Principles of Toxicology

Principles of Toxicology

Edited by Ned Burnett

hayle
medical
New York

Hayle Medical,
750 Third Avenue, 9th Floor,
New York, NY 10017, USA

Visit us on the World Wide Web at:
www.haylemedical.com

ISBN: 978-1-63241-528-8

The publisher's policy is to use permanent paper from mills that operate a sustainable forestry policy. Furthermore, the publisher ensures that the text paper and cover boards used have met acceptable environmental accreditation standards.

Printed in China.

Cataloging-in-Publication Data

Principles of toxicology / edited by Ned Burnett.
p. cm.
Includes bibliographical references and index.
ISBN 978-1-63241-528-8
1. Toxicology. 2. Poisoning. 3. Pharmacology. I. Burnett, Ned.
RA1211 .P75 2018
615.9--dc23

Table of Contents

Preface

Toxicology is the branch of chemistry which studies the affects of chemicals and pharmaceutical drugs on organisms. It focuses on the harmful effects of dosages and the extent to which an organism can be affected because of it. In the field of medicine, toxicology studies the impact of dosage on living organisms with respect to environmental factors, age, sex, etc. This book is compiled in such a manner, that it will provide in-depth knowledge about the field of toxicology. Those with an interest in the field of toxicology would find this book helpful.

This book has been the outcome of endless efforts put in by authors and researchers on various issues and topics within the field. The book is a comprehensive collection of significant researches that are addressed in a variety of chapters. It will surely enhance the knowledge of the field among readers across the globe.

It gives us an immense pleasure to thank our researchers and authors for their efforts to submit their piece of writing before the deadlines. Finally in the end, I would like to thank my family and colleagues who have been a great source of inspiration and support.

Editor

Anti-Heparanase Aptamers as Potential Diagnostic and Therapeutic Agents for Oral Cancer

Suzanne C. Simmons[1]◑, Hannaleena Jämsä[4,5]◑, Dilson Silva[2], Celia M. Cortez[2], Edward A. McKenzie[3], Carolina C. Bitu[4,5], Sirpa Salo[4], Sini Nurmenniemi[4], Pia Nyberg[4,5], Juha Risteli[6], Carlos E. B. deAlmeida[1,7], Paul E. C. Brenchley[8], Tuula Salo[4,5,9,10]*, Sotiris Missailidis[11]*

1 Department of Chemistry and Analytical Sciences, The Open University, Milton Keynes, United Kingdom, **2** Institute of Mathematics and Statistics, Rio de Janeiro State University, Rio de Janeiro, Brazil, **3** Manchester Institute of Biotechnology, University of Manchester, Manchester, United Kingdom, **4** Department of Diagnostics and Oral Medicine, Institute of Dentistry, University of Oulu, Oulu, Finland, **5** Medical Research Center and Oulu University Hospital, Oulu, Finland, **6** Institute of Diagnostics, Department of Clinical Chemistry, University of Oulu, Oulu, Finland, **7** Laboratório de Radiobiologia, Instituto de Radioproteção e Dosimetria, Rio de Janeiro, Brazil, **8** Renal Research Group, University of Manchester, Manchester, United Kingdom, **9** Graduate Program in Estomatopatologia, Piracicaba Dental School, University of Campinas, Piracicaba, São Paulo, Brazil, **10** Institute of Dentistry, University of Helsinki, Helsinki, Finland, **11** Institute of Biophysics Carlos Chagas Filho, Federal University of Rio de Janeiro, Rio de Janeiro, Brazil

Abstract

Heparanase is an endoglycosidase enzyme present in activated leucocytes, mast cells, placental tissue, neutrophils and macrophages, and is involved in tumour metastasis and tissue invasion. It presents a potential target for cancer therapies and various molecules have been developed in an attempt to inhibit the enzymatic action of heparanase. In an attempt to develop a novel therapeutic with an associated diagnostic assay, we have previously described high affinity aptamers selected against heparanase. In this work, we demonstrated that these anti-heparanase aptamers are capable of inhibiting tissue invasion of tumour cells associated with oral cancer and verified that such inhibition is due to inhibition of the enzyme and not due to other potentially cytotoxic effects of the aptamers. Furthermore, we have identified a short 30 bases aptamer as a potential candidate for further studies, as this showed a higher ability to inhibit tissue invasion than its longer counterpart, as well as a reduced potential for complex formation with other non-specific serum proteins. Finally, the aptamer was found to be stable and therefore suitable for use in human models, as it showed no degradation in the presence of human serum, making it a potential candidate for both diagnostic and therapeutic use.

Editor: Sophia N Karagiannis, King's College London, United Kingdom

Funding: Dr. Bonacossa was supported by a post-doctoral fellowship from Conselho Nacional de Pesquisa Científica - CNPq, on the scope of the programme Science with no Borders, MCTI, Brazil. Dr. Missailidis would like to acknowledge the financial support from Conselho Nacional de Pesquisa Científica - CNPq, on the scope of a senior visiting researcher programme at FIOCRUZ for part of this project. The research group of Prof. Salo was supported by the Academy of Finland, the Finnish Cancer Organisations, Finnish Dental Society Apollonia and the Oulu University Hospital KEVO grant. The funders had no role in study design, data collection and analysis, decision to publish, or preparation of the manuscript.

Competing Interests: The authors have declared that no competing interests exist.

* Email: sotirismissailidis@gmail.com (SM); tuula.salo@oulu.fi (TS)

◑ These authors contributed equally to this work.

Introduction

Heparanase is a β-1,4-endoglycosidase enzyme [1] that participates in extracellular matrix (ECM) degradation and remodeling [1]. The heparanase gene was first cloned in 1999 by the Vlodavsky and Parish groups in the seminal back to back Nature medicine papers [2,3].

The nascent polypeptide is a 543 amino acid pre-proenzyme, which after removal of the signal peptide sequence in the endoplasmic reticulum, undergoes proteolytic processing in late endosomes/lysosomes by cathepsin-L like proteases [4] at sites Glu109-Ser110 and Gln157-Lys158, yielding a N-terminal 8 kDa polypeptide, a C-terminal 50 kDa polypeptide and between them a 6 kDa linker polypeptide [3]. The 50 and 8 kDa polypeptides associate to form a heterodimeric active enzyme, whilst the 6 kDa linker is excised and degraded [5,6].

Heparanase activity is associated with activated leukocytes, mast cells, placental tissue and macrophages and the enzyme is secreted by activated CD4 + T cells [7,8,9], platelets [3], neutrophils and metastatic cells [10]. Upon secretion of heparanase from metastatic tumour cells, the enzyme hydrolyses the glycosidic bonds of heparan sulfate chains attached to proteoglycans to a product of 10–20 sugar units in length [11], leading to penetration of the endothelial cells of blood vessels and target organs by the tumor cell. Liberation of bound cytokines and growth factors sequestered by heparan sulfate chains in tissues [12] further facilitates growth of the tumour and promotes angiogenesis and proliferation of secondary tumours [13]. Levels of heparanase expression in tumour cells correlate with their metastatic potential; elevated levels of heparanase mRNA and protein have been found in cancer patients who show significantly shorter postoperative survival times than patients whose heparanase levels are normal [13,14].

Heparanase upregulation in cancer cells from myeloma, lymphoblastoid and breast cancer reflects in augmentation of exosome secretion with an enhanced content of syndecan-1, VEGF and HGF whose roles are closely related to tumor aggressiveness [15]. In addition to its function in cancer progression, heparanase enzyme also plays a major role in inflammation *per se* and carcinogenesis related to inflammatory process [16]. The enzyme has been detected in a variety of immune cells including T and B cells, macrophages, neutrophils and mast cells. It has been shown to mediate extravasation through the endothelial barrier via the remodeling of ECM heparan sulfate, which then allows trafficking to the sites of inflammation [10,17,18]. Heparanase expression has been linked to tumorigenesis in a number of different cancers, for example, acute myeloid leukaemia [19], bladder, brain [20], breast [21], colon [22], gastric [23], oesophageal [24], oral [25], pancreatic [14], and cervical cancer [26], suggesting that it may be a suitable target for drug therapy. Currently available inhibitors of heparanase include neutralizing antibodies [27], peptides [28] and small molecules [29,30].

A number of modified heparins and sulphated oligosaccharides have also been shown to be potent heparanase inhibitors with promising anti-tumour activities and have now advanced to the clinical testing stages. Examples of these include SST0001, M402, PI-88 and PG545. SST0001 is a fully N-acetylated modified heparin which lacks anti-coagulant activity and shown to be a selective heparanase inhibitor. It is currently in Phase I/II clinical trials for treatment of myeloma patients. M402 is an N-sulfated modified heparin that binds a wider range of growth factors compared to SST0001. This has progressed to Phase I/II clinical trials as a combination therapy with the chemotherapy agent gemcitabine for the treatment of metastatic pancreatic cancer. PI-88 is a sulphated polysaccharide with potent anti-angiogenic and anti-metastatic activity and with reduced unwanted anticoagulant activity. It has reached Phase III clinical trials for post-resection hepatocellular carcinoma. PG545 is a tetrasaccharide which has superior pharmacokinetic properties due to its high degree of lipophilicity. It has shown potent anti-tumour activity in PI88 resistant models, however, Phase I clinical trials in late 2010 were abandoned due to unexpected injection site reactions [31].

Aptamers are short DNA or RNA oligonucleotides developed for diagnostic and therapeutic use that display high binding affinity and specificity for target molecules [32–34]. The affinity of aptamers has been compared with that of antibodies (i.e. in the nanomolar range), but as aptamers are smaller (8–25 kDa compared to the 150 kDa size of antibodies), they can both penetrate tissues and be cleared from the plasma within minutes of intravenous administration without triggering an immune response, which can be useful when using them as diagnostic agents [35]. For therapeutic use, they are able to retain their function and binding characteristics upon modification with other moieties to improve their stability and solubility, whilst reducing their toxicity and plasma clearance [35–38][39–41]. Typically, aptamers are from 22 to 100 bases in length, and contain a region of variable sequence, flanked by known sequences, which are used for amplification and identification purposes. A large repertoire of different sequence combinations (typically in the region of 10^{15}) in the central domain creates many different folding arrangements, specificity and binding affinity for different molecules. Aptamers are typically produced based on the SELEX (systematic evolution of ligands by exponential enrichment) procedure [42], although a number of other selection methodologies are currently available [43–45].

Aptamers were previously generated against active human recombinant heparanase using a modified SELEX protocol and salt elution series. Selection yielded three aptamers, '1.5 M short', a 30 base truncated version of '1.5 M long' (73 bases), and '3.0 M' (55 bases). ELISAs and fluorescence titrations separated the two longer aptamers as showing higher affinity and recognition of heparanase in placental cells, whereas placental tissue staining favoured '1.5 M long'. This was confirmed in a Matrigel invasion assay using ovarian carcinoma cells previously shown to require heparanase for invasion [46]. Two additional aptamers, termed 'pink' and 'yellow' were selected against the linker peptide sequence of pro-heparanase, as these could have a function in inhibiting the formation of the active heterodimer enzyme, by blocking peptide protease excision.

In this study, efforts were made to further characterise the previously selected aptamers and to assess their potential as a diagnostic or therapeutic agent. The stability of the aptamer was assessed by incubation over different time points with human and mouse serum, and polyacrylamide gel electrophoresis used to determine the extent of its degradation by nucleases present in the serum. An additional invasion assay, in addition to the previous mouse EHS-tumour derived Matrigel invasion assay, was carried out using human uterine leiomyoma tissue and heparanase-expressing human oral squamous carcinoma cells (HSC-3), as this experiment represents a more authentic picture of what happens in human tissue. Furthermore, a cell cytotoxicity/cell proliferation assay was performed to verify that any inhibition of invasion observed was not a result of cytotoxicity on the part of the aptamers. Finally, the interactions of the aptamers with serum proteins was investigated to both verify specificity of the aptamers and study their potential transport by such proteins in the bloodstream. Increasing literature in the DNA aptamer field has demonstrated that these molecules have tremendous therapeutic potential in cancer therapy treatment and have already been used as so called escort molecules to deliver drugs into cancer cells (reviewed in [47]). AS1411 is an example of a DNA aptamer that has progressed to clinical trials testing. The aptamer in this case specifically targets nucleolin protein and has been trialed with metastatic, clear-cell, renal cell carcinoma patients who have been refractory to prior tyrosine kinase inhibitors [48].

Materials and Methods

Cell culture

Human tongue squamous carcinoma cells, HSC-3 (JCRB 0623; Osaka National Institute of Health Sciences, Osaka, Japan), were cultured in growth media: 50% DMEM 50% Ham's F-12 (Sigma Aldrich) and additionally supplemented with 50 µg/ml ascorbic acid, 250 ng/ml amphotericin B, 5µg/ml insulin (bovine pancreas), 0.4 ng/ml hydrocortisone and 10% heat-inactivated foetal bovine serum. The culture supplement was purchased from Sigma Aldrich. All cell cultures were carried out using pre-warmed reagents. Cells were incubated in 95% air/5% CO_2 at 37°C. Cells were passaged by removing media and washing with HBSS (Sigma Aldrich), then adding 3 ml 1× Trypsin-EDTA (Sigma Aldrich) and incubating for 5 minutes. The Trypsin-EDTA was inactivated by adding 7 ml growth media and removing any cell clumps. 1 ml cell suspension (plus 24 ml fresh growth media) was retained in the flask for maintenance of stocks and the remaining 9 ml was counted and used for experiments.

Organotypic invasion assay and analyses of the inhibitors on invasion

The organotypic invasion assay and the quantitation of results were performed as described in Nurmenniemi *et al.* (2009) [49]. Briefly, 7×10^5 HSC-3 cells suspended in media containing the appropriate aptamer (Unrelated, 1.5 M short, 1.5 M long, 3 M,

Pink and Yellow) or an antibody against heparanase (Hpa Ab, 0.7 mM). Also (2-{4-[(E)-3-(4-bromophenyl) acryloylamino]-3-fluorophenyl} benzooxazol-5-yl) acetic acid, abbreviated to BAFB, was used, as this was shown to have inhibitory effects upon heparanase in previous studies [29] (Table 1). Each aptamer or antibody was added at 1 μM to HSC-3 cell suspension in the beginning of the study and in the HSC-3 cell culture media throughout the experiment. Myoma disks without HSC-3 cells and HSC-3 cells without inhibitor were also included in the assay as controls. The disks were incubated for 14 days at 37°C with 5% CO_2 with the media containing the appropriate inhibitors. The media were collected, centrifuged, and fresh media with inhibitors were changed at days 4, 7, 10 and 14. From the collected media supernatants, the degradation products of myoma tissue type III collagen were analyzed using SP99 radioimmunoassay (RIA) for C-terminal telopeptide (IIICTP), and N-terminal telopeptide (IIINTP) indirect enzyme immunoassays (EIA) for N-terminal telopeptide, following the methods described in Nurmenniemi *et al.* ([49] for RIA and [50] for EIA). On day 14, the myoma disks were fixed in 4% paraformaldehyde and prepared for immuno-histological analysis. Six μm histological sections of myoma disks were stained with monoclonal pancytokeratin antibody (DAKO, clone AE1/AE3 at a 1:150 dilution) and viewed under a microscope at 100 × magnification. Nine representative images were taken from each of the three repeats of every treatment. Images were analyzed as described in Nurmenniemi *et al.* (2009) [49]. Differences in the invasion area and depth were evaluated using a Student's t-test and Mann-Whitney test and p-values less than 0.05 were considered statistically significant.

Cell proliferation assay

To determine the effect of the '1.5 M short' aptamer on HSC-3 cell proliferation, we used the CellTiter 96 AQueous Cell Proliferation Assay (Promega), an MTS assay. Approximately 1 × 10^4 cells were seeded in triplicate for in a 96-well plate with 1μM of the short aptamer. After 24, 48 and 72 h, 20 ml of CellTiter 96 Aqueous One Solution Reagent were added to each well and cells were incubated for 1 h at 37°C in a 5% CO_2 incubator. The absorbance recorded at 490 nm on a FLUOstar Optima plate reader was used as a representation of the relative number of living cells in culture.

Serum stability assay

Aptamers '1.5 M short', '1.5 M long' and '3.0 M' were incubated at a concentration of 5 μM with human and mouse serum for 30, 60, 120, 180, 240 and 300 minutes at 37°C. The reaction was then stopped by the addition of 100 mM EDTA and the products ran on a 12% native polyacrylamide gel, alongside a 25 bp DNA marker ladder. Gels were stained using ethidium bromide and viewed under UV light.

Serum albumin binding

Bovine Serum Albumin (BSA) was purchased from Sigma-Aldrich Ltd (Gillingham, UK, product code A7030 10 g). UV experiments were conducted on a Bio-Tek Uvikon XL with a Peltier Thermosystem for temperature control and stirring facility, connected to the PC utilizing Lab Power Junior software for data collection and analysis. Fluorimeter used was a Horiba Jobin Yvon Fluoromax-P equipped with a photon counter and Peltier system for temperature control and stirring facility, coupled to a PC utilizing Datamax software for spectral analysis. Initial measurements were taken to verify the presence or absence of fluorescent emission of both aptamers for excitation wavelength of 290 nm (selective for tryptophan residues) and emission wavelengths between 300 and 400 nm. Both aptamers were titrated in water and 10 mM phosphate buffer solutions pH 7.4, at 37°C. The 1.5 M short aptamer concentration varied from 0.3 to 8.0 μM, and the 1.5 M long aptamer varied from 0.5 to 8.0 μM, showing the intrinsic fluorescence of these aptamers. Both aptamers presented fluorescence emission spectra in this range. Earlier tests showed that aptamer concentrations ranging from 0.1 μg/ml to 8 μg/ml did not interfere in the evaluation of albumin quenching [51]. Quenching measurements were taken in 1 ml of 6 μM albumins in phosphate buffer pH 7.4. Emission spectra were registered from 300 to 400 nm wavelength, after a reaction time of 90 sec from each aptamer addition. Both emission and excitation bandwidth were set to 3 nm. Aptamer was added from a concentrated stock solution so that the volume increment was negligible. Experiments were performed at 37°C, pH 7.4.

To evaluate any existing primary and/or secondary inner filter effects (IFEs), correction procedures based on absorbance measurements of solutions were performed at excitation and emission wavelengths of albumin. This effect consists on the absorption of exciting and/or emitted radiation by dissolved species, including the fluorophore itself [52]. Absorbance measurement of aptamers/albumin solutions at excitation and emission wavelengths of albumin showed that inner filter effect caused by absorption of emitted radiation was negligible.

Results

The anti-heparanase aptamers inhibit carcinoma cell invasion

The invasion of HSC-3 cells was studied with a human myoma organotypic model [49] exposing the carcinoma cells to various aptamers (Unrelated, 1.5 M short, 1.5 M long, 3 M, Pink, and Yellow) or heparanase antibody (Hpa Ab) (Fig. 1A). The effects of these compounds compared to control (no inhibitor) on invasion area (calculated based on the μm-area of invasive cells) and depth of invasion (the distance from the lower surface of the noninvasive cell layer to the deepest invaded cell) were analyzed (Fig. 1B and C). Aptamer 1.5 M Short decreased significantly the total invasion

Table 1. The aptamer sequences used in this study.

Name	Sequence
1.5 M Long	GGGAGACAAGAATAAACGCTCAAATGG **ACTTTTGAATGTGGCAACAAATTCGACAGG** AGGCTCACAACAGGC
1.5 M Short	ACTTTTGAATGTGGCAACAAATTCGACAGG
Pink	TTGCTCCTTATAGAGCCGTCCGAGC
Yellow	CTAAAGTGCCTCACGCTGTTAACTC

In bold the sequence of the short aptamer, which is the part of the long aptamer that is structured.

area ($p = 0.0001$) similar to Hpa Ab, which was used as a positive invasion inhibitor control [27]. On the contrary, none of the other compounds had an effect on HSC-3 invasion (Figure 1). The invasion depth decreased significantly only after Hpa Ab treatment ($p = 0.0001$) (Figure 1C). Based on our previous findings, the degradation of type III collagen by HSC-3 cells measured with RIA peaks from days 7 to 11 [49]. Similarly, the media analyzed by RIA from day 4 did not show differences between any of the treatments (not shown). The media change upon termination of the experiment at day 14 showed only statistical significance from the treatment with no cells added ($p = 0.001$; not shown). Degradation of type III collagen without inhibitors was highest at days 7 (not shown) and 10 (Figure 2A–D). On day 10, the treatment with heparanase antibody inhibited significantly the degradation compared to unrelated control ($p = 0.07$ in RIA, and $p = 0.03$ in EIA). On day 10 there was a significant inhibition by 1.5 M Short ($p = 0.004$ in RIA, and $p = 0.007$ in EIA) and 1.5 M Long ($p = 0.02$ in RIA, and $p = 0.03$ in EIA) compared to HSC-3 control. However, 3 M, Pink, and Yellow aptamers, as well as BAFB showed no significant decrease in the amount of collagen degradation products, indicating that they were not successful inhibitors of invasion.

The short anti-heparanase aptamer does not exhibit any cytotoxicity

The cytotoxicity of the selected aptamers on HSC-3 cells was studied, to verify that the inhibition of invasion observed in the organotypic model was a result of the inhibition of the heparanase, as previously verified [46] and not cell cytotoxicity. The MTS assay was performed over 72 hrs, with a single addition of the aptamer in the beginning of the assay, and measurements over the period intervals of 24, 48 and 72 hrs. No change in cell viability and cell growth was observed between the cells where aptamer was added and the control (see Figure 3). Only the aptamers that showed inhibition of invasion were tested for cytotoxicity, to investigate if the inhibitory effect observed was due to cytotoxicity or inhibition of heparanase. Aptamers that do not inhibit cell invasion clearly have no effect on the cells and therefore there was no reason to be further studied for cytotoxicity.

The short anti-heparanase aptamer does not bind significantly to serum proteins

Short and long aptamers 1.5 M were initially titrated stepwise into water and phosphate buffer solution, pH 7.4 at 37°C to investigate their intrinsic fluorescence. Both aptamers have intrinsic fluorescence with peaks at 380 nm, which increases in

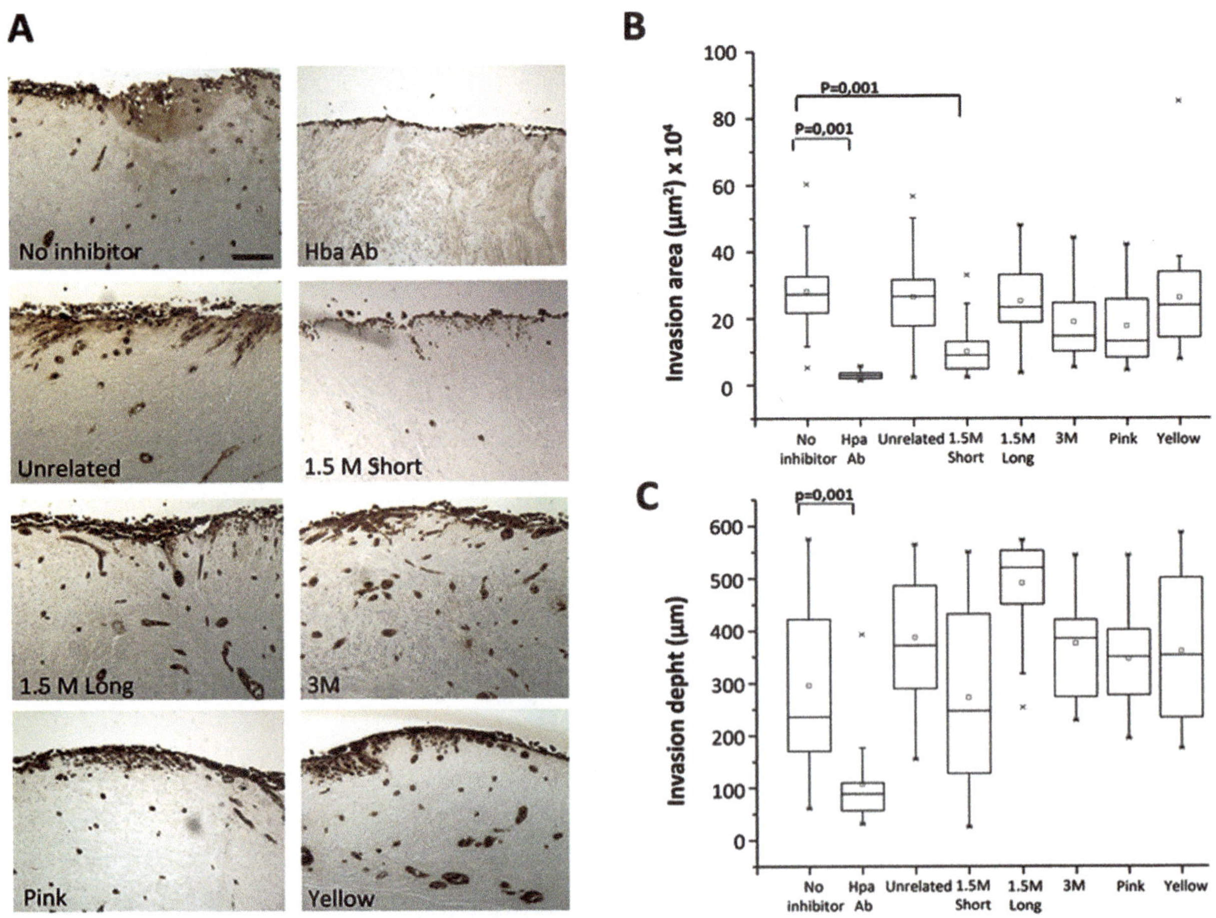

Figure 1. HSC-3 invasion in myoma discs. A: Paraffin-embedded 14-day myoma organotypic sections were stained for pancytokeratin marker AE1/AE3 to analyze HSC-3 invasion after various treatments: no inhibitor, Hpa Ab, (the polyclonal heparanase antibody as a positive control), unrelated aptamer, (selected against a target involved in Alzheimer's disease), anti-heparanase aptamers 1.5 M Short, 1.5 M long and 3 M; linker peptide aptamers Pink and Yellow. Scale bar is 100 µm. The differences in invasion area (B) and invasion depth (C) after various treatments ($n = 27$/treatment). The statistics were done as two-sample *t*-test and Mann-Whitney test.

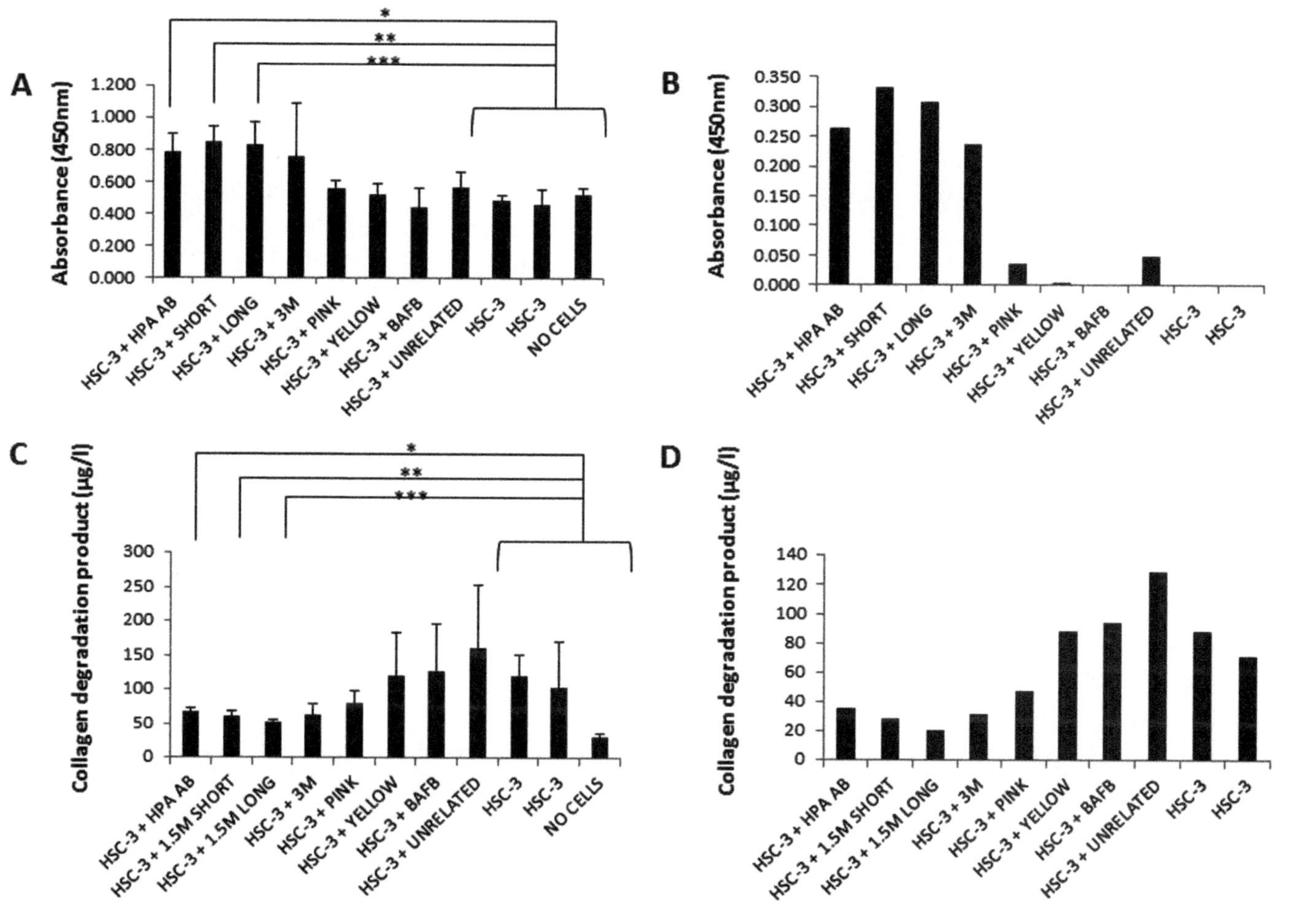

Figure 2. EIA and RIA assays. A: EIA (IIINTP indirect enzyme immunoassays) detecting N-terminal telopeptide from collagen type III degradation products at day 10 media change. Increasing absorbance means less collagen degradation product present. Hpa Ab ($p = 0.03$), 1.5 M Short ($p = 0.007$) and 1.5 M Long ($p = 0.03$) showed significant increase in absorbance compared to no inhibitor, suggesting they have inhibited the invasion of HSC-3 cells. B: The graph shows previous EIA values adjusted for negative control at day 10 media change. C: RIA (radioimmunoassay for type III collagen) detecting C-terminal telopeptide at day 10 media change. Increasing levels mean less collagen degradation product. Hpa Ab ($p = 0.07$), 1.5 M Short ($p = 0.004$) and 1.5 M Long ($p = 0.02$). D: RIA has confirmed the EIA assays showing significantly lower collagen degradation products than that for tissues without inhibitor added, indicating that they were successful inhibitors of invasion.

fluorescence intensity upon a corresponding increase of the aptamer concentration; the short showing less fluorescence than the long aptamer (Figure 4A and B). Water was used as a diluent to show that the aptamer and not phosphate buffer solution was the cause of the fluorescence, although the fluorescence for the short aptamer increased upon using phosphate buffer solution as the diluent. However, this was likely due to a difference in pH as in fluorescence spectroscopy pH changes have a significant influence on results.

The short aptamer was able to quench the fluorescence of HSA at 37°C by 1.5% (±0.03%) and 9% (±1.2%) at 1:100 and 1:10 molar ratios respectively, with quenching of 10% achieved by a molar concentration 9.1 times lower than HSA (Figure 5). Long aptamer is able to quench the fluorescence of HSA by 10% at a concentration 18.2 times lower than HSA, and by 2.4% (±0.1%) and 16% (±0.6%) at 1:100 and 1:10 molar ratios. The results show that both aptamers are able to quench the fluorescence of HSA, although the long aptamer was more effective. HSA quenching indicates that the aptamers reach sub domain IIA, where its single tryptophan is located. This tryptophan residue is located at site 214 in subdomain IIA, within which there is a large hydrophobic cavity with many arginine residues near the surface [53], which have been shown in different studies to serve as anchor points for aptamers [54].

To gain more information about the type of interaction occurring between the aptamers and HSA, UV spectrometry titrations were carried out by titrating increasing concentrations of short and long aptamers and 6 µM HSA diluted in phosphate buffer pH 7.4, as shown in Figure 5. The addition of both aptamers to phosphate buffer and HSA increased the overall absorbance, showing that the aptamer was responsible for this increase rather than HSA. The increase was more pronounced for the long aptamer over the short aptamer, and both produced a shift of the maximum absorbance to the left upon addition of increasing concentrations of aptamer.

The shift observed from the short aptamer (Figure 6A) moved 6 nm to the left, suggesting that only a slight conformational change in the protein was occurring [55] and therefore, HSA quenching by this aptamer is most likely due to dynamic quenching. However, in the case of the long aptamer (Figure 6B), not only was there a substantial shift in the maximal absorption by 20 nm to the left, but a complete change in the shape of the peak

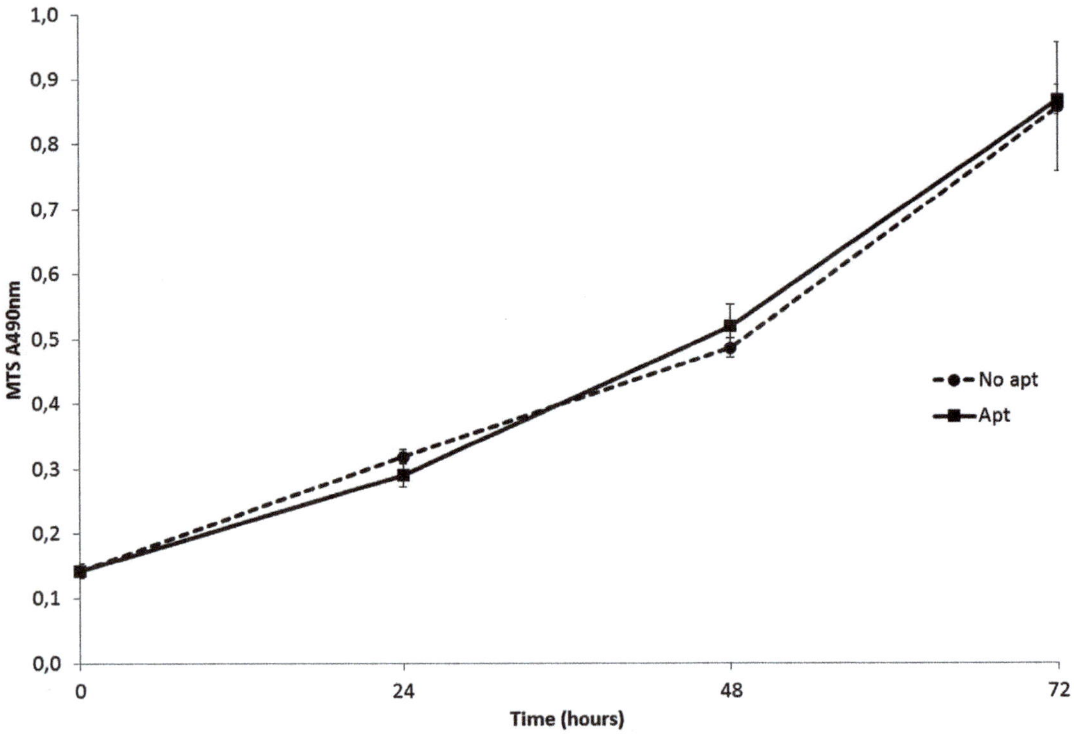

Figure 3. The MTS absorbance at 490 nm is shown over 24, 48 and 72 h in the presence and absence of the 1.5 M short aptamer. The presence of the aptamer at 1 mM concentration was found to have no effect on the cell growth in comparison with the control.

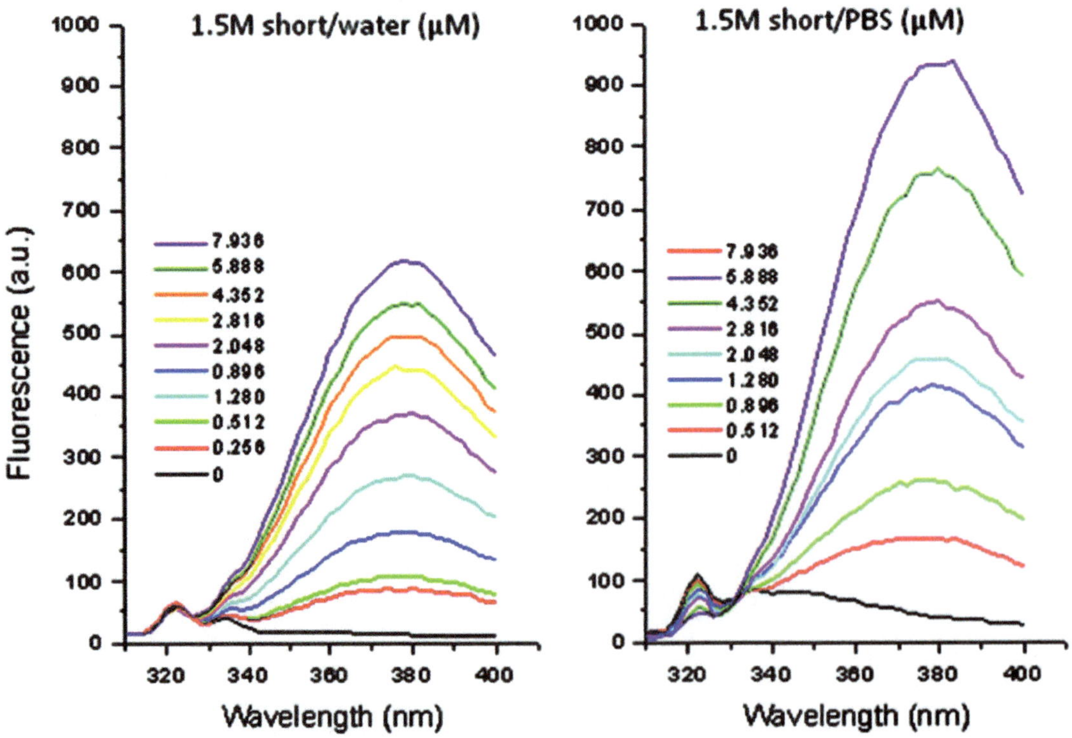

Figure 4. Fluorescence spectra of short aptamer in water and phosphate buffer solution (A) and PBS (B), at 37°C. Fluorescence increases upon increasing the concentration of aptamer in both phosphate and water, showing that although the fluorescence is higher in phosphate, the aptamer is in fact the cause and the pH difference in water and PBS is the most likely reason for the increase of fluorescence of the aptamer in PBS.

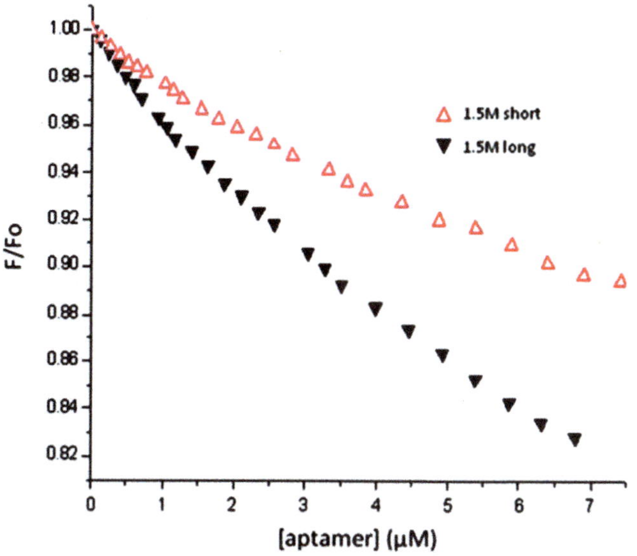

Figure 5. Stern-Volmer plots for HSA titrated by short and long apatmers 37°C. Excitation wavelength 290 nm, [HSA] = 6 μM. Excitation wavelength at 290 mM in a solution of sodium phosphate. Data is the mean of six values showing no greater standard deviation than 11%. The quenching effect is more considerable for long than short aptamer.

was observed, incorporating the peak at 260 nm of the aptamer, suggesting that one complex was formed and that the quenching was due to the static quenching phenomenon with long aptamers [55]. Thus, the UV titrations suggested that the short aptamer did not form a complex with HSA and that the interactions were due to dynamic quenching, whereas the long aptamer was suggested to form a ground state complex with HSA, in contrast with other aptamers previously studied, which also show specificity and complex formation only with their target protein [56]. This work has been expanded and interactions of the aptamers with serum proteins and the specific position of their interaction has been calculated and published separately, as it was not within the scope of this article [57].

Aptamers are stable in human serum

To assess the aptamers' suitability as therapeutic agents, it was necessary to have an understanding of how stable the unmodified aptamers would be in the body, as, in the bloodstream alone, there are many nucleases capable of degrading the aptamers. Thus, to verify the stability of the aptamers in human serum, we have characterized any degradation products by gel electrophoresis. Comparison of bands on the gels for 1.5 M short aptamer incubated for different time points with human and mouse serum, with that of aptamer only showed that 1.5 M short aptamer was not subject to nuclease degradation from human serum as the bands did not show any smearing or decrease in size or intensity compared to aptamer only, and hence, no breakdown of the aptamer into smaller fragments was observed (data not shown). With mouse serum, however, there was a decrease in primary band intensity at five hours' incubation time, suggesting that nucleases have degraded the aptamer by that time.

Discussion

In this study we have explored the potential of previously selected aptamers against heparanase as promising diagnostic and therapeutic agents against oral cancer. The aptamers were previously shown to have high affinity against heparanase and were functional in a Matrigel assay. On these initial studies, it was found that the longer aptamers had a higher affinity for heparanase and they had performed well in fluorescent microscopy and Matrigel invasion assays. However, when we examined these aptamers on the organotypic invasion assay and analysed their potential to block invasion, it was found that the short aptamer was far more capable of doing so, compared to its long counterparts. This was also verified by the analysis by RIA and EIA of the degradation products of myoma tissue, namely type III collagen C- and N-terminal telopeptide respectively. The 1.5 M Short and 1.5 M Long aptamers consist of the same variable region and in fact the short one is a truncated version of the long. However, it appears that although the long one has a slightly higher affinity, probably due to increased interactions between the protein and the primer parts of the aptamer, these resulted in reducing the ability of these aptamers to inhibit tissue invasion. The presence of various proteins in the actual tissue, as compared to the Matrigel experiment previously performed, may be the reason for this, as the long aptamer may form other interactions with such proteins, or the primer tails may have a steric hindrance effect on the tissue, which is not apparent in the simpler matrigel model. This, in fact, was confirmed by the study of the interactions between the two aptamers and serum proteins. In these studies it was found that the long aptamer formed a complex with human serum albumin, whereas the short aptamer did not form a complex and showed only a limited dynamic quenching. In a further study [57], we have modelled the interactions of the short and long aptamers with HSA and have identified that indeed the long aptamer forms a complex with serum albumins in a single binding site, close to Trp 214 of HSA or 212 of BSA, at the subdomain IIA of these proteins, in a positively charged cavity lined with lysine and arginine residues [57]. It has been demonstrated that the shorter aptamer species lacks the ability to form complexes with serum proteins and exhibits thus higher specificity for its target, which justifies our choice of using it in any further therapeutic or diagnostic development and is in agreement with the myoma data presented in this work. One further important feature of this study is the demonstration that post-SELEX modifications may be more beneficial for aptamer selection than initial counter-selection steps, where this is possible. In a series of studies with various methodologies of detection, aptamer affinity for their target has been compared to that for albumin. The majority of the exemplars for new aptamer-based detection methodologies are based on the thrombin aptamers. In a study of aptamer-enhanced laser desorption/ionization study, the thrombin-binding DNA aptamer was used for affinity capture of thrombin in MALDI-TOF-MS. This aptamer was shown to be capable to bind to thrombin in a thrombin/albumin mixture [58]. Similarly, aptamers have been shown to distinguish thrombin from albumin in a QCM experiment [59]. Another G-quadruplex based thrombin aptamer in cationic polythiophene protein detection arrays was also able to detect thrombin over albumin in the attomole range in less than one hour without any tagging of the target [60]. The thrombin aptamer has also been used in an electrochemical detection assay, where it has been able to separate thrombin from BSA, HSA, Lysozyme and immunoglobulin G [61].

Apart from the thrombin aptamers, other aptamers in detection assays have also been compared with albumin, or have shown specific binding in the presence of high concentration of albumins. In an electrochemical sensor, aptamers against lysozyme have been shown to detect lysozyme in a mixture of six proteins

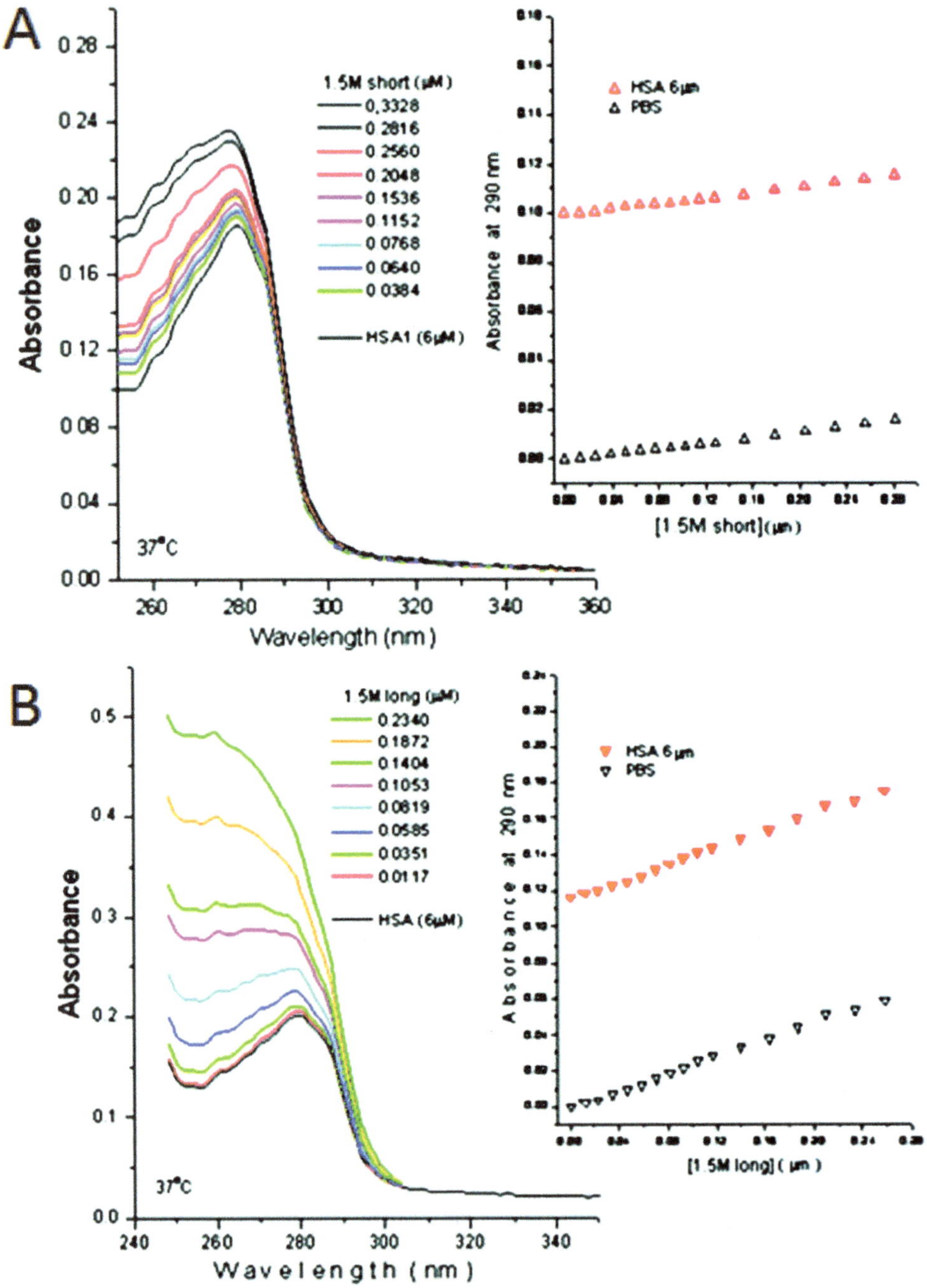

Figure 6. UV wavelength scan of HSA (left) and plot of PBS (right) titrated with 1.5 M short aptamer (A) and UV wavelength scan of HSA (left) and plot of PBS (right) titrated with 1.5 M long aptamer (B).

including albumin [62]. Immunoglobulin E has also been detected in serum over albumin [63], whereas an anti- *F. tularensis* aptamer cocktail, when tested in a sandwich Aptamer-Linked Immobilized Sorbent Assay (ALISA) and dot blot analysis, exhibited specificity in its ability to bind only to tularemia bacterial antigen from subspecies *japonica, holarctica* (also known as *palaearctica*) and *tularensis* but not to *Bartonella henselae*, nor to pure chicken albumin or chicken lysozyme, demonstrating the ability of this aptamer cocktail to function as a bacterial detection agent [64].

Depending on the aptamer species, some aptamers present cross-reactivity with serum albumins, whereas the majority of them are capable of distinguishing between the protein they have been selected for, and albumins. Thus, for example, when we investigated a number of KLK6 aptamers with serum albumins, we identified that the majority of the selected aptamers against that target were specific, but one of them had significant affinity for albumin [65]. In addition, it is important to note that the same aptamer may or may not form complexes with HSA or BSA, depending on their post-SELEX refinement. Thus, the heparanase aptamer of this study, when truncated for the binding site of the specific target protein, does not form a stable complex with serum proteins, whereas its longer counterpart that contains the flanking primers, not selected for specific binding, can do. This is important with respect to selection strategies, as, in the case of heparanase, we started the selection with a naïve library, containing all possible species, and the selection and studies clearly indicated that one aptamer species was the best candidate. This candidate would have been lost if a counter-selection against albumins had been performed at the beginning, as it also presents affinity for these proteins prior to truncation. However, with a simple truncation of the flanking primers, the aptamer gained the necessary specificity to be further developed for therapeutic and diagnostic applications.

Furthermore, the effect seen in the inhibition of the invasion assay could have been a result of a cytotoxic effect on the part of aptamers. This possibility was eliminated in a cytotoxicity assay, which clearly demonstrated that the aptamers did not show any cytotoxic effect on these cells after 72 hours of incubation, thus verifying that the inhibition of invasion was in fact due to inhibition of heparanase. Finally, aptamers were found to be stable in human serum even without any modification, making them potentially interesting therapeutic reagents on their own accord. This is important as such stability would reduce production costs of such an aptamer, if it were selected for subsequent therapeutic or diagnostic applications.

Author Contributions

Conceived and designed the experiments: SM TS CEBD JR DS CMC. Performed the experiments: SCS HJ CEBD DS EAM CB SS SN PN. Analyzed the data: SM CMC CEBD HJ ST. Contributed reagents/materials/analysis tools: EAM PB TS SM. Wrote the paper: SM DS CEBD TS EAM.

References

1. Eldor A, Bar-Ner M, Yahalom J, Fuks Z, Vlodavsky I (1987) Role of heparanase in platelet and tumor cell interactions with the subendothelial extracellular matrix. Semin Thromb Hemost 13: 475–488.
2. Vlodavsky I, Friedmann Y, Elkin M, Aingorn H, Atzmon R, et al. (1999) Mammalian heparanase: gene cloning, expression and function in tumor progression and metastasis. Nat Med 5:793–802
3. Hulett MD, Freeman C, Hamdorf BJ, Baker RT, Harris MJ, et al. (1999) Cloning of mammalian heparanase, an important enzyme in tumor invasion and metastasis. Nat Med 5: 803–809.
4. Abboud-Jarrous G, Rangini-Guetta Z, Aingorn H, Atzmon R, Elgavish S, et al. (2005) Site-directed mutagenesis, proteolytic cleavage, and activation of human proheparanase. J Biol Chem 280: 13568–13575.
5. Levy-Adam F, Miao HQ, Heinrikson RL, Vlodavsky I, Ilan N (2003) Heterodimer formation is essential for heparanase enzymatic activity. Biochem Biophys Res Commun.308: 885–91.
6. McKenzie E, Young K, Hircock M, Bennett J, Bhaman M, et al. (2003) Biochemical characterization of the active heterodimer form of human heparanase (Hpa1) protein expressed in insect cells. Biochem J 373: 423–35.
7. Adams DH, Shaw S (1994) Leucocyte-endothelial interactions and regulation of leucocyte migration. Lancet 343: 831–836.
8. Blotnick S, Peoples GE, Freeman MR, Eberlein TJ, Klagsbrun M (1994) T lymphocytes synthesize and export heparin-binding epidermal growth factor-like growth factor and basic fibroblast growth factor, mitogens for vascular cells and fibroblasts: Differential production and release by CD4+ and CD8+ T cells. Proc Natl Acad Sci U S A 91: 2890–2894.
9. Gilat D, Hershkoviz R, Goldkorn I, Cahalon L, Korner G, et al. (1995) Molecular behavior adapts to context: Heparanase functions as an extracellular matrix-degrading enzyme or as a T cell adhesion molecule, depending on the local pH. J Exp Med 181: 1929–1934.
10. Vlodavsky I, Eldor A, Haimovitz-Friedman A, Matzner Y, Ishai-Michaeli R, et al. (1992) Expression of heparanase by platelets and circulating cells of the immune system: Possible involvement in diapedesis and extravasation. Invasion Metastasis 12: 112–127.
11. Pikas DS, Li JP, Vlodavsky I, Lindahl U (1998) Substrate specificity of heparanases from human hepatoma and platelets. J Biol Chem 273: 18770–18777.
12. Lindahl U, Kusche-Gullberg M, Kjellen L (1998) Regulated diversity of heparan sulfate. J Biol Chem 273: 24979–24982.
13. Gohji K, Hirano H, Okamoto M, Kitazawa S, Toyoshima M, et al. (2001) Expression of three extracellular matrix degradative enzymes in bladder cancer. Int J Cancer 95: 295–301.
14. Koliopanos A, Friess H, Kleeff J, Shi X, Liao Q, et al. (2001) Heparanase expression in primary and metastatic pancreatic cancer. Cancer Res 61: 4655–4659.
15. Thompson CA, Purushothaman A, Ramani VC, Vlodavsky I, Sanderson RD (2013) Heparanase regulates secretion, composition, and function of tumor cell-derived exosomes. J Biol Chem 288: 10093–10099.
16. Meirovitz A, Goldberg R, Binder A, Rubinstein AM, Hermano E, et al. (2013) Heparanase in inflammation and inflammation-associated cancer. FEBS J 280: 2307–2319.
17. Li JP, Vlodavsky I (2009) Heparin, heparan sulfate and heparanase in inflammatory reactions. Thromb Haemost 102: 823–828.
18. Vlodavsky I, Beckhove P, Lerner I, Pisano C, Meirovitz A, et al. (2012) Significance of heparanase in cancer and inflammation. Cancer Microenviron 5: 115–132.
19. Vlodavsky I, Goldshmidt O, Zcharia E, Atzmon R, Rangini-Guatta Z, et al. (2002) Mammalian heparanase: Involvement in cancer metastasis, angiogenesis and normal development. Semin Cancer Biol 12: 121–129.
20. Marchetti D, Nicolson GL (2001) Human heparanase: A molecular determinant of brain metastasis. Adv Enzyme Regul 41: 343–359.
21. Maxhimer JB, Quiros RM, Stewart R, Dowlatshahi K, Gattuso P, et al. (2002) Heparanase-1 expression is associated with the metastatic potential of breast cancer. Surgery 132: 326–333.
22. Friedmann Y, Vlodavsky I, Aingorn H, Aviv A, Peretz T, et al. (2000) Expression of heparanase in normal, dysplastic, and neoplastic human colonic mucosa and stroma. evidence for its role in colonic tumorigenesis. Am J Pathol 157: 1167–1175.
23. Tang W, Nakamura Y, Tsujimoto M, Sato M, Wang X, et al. (2002) Heparanase: A key enzyme in invasion and metastasis of gastric carcinoma. Mod Pathol 15: 593–598.
24. Mikami S, Ohashi K, Usui Y, Nemoto T, Katsube K, et al. (2001) Loss of syndecan-1 and increased expression of heparanase in invasive esophageal carcinomas. Jpn J Cancer Res 92: 1062–1073.
25. Ikuta M, Podyma KA, Maruyama K, Enomoto S, Yanagishita M (2001) Expression of heparanase in oral cancer cell lines and oral cancer tissues. Oral Oncol 37: 177–184.
26. Zeng C, Ke ZF, Luo WR, Yao YH, Hu XR, et al. (2013) Heparanase overexpression participates in tumor growth of cervical cancer in vitro and in vivo. Med Oncol 30: 403–409.
27. He X, Brenchley PE, Jayson GC, Hampson L, Davies J, et al. (2004) Hypoxia increases heparanase-dependent tumor cell invasion, which can be inhibited by antiheparanase antibodies. Cancer Res 64: 3928–3933.
28. Vlodavsky I, Ilan N, Naggi A, Casu B (2007) Heparanase: Structure, biological functions, and inhibition by heparin-derived mimetics of heparan sulfate. Curr Pharm Des 13: 2057–2073.
29. Courtney SM, Hay PA, Buck RT, Colville CS, Phillips DJ, et al. (2005) Furanyl-1,3-thiazol-2-yl and benzoxazol-5-yl acetic acid derivatives: Novel classes of heparanase inhibitor. Bioorg Med Chem Lett 15: 2295–2299.
30. Ishida K, Hirai G, Murakami K, Teruya T, Simizu S, et al. (2004) Structure-based design of a selective heparanase inhibitor as an antimetastatic agent. Mol Cancer Ther 3: 1069–1077.
31. Pisano C, Vlodavsky I, Ilan N, Zunino F (2014) The potential of heparanase as a therapeutic target in cancer. Biochem Pharmacol http:dx.doi.org/10.10161/J.bcp.2014.02.010
32. Famulok M, Mayer G (1999) Aptamers as tools in molecular biology and immunology. Curr Top Microbiol Immunol 243: 123–136.
33. Gold L, Polisky B, Uhlenbeck O, Yarus M (1995) Diversity of oligonucleotide functions. Annu Rev Biochem 64: 763–797.
34. Osborne SE, Matsumura I, Ellington AD (1997) Aptamers as therapeutic and diagnostic reagents: Problems and prospects. Curr Opin Chem Biol 1: 5–9.
35. Hicke BJ, Marion C, Chang YF, Gould T, Lynott CK, et al. (2001) Tenascin-C aptamers are generated using tumor cells and purified protein. J Biol Chem 276: 48644–48654.
36. Cao Z, Tong R, Mishra A, Xu W, Wong GC, et al. (2009) Reversible cell-specific drug delivery with aptamer-functionalized liposomes. Angew Chem Int Ed Engl 48: 6494–6498.
37. Floege J, Ostendorf T, Janssen U, Burg M, Radeke HH, et al. (1999) Novel approach to specific growth factor inhibition in vivo: Antagonism of platelet-derived growth factor in glomerulonephritis by aptamers. Am J Pathol 154: 169–179.
38. Hicke BJ, Stephens AW, Gould T, Chang YF, Lynott CK, et al. (2006) Tumor targeting by an aptamer. J Nucl Med 47: 668–678.
39. Jellinek D, Green LS, Bell C, Lynott CK, Gill N, et al. (1995) Potent 2'-amino-2'-deoxypyrimidine RNA inhibitors of basic fibroblast growth factor. Biochemistry 34: 11363–11372.
40. Ruckman J, Green LS, Beeson J, Waugh S, Gillette WL, et al. (1998) 2'-fluoropyrimidine RNA-based aptamers to the 165-amino acid form of vascular endothelial growth factor (VEGF165). inhibition of receptor binding and VEGF-induced vascular permeability through interactions requiring the exon 7-encoded domain. J Biol Chem 273: 20556–20567.

41. Da Pieve C, Williams P, Haddleton DM, Palmer RM, Missailidis S (2010) Modification of thiol functionalized aptamers by conjugation of synthetic polymers. Bioconjug Chem 21: 169–174.
42. Ferreira CS, Matthews CS, Missailidis S (2006) DNA aptamers that bind to MUC1 tumour marker: Design and characterization of MUC1-binding single-stranded DNA aptamers. Tumour Biol 27: 289–301.
43. Missailidis S, Thomaidou D, Borbas KE, Price MR (2005) Selection of aptamers with high affinity and high specificity against C595, an anti-MUC1 IgG3 monoclonal antibody, for antibody targeting. J Immunol Methods 296: 45–62.
44. White R, Rusconi C, Scardino E, Wolberg A, Lawson J, et al. (2001) Generation of species cross-reactive aptamers using "toggle" SELEX. Mol Ther 4: 567–573.
45. Berezovski M, Musheev M, Drabovich A, Krylov SN (2006) Non-SELEX selection of aptamers. J Am Chem Soc 128: 1410–1411.
46. Simmons SC, McKenzie EA, Harris LK, Aplin JD, Brenchley PE, et al. (2012) Development of novel single-stranded nucleic acid aptamers against the pro-angiogenic and metastatic enzyme heparanase (HPSE1). PLoS One 7: e37938.
47. Scaggiante B, Dapas B, Farra R, Grassi M, Pozzato G, et al. (2013) Aptamers as targeting delivery devices or anti-cancer drugs for fighting tumors. Curr Drug Metab.14: 565–82.
48. Rosenberg JE, Bambury RM, Van Allen EM, Drabkin HA, Lara PN Jr, et al. (2014) A phase II trial of AS1411 (a novel nucleolin-targeted DNA aptamer) in metastatic renal cell carcinoma. Invest New Drugs 32:178–87.
49. Nurmenniemi S, Sinikumpu T, Alahuhta I, Salo S, Sutinen M, et al. (2009) A novel organotypic model mimics the tumor microenvironment. Am J Pathol 175: 1281–1291.
50. Nurmenniemi S, Koivula MK, Nyberg P, Tervahartiala T, Sorsa T, et al. (2012) Type I and III collagen degradation products in serum predict patient survival in head and neck squamous cell carcinoma. Oral Oncol 48: 136–140.
51. Gilbert JC, DeFeo-Fraulini T, Hutabarat RM, Horvath CJ, Merlino PG, et al. (2007) First-in-human evaluation of anti von willebrand factor therapeutic aptamer ARC1779 in healthy volunteers. Circulation 116: 2678–2686.
52. Puchalski M, Morra M, Wandruszka Rv (1991) Assessment of inner filter effect corrections in fluorimetry. Fresenius J Anal Chem 340: 341–344.
53. Kratochwil NA, Huber W, Muller F, Kansy M, Gerber PR (2002) Predicting plasma protein binding of drugs: A new approach. Biochem Pharmacol 64: 1355–1374.
54. Hermann T, Patel DJ (2000) Adaptive recognition by nucleic acid aptamers. Science 287: 820–825.
55. Lakowicz JR (2006) Principles of fluorescence spectroscopy. London: Springer. 954 p.
56. Cortez CM, Silva D, Silva CM, Missailidis S (2012) Interactions of aptamers with sera albumins. Spectrochim Acta A Mol Biomol Spectrosc 95: 270–275.
57. Silva D, Cortez CM, Silva CM, Missailidis S (2013) A fluorescent spectroscopy and modelling analysis of anti-heparanase aptamers-serum protein interactions. J Photochem Photobiol B, 127: 68–77.
58. Dick LW Jr, McGown LB (2004) Aptamer-Enhanced Laser Desorption/Ionization for Affinity Mass Spectrometry. Anal Chem, 76: 3037–3041.
59. Hianik T, Ostatna V, Zajacova Z, Stoikova E, Evtugyn G (2005) Detection of aptamer-protein interactions using QCM and electrochemical indicator methods. Bioorg Med Chem Lett 15: 291–295.
60. Béra Abérem M, Najari A, Ho H-A, Gravel J-F, Nobert P, et al. (2006), Protein Detecting Arrays Based on Cationic Polythiophene–DNA-Aptamer Complexes. Adv Mater 18: 2703–2707.
61. Wang Y, He X, Wang K, Ni X, Su J, et al. (2011) Electrochemical detection of thrombin based on aptamer and ferrocenylhexanethiol loaded silica nanocapsules. Biosens Bioelectronics 26: 3536–3541.
62. Kawde A-N, Rodriguez MC, Lee TMH, Wang J (2005) Label-free bioelectronics detection of aptamer-protein interactions. Electrochem Comm 7: 537–540.
63. Cole JR, Dick LW Jr, Morgan EJ, McGown LB (2007) Affinity Capture and Detection of Immunoglobulin E in Human Serum Using and Aptamer-Modified Surface in Matrix-Assisted Laser Desorption/Ionization Mass Spectroscopy. Anal Chem 79: 273–279.
64. Vivekananda J, Kiel JL (2006) Anti-Francisella tularensis DNA aptamers detect tularemia antigen from different subspecies by Aptamer-Linked Immobilized Sorbent Assay. Lab Investig 86: 610–618.
65. Arnold S, Pampalakis G, Kantiotou K, Silva D, Cortez CM, et al. (2012) One round of SELEX for the generation of DNA aptamers directed against KLK6. Biol Chem 393: 343–353.

2

Early versus Deferred Treatment for Smoldering Multiple Myeloma

Minjie Gao[1,☯], Guang Yang[1,☯], Van S. Tompkins[2], Lu Gao[1], Xiaosong Wu[1], Yi Tao[1], Xiaojing Hu[1], Jun Hou[1], Ying Han[1], Hongwei Xu[3], Fenghuang Zhan[3]*, Jumei Shi[1]*

1 Department of Hematology, Shanghai Tenth People's Hospital, Tongji University School of Medicine, Shanghai, China, **2** Department of Pathology, University of Iowa Carver College of Medicine, Iowa City, Iowa, United States of America, **3** Department of Internal Medicine, University of Iowa Carver College of Medicine, Iowa City, Iowa, United States of America

Abstract

Purpose: Whether patients with smoldering multiple myeloma (SMM) needed to receive early interventional treatment remains controversial. Herein, we conducted a meta-analysis comparing the efficacy and safety of early treatment over deferred treatment for patients with SMM.

Methods: MEDLINE and Cochrane Library were searched to May 2014 for randomized controlled trials (RCTs) that assessed the effect of early treatment over deferred treatment. Primary outcome measure was mortality, and secondary outcome measures were progression, response rate, and adverse events.

Results: Overall, 5 trials including 449 patients were identified. There was a markedly reduced risk of disease progression with early treatment (Odds Ratio [OR] = 0.13, 95% confidence interval [CI] = 0.07 to 0.24). There were no significant differences in mortality and response rate (OR = 0.85, 95% CI = 0.45 to 1.60, and OR = 0.63, 95% CI = 0.32 to 1.23, respectively). More patients in the early treatment arm experienced gastrointestinal toxicities (OR = 10.02, 95%CI = 4.32 to 23.23), constipation (OR = 8.58, 95%CI = 3.20 to 23.00) and fatigue or asthenia (OR = 2.72, 95%CI = 1.30 to 5.67). No significant differences were seen with the development of acute leukemia (OR = 2.80, 95%CI = 0.42 to 18.81), hematologic cancer (OR = 2.07, 95%CI = 0.43 to 10.01), second primary tumors (OR = 3.45, 95%CI = 0.81 to 14.68), nor vertebral compression (OR = 0.18, 95%CI = 0.02 to 1.59).

Conclusions: Early treatment delayed disease progression but increased the risk of gastrointestinal toxicities, constipation and fatigue or asthenia. The differences on vertebral compression, acute leukemia, hematological cancer and second primary tumors were not statistically significant. Based on the current evidence, early treatment didn't significantly affect mortality and response rate. However, further much larger trials were needed to provide more evidence.

Editor: Zhenyu Li, University of Kentucky, United States of America

Funding: This study was supported by grants from the National Natural Science Foundation of China (No. 81372391, 81071856 to J.S., 81000199 to X.H., 81250110552 to T.V., and 81228016 to F.Z.), Shanghai Science and Technology Program (No. 12410705100 to J.S.), Shanghai Tenth People's Hospital Funds (No. 040113015 to J.S.). The funders conceived and designed the experiments, contributed reagents/materials/analysis tools, drafted the paper, and read and approved the final paper.

Competing Interests: The authors have declared that no competing interests exist.

* Email: shijumei@hotmail.com (JMS); fenghuang-zhan@uiowa.edu (FHZ)

☯ These authors contributed equally to this work.

Introduction

Multiple myeloma (MM) is a cancer of plasma cells that accumulate in the bone marrow and produce a monoclonal protein [1]. It is the second most common hematologic cancer, accounting for 15–20% of deaths from hematologic cancer and about 2% of deaths from cancer [1,2]. Advances in the basic understanding of MM and the introduction of novel agents including thalidomide, lenalidomide and bortezomib, have delayed progression and prolonged patients' survival. Despite these advances, unfortunately there is still no cure for MM, and the goal of therapy is to prolong survival [3,4]. Prognosis of patients with MM depends mainly on the stage of the disease [5].

According to the International Myeloma Working Group guidelines, MM-related neoplastic conditions have been categorized into monoclonal gammopathy of undetermined significance (MGUS), smoldering (asymptomatic) myeloma (SMM), solitary plasmacytoma of bone, and symptomatic multiple myeloma [6]. Diagnosed symptomatic multiple myeloma falls into one of three stages, IB, II, or IIIA/B, whereas SMM is identified as stage IA myeloma as determined by Durie/Salmon staging [6]. SMM patients are at high risk of progressing to symptomatic myeloma, with a median time to progression of approximately 1–2 years [7]. This intermediate condition is characterized by monoclonal protein concentrations of more than or equal to 3 g/dL and/or bone marrow plasma cells that are more than or equal to 10% without myeloma related organ or tissue impairment [8–10]. 54% of SMM patients will progress to frank multiple myeloma 5 years after initial diagnosis. This value exceeds 70% 15 years post-diagnosis [11,12].

Historically, the majority of patients with SMM are followed without therapy until symptoms developed [13,14]. The main reason for this conservative approach was that previous options for MM therapy were limited and toxic and unable to improve survival benefit [5,15] based on the conclusions from three small, randomized controlled trials prior to 2003 [16–18]. These trials compared early treatment with melphalan and prednisone with observation and demonstrated a significant delay in time to progression. However, this did not translate into better overall survival. Because all the trials included in the He et al. meta-analysis [5] used non-targeted agents (melphalan and prednisone) and had small sample sizes, the treatment of SMM remained controversial [12]. A better understanding of the biology of MM has led to a wealth of novel agents to treat it, providing for the possibility that at least some of these new agents may be used to effectively intervene at early stages of MM. Investigators have since begun to carry out clinical trials to evaluate novel drugs and new combinations of existing treatments in SMM. From the year 2003 onwards, two phase randomized, controlled trials (RCTs) have been completed [19,20] and one phase III RCT with 370 SMM patients is ongoing. Immuno-modulatory agents (IMiDs) were used in each of these three more recent RCTs.

To gain a better, more complete and current understanding of the efficacy and safety of early treatment over deferred treatment for SMM patients, we conducted a meta-analysis based on data from the five completed phase III RCTs. Our analysis includes the three trials [16–18] previously used by He et al. [5] but also include the two more recent trials that used IMiDs [19,20]. In performing our meta-analysis we address whether early treatment of SMM with the newer, targeted agents (IMiDs) have improved clinical benefit over the non-targeted agents (melphalan and prednisone), and whether increasing the sample size provides more reliable data for early treatment regardless of agent type.

Methods

Search strategy

We used MEDLINE and Cochrane Library to locate all relevant studies published up to May 2014. The search criterion was listed in Table S1 and Table S2. We used 'the related articles' function in PubMed to identify other potentially relevant articles. Further, we searched the ClinicalTrials.gov registry. We also checked all the references of retrieved articles and asked for additional data and explanations when key information relevant to the meta-analysis was missing. The data was collected only from published, peer-reviewed papers.

Selection criteria

Studies included in our analyses were required to be RCTs that compared early versus deferred treatment for SMM. The treatment strategy and the criteria used for selecting patients also needed to be reported. Further, every RCT must have either reported clinical outcomes or safety of the treatments. The eligibility of each study was assessed independently by two investigators.

Data extraction and methodological quality assessment

The quality of trials was evaluated by two independent reviewers by examining the adequacy of the allocation randomization, allocation concealment, blinding, data analysis, withdrawals and dropouts and power analysis. Two reviewers performed data extraction independently based on selection criteria. If a disagreement arose, agreement was achieved through consultation with a third reviewer.

Outcomes Assessments

Our primary outcome for this meta-analysis was mortality. Secondary outcomes included progression, response rate and adverse events. Mortality was defined as time from randomization to death from any cause. Progression was defined as time from randomization to documentation of progression. Safety outcomes included the incidence of adverse events, specifically gastrointestinal toxicities, constipation, fatigue or asthenia, vertebral compression, acute leukemia, hematological cancer and second primary tumor.

Statistical analysis

The effect of treatment for each study was expressed as Odds Ratio [OR] of the early treatment arm over deferred treatment arm, with 95% confidence interval [CI]. A random-effect model was obtained to conduct a meta-analysis of all the relevant RCTs to get a conservative conclusion. Heterogeneity was assessed by both chi-squared test and I^2 statistics. Statistically significant heterogeneity was defined as p-value <0.10 or I^2 statistic$>50\%$. Revman software (5.2) was used to perform all calculations.

Results

1 Selection of the trials

Our initial search yielded 279 potentially relevant studies, of which 90 studies were duplicated and 176 studies were deemed ineligible after screening titles and abstracts (Figure 1). The full text of the remaining 13 studies was reviewed in full, of which five randomized controlled trials fully met the inclusion criteria [16–20]. Excluded full-text studies, with the reasons for exclusion, were listed in Table S3.

2 Description of trials

An outline of the five trials is provided in Table 1. The trial results were published between 1993 and 2013. A total of 449 patients with SMM were included. Three of the trials used melphalan and prednisone as early treatment agents [16–18]. Two of the trials used IMiDs as early treatment agents [19–20]. The methodological quality of the 5 trials is summarized in Table 2. Two trials described the methods of randomization and allocation concealment [17,18]. Three trials used an intention-to-treat analysis [16,19,20]. None of the five trials described power analysis and reported double blinding of the participants and outcome assessors. All trials described withdrawls and dropouts [16–20].

3 Mortality

In the early treatment and deferred treatment arms, 223 and 226 patients were evaluable for mortality, respectively. In total, there were 77 deaths in the early treatment arm and 81 deaths in the deferred treatment arm. As shown in Figure 2, no statistically

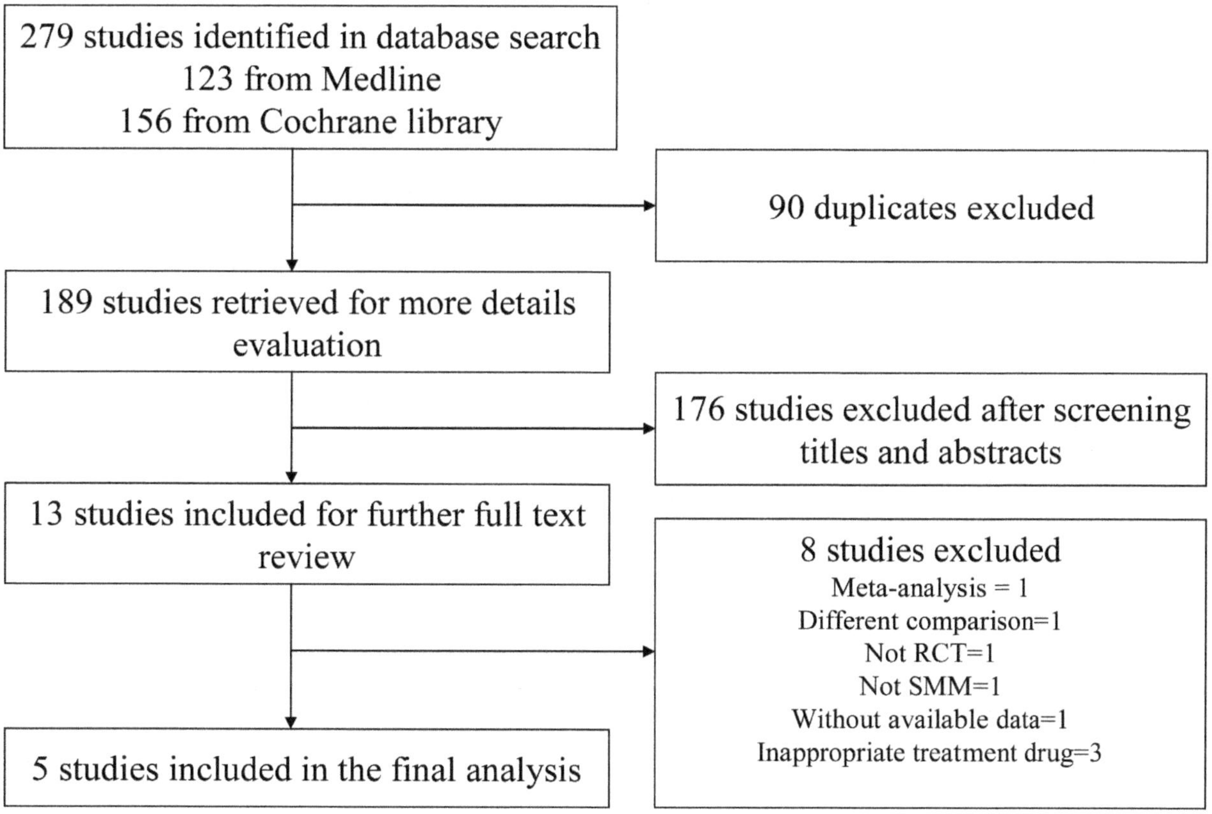

Figure 1. Flowchart of the selection of studies.

significant heterogeneity was observed ($p = 0.10$, $I^2 = 49\%$), and there was no significant difference in overall mortality between the two arms (pooled OR = 0.85, 95%CI = 0.45 to 1.60). When stratifying the data according to treatment type, neither alkylators nor IMiDs had a significantly reduced mortality. There was evidence of significant heterogeneity in the IMiDs group ($I^2 = 52\%$).

4 Progression

In the early and deferred treatment arms, 198 and 201 patients were evaluable for disease progression, respectively. Twenty-seven (27) patients progressed to MM in the early treatment arm compared to 108 in the deferred treatment arm. As shown in Figure 3, there was a markedly reduced risk of disease progression with early treatment (pooled OR = 0.13, 95%CI = 0.07 to 0.24).

Table 1. Characteristics of studies fulfilling inclusion criteria in the meta-analysis.

Author [year]	Disease	Early treatment defined as	Deferred treatment defined as	No. of enrolled/ analyzed patients	Intervention
Hjorth [1993]	SMM	Immediate treatment on diagnosis/randomization	Observation until symptomatic disease progression	E: 25/25 D: 25/25	M: 0.25 mg/kg P: 2 mg/kg d1-4 of 6 w intervals
Riccardi [1994]	SMM	Immediate treatment on diagnosis/randomization	Observation until symptomatic disease progression	E:38/34 D:40/40	M: 0.21 mg/kg d1-4 of 6 w intervals P: 0.5 mg/kg d1-10 of 6 w intervals
Riccardi [2000]	SMM	Immediate treatment on diagnosis/randomization	Observation until symptomatic disease progression	E:75/72 D:70/66	M: 0.21 mg/kg d1-4 of 6 w intervals P: 0.5 mg/kg d1-10 of 6 w intervals
Witzig [2013]	SMM	Immediate treatment on diagnosis/randomization	Observation until symptomatic disease progression	E:35/35 D:33/33	ZLD: 4 mg/d Thal: 200 mg/d a 28 d cycle
Mateos [2013]	High-risk SMM	Immediate treatment on diagnosis/randomization	Observation until symptomatic disease progression	E:57/57 D:62/62	Induction (L: 25 mg/d d1-21 Dex: 20 mg/d d1-4,12-15 4 w intervals, 9 cycles) Maintenance (L: 10 mg/d d1-21, a 28 d cycle, 2 y)

MM: multiple myeloma; SMM: smouldering myeloma; M: melphalan; P: prednisone; ZLD: zoledronic acid; Thal: thalidomide; L: lenalidomide; Dex: dexamethasone; E: early treatment arm; D: deferred treatment arm; d: day; w: week; y: year.

Table 2. Methodological quality assessment of included trial.

Author [year]	Allocation generation	Allocation concealment	Double blinding	ITT	Withdrawls and dropouts described	Power analysis described
Hjorth [1993]	Unclear	Unclear	No	Yes	Yes	No
Riccardi [1994]	Computer generated	Adequate	No	No	Yes	No
Riccardi [2000]	Computer generated	Adequate	No	No	Yes	No
Witzig [2013]	Unclear	Unclear	No	Yes	Yes	No
Mateos [2013]	Unclear	Unclear	No	Yes	Yes	No

ITT: intention-to-treat.

No statistically significant heterogeneity was observed ($p = 0.23$, $I^2 = 31\%$). When divided by the treatment type, early treatment with either alkylators or IMiDs significantly delayed disease progression. There was evidence of significant heterogeneity in the alkylators group ($I^2 = 69\%$).

5 Response Rate

In the early and deferred treatment arms, 100 and 56 patients were evaluable for response rate, respectively. There were 43 responses in the early treatment arm and 31 responses in the deferred treatment arm. As shown in Figure 4, no heterogeneity was observed ($p = 0.45$, $I^2 = 0\%$). There was no significant difference in response rate between the two arms (pooled OR = 0.63, 95%CI = 0.32 to 1.23).

6 Adverse events

6.1 Acute leukemia. Acute leukemia was reported in three trials. 4 of 131 patients in the early treatment arm and 1 of 131 patients in the deferred arm developed acute leukemia. No heterogeneity was observed ($p = 0.59$, $I^2 = 0\%$). The incidence of acute leukemia in the early treatment arm did not differ significantly from that in the deferred treatment arm (pooled OR = 2.80, 95%CI = 0.42 to 18.81) (Figure 5).

6.2 Hematologic cancer. Hematologic cancer was reported in four studies. 5 of 188 patients in the early treatment arm and 2 of 193 patients in the deferred arm developed hematologic cancer. No heterogeneity was observed ($p = 0.74$, $I^2 = 0\%$). The incidence of hematologic cancer in the early treatment arm did not significantly differ from that in the deferred treatment arm (pooled OR = 2.07, 95%CI = 0.43 to 10.01) (Figure. 5).

6.3 Second primary tumors. Second primary tumors were reported in four studies. 8 of 188 patients in the early treatment arm and 2 of 193 patients in the deferred arm developed second primary tumors. No heterogeneity was observed ($p = 0.82$, $I^2 = 0\%$). The incidence of second primary tumors did not significantly differ between the early and deferred treatment arms (pooled OR = 3.45, 95%CI = 0.81 to 14.68) (Figure 5).

6.4 Vertebral compression. Vertebral compression was reported in two trials. None of the 97 patients in the early treatment arm, but 4 of 91 patients in the deferred treatment arm

Study or Subgroup	Early treatment Events	Early treatment Total	Deferred treatment Events	Deferred treatment Total	Weight	Odds Ratio M-H, Random, 95% CI
1.1.1 Alkylators						
Hjorth 1993	17	25	12	25	17.6%	2.30 [0.73, 7.27]
Riccardi 1994	6	34	12	40	18.3%	0.50 [0.16, 1.52]
Riccardi 2000	41	72	35	66	28.3%	1.17 [0.60, 2.29]
Subtotal (95% CI)		**131**		**131**	**64.2%**	**1.11 [0.53, 2.31]**
Total events	64		59			
Heterogeneity: Tau² = 0.19; Chi² = 3.55, df = 2 (P = 0.17); I² = 44%						
Test for overall effect: Z = 0.27 (P = 0.79)						
1.1.2 IMiDs						
Mateos 2013	4	57	13	62	16.9%	0.28 [0.09, 0.93]
Witzig 2013	9	35	9	33	18.9%	0.92 [0.31, 2.71]
Subtotal (95% CI)		**92**		**95**	**35.8%**	**0.53 [0.17, 1.67]**
Total events	13		22			
Heterogeneity: Tau² = 0.36; Chi² = 2.08, df = 1 (P = 0.15); I² = 52%						
Test for overall effect: Z = 1.09 (P = 0.28)						
Total (95% CI)		**223**		**226**	**100.0%**	**0.85 [0.45, 1.60]**
Total events	77		81			
Heterogeneity: Tau² = 0.25; Chi² = 7.87, df = 4 (P = 0.10); I² = 49%						
Test for overall effect: Z = 0.50 (P = 0.61)						
Test for subgroup differences: Chi² = 1.13, df = 1 (P = 0.29), I² = 11.9%						

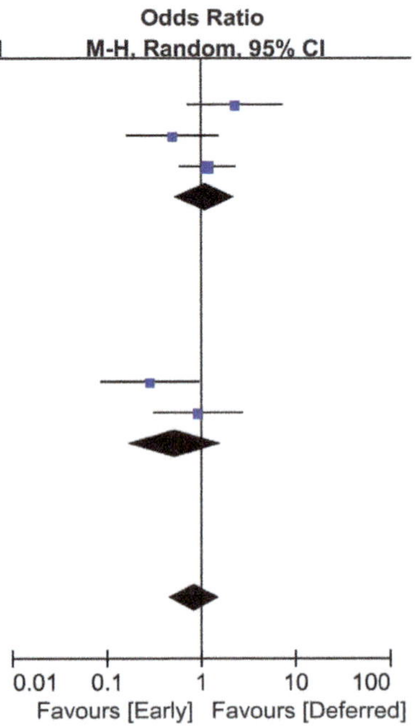

Figure 2. Mortality odds ratios of early treatment versus deferred treatment. The 95% CIs are shown. Squares represent the odds ratios and 95% CI for alkylators or IMiDs and finally for the overall estimate. CIs, confidence intervals; IMiDs, immuno-modulatory agents.

Study or Subgroup	Early treatment Events	Early treatment Total	Deferred treatment Events	Deferred treatment Total	Weight	Odds Ratio M-H, Random, 95% CI
1.2.1 Alkylators						
Riccardi 1994	4	34	12	40	19.6%	0.31 [0.09, 1.08]
Riccardi 2000	5	72	34	66	25.8%	0.07 [0.03, 0.20]
Subtotal (95% CI)		**106**		**106**	**45.4%**	**0.14 [0.03, 0.61]**
Total events	9		46			
Heterogeneity: Tau² = 0.77; Chi² = 3.27, df = 1 (P = 0.07); I² = 69%						
Test for overall effect: Z = 2.63 (P = 0.009)						
1.2.2 IMiDs						
Mateos 2013	13	57	47	62	33.0%	0.09 [0.04, 0.22]
Witzig 2013	5	35	15	33	21.5%	0.20 [0.06, 0.64]
Subtotal (95% CI)		**92**		**95**	**54.6%**	**0.12 [0.06, 0.25]**
Total events	18		62			
Heterogeneity: Tau² = 0.01; Chi² = 1.04, df = 1 (P = 0.31); I² = 4%						
Test for overall effect: Z = 5.84 (P < 0.00001)						
Total (95% CI)		**198**		**201**	**100.0%**	**0.13 [0.07, 0.24]**
Total events	27		108			
Heterogeneity: Tau² = 0.13; Chi² = 4.32, df = 3 (P = 0.23); I² = 31%						
Test for overall effect: Z = 6.33 (P < 0.00001)						
Test for subgroup differences: Chi² = 0.03, df = 1 (P = 0.86), I² = 0%						

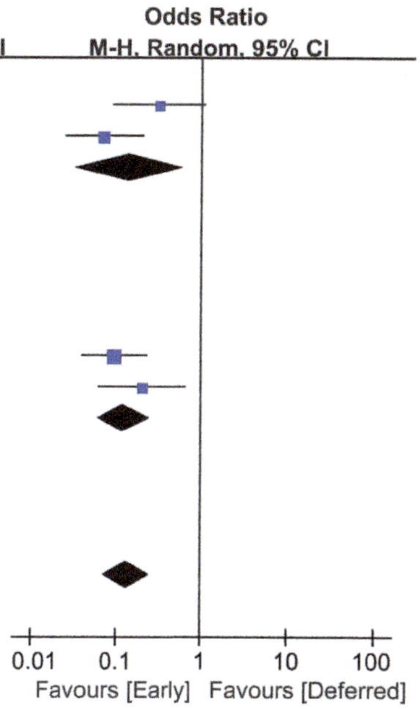

Figure 3. Progression rate odds ratios of early treatment versus deferred treatment. The 95% CIs are shown. Squares represent the odds ratios and 95% CI for alkylators or IMiDs and finally for the overall estimate. CIs, confidence intervals; IMiDs, immuno-modulatory agents.

suffered from vertebral compression. No heterogeneity was observed ($p = 0.99$, $I^2 = 0\%$). The incidence of vertebral compression in the early treatment arm did not significantly differ from that in the deferred treatment arm (pooled OR = 0.18, 95%CI = 0.02 to 1.59) (Figure 5).

6.5 Gastrointestinal toxicities. Gastrointestinal toxicities, including both constipation and diarrhea, were reported in only two the trials. Both studies reported the incidence of constipation, but only one reported the incidence of diarrhea. Forty-six (46) of 92 patients in the early treatment arm and 9 of 95 patients in the deferred treatment arm suffered from gastrointestinal toxicities. No heterogeneity was observed ($p = 0.46$, $I^2 = 0\%$). Patients in the early treatment arm experienced a greater incidence of gastrointestinal toxicities (pooled OR = 10.02, 95%CI = 4.32 to 23.23) (Figure 5).

6.6 Constipation. 32 of 92 patients in the early treatment arm and 7 of 95 patients in the deferred treatment arm suffered from constipation when only constipation among gastrointestinal toxicities was evaluated. No heterogeneity was observed ($p = 0.65$, $I^2 = 0\%$). We found significant differences between the two arms, with more patients in early treatment arm experiencing greater incidence of constipation (pooled OR = 8.58, 95%CI = 3.20 to 23.00) (Figure 5), but all the cases were grade 1 or 2.

6.7 Fatigue or asthenia. Fatigue or asthenia was reported in two trials. 42 of 92 patients in the early treatment arm and 26 of 95 patients in the deferred treatment arm experienced fatigue or asthenia. No heterogeneity was observed ($p = 0.58$, $I^2 = 0\%$). We found significant differences between the two arms, with a greater incidence of fatigue or asthenia in the early versus deferred treatment arms (pooled OR = 2.72, 95%CI = 1.30 to 5.67) (Figure 5). Fatigue or asthenia at grade 1 or 2 was reported in 37 of 92 patients in the early treatment arm and 23 of 95 patients in the deferred treatment arm. Higher grade fatigue or asthenia (3 or 4) was reported in 5 of 92 patients in the early treatment arm and 3 of 95 patients in the deferred treatment arm.

Discussion

The meta-analysis by He et al. in 2003 [5] evaluated three RCTs that used melphalan and prednisone to treat patients with SMM. Researchers have since questioned whether their conclusion that early treatment did not decrease mortality was actually a false-negative result. This was based on the fact that they only had

Study or Subgroup	Early treatment Events	Early treatment Total	Deferred treatment Events	Deferred treatment Total	Weight	Odds Ratio M-H, Random, 95% CI
Hjorth 1993	13	25	12	22	33.7%	0.90 [0.29, 2.85]
Riccardi 2000	30	75	19	34	66.3%	0.53 [0.23, 1.19]
Total (95% CI)		**100**		**56**	**100.0%**	**0.63 [0.32, 1.23]**
Total events	43		31			
Heterogeneity: Tau² = 0.00; Chi² = 0.56, df = 1 (P = 0.45); I² = 0%						
Test for overall effect: Z = 1.35 (P = 0.18)						

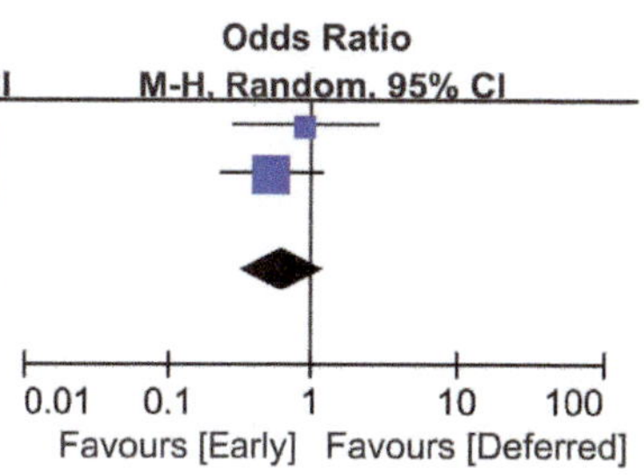

Figure 4. Response rate odds ratios of early treatment versus deferred treatment. Squares on the odds ratios plot are proportional to the weight of each study, which is based on the M-H method. Odds ratios are presents with 95% CIs. CIs, confidence intervals.

Study or Subgroup	Early treatment Events	Early treatment Total	Deferred treatment Events	Deferred treatment Total	Weight	Odds Ratio M-H, Random, 95% CI
1.3.1 Acute leukemia						
Riccardi 1994	0	34	0	40		Not estimable
Hjorth 1993	2	25	0	25	38.1%	5.43 [0.25, 118.96]
Riccardi 2000	2	72	1	66	61.9%	1.86 [0.16, 20.97]
Subtotal (95% CI)		**131**		**131**	**100.0%**	**2.80 [0.42, 18.81]**
Total events	4		1			
Heterogeneity: Tau² = 0.00; Chi² = 0.29, df = 1 (P = 0.59); I² = 0%						
Test for overall effect: Z = 1.06 (P = 0.29)						
1.3.2 Hematologic cancer						
Mateos 2013	0	34	0	40		Not estimable
Hjorth 1993	2	25	0	25	26.0%	5.43 [0.25, 118.96]
Riccardi 2000	1	57	1	62	31.8%	1.09 [0.07, 17.83]
Riccardi 1994	2	72	1	66	42.2%	1.86 [0.16, 20.97]
Subtotal (95% CI)		**188**		**193**	**100.0%**	**2.07 [0.43, 10.01]**
Total events	5		2			
Heterogeneity: Tau² = 0.00; Chi² = 0.59, df = 2 (P = 0.74); I² = 0%						
Test for overall effect: Z = 0.91 (P = 0.36)						
1.3.3 Second primary tumors						
Riccardi 1994	0	34	0	40		Not estimable
Hjorth 1993	2	25	0	25	22.0%	5.43 [0.25, 118.96]
Riccardi 2000	2	72	1	66	35.6%	1.86 [0.16, 20.97]
Mateos 2013	4	57	1	62	42.4%	4.60 [0.50, 42.47]
Subtotal (95% CI)		**188**		**193**	**100.0%**	**3.45 [0.81, 14.68]**
Total events	8		2			
Heterogeneity: Tau² = 0.00; Chi² = 0.40, df = 2 (P = 0.82); I² = 0%						
Test for overall effect: Z = 1.68 (P = 0.09)						
1.3.4 Vertebral compression						
Hjorth 1993	0	25	2	25	49.5%	0.18 [0.01, 4.04]
Riccardi 2000	0	72	2	66	50.5%	0.18 [0.01, 3.78]
Subtotal (95% CI)		**97**		**91**	**100.0%**	**0.18 [0.02, 1.59]**
Total events	0		4			
Heterogeneity: Tau² = 0.00; Chi² = 0.00, df = 1 (P = 0.99); I² = 0%						
Test for overall effect: Z = 1.54 (P = 0.12)						
1.3.5 Gastrointestinal toxicity						
Mateos 2013	24	57	3	62	43.6%	14.30 [4.00, 51.11]
Witzig 2013	22	35	6	33	56.4%	7.62 [2.49, 23.32]
Subtotal (95% CI)		**92**		**95**	**100.0%**	**10.02 [4.32, 23.23]**
Total events	46		9			
Heterogeneity: Tau² = 0.00; Chi² = 0.54, df = 1 (P = 0.46); I² = 0%						
Test for overall effect: Z = 5.37 (P < 0.00001)						
1.3.6 Constipation						
Mateos 2013	10	57	1	62	22.3%	12.98 [1.60, 104.99]
Witzig 2013	22	35	6	33	77.7%	7.62 [2.49, 23.32]
Subtotal (95% CI)		**92**		**95**	**100.0%**	**8.58 [3.20, 23.00]**
Total events	32		7			
Heterogeneity: Tau² = 0.00; Chi² = 0.20, df = 1 (P = 0.65); I² = 0%						
Test for overall effect: Z = 4.27 (P < 0.0001)						
1.3.7 Fatigue or Asthenia						
Witzig 2013	27	35	20	33	48.8%	2.19 [0.76, 6.29]
Mateos 2013	15	57	6	62	51.2%	3.33 [1.19, 9.32]
Subtotal (95% CI)		**92**		**95**	**100.0%**	**2.72 [1.30, 5.67]**
Total events	42		26			
Heterogeneity: Tau² = 0.00; Chi² = 0.31, df = 1 (P = 0.58); I² = 0%						
Test for overall effect: Z = 2.66 (P = 0.008)						

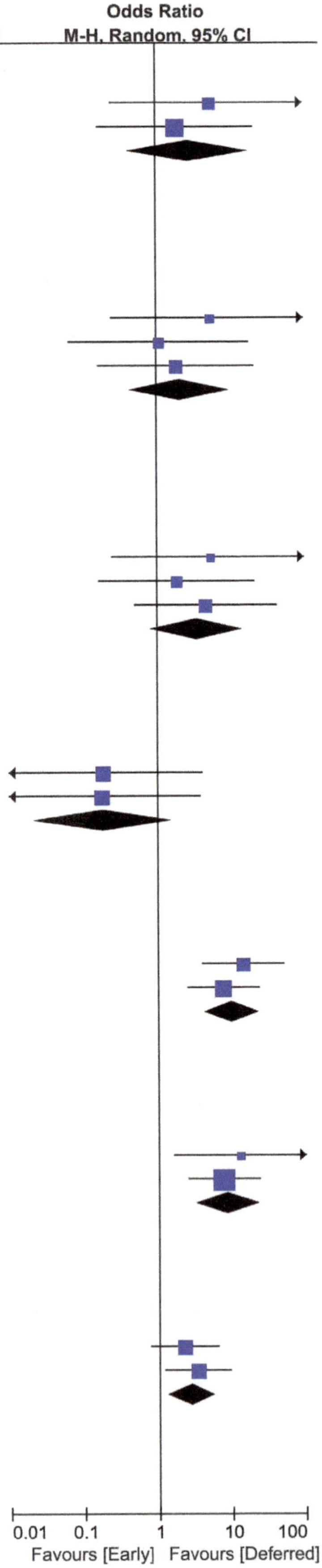

Figure 5. Adverse events rate odds ratios of early treatment versus deferred treatment. Squares on the odds ratios plot are proportional to the weight of each study, which is based on the M-H method. Odds ratios are presents with 95% CIs. CIs, confidence intervals.

262 patients in their analysis, which was only 70% of the requisite 350 patients or more that were needed to reliably detect 15% mortality difference based on the Pogue and Yusuf formula they used. In our analysis, 449 patients were included, providing us more statistical power to evaluate this difference. However, even with a greater number of patients, there was no survival benefit in treating SMM patients early. This held true when the data were combined or when we analyzed the trials by treatment group, alkylators and IMiDs, respectively. Notably, doing this puts us under the same criticism of the work by He et al. because of the reduced number of patients in each group. More large-scale trials with IMiDs and newly emerging therapies are needed to provide more evidence.

Mhaskar et al. [20] performed a meta-analysis of 10 RCTs comparing efficacy of early versus late first-line treatments for cancer. They conducted subgroup analysis based on cancer type, including MM. Their subgroup analyses for MM showed no difference in survival and response rate. But there was statistically significant difference in progression-free-survival. These results were consistent with ours. But they did not compare safety of early versus late first-line treatments for MM, which is one important part in our analysis.

Only the Mateos trial [21], among all trials included here, showed that mortality was significantly lower in the early treatment arm compared to that of the deferred treatment arm (7% versus 21%; OR = 0.28, 95%CI = 0.09 to 0.93). Interestingly, although all the patients included in all the trials we analyzed had SMM, the patients in the Mateos trial included a large number of high risk SMM. The addition of the high risk SMM patients could contribute to the different clinical outcomes. This is, however, confounded because there has not been a consistent definition for high risk SMM [22–24], and none of the other trials selected patients with high risk SMM, so we could not address this question. Moving forward, it will be important not only to evaluate early treatment for patients with high-risk SMM, but also to consistently define high-risk SMM.

Along with no reduction in mortality, we also found that early treatment with alkylators did not significantly improve response rates. Notably, these data were available for only two trials. Since the response rate of the deferred treatment arm in these trials with IMiDs were not reported, we were not sure whether early treatment with IMiDs could improve response rates, comparatively.

Despite there being no difference in mortality or response rates between the early and deferred treatment arms, our meta-analysis demonstrated that early treatment decreased the risk of disease progression. This also held true when we stratified the trials by treatment with either alkylators or IMiDs. The fact that overall mortality is not reduced but progression is delayed indicates that the early treatment group survives for a shorter amount of time than the deferred treatment group once the disease has progressed. This suggests that early treatment might actually predispose toward a more aggressive, more difficult to treat MM disease once it has progressed. The addition of molecular analyses at both diagnosis and progression may help to explain this phenomenon.

In addition to efficacy, safety is an equally important consideration for whether to subject patients to early treatment. Only vertebral compression and acute leukemia were analyzed in the meta-analysis of 2003. Apart from these adverse events, we also extracted data on gastrointestinal toxicity, hematological cancer, second primary tumor, and fatigue or asthenia. Other adverse events, such as neuropathy, anemia, rash and infection were not consistently reported, and were not considered in this analysis. Although there was a trend for increased incidence of acute leukemia, hematologic cancer, second primary tumors and vertebral compression, our meta-analysis revealed that there was no statistically significant difference between the early and deferred treatment arms. There were few cases that reported these adverse events and follow-up time was limited. Therefore, our findings should be further verified with more RCTs and longer follow-up times. Analyses of gastrointestinal toxicities showed that early treatment with IMiDs was associated with an increased incidence. When only constipation was evaluated, we found an increased incidence for the early IMiD treatment group, but all the cases were moderate. Data on gastrointestinal toxicities were not available for trials that used alkylators. Data were integrated regarding fatigue and asthenia because they were frequently reported together. Early treatment with IMiDs appeared to increase the incidence of fatigue or asthenia, but most cases were moderate. Such data were not available for studies that used alkylators. Our findings call attention to gastrointestinal toxicities and fatigue or asthenia as the predominant safety concerns associated with treating SMM patients with IMiDs.

There were a number of limitations of this meta-analysis. First, all individual studies used in our analyses had small sample sizes and did not have predetermined power analyses. Due to the small sample sizes, the data on vertebral compression, acute leukemia, hematological cancer and second primary tumor were not interpretable. Second, only two of the included studies reported an adequate method for randomized allocation and allocation concealment. Third, sample sizes were also too small for reliable subgroup analyses for IMiDs versus alkylators. Fourth, separate subgroup analysis was not possible on low- versus high-risk SMM. Finally, statistically significant heterogeneity was seen in subgroup of progression ($I^2 = 69\%$) and in subgroup of mortality ($I^2 = 52\%$). This might be derived largely from potential differences in the proportion of different SMM risk stratification, non-uniform reporting of clinical parameters, and variability in clinical factors.

Despite these limitations, our meta-analysis demonstrated that early treatment did indeed delay disease progression, but delaying progression did not translate into an improved response rate or decreasing mortality. However, further RCTs were needed to determine if early treatment of SMM with the newer, targeted agents and/or patients characteristics were more likely to improve response rate and overall survival. Early treatment also resulted in an increased risk of gastrointestinal toxicities and fatigue or asthenia.

Supporting Information

Table S1 Search criterion of Medline (via Pubmed, from inception to May 16, 2014).

Table S2 Search criterion of Cochrane Library (from inception to May 17, 2014).

Table S3 Characteristics of excluded full-text studies.

Checklist S1 PRISMA checklist.

Author Contributions

Conceived and designed the experiments: JmS FhZ MjG. Performed the experiments: MjG GY. Analyzed the data: MjG GY. Contributed reagents/materials/analysis tools: VST LG XsW YT XjH JH YH HwX. Wrote the manuscript: JmS FhZ MjG GY. Obtaining funding: JmS FhZ XjH VST. Read and approved the final paper: MjG GY VST LG XsW YT XjH JH YH HwX JmS FhZ.

References

1. Raab MS, Podar K, Breitkreutz I, Richardson PG, Anderson KC (2009) Multiple myeloma. Lancet 374: 324–339.
2. Smith D, Yong K (2013) Multiple myeloma. BMJ 346: f3863.
3. Dolloff NG, Talamo G (2013) Targeted therapy of multiple myeloma. Adv Exp Med Biol 779: 197–221.
4. Harousseau JL (2010) Optimising patient outcomes in myeloma. Cancer Treat Rev 36 Suppl 2: S33–S35.
5. He Y, Wheatley K, Clark O, Glasmacher A, Ross H, et al. (2003) Early versus deferred treatment for early stage multiple myeloma. Cochrane Database Syst Rev D4023.
6. Palumbo A, Sezer O, Kyle R, Miguel JS, Orlowski RZ et al. (2009) International Myeloma Working Group guidelines for the management of multiple myeloma patients ineligible for standard high-dose chemotherapy with autologous stem cell transplantation. Leukemia 23: 1716–1730.
7. Rajkumar SV, Gertz MA, Lacy MQ, Dispenzieri A, Fonseca R, et al.(2003) Thalidomide as initial therapy for early-stage myeloma. Leukemia 17: 775–779.
8. Fouquet G, Guidez S, Herbaux C, Demarquette H, Leleu X (2013) [Smoldering multiple myeloma.]. Rev Med Interne.
9. Waxman AJ, Kuehl M, Balakumaran A, Weiss B, Landgren O (2010) Smoldering (asymptomatic) multiple myeloma: revisiting the clinical dilemma and looking into the future. Clin Lymphoma Myeloma Leuk 10: 248–257.
10. Kyle RA, Rajkumar SV (2007) Monoclonal gammopathy of undetermined significance and smouldering multiple myeloma: emphasis on risk factors for progression. Br J Haematol 139: 730–743.
11. Kyle RA, Remstein ED, Therneau TM, Dispenzieri A, Kurtin PJ, et al. (2007) Clinical course and prognosis of smoldering (asymptomatic) multiple myeloma. N Engl J Med 356: 2582–2590.
12. Ghobrial IM (2013) Revisiting treatment paradigms in high-risk smoldering multiple myeloma: out with the old, in with the new? Leuk Lymphoma 54: 2328–2330.
13. Korde N, Kristinsson SY, Landgren O (2011) Monoclonal gammopathy of undetermined significance (MGUS) and smoldering multiple myeloma (SMM): novel biological insights and development of early treatment strategies. Blood 117: 5573–5581.
14. Blade J, Dimopoulos M, Rosinol L, Rajkumar SV, Kyle RA (2010) Smoldering (asymptomatic) multiple myeloma: current diagnostic criteria, new predictors of outcome, and follow-up recommendations. J Clin Oncol 28: 690–697.
15. Rajkumar SV (2012) Preventive strategies in monoclonal gammopathy of undetermined significance and smoldering multiple myeloma. Am J Hematol 87: 453–454.
16. Hjorth M, Hellquist L, Holmberg E, Magnusson B, Rodjer S, et al. (1993) Initial versus deferred melphalan-prednisone therapy for asymptomatic multiple myeloma stage I–a randomized study. Myeloma Group of Western Sweden. Eur J Haematol 50: 95–102.
17. Riccardi A, Ucci G, Luoni R, Brugnatelli S, Mora O, et al. (1994) Treatment of multiple myeloma according to the extension of the disease: a prospective, randomised study comparing a less with a more aggressive cystostatic policy. Cooperative Group of Study and Treatment of Multiple Myeloma. Br J Cancer 70: 1203–1210.
18. Riccardi A, Mora O, Tinelli C, Vanlentini D, Brugnatelli S, et al. (2000) Long-term survival of stage I multiple myeloma given chemotherapy just after diagnosis or at progression of the disease: a multicentre randomized study. Cooperative Group of Study and Treatment of Multiple Myeloma. Br J Cancer 82: 1254–1260.
19. Witzig TE, Laumann KM, Lacy MQ, Hayman SR, Dispenzieri A et al. (2013) A phase III randomized trial of thalidomide plus zoledronic acid versus zoledronic acid alone in patients with asymptomatic multiple myeloma. Leukemia 27: 220–225.
20. Mhaskar AR, Quinn G, Vadaparampil S, Djulbegovic B, Gwede CK, et al. (2010) Timing of first-line cancer treatments - early versus late - a systematic review of phase III randomized trials. Cancer Treat Rev 36: 621–628.
21. Mateos MV, Hernandez MT, Giraldo P, de la Rubia J, de Arriba F, et al. (2013) Lenalidomide plus dexamethasone for high-risk smoldering multiple myeloma. N Engl J Med 369: 438–447.
22. Rajkumar SV, Larson D, Kyle RA (2011) Diagnosis of smoldering multiple myeloma. N Engl J Med 365: 474–475.
23. Larsen JT, Kumar SK, Dispenzieri A, Kyle RA, Katzmann JA, et al. (2013) Serum free light chain ratio as a biomarker for high-risk smoldering multiple myeloma. Leukemia 27: 941–946.
24. Bianchi G, Kyle RA, Larson DR, Witzig TE, Kumar S, et al. (2013) High levels of peripheral blood circulating plasma cells as a specific risk factor for progression of smoldering multiple myeloma. Leukemia 27: 680–685.

3

A Small Library of Synthetic Di-Substituted 1, 4-Naphthoquinones Induces ROS-Mediated Cell Death in Murine Fibroblasts

Oscar Ramirez[1,2¤a], **Laura B. Motta-Mena**[1¤b], **Amanda Cordova**[1], **Kristine M. Garza**[1,2]*

1 Dept. of Biological Sciences, University of Texas at El Paso, El Paso, Texas, United States of America, **2** Border Biomedical Research Center, University of Texas at El Paso, El Paso, Texas, United States of America

Abstract

Synthesis of compound libraries and their concurrent assessment as selective reagents for probing and modulating biological function continues to be an active area of chemical biology. Microwave-assisted solid-phase Dötz benzannulation reactions have been used to inexpensively synthesize 2, 3-disubstituted-1, 4-naphthoquinone derivatives. Herein, we report the biological testing of a small library of such compounds using a murine fibroblast cell line (L929). Assessment of cellular viability identified three categories of cytotoxic compounds: no toxicity, low/intermediate toxicity and high toxicity. Increased levels of Annexin-V-positive staining and of caspase 3 activity confirmed that low, intermediate, and highly toxic compounds promote cell death. The compounds varied in their ability to induce mitochondrial depolarization and formation of reactive oxygen species (ROS). Both cytotoxic and non-cytotoxic compounds triggered mitochondrial depolarization, while one highly cytotoxic compound did not. In addition, all cytotoxic compounds promoted increased intracellular ROS but the cells were only partially protected from compound-induced apoptosis when in the presence of superoxide dismutase, catalase, or ascorbic acid suggesting utilization of additional pro-death mechanisms. In summary, nine of twelve (75%) 1, 4-naphthoquinone synthetic compounds were cytotoxic. Although the mitochondria did not appear to be a central target for induction of cell death, all of the cytotoxic compounds induced ROS formation. Thus, the data demonstrate that the synthesis regime effectively created cytotoxic compounds highlighting the potential use of the regime and its products for the identification of biologically relevant reagents.

Editor: Shama Ahmad, University of Colorado, Denver, United States of America

Funding: This work was supported by the University of Texas at El Paso Border Biomedical Research Center, funded by the National Institutes of General Medical Sciences – Research Centers at Minority Institutions [G12RR08124]. L.B.M.M. was supported by the National Science Foundation-funded University of Texas at El Paso Model Institution of Excellence Research Experience for Undergraduates Program. K.M.G. and this work was supported by the National Institute of General Medical Sciences SCORE Program [2S06GM1008012]. Undergraduate students who contributed to this work were supported by the National Science Foundation-funded Louis Stokes Alliance for Minority Participation Program [HRD1202008] (B.B) and by the National Institutes of Health-funded Bridges to the Baccalaureate Program [R25GM049011-14] (S.G.). The funders had no role in study design, data collection and analysis, decision to publish, or preparation of the manuscript.

Competing Interests: The authors have declared that no competing interests exist.

* Email: kgarza@utep.edu

¤a Current address: Section of Hematology, Yale University, New Haven, CT, United States of America
¤b Current address: Dept. of Biophysics, Univ. of Texas Southwestern Medical Center, Dallas, TX, United States of America

Introduction

Quinones are aromatic compounds naturally present in bacteria and eukaryotes. They are often involved in the biochemistry of energy production and serve as vital links in electron transport in the form of ubiquinones [1]. This biological activity is related to the acceptance of one and/or two electrons to form the corresponding radical anion or dianion species (electrophiles). Quinones are also natural defensive products made by plants and have been employed as anti-fungal agents, broad-spectrum anti-bacterials, and anti-malarial drugs [2–4]. Moreover, extensively substituted anthroquinones or *p*-benzoquinones or naphthoquinones with reactive or heterocyclic groups are effective anti-cancer agents [5,6] forming one of the largest classes of cytotoxic agents used therapeutically against cancer.

Quinones are particularly effective at inducing apoptosis [7,8] and as such provide a rich source of unique cytotoxic reagents that can be exploited. Of note, are the anti-malarial naphthoquinones, particularly hydroxyl-1, 4-naphthoquinone (atavoquone), which is used in combination with proguanil (known as Malarone) for the prevention and treatment of malaria [9,10]. In an effort to create 1, 4-naphthoquinone libraries a novel organic synthesis regime was developed by Shanmugasundaram et. al using microwave-assisted solid-phase Dötz benzannulation reactions [11]. This was the first report to use solid-supported Dötz benzannulation reaction and the subsequent oxidative cleavage process to generate derivatives of this class of quinones.

Here, twelve different 2, 3-disubstituted-1, 4-naphthoquinones synthesized using this regime were screened for biological activity against the murine fibroblast cell line, L929. The L929 cell line was chosen to serve as the adherent cell assay model due to its ease of use and its frequent use in toxicity assays for various agents [12–14]. The data demonstrates that the majority of the naphthoquinones studied were cytotoxic and promoted the induction of ROS formation.

Materials and Methods

Compounds

2, 3-disubstituted naphthoquinones were synthesized on solid support utilizing the Dotz reaction with solid supported Fischer carbine complexes as described [11]. The compounds were analyzed by gas chromatography and ranged from 95–99% purity. They were dissolved in DMSO at 20 mg/ml and stored frozen at −20°C.

Cell culture

The murine fibroblast cell line, L929, was purchases from ATCC (NCTC clone 929; L cell, L-929 derivative of Strain L; ATCC CCL-1). The cells were grown in Dulbecco's Modified Eagles Medium (DMEM) supplemented with 10% fetal bovine serum, 100 Units/ml Penicillin, 100 μg/ml Streptomycin, and 1X Glutamax (all cell culture reagents were purchased from Invitrogen Corp., Grand Island, NY). The cells were grown at 37°C, 5% CO_2 and were used for experiments when in expansion phase of culture (at approximately 70% confluency).

MTT viability assay

The MTT (3-(4, 5-dimethylthiazol-2-yl)-2, 5-diphenyl tetrazolium bromide) viability assays were carried out as per manufacturers' instructions (Sigma). The L929 cells were plated at 50,000 cells per well in a 96-well flat-bottom plate (Becton-Dickinson Labware, Franklin Lakes, NJ). The cells were cultured with decreasing concentrations of the compounds (in duplicate), beginning at 0.50 mg/ml and decreasing by half over eleven additional dilutions. As controls, cells remained untreated (media alone) or were treated with equivalent dilutions of dimethylsulfoxide (DMSO, vehicle control) (Sigma). Following a 48 h incubation, the MTT reagent was added at 10% of the total volume per well. Following a 4 h incubation, the formed crystals were solubilized by removing approximately ¾ of the supernatant and adding 50 μl of solvent (10% Triton × and 1% 12 N HCl in isopropanol) (Sigma). All wells were vigorously pipetted (without forming bubbles) to help dissolve the crystals. The plates were then read in a 96-well plate spectrophotometer at 570 nm.

Microscopy of compound-treated cells

Cells were grown at a density of 1×10^5 cells per well in a four-well chambered slide (Nalgene Nunc, Rochester, NY) in the presence of medium alone, the indicated compound (IC_{50}), DMSO, or H_2O_2. The cells were cultured for 24 h at 37°C, 5% CO_2. After the 24 h incubation, the cells were observed at 40× using a Zeiss Axiovert 200 microscope. Digital images were captured using the AxioCam HR digital camera and were processed with Axiovision Software, Version 4.1.

Flow cytometry assessment of apoptosis

L929 cells (500,000 cells/well) were cultured in 12-well plates for 48 h in the presence of medium, DMSO, or compound (IC_{50}). The cells were then washed in FACS buffer (phosphate buffered saline +2% fetal bovine serum), resuspended in 50 μl of buffer and then incubated with 30% normal mouse serum (Sigma) to prevent non-specific staining, for 15 min. at room temperature. The cells were then treated with 5 μl of FITC-conjugated annexin V (1 mg/ml stock) and 2 μl of propidium iodide (5 mg/ml stock). The cells were then mixed gently and incubated at room temperature for 15 minutes in the dark followed by an analysis by flow cytometry

Compound	Position R_1	Position R_2	Position R_3
2	Ph	Ph	H
3	CH_3	CH_3	H
4	OH	$CH_2CH{=}C(CH_3)_2$	H
5	CH_3	H	OH
6	H	H	H
7	H	H	OH
11	$CH_2CH_2CH_3$	H	H
12	CH_2CH_3	CH_3	H

core quinone structure

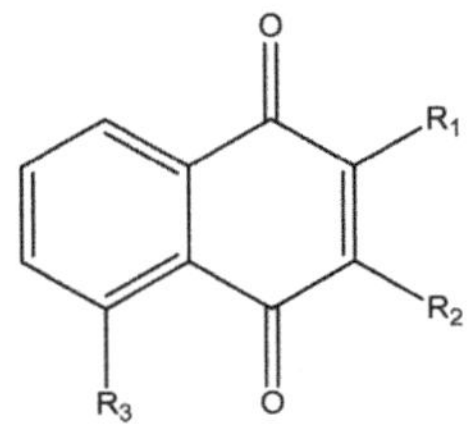

1

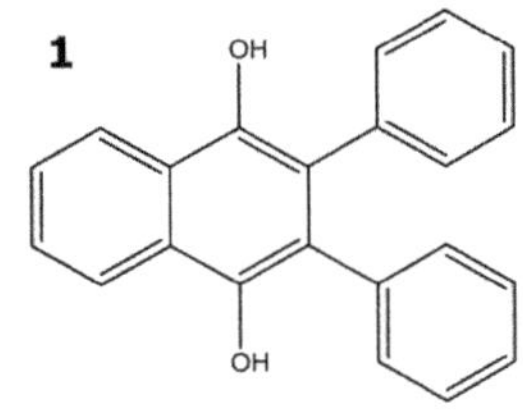

8

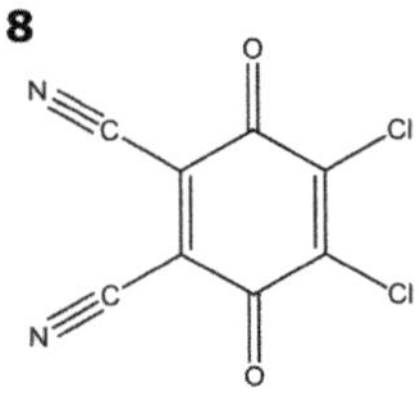

10

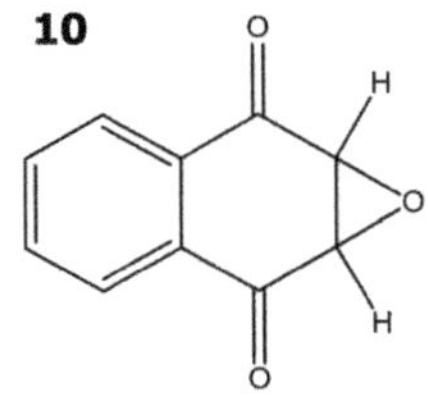

Figure 1. Structures of the synthetic 2, 3-disubstituted-1, 4-naphthoquinone library. All compound structures are based on the core structure (upper left), except for **1**, **8**, and **10**. Compound **9** is vitamin K1.

within the hour. 10,000 events per sample were collected and analyzed by EXPO32 software.

Detection of active caspase-3

The assay was performed as per the manufacturers' instructions (BioSource International, Camarillo, CA). The L929 cells were plated at 3×10^6 cells per well in 6-well plates. The cells were then treated with the indicated compounds for 24 h. After treatment with the compounds, the cells were centrifuged, washed twice with 1× PBS, were re-suspended in 50 μL of chilled Cell Lysis Buffer and were incubated for 10 min. on ice. The samples were then centrifuged and the cytosolic extracts were then transferred to 96-well flat-bottom plates. Reaction buffer (5 μL of 1.0 M DTT to 500 μL of 2× Reaction buffer), made just prior to use, and was added to each well. DEVD-pNA substrate (5 μL) was then added to each well and the samples were incubated in the dark for an additional 2 h. The plates were then read on a spectrophotometer at a wavelength of 405 nm.

Detection of mitochondrial membrane depolarization

L929 cells (5×10^4 cells/well) were cultured in 12-well plates for 48 h in the presence of medium, DMSO, camptothecin (4 μM, positive control) or compound (IC_{50}). The cells were then washed in FACS buffer resuspended in 100 μl of buffer. DePsipher (5, 5′6,

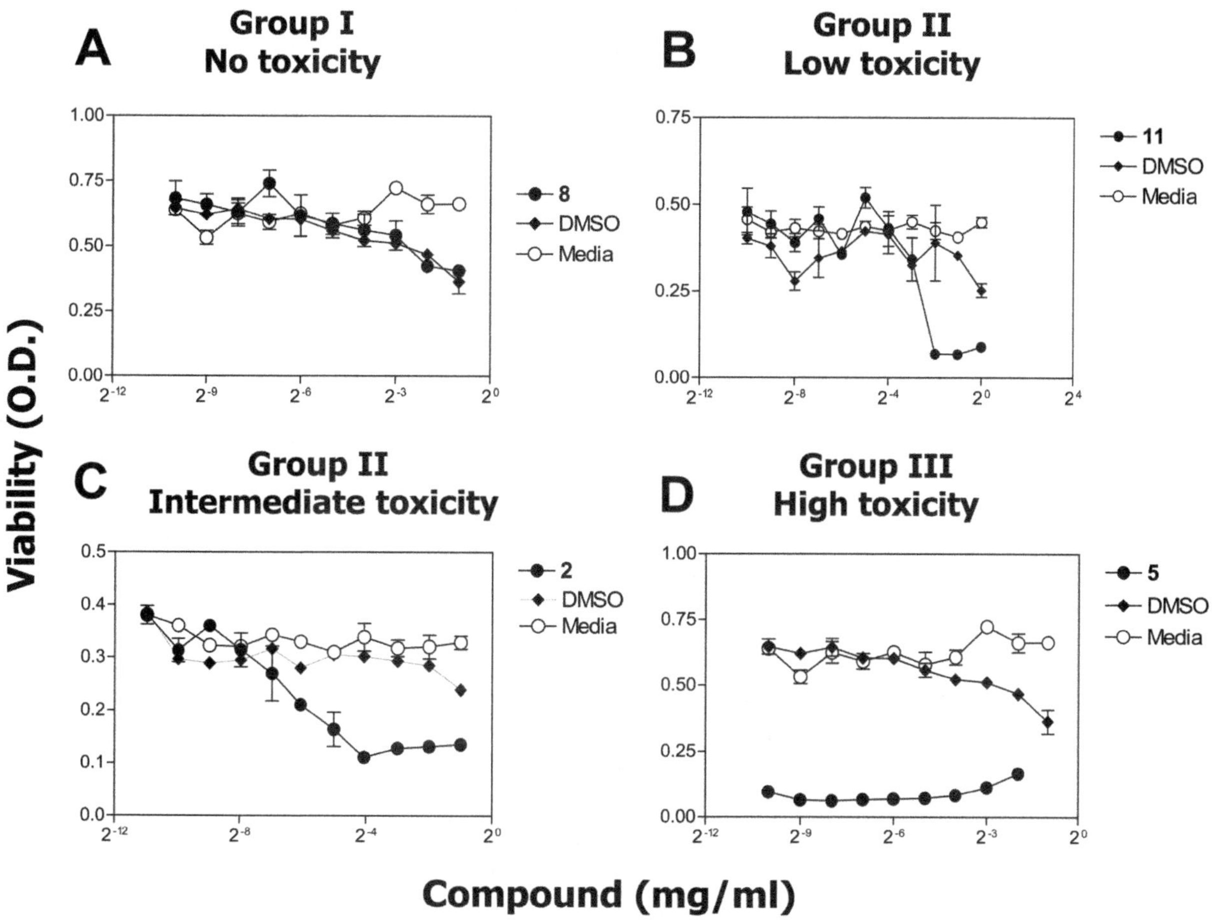

Figure 2. The synthetic naphthoquinone compounds vary in their ability to inhibit cellular viability. L929 cells were cultured for 48 hrs in the absence or presence of decreasing concentrations of vehicle control (DMSO) or decreasing concentrations of the indicated compound. The cells were then assessed for viability by a MTT assay. The compounds presented with no toxicity (Group I), low to intermediate toxicity (Group II), and high toxicity (Group III) (as described in the text). Compounds **2**, **5**, **8**, and **11** are representative examples of each group. The data is presented as the mean ± SEM of triplicate wells and is one of three representative experiments.

6′-tetrachloro-1, 1′, 3, 3′-tetraethylbenzimidazolylcarbocyanine iodide) (5 μl per sample) (Trevigen, Inc., Gaithersburg, MD) was then added. The cells were incubated at room temp., in the dark for a minimum of 15 min. The cells were assayed within the hour by flow cytometry. 10,000 events per sample were collected and analyzed by EXPO32 software.

Detection of reactive oxygen species (ROS) by DHCF-DA

Cells were plated at 50,000 cells per well in 96-well plates, protected or unprotected by 10 μM ascorbic acid. Following a 4 h incubation period, the cells were then treated with the indicated compounds at the IC_{50} or with H_2O_2 as a positive control. Twenty-four hours later, the plates were centrifuged and the supernatants were collected. The adherent cells were treated with 20 μL of 0.25% trypsin (Invitrogen) for 2 min. The cells were then removed and transferred to black-walled 96-well plates to prevent bleeding of fluorescence between sample wells. Dichlorofluorescein-Diacetate (10 μM) (Invitrogen - Molecular Probes), was then added to each well. The plates were then immediately read on a fluorescent plate reader using excitation wavelength of 485 nm and emission wavelength of 538 nm. Fluorescence was determined every 10 min over a 320 min. period.

Ascorbic acid protection assay

Prior to the viability assay, the cells were treated for 24 hrs with 10 μM ascorbic acid (Sigma). The cells were then thoroughly washed with Hank's Balanced Salt Solution (HBSS (Invitrogen) and the MTT viability assay was conducted as described above.

Catalase and super oxide dismutase (SOD) protection assay

L929 cells were plated at 50,000 cells/well in a 96-well flat-bottom plate. Catalase (1 mg/ml), SOD (0.5 mg/ml), or combinations of the two were added to the cells. The cells were also treated with the indicated compounds at a concentration that inhibited 50% of cellular viability or with media or DMSO as controls. Following a 48 h incubation period, cellular viability was assessed as described above (MTT assay).

Statistical analysis

Statistical significance was analyzed with Student's t-test. The level of significance was set at 0.05.

Table 1. Summary of compounds assessed for inhibition of viability.[a]

Compound	IC_{50}[b]
GROUP I – No Toxicity	
1:diphenyl phenol	No inhibition
8:2, 3-dichloro-5, 6-dicyano-1, 4-benzoquinone	No inhibition
9:2-methyl-3-phytyl-1, 4-naphthoquinone (Vitamin K1)	No inhibition
GROUP II – Low to Intermediate Toxicity	
2:2, 3-diphenyl-1, 4-napthoquinone	0.030 mg/ml = 96.7 μM
3:2, 3-dimethyl-1, 4-napthoquinone	0.008 mg/ml = 42.9 μM
4:2-hydroxy-3-(3-methyl-2 butenil)-1, 4 naphthoquinone	0.015 mg/ml = 62.0 μM
10:1, 4-napthoquinone epoxide	0.004 mg/ml = 23.0 μM
11:2-propyl-1, 4-napthoquinone	0.060 mg/ml = 300 μM
12:2-ethyl-3-methyl-1, 4-napthoquinone	0.030 mg/ml = 150 μM
GROUP III – High Toxicity	
5:2-methyl-5-hydroxy-1, 4-napthoquinone (plumbagin)	<0.001 mg/ml = <5.3 μM
6:1, 4-napthoquinone	<0.001 mg/ml = <6.3 μM
7:5-hydroxy-1, 4-napthoquinone (jugalone)	<0.001 mg/ml = <5.7 μM

[a]Viability was assessed by an MTT assay at 48 hrs of incubation over a series of doubling dilutions.
[b]Concentration of compound that induces 50% inhibition of survival.

Results

Compound effects on cellular viability and morphology

A total of twelve compounds were screened (Figure 1 and Table 1) for effects on a murine fibroblast cell line (L929). To determine if the compound library of 2, 3-disubstituted-1, 4-naphthoquinoneswere cytotoxic, the L929 cells were cultured in the presence of ten two-fold dilutions of each compound and were measured for viability. Cytotoxicity was measured as an inverse function of cell survival using a colorimetric viability assay. As indicated in Figure 2, the effects of the different compounds were placed into three groups with respect to cytotoxicity: (1) no cytotoxicity (no observed difference from DMSO vehicle control) (Fig. 2A); (2) low to intermediate cytotoxicity (the compound inhibited cell survival in a dose-responsive manner) (Fig. 2B and 2C); and (3) high cytotoxicity (>50% inhibition of cell survival at the lowest concentration examined) (Fig. 2D). The concentration of each compound that induced 50% inhibition of survival (IC_{50}) is listed in Table 1. Of note are the compounds used as controls: compound **5** (plumbagin) and **7** (juglone), both which are naturally occurring. Plumbagin is found in the roots of the Plumbago plant; juglone is found in the leaves, roots and bark of plants in the Juglandaceae family, particularly the black walnut. Both compounds have been found to be quite cytotoxic in a number of model systems [15–17] including our L929 murine fibroblast cell line, verifying that the new synthetic scheme being used to produce the naphthoquinones accurately generates cytotoxic compounds.

To assess the effect of our small library of compounds on the morphology of the cells, the fibroblasts were cultured in 4-chamber well slides in the presence of each compound at the IC_{50}. Following a 24 hr incubation, the slides were assessed by light microscopy. Untreated cells presented with two general morphologies: adherent-fibrous and semi-adherent rounded, which is typical of this cell line in different phases of the cell cycle (Fig. 3A). Similar morphologies were seen for cells treated with the vehicle control (Fig. 3B). In contrast, hydrogen peroxide (H_2O_2) treated cells were rounded, granular, and vesiculated, which is typical of dying cells (Fig. 3F). At a single concentration (IC_{50}), the compounds either had no effect on the cells, as represented by compound **9** (Fig. 3E), or were detrimental to cell survival, as represented by compounds **2** and **5** (plumbagin) (Fig. 3C and 3D), further demonstrating that members of our small compound library were cytotoxic.

Cytotoxic compounds induces phosphotidyl serine externalization and caspase 3 activation

We next assessed possible mechanisms the cytotoxic compounds (intermediate and high) could be utilizing to inhibit cell survival. We first assessed the loss of plasma membrane phosphatidylserine (PS) asymmetry. The appearance of PS at the cell surface, often considered a hallmark of apoptosis, has been said to prepare the dying cell to be phagocytosed [18]. We employed Annexin V staining to ascertain the extent of PS externalization upon treatment with the compounds. The L929 cells were treated with the IC_{50} for the intermediate cytotoxic compounds or with the lowest concentration tested for the highly cytotoxic compounds and were then stained with FITC-conjugated Annexin V in combination with propidium iodide (PI) and assessed by flow cytometry (Figure 4A). Very little necrosis was induced by any of the compounds (<10% above background, data not shown). In contrast, all the cytotoxic compounds (**2**, **3**, **4**, **5**, **6**, 7, **9**, **10**, and **11**) promoted externalization of PS significantly above that induced by medium alone. Compound 5, plumbagin, served as a positive control [15,16].

Caspase 3 is thought to play a central role in the execution phase of apoptosis in that it is indispensable for several hallmark features of the process [19]. Although it has been reported that the presence of cleaved caspase 3 can be associated with non-lethal biological processes [20], the highly cytotoxic compound **5** and **7** (plumbagin and jugalone) have been shown to act through caspase 3 [21]. The ability of the cytotoxic compounds to promote activation of caspase 3 in treated L929 cells was therefore also assessed. All of the cytotoxic compounds activated caspase 3. The data is presented in Figure 4B, using compounds **2**, **6**, and **9** as

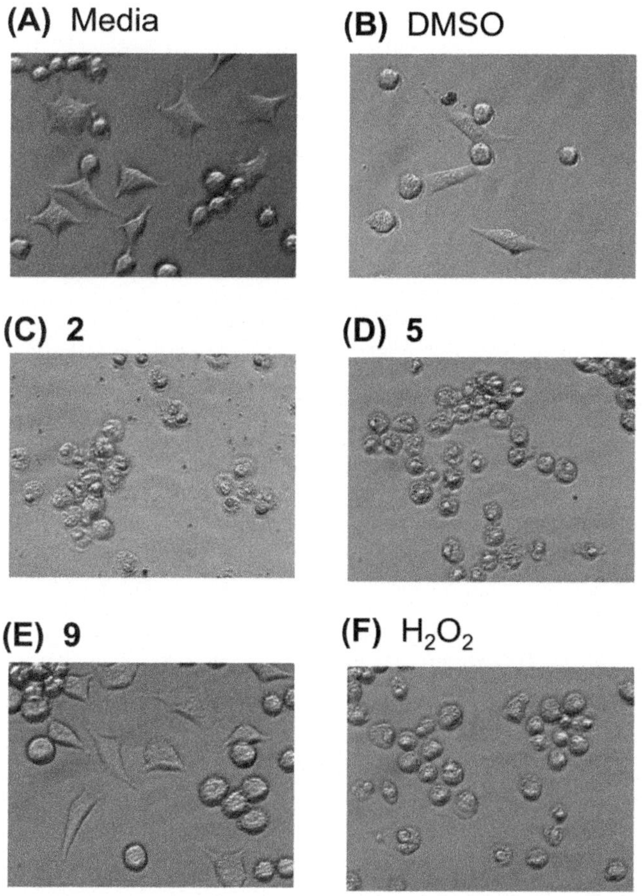

Figure 3. The cytotoxic compounds promote abnormal cellular morphology. The cellular morphology of L929 cells, grown in chamber well slides for 24 hrs in the presence of media alone (A), vehicle control (DMSO) (B), compound **2** (C), **5** (D), **9** (E), or H_2O_2 (F), was assessed by light microscopy (40 X). The cells were treated with the IC_{50} of each compound, an equivalent volume of DMSO (vehicle control), or 600 μM H_2O_2 (positive control).

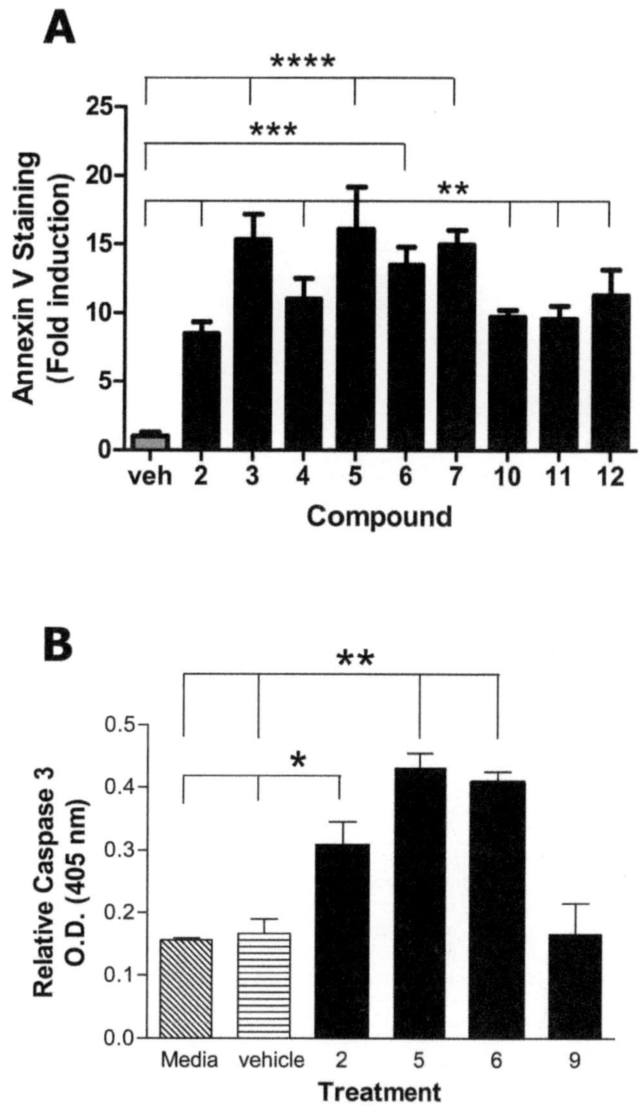

Figure 4. The cytotoxic compounds promote PS externalization and activate caspase 3. L929 cells were cultured for 48 hrs in the presence of media, vehicle control (DMSO), or the indicated compound (concentration = IC_{50}) (compound **5**, plumbagin, served as an internal positive control). The cells were harvested, washed, stained with Annexin V-FITC and were assessed by flow cytometry within 30 minutes of staining (A). Percentage of apoptotic non-treated cells (background) was approximately 10–20%. The data is plotted as the fold increase in apoptosis above that induced by the vehicle control. In parallel experiments, cells grown in the presence of media, vehicle control (DMSO) or the indicated compound were assessed for the presence of active caspase 3 using a colorimetric assay kit (BioSource) (B). Although all compounds were tested, compounds **2**, **6**, and **9** are presented as representative compounds for intermediate, high, or no toxicity, respectively. Compound **5** (plumbagin) served as an internal positive control. Data is presented as the mean ± SEM of duplicate wells and is one of three representative experiments. (A) ** $p \leq 0.001$, *** $p \leq 0.0004$, **** $p \leq 0.0001$. (B) * $p \leq 0.049$, ** $p \leq 0.0076$.

representative compounds for the three cytotoxicity categories (intermediate, high, and no cytotoxicity, respectively), and compound **5** (plumbagin) as a control, Interestingly, the low/intermediate cytotoxic compounds (represented by compound **2**) activated caspase 3 equivalently to the highly toxic compounds (represented by compound **6**) and to that of the compound **5** positive control. This suggests that although caspase 3 is activated, the extent of activation was not proportional to the extent of cytotoxicity.

Some of the cytotoxic compounds promote mitochondrial depolarization

To determine if the cytotoxic compounds were promoting cell death through actions on mitochondrial membrane potential, the DePsipher Kit from Trevigen was used. DePsipher is a unique cationic dye that indicates the loss of the mitochondrial potential. The dye readily enters cells and fluoresces bright red in its multimeric form within healthy mitochondria. If the mitochondrial membrane potential were to collapse, the DePsipher reagent cannot accumulate within the mitochondria. In these cells, the reagent remains in the cytoplasm as a green fluorescent monomeric form. As shown in Figure 5A, all the cytotoxic compounds, with the exclusion of compound **5**, promoted significant changes in mitochondrial membrane potential. Compound **5** (plumbagin) is of particular interest because it is one of the most cytotoxic compounds (Fig. 2) yet its mode of action is different than any of the other cytotoxic compounds tested as is demonstrated by the flow cytometry profile (Fig. 5B).

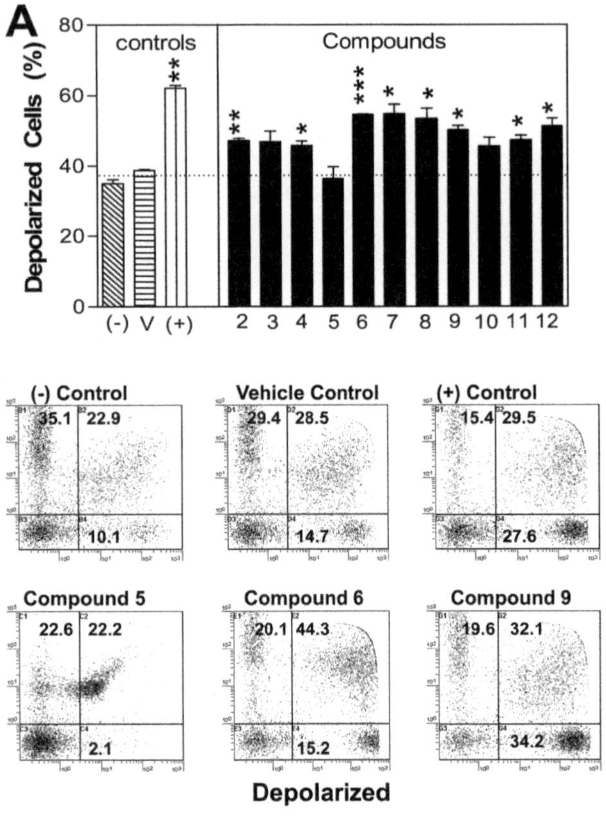

Figure 5. The synthetic naphthoquinone compounds vary in their ability to induce mitochondrial depolarization. L929 cells were in the presence of media (-), DMSO vehicle control (V), a positive control (+) or the indicated compound (concentration = IC_{50}). Compound **7** (jugalone) functioned as an internal control. The cells were harvested, washed, stained with DePsipher to assess mitochondrial depolarization and were assessed by flow cytometry within 30 minutes of staining. The upper and lower right hand quadrants of the dot plots (represented in panel B) together constitute the percentage of cells that underwent mitochondrial depolarization. Data is plotted as the mean ± SEM percentage of depolarized cells from duplicate samples (A) or as representative dot plots (B). The data is taken from one of two representative experiments. * $p \leq 0.036$, ** $p \leq 0.0066$, *** $p \leq 0.0002$.

Cytotoxic compounds promote production of reactive oxygen species

The mechanism of cytotoxicity by naphthoquinones varies due to differences in structures, diverse pharmacological effects, and different assay systems. However, the most commonly utilized mechanism appears to be the promotion of reduction-oxidation reactions [1]. The compounds are proposed to catalytically cycle and generate oxidative radicals such as hydrogen peroxide and superoxide, which then damage the cell. To assess whether our library of compounds promoted oxidative stress, L929 cells were treated with the representative compounds (**2**, **5**, and **9**) and were assessed by a fluorometric assay for the formation of reactive oxygen species (ROS) (Fig. 6A). The vehicle control (DMSO-treated) and compound **9**-treated cells induced ROS equal to that generated in non-treated cells (medium). The intermediate cytotoxic compound (**2**) and the highly cytotoxic compound (**5**) both induced levels of ROS equal to that induced by H_2O_2 (although not at the same rate).

To further assess the formation of oxidative radicals upon treatment with the cytotoxic naphthoquinone compounds, the compound-treated cells were concomitantly treated with the antioxidant proteins superoxide dismutase (SOD), catalase, or a combination of the two. SOD converts superoxide to hydrogen peroxide and catalase converts hydrogen peroxide to water and free oxygen, thus protecting the cells from potential ROS damage. The cells were treated with three representative compounds (**2**, **5**, and **9**) in the absence or presence of the oxidative radical inhibitors. Following a 48 hr incubation, the cells were assessed for viability.

As shown in Figure 6B, the addition of catalase partially protected compound **2**-treated cells, enhancing cell survival in the presence of the compound. The addition of SOD did not protect cells against compound **2** and the combination of catalase and SOD did not protect the cells against compound **2** beyond that provided by catalase alone. The addition of catalase, SOD, or a combination of the two also did not protect cells treated with the highly cytotoxic compound **5**. In parallel experiments, the cells to be treated with the compounds were first pre-loaded for 4 h with ascorbic acid. Ascorbic acid is a small molecule which sacrificially reacts with and eliminates oxidative radicals, and was used as a non-protein control for protection against ROS. As shown in Figure 6B, the cells treated with either compound **2** or **5** were partially protected from cell death when pre-treated with ascorbic acid. Thus, the data demonstrates that the cytotoxic naphthoquinone compounds induce oxidative radicals however, this is unlikely the only mechanism by which the compounds induce cell death.

Discussion

Twelve 2, 3-1, 4-naphthoquinone compounds produced by a novel organic synthetic scheme were screened for cytotoxicity against a murine fibroblast cell line. Two of the compounds (plumbagin = compound **5** and juglone = compound **7**), which occur naturally in plants as chemicals for defense, were used as internal controls for synthesis of biologically relevant 1, 4-naphthoquinone derivatives. The scheme utilized was unique in that it was the first report to demonstrate the use of the Dötz benzannulation of Fischer carbene complexes with alkynes to form substituted phenols and was the first report to apply these reactions to solid-phase organic synthesis [11]. The goal of the current study was to determine if the products generated by this process possessed the biological activity commonly associated with naphthoquinones.

The small library of twelve compounds presented with three categories of cytotoxicity: no cytotoxicity, low/intermediate cytotoxicity, and high cytotoxicity. Through further analysis, the low/intermediate, and highly cytotoxic compounds were determined to promote cell death as demonstrated by membrane PS externalization and by the activation of caspase 3. Published reports have established that naphthoquinone compounds with similar structures to those of our library are commonly used in experimental models of oxidative stress induction [22–24]. Moreover, both plumbagin and jugalone have been shown to induce oxidative stress as well [23,25–27]. We therefore sought to assess whether ROS production associated with cytotoxicity. Indeed, the formation of ROS was induced by all of the cytotoxic compounds. However, compound-treated cells were only partially protected by the anti-oxidant enzymes SOD and/or catalase or the ROS scavenger ascorbic acid suggesting the potential role of additional cell-death inducing properties of the cytotoxic compounds.

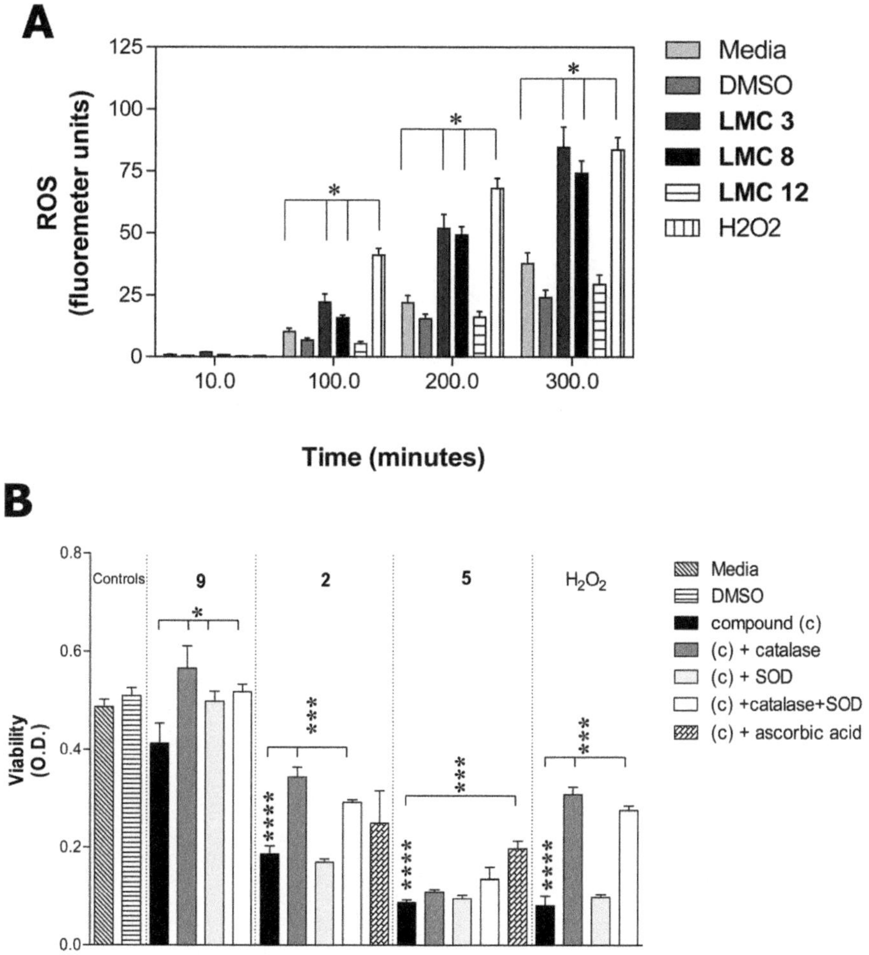

Figure 6. Compound-induced cell death associates with formation of ROS. L929 cells were cultured in the presence of media, vehicle control (DMSO), or the indicated compound (concentration = IC_{50}). At 24 hrs, the cultures were assessed for the formation of intracellular reactive oxygen species (ROS) utilizing a fluorometric assay (Molecular Probes) over a 350 minute time period (A). The assay was conducted for all twelve naphthoquionones, however the only representative compounds are shown. The data is presented as the mean ± SEM of triplicate wells and is taken from one of three representative experiments. In separate experiments, the cells incubated with the compounds were also treated with catalase (1 mg/ml), superoxide dismutase (SOD, 0.5 mg/ml), or a combination of the two. A separate group of cells were pre-treated for 4 hours with 10 μM ascorbic acid, were washed and then grown in the presence of media, vehicle control (DMSO), or the indicated compound (concentration = IC_{50}). At 48 hrs, all groups of cells were assessed for viability (B). The data is presented as the mean ± SEM of five replicate wells and is taken from one of three representative experiments. (A) * $p < .001$. (B). * $p \leq 0.043$, *** $p \leq 0.0004$, **** $p \leq 0.0002$, (each compound treatment, solid black, is compared to the DMSO vehicle control).

The additional cytotoxic mechanism induced by our compound library is particularly underscored by compound **5** (plumbagin), which was shown to promote the generation of ROS but does not activate mitochondrial depolarization (nor are its actions protected by SOD and/or catalase or ascorbic acid). This is in contrast to what has been seen in other mammalian cell model systems [26,28], where changes in mitochondrial membrane potential were observed. Plumbagin may instead be inducing cell cycle arrest through effects on cell cycle progression mediators and pro-survival molecules [29,30]. Although not tested here, it is possible that our compounds are also interacting with glutathione (GSH) or other thiol-containing proteins. Glutathione GSH, generated from the reduction of glutathione disulfide (GSSG), is a critical molecule in resisting oxidative stress acting as a scavenger for hydroxyl radicals, singlet oxygen, and various electrophiles [31,32]. In this capacity, it is possible that the electrophilic disubstituted 1, 4-naphthoquinone compounds interact with GSH inhibiting its ability to scavenge ROS. This may explain the partial protection from ROS formation imparted by SOD, catalase, and ascorbic acid upon treatment with the compounds. Indeed, quinones and naphthoquinones, including jugalone, have been shown to readily cause depletion of intracellular GSH [23,24,33,34].

In summary, our small library of compounds has yielded a group of synthetically generated naphthoquinones that mediate cytotoxicity. Induction of cell death by the cytotoxic compounds is associated with PS externalization, activation of caspase 3, and induction of ROS formation. Our data is in agreement with reports demonstrating that compounds with structures similar to our compounds are cytotoxic and induce cellular oxidative stress [22–29]. Of significance is that in our study, 75% of the synthetic compounds possess pro-cell death activity as determined by a number of assays in a well-defined eukaryotic cell system, offering a number of reagents for future study and highlighting the effectiveness of the unique organic synthetic scheme to generate biologically active naphthoquinone compounds.

Acknowledgments

We would like to sincerely thank Dr. Luis E. Martinez (formerly of the Dept. of Chemistry, University of Texas at El Paso; now the Director of the Center of Innovation and Entrepreneurship at Trinity University) for providing us with the small naphthoquinone library. We would also like to thank Bethany Barrow and Susan Garcia, UTEP undergraduate students who assisted O.R. with the cytotoxicity assays. Lastly, we thank Dr. Kyle Johnson (Dept. of Biological Sciences, University of Texas at El Paso) and Dr. Jacen S. Maier-Moore (Dept. of Clinical Laboratory Sciences) for critical review of the manuscript.

Author Contributions

Conceived and designed the experiments: OR KMG. Performed the experiments: OR LBMM AC. Analyzed the data: OR LBMM AC KMG. Contributed reagents/materials/analysis tools: KMG. Contributed to the writing of the manuscript: OR KMG.

References

1. Bentinger M, Tekle M, Dallner G (2010) *Coenzyme Q – Biosynthesis and functions*. Biochemical and Biophysical Research Communications 396(1): p. 74–79.
2. Kim YM, Lee CH, Kim HG, Lee HS (2004) *Anthraquinones Isolated from Cassia tora (Leguminosae) Seed Show an Antifungal Property against Phytopathogenic Fungi*. Journal of Agricultural and Food Chemistry 52(20): p. 6096–6100.
3. Tasdemir D, Brun R, Yardley V, Franzblau SG, Ruedi P (2006) *Antituberculotic and Antiprotozoal Activities of Primin, a Natural Benzoquinone: In vitro and in vivo Studies*. Chemistry & Biodiversity 3(11): p. 1230–1237.
4. Ui H, Ishiyama A, Sekiguchi H, Namatame M, Nishihara A, et al. (2007) *Selective and Potent In Vitro Antimalarial Activities Found in Four Microbial Metabolites*. The Journal of Antibiotics 60(3): p. 220–222.
5. Brondani DJ, Nascimento CR, de M Moreira DR, Lima Leite AC, de Souza IA, et al. (2007) *Synthesis and Antitumour Activity of the Primin (2-methoxy-6-n-pentyl-1, 4-benzoquinone) and Analogues*. Medicinal Chemistry 3(4): p. 369–372.
6. Verma RP (2006) *Anti-Cancer Activities of 1, 4-Naphthoquinones: A QSAR Study*. Anti-Cancer Agents in Medicinal Chemistry - Anti-Cancer Agents 6(5): p. 489–499.
7. O'Brien PJ (1991) *Molecular mechanisms of quinone cytotoxicity*. Chemico-Biological Interactions 80(1): p. 1–41.
8. Bentinger M, Tekle M, Dallner G (2010) *Coenzyme Q – Biosynthesis and functions*. Biochemical and Biophysical Research Communications 396(1): p. 74–79.
9. Patel SN, Kain KC (2005) *Atovaquone/proguanil for the prophylaxis and treatment of malaria*. Expert review of anti-infective therapy 3(6): p. 849–861.
10. Nakato H, Vivancos R, Hunter P (2007) *A systematic review and meta-analysis of the effectiveness and safety of atovaquone proguanil (Malarone) for chemoprophylaxis against malaria*. Journal of antimicrobial chemotherapy 60(5): p. 929–936.
11. Shanmugasundaram M, Garcia-Martinez I, Qingyi L, Estrada A, Martinez NE, et al. (2005) *Microwave-assisted solid-phase Dötz benzannulation reaction: a facile synthesis of 2, 3-disubstituted-1, 4-naphthoquinones*. Tetrahedron letters 46(44): p. 7545–7548.
12. Chen J, Jiang J, Zhang F, Yu H, Zhang J (2004) *Cytotoxic effects of environmentally relevant chlorophenols on L929 cells and their mechanisms*. Cell Biology and Toxicology 20(3): p. 183–196.
13. Ozdemir KG, Yilmaz H, Yilmaz S (2009) *In vitro evaluation of cytotoxicity of soft lining materials on L929 cells by MTT assay*. Journal of Biomedical Materials Research Part B: Applied Biomaterials 90(1): p.82–86.
14. Süloğlu AK, Karacaoğlu E, Koçkaya EA, Selmanoğlu G, Loğoglu E (2014) *Cytotoxic effects of a novel thiola benzen derivative 2, 4-dithiophenoxy-1-iodo-4-bromobenzene ($C_{18}H_{12}S_2IBr$)in L929 cells*. International Journal of Toxicology 33(4): p. 319–324
15. Babich H, Stern A (1993) *In Vitro Cytotoxicities of 1, 4-Naphthoquinone and Hydroxylated 1, 4-Naphthoquinones to Replicating Cells*. Journal of Applied Toxicology 13(5): p. 353–358.
16. Inbaraj JJ, Chignell CF (2003) *Cytotoxic Action of Juglone and Plumbagin: A Mechanistic Study Using HaCaT Keratinocytes*. Chemical Research in Toxicology 17(1): p. 55–62.
17. Aithal BK, Kumar MR, Rao BN, Udupa N, Rao BS (2009) *Juglone, a naphthoquinone from walnut, exerts cytotoxic and genotoxic effects against cultured melanoma tumor cells*. Cell Biology International 33(10): p. 1039–1049.
18. Balasubramanian K, Mirnikjoo B, Schroit AJ (2007) *Regulated externalization of PS at the cell surface: implications for apoptosis*. J Biol Chem 282(25): 18357–18364;
19. Porter AG, Jänicke RU (1999) *Emerging roles of caspase-3 in apoptosis*. Cell death and differentiation. 6(2): 99–104.
20. Kroemer G, Galluzzi L, Vandenabeele P, Abrams J, Alnemri ES (2009) *Classification of cell death: recommendations of the Nomenclature Committee on Cell Death*. Cell Death Differentiation 16(1): 3–11.
21. Seshadri PA, Rajaram A, Rajaram R (2011) *Plumbagin and jugalone induce caspase-3-dependent apoptosis involving the mitochondria through ROS generation in human peripheral blood lymphocytes* 51(11): 2090–2107.
22. Watanabe N, Forman HJ (2003) *Autoxidation of extracellular hydroquinones is a causative event for the cytotoxicity of menadione and DMNQ in A549-S cells*. Arch Biochem Biophys 411 (1) 145–157.
23. Castro FA, Mariani D, Panek AD, Eleutherio EC, Pereira MD (2008) *Cytotoxicity mechanism of two naphthoquinones (menadione and plumbagin) in Saccharomyces cerevisiae*. PLoS One3(12): e3999.
24. Di Rosso ME, Barreiro Arcos ML, Elingold I, Sterle H, Baptista Ferreira S (2013) *Novel o-naphthoquinones induce apoptosis of EL-4 T lymphoma cells through the increase of reactive oxygen species*. Toxicology In Vitro 27(7)2094–2104.
25. Hassan HM, Fridovich I (1979) *Intracellular production of superoxide radical and of hydrogen peroxide by redox active compounds*. Archives of Biochemistry and Biophysics. 196: 385–395.
26. Srinivas P, Gopinath G, Banerji A, Dinakar A, Srinivas G (2004) *Plumbagin induces reactive oxygen species, which mediate apoptosis in human cervical cancer cells*. Molecular carcinogenesis 40(4): p. 201–211.
27. Babula, P. Vojtěch Adamb, René Kizekb, Zdeněk Sladkýa, Ladislav Have (2009) *Naphthoquinones as allelochemical triggers of programmed cell death*. Environmental and Experimental botany 65: 330–337.
28. Chen CA, Chang HH, Kao CY, Tsai TH, Chen YJ (2009) *Plumbagin, Isolated from Plumbago zeylanica, Induces Cell Death through Apoptosis in Human Pancreatic Cancer Cells*. Pancreatology 9(6): p. 797–809.
29. Hsu YL, Cho CY, Kuo PL, Huang YT, Lin CC (2006) *Plumbagin (5-hydroxy-2-methyl-1, 4-naphthoquinone) induces apoptosis and cell cycle arrest in A549 cells through p53 accumulation via c-Jun NH2-terminal kinase-mediated phosphorylation at serine 15 in vitro and in vivo*. The journal of pharmacology and experimental therapeutics 318(2): p. 484–494.
30. Wang CC, Chiang YM, Sung SC, Hsu YL, Chang JK, et al. (2008) *Plumbagin induces cell cycle arrest and apoptosis through reactive oxygen species/c-Jun N-terminal kinase pathways in human melanoma A375.S2 cells*. Cancer Letters 259(1): p. 82–98.
31. Watanabe N, Dickinson DA, Liu RM, Forman HJ (2004) *Quinones and Glutathione Metabolism*. Methods in Enzymology 378: p. 319–340.
32. Deponte M (2013) *Glutathione catalysis and the reaction mechanisms of glutathione-dependent enzymes* 1830(5): 3217–3266.
33. Ross D, Thor H, Orrenius S, Moldeus P (1985) *Interactoin of menadione (2-methyl-1, 4-naphthoquinone) with glutathione (liver cells)* Chemico-Biological Interactions 55: 177–184.
34. Doherty MD, Rodgers A, Cohen GM (1987) *Mechanisms of toxicity of 2- and 5-hydroxy-1, 4-naphthoquinone; absence of a role for redox cycling in the toxicity of 2-hydroxy-1, 4-naphthoquinone to isolated hepatocytes*. Journal of Applied Toxicology. 7(2): 123–129.

4

Nutritional Advice in Older Patients at Risk of Malnutrition during Treatment for Chemotherapy: A Two-Year Randomized Controlled Trial

Isabelle Bourdel-Marchasson[1,2,3*], Christelle Blanc-Bisson[1], Adélaïde Doussau[4,5], Christine Germain[4], Jean-Frédéric Blanc[6,7], Jérôme Dauba[8], Cyril Lahmar[8], Eric Terrebonne[9], Cédric Lecaille[10], Joël Ceccaldi[11], Laurent Cany[12], Sandrine Lavau-Denes[13], Nadine Houede[14], François Chomy[14], Jessica Durrieu[1], Pierre Soubeyran[14,7], Pierre Senesse[15,16], Geneviève Chene[4,5], Mariane Fonck[14]

1 CHU Bordeaux, Clinical Gerontology, Bordeaux, France, **2** CNRS, RMSB, UMR 5536, Bordeaux, France, **3** Université Bordeaux, RMSB, UMR 5536, Bordeaux, France, **4** CHU Bordeaux, Unité d'épidémiologie clinique, Bordeaux, France, **5** INSERM, CIC-EC7, Bordeaux, France, **6** CHU Bordeaux, Hépatologie, gastroentérologie and oncologie digestive, Bordeaux, France, **7** Université Bordeaux, Bordeaux, France, **8** Centre Hospitalier Layné-Mont de Marsan, Service Oncologie, Mont de Marsan, France, **9** CHU Bordeaux, Service Hépato-gastro et oncologie digestive, Bordeaux, France, **10** Polyclinique Bordeaux Nord, Service hépato gastroentérologie, Bordeaux, France, **11** Centre Hospitalier de Libourne, Libourne, France, **12** Clinique Francheville, Périgueux, France, **13** CHU Limoges, Unité de recherche clinique en oncologie médicale, Limoges, France, **14** Institut Bergonié, Departement d'oncologie médicale, Bordeaux, France, **15** Universités Montpellier 1 et 3, Laboratoire Epsylon EA 4556, Montpellier, France, **16** CLCC du Val d'Aurelle, Unité de Nutrition clinique et de gastroentérologie, Montpellier, France

Abstract

Objective: We tested the effect of dietary advice dedicated to increase intake in older patients at risk for malnutrition during chemotherapy, versus usual care, on one-year mortality.

Method: We conducted a multicentre, open-label interventional, stratified (centre), parallel randomised controlled trial, with a 1:1 ratio, with two-year follow-up. Patients were aged 70 years or older treated with chemotherapy for solid tumour and at risk of malnutrition (MNA, Mini Nutritional Assessment 17–23.5). Intervention consisted of diet counselling with the aim of achieving an energy intake of 30 kCal/kg body weight/d and 1.2 g protein/kg/d, by face-to-face discussion targeting the main nutritional symptoms, compared to usual care. Interviews were performed 6 times during the chemotherapy sessions for 3 to 6 months. The primary endpoint was 1-year mortality and secondary endpoints were 2-year mortality, toxicities and chemotherapy outcomes.

Results: Between April 2007 and March 2010 we randomised 341 patients and 336 were analysed: mean (standard deviation) age of 78.0 y (4·9), 51.2% male, mean MNA 20.2 (2.1). Distribution of cancer types was similar in the two groups; the most frequent were colon (22.4%), lymphoma (14.9%), lung (10.4%), and pancreas (17.0%). Both groups increased their dietary intake, but to a larger extent with intervention ($p<0.01$). At the second visit, the energy target was achieved in 57 (40.4%) patients and the protein target in 66 (46.8%) with the intervention compared respectively to 13 (13.5%) and 20 (20.8%) in the controls. Death occurred during the first year in 143 patients (42.56%), without difference according to the intervention ($p = 0.79$). No difference in nutritional status changes was found. Response to chemotherapy was also similar between the groups.

Conclusion: Early dietary counselling was efficient in increasing intake but had no beneficial effect on mortality or secondary outcomes. Cancer cachexia antianabolism may explain this lack of effect.

Trial Registration: ClinicalTrials.gov NCT00459589

Editor: Vincent Wong, The Chinese University of Hong Kong, Hong Kong

Funding: This work was supported by the National Hospital Program of Clinical Research (Programme Hospitalier de Recherche Clinique 2006) (46%), La Ligue contre le cancer (52%) and AMGEN (2%) and sponsored by the university hospital of Bordeaux (CHU Bordeaux). The funders had no role in study design, data collection and analysis, decision to publish, or preparation of the manuscript.

Competing Interests: The authors have declared that no competing interests exist.

* Email: isabelle.bourdel-marchasson@chu-bordeaux.fr

Introduction

Weight loss in patients with cancer has long been linked to poor prognosis [1]. It is therefore recommended to assess nutritional status in patients undergoing nonsurgical oncologic treatment [2,3]. Dietary counselling has been proposed in malnourished patients with gastrointestinal cancers during chemotherapy treatment [4].

A randomised controlled trial of a nutritional intervention was conducted in 358 adults with weight loss and treated with chemotherapy for metastatic or locally advanced digestive or non-small cell lung cancer [5]. The aim was to assess the effects of nutritional advice and the prescription of an oral nutritional supplement, either alone or combined for 6 weeks. None of the interventions produced any benefit on outcomes including one-year mortality and quality of life. The lack of effect of the interventions may be due to the fact that cancers were advanced in this study. Another similar trial of dietary support during 12 weeks of chemotherapy reported no benefit in 192 patients with non-small-cell lung and colon cancer [6]. A meta-analysis of nutritional intervention trials in cancer patients in very heterogeneous situations (nutrition, cancer, treatments) did not show any benefit of these interventions on mortality [7].

All these studies involved adult populations but without any specific analysis of older patients. Nevertheless, cancers are more and more frequent in older people and in 2005 it was estimated that 56% of new cancers occurred in people older than 65 y [8]. One-year mortality increased by two-fold in older patients with cancer undergoing chemotherapy who were malnourished or at risk of malnutrition according to the MNA© (Mini Nutritional Assessment) [9]. Very few patients were malnourished (3 out of 202) so the relationships mainly concerned those at risk. In a similar study including more malnourished patients (MNA<17, 13.8%), the relative risk (RR) of one-year mortality was higher (2.77) for those with an MNA©<24 [10]. This suggests an effect related to the severity of nutritional impairment according to the MNA©. This specific tool for older patients includes items directly associated to nutrition, such as anthropometric measures and nutritional intake, together with health-related quality of life measures such as comorbidities, mental health, autonomy and subjective health [11], was shown correlated with cancer cachexia features [12] and was one of the independent predictor for chemotherapy toxicity [13]. MNA is thus the consensual tool to assess malnutrition in older patients with cancer [3,14]. In older patients, the increased risk of mortality associated with a lower MNA© score may be due to factors other than cancer. Thus, nutritional support targeted on gerontological assessment may better address the needs of older patients with cancer [15]. No clinical trial has yet tested the effect of nutritional intervention in older patients with cancer who were malnourished or at risk of malnutrition according to the MNA© scale.

In the present randomised clinical trial involving older patients at risk of malnutrition, we assessed the effect of dietary counselling implemented from the start of chemotherapy for solid tumours and lymphoma on one-year and subsequent mortality and on severe toxicities.

Methods

The protocol for this trial and supporting CONSORT checklist are available as supporting information; see Checklist S1 and Protocol S1.

Study design

This study was a multicenter superiority randomized controlled trial of patients with cancer receiving chemotherapy in 12 public and private settings in South-West France, comparing Usual Care to Usual Care + Nutritional Intervention in two parallel arms. Patients were enrolled between April 2007 and March 2010 and follow-up ended in April 2012. The randomization list was prepared and stored by the clinical trial unit biostatistician. Randomization was centralized by internet, with a 1:1 ratio, stratified on recruitment centre. Informed written consent was retrieved from patients after eligibility criteria were checked. All assessments and dietary interviews were performed in the cancer treatment setting. The institutional Review Board of South-West France and Overseas French departments, France, approved the study protocol. The trial was recorded with ClinicalTrials.gov, number NCT00459589.

Patient selection

Patients older than 70 y with lymphoma or carcinoma with an indication of chemotherapy and a Karnofsky index higher than 50% were screened for participation. Eligible carcinomas were from the colon, stomach, pancreas and biliary tract, ovary, prostate, bladder, and lung. Lymphoma types were any B cell lymphoma, any T lymphoma, low malignancy lymphomas such as follicular, lymphoplasmacytic, lymphocytic, mantle, MALT, and other marginal zone lymphoma. Patients with a carcinoma of unproven origin but compatible with any tumour in the above-mentioned list could be included if a chemotherapy was planned.

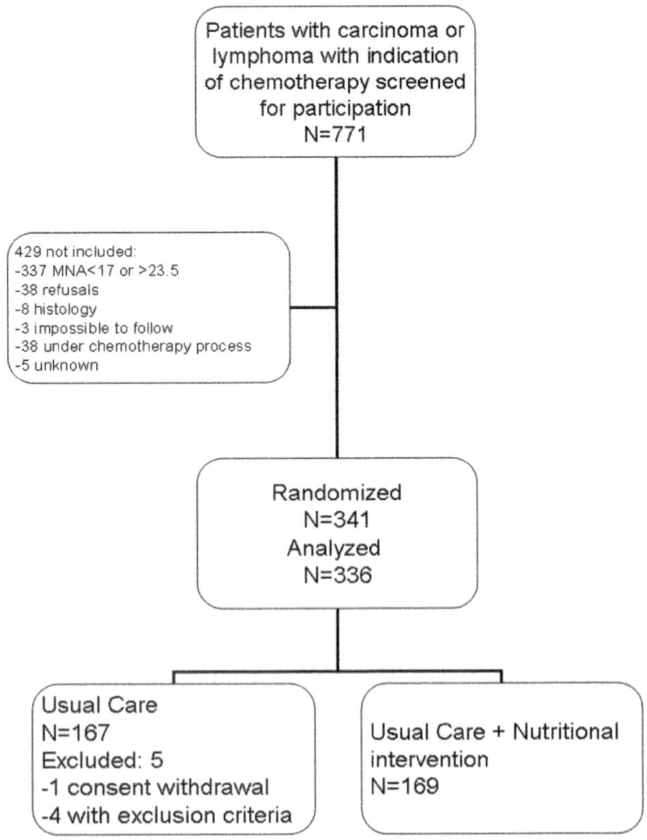

Figure 1. Flowchart of participant progression through a randomized controlled trial of nutritional intervention in older patients at risk of malnutrition.

Table 1. Main baseline characteristics of participants in the trial.

	Usual Care	Usual Care + Nutritional Intervention
	N = 167	N = 169
Full MNA, mean (SD)	20.4 (2.1)	20.1 (2.0)
Age, years, mean (SD)	78.3 (4.7)	77.7 (5.2)
Gender, male % (n)	54.5 (91)	47.9 (81)
ECOG PS 0–1% (n)	71.6 (88)	78.4 (91)
Cancer % (n)		
Colon	19.2 (32)	25.6 (17)
Stomach	8.4 (14)	6.5 (11)
Pancreas and cholangiocarcinoma	17.4 (29)	19.7 (33)
Non-small cell lung	10.8 (18)	10.1 (17)
Prostate	5.4 (9)	2.4 (4)
Bladder	7.8 (13)	4.2 (7)
Ovary	7.8 (13)	7.1 (12)
Breast	7.2 (12)	9.5 (16)
Lymphoma	16.2 (27)	13.7 (23)
Metastasis (carcinoma), % (n)		
None	30.0 (42)	35·6 (52)
Mx (unknown at chemotherapy start)	8.6 (12)	8.2 (12)
IPI score 2–3 (lymphoma) % (n)	40·7 (11)	57.1 (12)
First line chemotherapy, % (n)	83.2 (139)	84.6 (143)
Weight change, % of usual body weight, mean (SD)	−8.6 (7.9)	−8.9 (6.6)
Lymphocytes/mm^3, mean (SD)	1.6 (1.9)	1.6 (2.5)
Haemoglobin, g/100 ml, mean (SD)	12.0 (1.6)	11.8 (1.7)
Serum albumin, g/l, mean (SD)	36.8 (6.2)	36.9 (6.9)
C-reactive protein, mg/l, mean (SD)	34.7 (64.7)	34.1 (42.2)

ECOG PS: Eastern Cooperative Oncology Group performance status.

Patients could receive a first to third line of routine chemotherapy. Patients with cerebral metastasis were not eligible. Any inability to take part in follow-up according to the schedule of the study for geographical or treatment-related reasons was an exclusion criterion. The full MNA© was used to screen patients at risk of malnutrition. The full MNA© is an 18-item questionnaire including anthropometric, general, dietary and subjective assessments; the maximum score is 30 indicating the best nutritional status and a score below 17 indicating malnutrition [11]. To be included in the randomized trial, subjects had to be at risk of malnutrition with a full MNA© in the 17 to 23·5 point range.

Intervention

The Usual Care group (UC) received the nutritional care routinely given in the cancer treatment settings and there were no restrictions for dietary advice, oral supplements or prescription of artificial nutrition. The Usual Care+Nutritional Intervention (UC+NI) group received usual care and nutritional intervention.

Nutritional intervention began the first day of chemotherapy. The study dietician, who did not belong to the staff of the cancer treatment setting, provided dietary advice with the aim of achieving an energy intake of 30 kCal/kg body weight/d and 1.2 g protein/kg/d [2]. Counselling was based on face-to face-interviewing and dietary advice cards, and involved caregivers or relatives if possible [16]. Actual dietary intake and gerontological assessment routinely applied [15] were taken into account to adapt the advice to each patient. The eight dietary cards addressed the issues of: 1-dietary balance, 2-loss of appetite and enrichment, 3-diarrhoea and constipation, 4-nausea and vomiting, 5-taste disturbances, 6-oral pain and feeling of dryness, 7- swallowing disorders, 8- diabetes. Dietary advice was complemented by prescription of an oral supplement if pertinent in order to increase intake. Six face-to-face visits were planned during chemotherapy sessions and contact was made by phone in the event of an interval between visits longer than two weeks due to the schedule of the chemotherapy. Intervention lasted 3 to 6 months according to the chemotherapy schedule of each patient. All study dieticians received training for the study, which was also given if a new dietician entered it. The duration of the intervention was 3 to 4 months according to the duration of chemotherapy. None of the staff in cancer treatment centres were aware of the content of the nutritional intervention.

Dietary intake assessment

Patients in both groups were instructed to fill in a one-day dietary record of the day before the visit. The UC patients gave their record to the staff of the centre and the UC+IC gave theirs to the study dietician who needed it to adapt counselling. Any intake was taken into account, including oral nutritional supplements and artificial nutrition. Nutrisoft Billnut software (Tours, France) was used to calculate intake by the coordinating centre [17]. Percent of patients achieving the goal in energy or protein intake were shown.

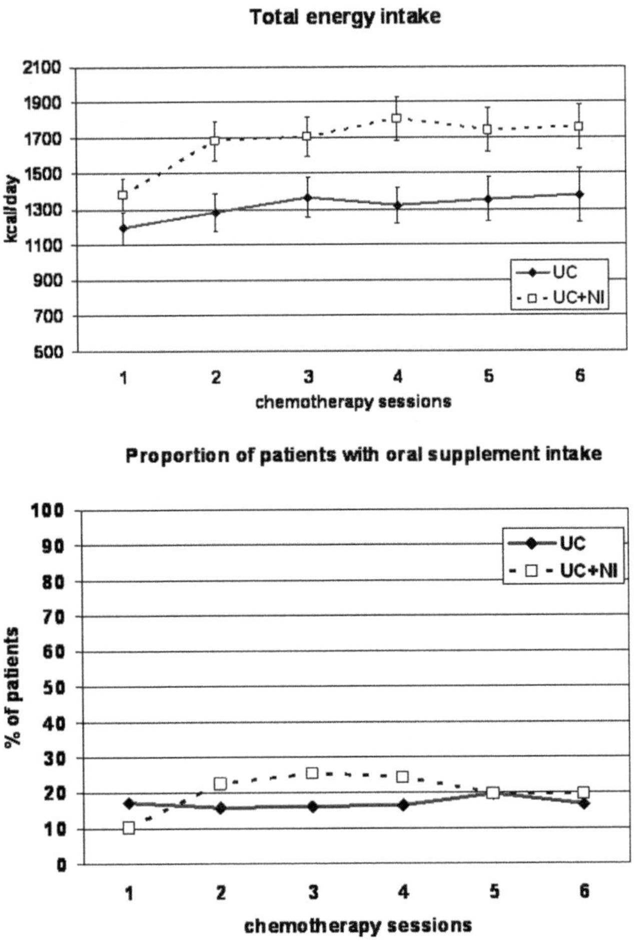

Figure 2. Dietary intake the day before each cycle during the chemotherapy period. Data are presented as mean and 95% CI, or proportion. Total dietary intake was analyzed with mixed models: increase of total intake at the second visit in both groups (UC+NI, $P< 0.0001$; **UC, $P = 0.02$), with higher increased in UC+NI compared to UC, $P<0.01$.

Outcomes

The main outcome was one-year mortality recorded in centres. Causes of death were recorded. General practitioners and patients' relatives were contacted if needed.

Chemotherapy management (dosage, changes and arrest), grade 3–4 toxicities including severe infections, weight changes, prescription of enteral or parenteral nutrition, and hospitalization for reasons other than chemotherapy were considered as secondary outcomes. These data were collected by investigators during the chemotherapy period. Two-year mortality was also assessed.

Determination of sample size

A reduction of 10% in one-year mortality was considered as clinically significant, considering a mortality rate in the UC group of 50% [9]. On the basis of the O'Brien and Fleming rule [18], an alpha 5% risk and a 80% power, with an interim analysis, and loss of follow-up expected lower than 5%, 820 patients had to be enrolled. The interim analysis was planned when half of the participants were evaluated for the main outcome, but it was not performed because recruitment was lower than expected and the study was stopped before.

Statistics

Analyses were performed on an intention-to-treat basis. Comparisons of dietary intake between groups were performed using a mixed linear model accounting for repeated measures. The Kaplan-Meier method was used to estimate survival probabilities and mortality was compared using the Wald Chi2 test in a Cox model adjusted on the recruiting centre of the participants. Proportions were used to describe qualitative variables of the secondary outcomes. Comparisons were made with the Chi2 test. For the outcomes that were recorded at several visits such as infections, the outcome was considered absent when the data was not available in the main analysis. In a robustness analysis, we used the missing = failure strategy. Quantitative variables were described with mean and standard deviation (SD) or Inter-Quartile Interval (IQR) and compared using Student's t test or the Wilcoxon test according to the distribution of the analysed variable.

Results

Patient flow

Among 771 screened patients, 341 were randomized in the trial (**Figure** 1) and 336 were analyzed: mean (standard deviation, SD) age 78.0 y (SD 4.9), 51.2% male, mean MNA 20.2 (SD 2.1). Distribution of cancer types was similar in the two groups; the most frequent were colon (22.4%), lymphoma (14.9%), lung (10.4%) and pancreas (17.0%). Chemotherapy was first line in 83.9%. The trial was stopped before full recruitment at the end of the inclusion period due to an insufficient rate of inclusions. The slow recruitment was attributed to the existence of competitive trials for chemotherapy agents. The resulting power of the analysis was estimated 40% instead of 80% to detect a one-year mortality difference of 10% or was 80% to detect a 15% difference between the two groups. Among patients identified as eligible, reasons for non-inclusion were mainly chemotherapy treatments already started. Five more subjects were excluded from the analyzed sample due to one withdrawal of consent and 4 major deviations from the eligibility criteria in the UC group (head and neck tumour and chronic lymphatic leukaemia, or 5th line of chemotherapy). Five patients with minor eligibility deviation (MNA<17 and one>23·5 and two with cardia cancer or endocrine carcinoma of the pancreas) were kept in the analysed sample after scientific committee approval. The resulting analyzed sample included 336 subjects. Groups were balanced for age, sex ratio, cancer characteristics, routine biochemical analyses and blood cell count (Table 1). Two participants from the UC+NI group refused to continue the intervention during the follow-up but were followed up for mortality. In the UC+NI group 877 dietician-visits were performed in relation to 990 planned visits so compliance with the intervention was estimated at 88.6%. There were also 450 phone contacts. None of the participants were lost to follow-up.

Dietary intake

The mixed model was applied in 1248 records, corresponding to 310 patients. At baseline, dietary intake was higher in the UC+NI group compared to the UC group (difference of 178 kcal/day, $p<0.01$). In both groups, dietary intake increased between visit 1 and visit 2 (UC+NI+328 kcal/day, $p<0.0001$; UC+132 kcal/day, $p = 0.02$) but the difference was higher in the UC+NI group than in the UC group ($p<0.01$) (**Figure** 2). At the visit 2, 57 (40.4%)

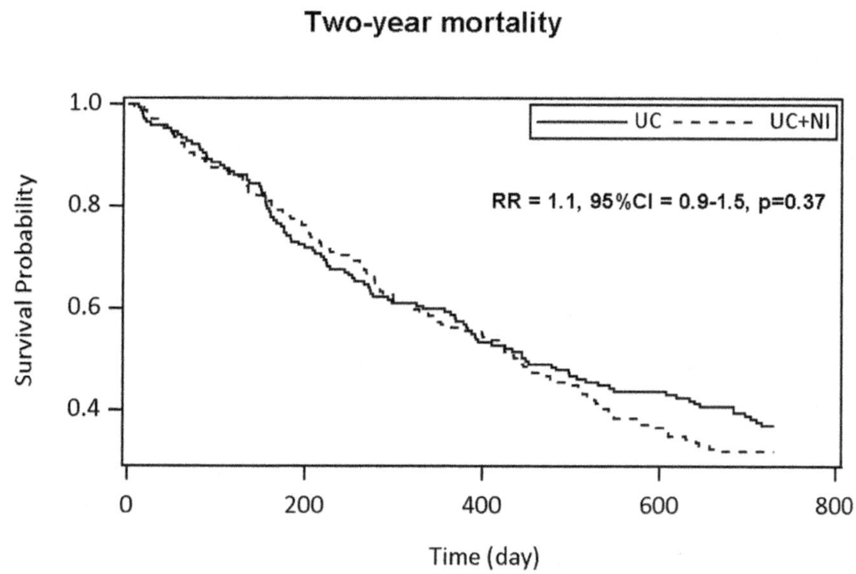

Figure 3. Two-year mortality according to groups UC and UC+NI. N = 336. Comparisons were performed with Cox model adjusted on recruiting centres.

patients in the UC+NI group compared to 13 (13.5%) in the UC group achieved the goal of 30 Kcal/kg/d or more and 66 (46.8%) in UC+NI group compared to 20 (20.8%) the goal of 1.2 g protein/kg/d.

In the UC group, 17.2% (23 patients) received a supplement at the first visit and the maximum rate during the follow-up was 19.6% (19 patients) (visit 5). In the UC+NI group, 10.3% (17 patients) received an oral nutritional supplement at the first visit and the maximum rate during the follow-up was 25.5% (42 patients) (visit 3).

Outcomes

One-year and two-year mortality were similar in both groups (**Figure** 3, respectively R = 1.1, 95%CI = 0.8–1.5, p = 0.74, and RR = 1.1, 95%CI = 0.9–1.5, p = 0.37). The main declared cause of death was cancer disease (Table 2). There was no difference in distribution of cause of death according to the groups. There was no lost of follow-up patient for the main outcome.

Chemotherapy management and outcomes were similar in both groups (Table 2). There were more UC patients with grade 3–4 infections than UC+NI ones (Table 2, p = 0.03). However, a robustness analysis was performed due to the existence of missing data and did not confirm the difference in the incidence of severe infections. The rate of weight change and other secondary outcomes were similar in both groups.

Discussion

This large randomised controlled trial investigated the effect of nutritional support in older patients treated by chemotherapy for cancer. Despite an increase in dietary intake that was higher in patients with dietary counselling, no improvement was noted in one- and two-year mortality in older subjects at risk of malnutrition and undergoing chemotherapy.

However, a lower rate of serious infections during chemotherapy was observed. This result should be interpreted with caution as it was not confirmed by the robustness analysis. A decreased rate of severe infections may have favourably impacted the prognosis but this event was relatively rare. The potential beneficial effect of nutritional intervention on this outcome was not sufficient to modify the survival of the group.

The trial was stopped before completion of the planned inclusions. However, it is unlikely that the lack of an observed effect of intervention is due to lack of power. In the whole sample, a trend of an increase in two-year mortality in the intervention group was seen. However, even if we had reached the planned sample size, such mortality rates in both arms would not have provided a significant difference in survival. This absence of effect on mortality and other outcomes was not due to unbalanced characteristics of patients between groups according to cancer disease, chemotherapy or nutritional baseline assessment. Furthermore, within each tumour type, one and two-year mortality rates were similar. Any non-drug intervention involving a person such as a dietician cannot be blind and patients and practitioners were aware of the allocation. Despite these limitations, the main result is reliable, all the more in that mortality is a very robust outcome and there were no patient lost to follow-up.

In adult patients with gastro-intestinal cancer, the increased mortality risk in patients with weight loss prior to chemotherapy has been attributed to decreased chemotherapy doses and increased toxicity rates [19]. The present findings do not support this hypothesis since we found no change in chemotherapy outcomes with nutritional support. However, we did not include malnourished patients (MNA© < 17) or those with a very poor prognosis according to their Karnosky index in order to better focus on nutritional risk. Thus, we cannot draw any firm conclusions about the effect of malnutrition on chemotherapy outcomes.

In patients with cancer treated with chemotherapy, nutritional interventional trials are scarce and none have been devoted to older patients. A retrospective study showed that patients with colorectal cancers who benefited from dietary counselling improved their nutritional status and had better survival [20]. However, randomized controlled trials did not show in adult patients with cancer an effect of nutritional interventions on mortality [5,7].

Table 2. Two-year mortality and other outcomes.

	Usual Care	Usual Care + Nutritional Intervention	P
	N = 167	N = 169	
One-year mortality, % (n)	41.3 (69)	43.8 (74)	0.74
Two-year mortality, % (n)	62.9 (105)	68.0 (115)	0.37
Mortality during chemotherapy, % (n)	3.7 (4)	6.4 (7)	
One-year mortality according to tumour origin, % of subject in tumour category, (n)			
Colon	18.8 (6)	30.2 (13)	
Stomach	50.0 (7)	54.5 (6)	
Pancreas and cholangiocarcinoma	62.1 (18)	66.7 (22)	
Non-small cell lung	77.8 (14)	76.5 (13)	
Prostate	22.2 (2)	75.0 (3)	
Bladder	53.8 (7)	57.1 (4)	
Ovary	15.4 (2)	8.3 (1)	
Breast	33.3 (4)	50.0 (8)	
Lymphoma	33.3 (9)	17.4 (4)	
Two-year mortality according to tumour origin, % of subject in tumour category, (n)			
Colon	40.6 (13)	48.8 (21)	
Stomach	78.6 (11)	81.8 (9)	
Pancreas and cholangiocarcinoma	75.9 (22)	87.9 (29)	
Non-small cell lung	94.4 (17)	94.1 (16)	
Prostate	88.9 (8)	100.0 (4)	
Bladder	84.6 (11)	71.4 (5)	
Ovary	15.4 (4)	41.7 (5)	
Breast	58.3 (7)	62.5 (10)	
Lymphoma	44.4 (12)	60.9 (14)	
Cause of one-year death, % (n)			
Cancer disease	89.9 (62)	86.5 (64)	
Chemotherapy toxicities	2.9 (2)	5.4 (4)	
Other	7.2 (5)	8.2 (6)	
Cause of two-year death, % (n)			
Cancer disease	93.3 (98)	90.4 (104)	
Chemotherapy toxicities	1.9 (2)	3.5 (4)	
Other	4.8 (5)	6.1 (7)	
Cause of two-year death, % (n)			
Cancer disease	93.3 (98)	90.4 (104)	
Chemotherapy toxicities	1.9 (2)	3.5 (4)	
Other	4.8 (5)	6.1 (7)	
Chemotherapy management % (n)			
At least one time chemotherapy not administrated,	65.0 (106)	64.8 (107)	0.97
At least one time change of protocol	62.0 (101)	64.2 (106)	0.67
Chemotherapy result at end of treatment, % (n)			
Full remission	11.1 (16)	7.0 (10)	0.39
Partial remission	23.6 (34)	30.8 (44)	
Stabilization	40.3 (58)	40.6 (58)	
Progression	25.0 (36)	21.7 (31)	
Weight change at end of chemotherapy, kg			
IQR	−4.0; 1	−5.0; 1	0.59
Proportion of patients with weight loss % (n)	57.7 (75)	56.9 (74)	
Change in serum albumin at end of treatment, g/l, IQR	−8.0; 1.4	−7.6; 3.0	0.59

Table 2. Cont.

	Usual Care	Usual Care + Nutritional Intervention	P
	N = 167	N = 169	
Incident fall, pressure ulcer or fracture, % (n)	6.0 (9)	5.6 (8)	0.87
Infection grade 3–4, % (n)	10.4 (7)	4.2 (7)	0.03
Hospitalisation, % (n)	34.4 (56)	29.1 (48)	0.31
Enteral/parenteral nutrition, % (n)	9.2 (15)	5.5 (9)	0.19

The absence of difference in survival between the intervention groups in the present study might be explained by two main hypotheses. First, the intervention might not sufficiently increase dietary intakes; second, the nutritional intervention might not be efficient to impact mortality in these patients.

In the present study, dietary counselling resulted in a positive effect on dietary intake in both groups after the start of chemotherapy. However, the targets of 30 kcal/kg/d or 1.2 g protein/d were not achieved in more than half of the patients, questioning the efficacy of dietary counselling in patients with cancer. On the other side, the patients in the UC group may have received more attention regarding their dietary intake than usual. This could have been a source of contamination bias, resulting in a lower difference in intakes between groups. The improvement in intake in both groups at cycle 2 might also be due to an improvement in health status owing to the global care given to these patients. The assessment of intake was different between the groups: in UC patients it was a self-report questionnaire with no help from a dietician; in UC+NI ones, the study dietician commented and improved the quality of the record with the patient during the interview. However, the baseline difference could be interpreted as a proxy of the measurement bias due to the assessment procedure, and the linear mixed model showed that, despite the baseline difference, the increase in intake was significantly higher in the intervention group between cycle 1 and 2. The other possible limitation regarding dietary intake assessment was the schedule of the records. We assessed the period distant from the chemotherapy session, so we probably assessed the highest intake during the period between two cycles. We made no assessment of the effects of chemotherapy on appetite just after the treatment. However, chemotherapy management was similar in both groups and the effects on appetite were likely the same. Two randomized studies previously showed an increase of actual dietary intake with dietary counselling in patients with solid tumours [21],[22].

Despite the higher dietary intake found in the intervention group, there were no differences in weight change between the groups. This finding is similar to the conclusion of other randomised control trials performed in adults [5],[22] in which no weight change difference was seen between the groups. The sample was not large enough to explore subgroups responses to intervention on intention to treat basis, all the more that they had not been planned in the trial design. It was thus not possible to identify if any, a sub-group of patients on the basis of their baseline characteristics, able to take advantage on the intervention, for example those who did or not lose weight prior to chemotherapy. In the already described RCT performed in adults selected on the basis of weight loss, irrespective of the intervention group, those who gained weight had better prognosis [5]. This suggested that weight gain was related to favourable response to cancer treatment and not to the tested intervention.

The increase in dietary intake was not efficient to prevent weight loss in our study. As in younger adults, [5] weight loss in these patients was probably mainly due to cancer disease within the framework of cancer cachexia and very unlikely due to other causes, as we hypothesized when constructing the trial.

The reason why increased energy intake fails to change body mass may involve the activation of anti-anabolic pathways during cancer cachexia [23,24]. It is likely that pharmaconutrition targeting cachexia related metabolism alterations should be tested. This trial cannot however provide hypothesis to drive pharmaco-nutrition related to cancer cachexia.

In patients receiving chemotherapy, increase in dietary intake just before treatment may also not be favourable. Indeed, it has been reported that short-term fasting (48 hours) can protect normal cells but not cancer cells from chemotherapy agents in mice and in cell culture [25]. An interventional study in patients undergoing chemotherapy for breast or prostate cancer is ongoing to test the effects of a controlled low calorie diet on the side effects and response to treatment (NCT01802346).

The situation in patients receiving radiotherapy might be different from that in those with chemotherapy. In head and neck cancer patients, several trials found an improvement in dietary intake [26], quality of life [27], and nutritional status [28]. It is noticeable that we also found no differences in quality of life items (data not shown) among the intervention groups in surviving patients at the end of chemotherapy. A trial of dietary counselling or oral supplementation in 111 patients with colon cancer treated with radiotherapy showed the superiority of dietary counselling compared to oral nutritional supplements and to control on outcomes during treatment [29]. Mortality was not analyzed in these patients with good prognosis. Nutritional support in dysphagic patients is probably mandatory unlike in patients with anorexia owing to the cancer disease itself. In dysphagic patients with oesophageal cancer, enteral support led to the same prognosis as in patients with the same cancer but without swallowing problems [30]. The question of the prognostic value of anorexia and subsequent weight loss related to cancer cachexia is thus crucial. On one hand, it is very difficult to overcome anorexia; on the other, the intensity of anorexia may reflect the severity of the cancer disease [31].

In conclusion, individual dietary counselling in older patients at risk for malnutrition during their chemotherapy treatment for cancer was associated with an increase in dietary intake but did not decrease mortality.

Supporting Information

Consort S1 INOGAD Consort.

Consort S2 INOGAD CONSORT Abstract.

Protocol S1 Protocol extract.

Protocol S2 Protocol full in French.

Acknowledgments

We are indebt to Mrs Catherine Maldonado, datamanager, and Mr Guillaume Dupouy, informatician.

Author Contributions

Conceived and designed the experiments: IBM GC MF P. Senesse P. Soubeyran. Performed the experiments: CBB JFB J. Dauba C. Lahmar ET C. Lecaille JC LC SLD NH FC J. Durrieu. Analyzed the data: AD CG IBM. Wrote the paper: IBM AD CG.

References

1. Dewys WD, Begg C, Lavin PT, Band PR, Bennett JM, et al. (1980) Prognostic effect of weight loss prior to chemotherapy in cancer patients. Eastern Cooperative Oncology Group. Am J Med 69: 491–497.
2. Arends J, Bodoky G, Bozzetti F, Fearon K, Muscaritoli M, et al. (2006) ESPEN Guidelines on Enteral Nutrition: Non-surgical oncology. Clin Nutr 25: 245–259.
3. Senesse P, Bachmann P, Bensadoun RJ, Besnard I, Bourdel-Marchasson I, et al. (2014) Clinical nutrition guidelines of the French Speaking Society of Clinical Nutrition and Metabolism (SFNEP): Summary of recommendations for adults undergoing non-surgical anticancer treatment. Dig Liver Dis 46: 867–874.
4. Senesse P, Assenat E, Schneider S, Chargari C, Magne N, et al. (2008) Nutritional support during oncologic treatment of patients with gastrointestinal cancer: who could benefit? Cancer Treat Rev 34: 568–575.
5. Baldwin C, Spiro A, McGough C, Norman AR, Gillbanks A, et al. (2011) Simple nutritional intervention in patients with advanced cancers of the gastrointestinal tract, non-small cell lung cancers or mesothelioma and weight loss receiving chemotherapy: a randomised controlled trial. J Hum Nutr Diet 24: 431–440.
6. Evans WK, Nixon DW, Daly JM, Ellenberg SS, Gardner L, et al. (1987) A randomized study of oral nutritional support versus ad lib nutritional intake during chemotherapy for advanced colorectal and non-small-cell lung cancer. J Clin Oncol 5: 113–124.
7. Baldwin C, Spiro A, Ahern R, Emery PW (2012) Oral nutritional interventions in malnourished patients with cancer: a systematic review and meta-analysis. J Natl Cancer Inst 104: 371–385.
8. Yancik R (2005) Population aging and cancer: a cross-national concern. Cancer J 11: 437–441.
9. Aaldriks AA, Maartense E, le Cessie S, Giltay EJ, Verlaan HA, et al. (2011) Predictive value of geriatric assessment for patients older than 70 years, treated with chemotherapy. Crit Rev Oncol Hematol 79: 205–212.
10. Soubeyran P, Fonck M, Blanc-Bisson C, Blanc JF, Ceccaldi J, et al. (2012) Predictors of early death risk in older patients treated with first-line chemotherapy for cancer. J Clin Oncol 30: 1829–1834.
11. Guigoz Y, Lauque S, Vellas BJ (2002) Identifying the elderly at risk for malnutrition. The Mini Nutritional Assessment. Clin Geriatr Med 18: 737–757.
12. Gioulbasanis I, Georgoulias P, Vlachostergios PJ, Baracos V, Ghosh S, et al. (2011) Mini Nutritional Assessment (MNA) and biochemical markers of cachexia in metastatic lung cancer patients: interrelations and associations with prognosis. Lung Cancer 74: 516–520.
13. Extermann M, Boler I, Reich RR, Lyman GH, Brown RH, et al. (2012) Predicting the risk of chemotherapy toxicity in older patients: the Chemotherapy Risk Assessment Scale for High-Age Patients (CRASH) score. Cancer 118: 3377–3386.
14. Extermann M, Hurria A (2007) Comprehensive geriatric assessment for older patients with cancer. J Clin Oncol 25: 1824–1831.
15. Blanc-Bisson C, Fonck M, Rainfray M, Soubeyran P, Bourdel-Marchasson I (2008) Undernutrition in elderly patients with cancer: target for diagnosis and intervention. Crit Rev Oncol Hematol 67: 243–254.
16. Durrieu J, Blanc-Bisson C, Fonck M, Béchade D, Becouarn Y, et al. (2011) "Eight dietary advice cards to address the needs of elderly cancer patients treated with palliative chemotherapy: INOGAD study". Huit fiches pratiques de conseil nutritionnel pour répondre aux besoins des patients âgés au cours de leur traitement de chimiothérapie: Étude INOGAD. Oncologie 13: 83–89.
17. Bourdel-Marchasson I, Barateau M, Rondeau V, Dequae-Merchadou L, Salles-Montaudon N, et al. (2000) A multi-center trial of the effects of oral nutritional supplementation in critically ill older inpatients. GAGE Group. Groupe Aquitain Geriatrique d'Evaluation. Nutrition 16: 1–5.
18. O'Brien PC, Fleming TR (1979) A multiple testing procedure for clinical trials. Biometrics 35: 549–556.
19. Andreyev HJ, Norman AR, Oates J, Cunningham D (1998) Why do patients with weight loss have a worse outcome when undergoing chemotherapy for gastrointestinal malignancies? Eur J Cancer 34: 503–509.
20. Dobrila-Dintinjana R, Trivanovic D, Zelic M, Radic M, Dintinjana M, et al. (2013) Nutritional Support in Patients with Colorectal Cancer during Chemotherapy: Does it Work? Hepatogastroenterology 60: 475–480.
21. Ovesen L, Allingstrup L, Hannibal J, Mortensen EL, Hansen OP (1993) Effect of dietary counseling on food intake, body weight, response rate, survival, and quality of life in cancer patients undergoing chemotherapy: a prospective, randomized study. J Clin Oncol 11: 2043–2049.
22. Uster A, Ruefenacht U, Ruehlin M, Pless M, Siano M, et al. (2013) Influence of a nutritional intervention on dietary intake and quality of life in cancer patients: a randomized controlled trial. Nutrition 29: 1342–1349.
23. Lenk K, Schuler G, Adams V (2010) Skeletal muscle wasting in cachexia and sarcopenia: molecular pathophysiology and impact of exercise training. J Cachexia Sarcopenia Muscle 1: 9–21.
24. Prado CM, Sawyer MB, Ghosh S, Lieffers JR, Esfandiari N, et al. (2013) Central tenet of cancer cachexia therapy: do patients with advanced cancer have exploitable anabolic potential? Am J Clin Nutr 98: 1012–1019.
25. Raffaghello L, Safdie F, Bianchi G, Dorff T, Fontana L, et al. (2010) Fasting and differential chemotherapy protection in patients. Cell Cycle 9: 4474–4476.
26. Isenring EA, Bauer JD, Capra S (2007) Nutrition support using the American Dietetic Association medical nutrition therapy protocol for radiation oncology patients improves dietary intake compared with standard practice. J Am Diet Assoc 107: 404–412.
27. Isenring EA, Capra S, Bauer JD (2004) Patient satisfaction is rated higher by radiation oncology outpatients receiving nutrition intervention compared with usual care. J Hum Nutr Diet 17: 145–152.
28. van den Berg MG, Rasmussen-Conrad EL, Wei KH, Lintz-Luidens H, Kaanders JH, et al. (2010) Comparison of the effect of individual dietary counselling and of standard nutritional care on weight loss in patients with head and neck cancer undergoing radiotherapy. Br J Nutr 104: 872–877.
29. Ravasco P, Monteiro-Grillo I, Vidal PM, Camilo ME (2005) Dietary counseling improves patient outcomes: a prospective, randomized, controlled trial in colorectal cancer patients undergoing radiotherapy. J Clin Oncol 23: 1431–1438.
30. Bozzetti F, Cozzaglio L, Gavazzi C, Bidoli P, Bonfanti G, et al. (1998) Nutritional support in patients with cancer of the esophagus: impact on nutritional status, patient compliance to therapy, and survival. Tumori 84: 681–686.
31. Argiles JM (2005) Cancer-associated malnutrition. Eur J Oncol Nurs 9 Suppl 2: S39–50.

5

Chemotherapy or Targeted Therapy as Second-Line Treatment of Advanced Gastric Cancer

Roberto Iacovelli[1,2]*, Filippo Pietrantonio[1], Alessio Farcomeni[3], Claudia Maggi[1], Antonella Palazzo[1,2], Francesca Ricchini[1], Filippo de Braud[1], Maria Di Bartolomeo[1]

1 Department of Medical Oncology, Fondazione IRCCS Istituto Nazionale dei Tumori, Milan, Italy, **2** PhD Program, Department of Radiology, Oncology and Human Pathology, Sapienza University of Rome, Rome, Italy, **3** Department of Public Health and Infectious Diseases, Statistics Section, Sapienza University of Rome, Rome, Italy

Abstract

Chemotherapy is a cornerstone in treatments of gastric cancer, but despite its benefit, less than 60% of patients receive salvage therapy in clinical practice. We performed a systematic review and meta-analysis based on trial data on the role of second-line treatment of advanced gastric cancer. MEDLINE/PubMed and Cochrane Library were searched for randomized phase III trials that compared active therapy to best supportive care in advanced gastric cancer. Data extraction was conducted according to the PRISMA statement. Summary HR for OS was calculated using a hierarchical Bayesian model and subgroup analysis was performed based on baseline Eastern Cooperative Oncology Group Performance Status (ECOG) performance status (0 vs. 1 or more). A total of 1,407 patients were evaluable for efficacy, 908 were treated in the experimental arms, with chemotherapy (231 pts) or with targeted therapies (677 pts). The risk of death was decreased by 18% (HR = 0.82; 95% CI, 0.79–0.85; posterior probability HR≥1: <0.00001) with active therapies. Chemotherapy and ramucirumab were able to decrease this risk by 27% and 22%, respectively. No differences were found between chemotherapy and ramucirumab. In patients with ECOG = 0 a greater benefit was found for chemotherapy with a reduction of the risk of death by 43% and no benefits were found for ramucirumab or everolimus. In patients with ECOG = 1 or more a significant reduction of the risk of death by 32% was reported in patients treated with ramucirumab, even if no significant difference was reported between chemotherapy and ramucirumab. This analysis reports that active and available therapies are able to prolong survival in patients with advanced gastric cancer with a different outcome based on initial patient's performance status. New trials based on a better patient stratification are awaited.

Editor: Qing-Yi Wei, Duke Cancer Institute, United States of America

Funding: The authors received no specific funding for this work.

Competing Interests: The authors have declared that no competing interests exist.

* Email: roberto.iacovelli@istitutotumori.mi.it

Introduction

Gastric cancer (GC) is the third leading cause of cancer death in both sexes worldwide (8.8% of the total), with the highest estimated mortality rates in Eastern Asia and the lowest in Northern America. High mortality rates are also present in both sexes in Central and Eastern Europe, and in Central and South America [1].

Systemic chemotherapy is a cornerstone in treatments of GC both in locally-advanced and metastatic disease. Although no standard regimen for the first-line chemotherapy have been set up on a global scale, its use is associated with a consistent reduction of the risk of death over best supportive care and the combination chemotherapy with cisplatin and fluoropyrimidine (5-FU) improves survival compared to single-agent 5-FU [2].

Considering the new treatment in first-line therapy options, whereas trastuzumab with standard fluoropyrimidine/cisplatin regimen is recommended in patients with HER-2 positive tumors, two and three-drug regimens including 5-FU, cisplatin, with or without an antracyclines, as well as irinotecan or docetaxel-containing regimens are reasonable treatment options for HER-2 negative patients [3,4].

Despite the majority of patients receive first-line therapy, the analysis of patients enrolled in upfront clinical trials revealed that the attitude towards second-line chemotherapy differ between European and Japanese studies, with a percentage of 14% in REAL-2 and 75% in the SPIRITS study [5,6].

Regarding patients treated in clinical practice, these percentages are even lower - with only about 45% receiving a salvage treatment compared to Japanese clinical studies. Despite the low number of patients treated in the second-line setting and intrinsic biases, the outcomes of patients receiving a salvage treatment seemed to be influenced positively, with survival times exceeding one year [7,8].

In recent years, several studies reported that second-line chemotherapy is not the only effective strategy able to increase survival in patients with advanced GC focusing on vascular endothelial growth factor receptor (VEGFR) as new target. In particular, VEGFR-2 is over-expressed in GC tissue as compared to normal mucosa and in presence of lymph nodal metastases [9]. Recent trials reported such as a monoclonal antibody – ramucirumab – or a tyrosine kinase inhibitor – apatinib – against the VEGFR-2 are able to increase the progression free survival and the overall survival in patients treated with one or two previous line of therapies [10,11].

The aim of this meta-analysis was to estimate the effect of second-line treatment of GC and to analyze the differential role of chemotherapy or targeted agents. We also investigated if different strategies have the same role in patients with different performance status, with the intent to find the best strategy for second-line treatment of this tumor.

Methods

Definition of the outcome

For each trial, chemotherapy or targeted therapy as single agents were considered as the experimental treatment and the placebo or the best supportive care (BSC) as the control one. Results were reported for the entire cohort and by type of treatments (chemotherapy, everolimus, ramucirumab), separately.

Selection of the studies

We reviewed MEDLINE/PubMed and Cochrane Library for citations from January 2004 to February 2014. The search criteria were limited to articles published in English language and phase III clinical trials using appropriate filters available on PubMed. The entry term for the search was "gastric tumor". During the selection process, search was further restricted to randomized controlled trials in which chemotherapy agents or targeted agent were used as second-line of therapy after first-line platinum- and fluoropyrimidine-based combination therapy over placebo or best supportive care for treatment of advanced gastric cancer. If more than one publication was found for the same trial, the most recent was considered for analysis.

Study quality was assessed by using the Jadad seven-item scale that included randomization, double blinding and withdrawals; the final score was reported between 0 and 5 [12].

Data extraction

Data's extraction was conducted independently by two co-authors (R.I. and C.M.) according to the Preferred Reporting Items for Systematic review and Meta-Analysis (PRISMA) statement (**Checklist S1**) [13]; any discrepancies were resolved by consensus between the two authors. The data obtained for each trial were reported in the presented tables, these were: first author's name, year of publication, trial phase, the number of patients evaluable, the number of arms, drugs used in the experimental and in the control arm, dosage, rate of patients with Eastern Cooperative Oncology Group Performance Status (ECOG-PS) of 0, 1 or 2; and median overall survival (OS) with the relative hazard ratio and 95% CI.

Statistical method

The HR for OS with the relative 95% CIs was extracted from each study. Summary HR for OS was calculated using a hierarchical Bayesian model [14], where logarithm of study HR was assumed to be normally distributed [15], each study effect was assumed to arise from a Gaussian centered on a study-specific log-HR and the extracted standard error, inflated by 25% to obtain a conservative statement. The study-specific log-HR was assumed to be Gaussian, centered on a pooled log-HR, which is the main object of interest. An informative prior is used for the variance of the pooled HR, as an inverse Gamma centered on an estimator obtained with a moment-based approach (inflated by 25% to obtain a conservative statement). Indirect comparisons were conducted by means of a similar model, assuming an additive shift for the difference of effects on the log-HR scale.

A two-tailed p-value of less than 0.05 was considered statistically significant. All data were collected using Microsoft Office Excel 2007; statistical analyses were performed using R software [16].

Results

The electronic search revealed 72 citations. After screening, 49 articles were eliminated because 23 were studies on adjuvant or neoadjuvant therapy, 6 were surgical studies and 20 were related to other aspects of gastric cancer. Among the remaining 23 studies on patients with advanced disease, 18 were eliminated because treatments were administered as first-line. At the end of the review process, only five articles were included in the meta-analysis because of their adequate quality and availability of data (**Figure 1**) [11,17–20]. Among these, only four were considered positive because reached the primary end-point [11,18–20]. The characteristics of each study are presented in **Table 1**.

Overall population

A total of 1,424 patients were available for this trial-based meta-analysis and among these 1,407 were evaluable for efficacy. The majority of patients had an ECOG performance status of 0 or 1, being the 95% and 83% in the experimental and control arm, respectively. A total of 908 patients were treated in the experimental arm, and among these 231 received chemotherapy and 677 received targeted therapies. In the experimental arm, the type of chemotherapy used was docetaxel (150 patients) or irinotecan (81 patients), while the targeted therapies were the anti-VEGFR-2 therapy ramucirumab (238 patients) or the mammalian target of rapamycin inhibitor (mTORi) everolimus (439 patients). In the control arm, patients were all treated with BSC with or without placebo.

In the overall population, the active treatment decreases the risk of death by 18% ($HR = 0.82$; 95% CI 0.79–0.85; posterior probability of $HR \geq 1$: <0.00001) (**Table 2**). When analysis was limited to studies reaching the primary end point the reduction of the risk of death was by 27% ($HR = 0.73$; 95% CI 0.61–0.86; posterior probability of $HR \geq 1$: $p < 0.00001$), as reported in **Table 2**.

Survival by type of therapy

When populations were divided based on type of therapy (chemotherapy vs. anti-VEGFR vs. mTORi), chemotherapy was able to decrease the risk of death by 27% ($HR = 0.73$; 95% CI, 0.58–0.96; posterior probability of $HR \geq 1$: 0.00942). On the other hand, treatment with ramucirumab was able to decrease the risk of death by 22% ($HR = 0.78$; 95% CI, 0.60–1.00), while no significant effect on OS was seen with everolimus ($HR = 0.90$; 95% CI, 0.75–1.08).

Survival by performance status

When populations were divided based on ECOG-PS, the 461 patients with ECOG 0 had a greater benefit when treated with chemotherapy over BSC, with a reduction of the risk of death by 43% ($HR = 0.57$; 95% CI, 0.36–0.91; posterior probability of $HR \geq 1$: 0.0092). In this group of patients, no benefit was found for ramucirumab or everolimus over BSC; indirect comparison found a better outcome for patients treated with chemotherapy

Table 1. Characteristic of the studies included in the analysis.

Author	Year	Experimental arm					Control arm					OS			Jadad' Score
		N° Pts	Therapy	ECOG 0	ECOG 1	ECOG ≥2	N° Pts	Therapy	ECOG 0	ECOG 1	ECOG ≥2	Exp Arm	Ctr arm	HR (95% CI)	
Ford HER et al.	2014	84	Docetaxel	28%	55%	17%	84	BSC	26%	60%	14%	5.2	3.6	0.67 (0.49–0.92)	2
Fuchs CS et al.	2014	238	Ramucirumab	28%	72%	-	117	Pbo	31%	85%	1%	5.2	3.8	0.78 (0.60–1.00)	5
Ohtsu A et al.	2013	439	Everolimus	33%	61%	6%	217	Pbo	32%	55%	12%	5.4	4.3	0.90 (0.75–1.08)	5
Kang JH et al.	2012	133	Chemotherapy	54%	46%	-	69	BSC	52%	48%	-	5.3	3.8	0.66 (0.48–0.89)	2
Thuss-P. PC et al.	2011	21	Irinotecan	81%		19%	19	BSC	74%		26%	4.0	2.4	0.48 (0.25–0.92)	2

compared to ramucirumab (posterior probability of chemotherapy worse than or equivalent to ramucirumab: 0.00439) (**Table 3**).

In the 912 patients with ECOG-PS = 1 or more, a trend for greater efficacy was confirmed for those treated with chemotherapy compared to patients treated with BSC, but difference was not strong (HR = 0.80; 95% CI, 0.34–1.89; posterior probability of HR≥1: 0.07). In the same group of patients, a significant effect was found for those treated with ramucirumab over BSC with a reduction of the risk of death by 32% (HR = 0.68; 95% CI, 0.51–0.92; p = 0.04) (**Table 4**). Indirect comparison did not report differences between patients treated with chemotherapy or ramucirumab (posterior probability of chemotherapy worse than or equivalent to ramucirumab: 0.7622).

In summary, regardless of treatment, very little evidence was found for efficacy in patients with ECOG-PS = 0 (HR = 0.88; 95% CI, 0.61–1.28; posterior probability of HR≥1: 0.174), due to the fact that ramucirumab and everolimus do not report a significant decrease of the risk of death, while a benefit was found for chemotherapy. On the other hand, a mild evidence of efficacy was found for patients with PS = 1 or more, with a reduction of 21% (HR = 0.79; 95% CI, 0.64–0.98; posterior probability of HR≥1: 0.015). Indirect comparison indicated that any active therapy over BSC was more effective on patients with ECOG-PS = 1 or more vs. ECOG-PS = 0 (posterior probability of HR in ECOG = 0 better than HR in ECOG = 1: <0.0001), suggesting that patients with symptomatic disease should not be immediately excluded by further lines of therapy.

Quality of the studies

Jadad' scores for each trial are listed in Table 1; the mean score was 3.2, confirming the good-quality of the included trials.

Discussion

First-line treatment of advanced GC with modern regimens confers a benefit of OS exceeding ten months. Even if the addition of trastuzumab to cisplatin-based chemotherapy significantly improved OS in HER-2 positive GC [21], the outcome of the majority of patients is still poor and disease progression invariably occurs.

In metastatic phase, the role of second-line was largely debated because the risk to expose patients to treatment toxicity is high due to performance status deterioration and disease-related symptoms.

Up to date, this analysis is the largest to report that second-line treatment is able to decrease the risk of death by 18%, with a more evident effect in favor of chemotherapy reaching a risk reduction of 27%. Despite the higher absolute benefit of chemotherapy, we are unable to find a relative superiority of this strategy over the targeted agent ramucirumab.

Studies included in this analysis reported that second-line chemotherapy increased the median OS of about two months as compared to BSC. In this context, the choice of the best drugs to use according to efficacy, toxicity and individual patients characteristics remain a open issue. A recent meta-analysis showed that different drugs such as docetaxel or irinotecan, or different administration schedules did not have any influence on outcomes [22]. In facts, the objective response rate and the disease control rate were similar and the decreased risks of death were 29% for docetaxel (HR = 0.71; 95% CI, 0.56–0.90) and 45% for irinotecan (HR = 0.55; 95% CI, 0.40–0.77) over BSC, respectively [22]. Despite this difference, a prospective phase III study comparing weekly paclitaxel to irinotecan as second-line of therapy in GC patients did not show significant differences. The results were a

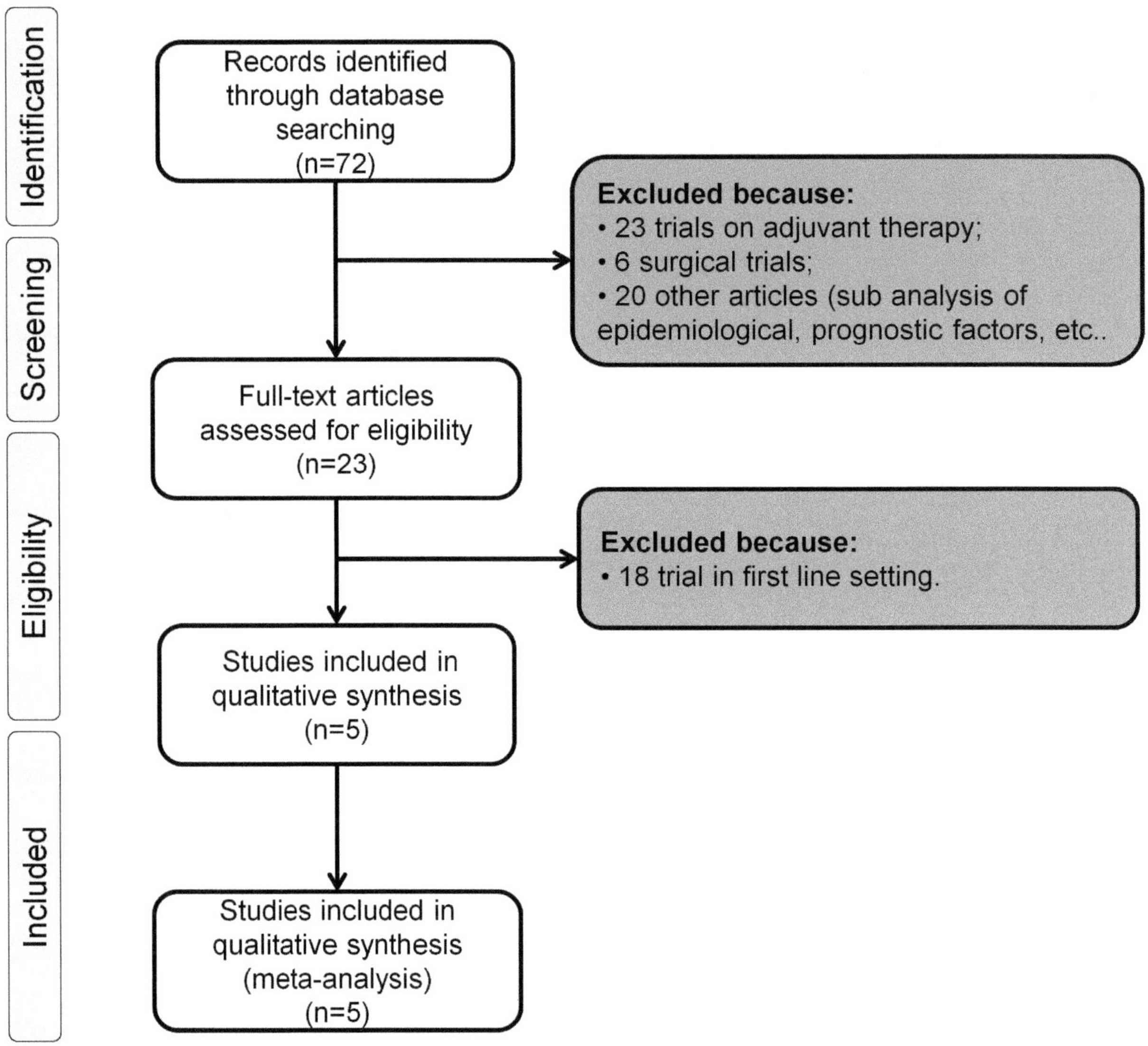

Figure 1. Flowchart of search process.

median OS of 8.4 and 9.5 months (HR 1.132; 95% CI, 0.86–1.49; p = 0.38), a median PFS of 2.3 and 3.6 months (HR 1.14; 95% CI, 0.88–1.49; p = 0.33), and an ORR of 13.6% and 20.9% (p = 0.20) for irinotecan and paclitaxel, respectively [23]. Another phase II study that compared the a novel liposomal formulation of irinotecan (PEP02) to standard irinotecan or docetaxel confirmed

Table 2. Overall Survival in overall population and based on type of studies.

Type of study	Trial	Year	N° of Patients		HR (95% CI)
			Experim. Arm	Control Arm	
Positive studies	Thuss-P. PC et al.	2011	21	19	0.48 (0.25–0.92)
	Kang JH et al.	2012	133	69	0.66 (0.48–0.89)
	Fuchs CS et al.	2014	238	117	0.78 (0.60–1.00)
	Ford HER et al.	2014	84	84	0.67 (0.49–0.92)
	Subtotal		*476*	*289*	*0.73 (0.61–0.86)*
Negative studies	Ohtsu A et al.	2013	439	217	0.90 (0.75–1.08)
TOTAL			**915**	**506**	**0.82 (0.79–0.85)**

Table 3. Overall Survival by type of therapy in patients with ECOG performance status of 0.

Type of therapy	Trial	Year	N° of Patients		HR (95% CI)
			Experim. Arm	Control Arm	
Chemotherapy	Kang JH et al.	2012	72	36	0.59 (0.38–0.90)
	Ford HER et al.	2014	22	19	0.48 (0.24–0.95)
	Subtotal		*94*	*55*	*0.57 (0.36–0.91)*
mTOR inhibitor	Ohtsu A et al.	2013	144	70	1.14 (0.81–1.61)
VEGFR inhibitor	Fuchs CS et al.	2014	67	31	1.07 (0.64–1.81)
TOTAL			**305**	**156**	**0.88 (0.61–1.28)**

the lack of any significant advantage in favor of any therapy when compared to other ones [24].

If the role of chemotherapy in second-line treatment of GC was clearly demonstrated, the role of combination chemotherapy over single agent remain an open issue. At least two studies tested this hypothesis comparing two different regimens. The first one compared the activity of irinotecan and 5-FU (FOLFIRI regimen) over irinotecan alone, reporting no significant difference in term of response rate (20.0 vs. 17.2%; $p = 0.525$), median PFS (3.0 vs. 2.2 months; $p = 0.481$) or OS (6.7 vs. 5.8 months; $p = 0.514$) [25]. The second one comparing docetaxel +5′DFUR to docetaxel alone, reported a significant survival benefit for the combination (7.6 vs. 4.0; $p<0.05$) [26], increasing the confusion in this area. Probably, new large phase III studies will be indispensable in order to better understand the role of combination chemotherapy in the salvage setting.

Until now, two evidence-based strategies are available for patients after first-line of therapy: single agent chemotherapy or targeted therapy such as ramucirumab. Even if the decreased risk of death seems to be higher with the use of chemotherapy in patients with ECOG-PS = 0 and for ramucirumab in patients with ECOG-PS = 1 or more, considering the nature of this analysis, no definitive conclusion may lead in favor of one strategy or the other, while a prospective trial may better address to this fascinating question.

It is noteworthy to mention the results of a recent placebo-controlled, randomized phase III trial that compared weekly paclitaxel and ramucirumab over weekly paclitaxel alone (RAINBOW study). For the first time, in the second-line setting, the addition of a targeted agent to standard chemotherapy demonstrated a significant survival advantage increasing the median OS from 7.36 to 9.63 months ($p = 0.0169$) and the median PFS from 2.9 to 4.4 months [27]. This positive outcome was reported both in PS = 0 and PS = 1 patients even if a statistical significant benefit was reached only in the PS = 1 patients [27].

Both studies confirm such as the anti-VEGFR2 monoclonal antibody seems to be more active compared to anti-VEGF ones or to anti-VEGFR tyrosine kinase inhibitors. Unfortunately, the reason for these differences is a current challenge and further studies may elucidate the pharmacological differences and probably to improve clinical outcome [28].

Considering the different toxicity profile – of any grade – we account such as ramucirumab was mainly characterized by: fatigue (36%), abdominal pain (29%), decrease of appetite (24%) and vomiting (20%) and by hematological toxicities such as arterial hypertension (16%) and bleeding or hemorrhage (13%) [11]. On the other hand, docetaxel was characterized by the hematological ones such as anemia (28%), neutropenia (16%) other than by specific gastrointestinal (86%), dermatological (42%), or neurological (30%) toxicities [18,20]. Moreover, the choice of irinotecan seems not improve the toxicity profile of taxanes-based chemotherapy as reported by comparative studies [19,23,24].

Even if no definitive data are available about the best treatment strategy for these patients, it is evident that patients' clinical conditions and co-morbidities, as well as the residual toxicities and the magnitude of benefits from first-line treatment, may all have a role in the choice of second-line therapy.

Table 4. Overall Survival by type of therapy in patients with ECOG performance status of 1 or more.

Type of therapy	Trial	Year	N° of Patients		HR (95% CI)
			Experim. Arm	Control Arm	
Chemotherapy	Kang JH et al.	2012	61	33	0.72 (0.46–1.13)
	Ford HER et al. PS = 1	2014	45	50	0.80 (0.53–1.21)
	Ford HER et al. PS = 2	2014	13	12	0.81 (0.36–1.82)
	Subtotal		*119*	*95*	*0.80 (0.34–1.89)*
mTOR inhibitor	Ohtsu A et al. PS = 1	2013	269	120	0.86 (0.58–1.08)
	Ohtsu A et al. PS = 2	2013	25	27	1.43 (0.82–2.48)
	Subtotal		*294*	*147*	*0.92 (0.70–1.23)*
VEGFR inhibitor	Fuchs CS et al.	2014	171	86	0.68 (0.51–0.91)
TOTAL			**584**	**328**	**0.79 (0.64–0.98)**

In this study, we analyzed the activity of different type of therapy in relationship with patient' performance status at the start of treatment and we found out that the benefit of chemotherapy was more evident in asymptomatic patients with ECOG = 0, with a reduction of the risk of death by 43% (HR = 0.57) as compared to patients with symptomatic disease (HR = 0.80). In this population with suboptimal performance status, the benefit of any type of therapy in terms of decreased risk of death is only by 21%, although it may be increased to 32% with the use of ramucirumab. Probably, a good baseline clinical condition increases the tolerability to chemotherapy and to its related toxicity, on the other hand, the use of a less-toxic approach might be preferred in patients with worse clinical conditions in order to improve quality of life.

This analysis may be influenced by several factor such as the low number of patients – less than 35% – with ECOG = 0 compared to patients with ECOG = 1 or more in the included studies.

Nevertheless, some other limitations may affect these results. First, and foremost, this is a trial-level meta-analysis based on studies and not on individual patient data. Confounding variables such as patient co-morbidities, extent of disease and differences in other possible prognostic factors could not be incorporated into such an analysis. Second, all the included studies were conducted in selected patients with adequate organ function and no severe co-morbidities at the time of study entry. Third, the data on the correlation between ECOG PS and outcomes derived from subgroup analyses of published studies.

Conclusions

Finally, this study confirms a significant benefit in terms of OS when active second-line treatments are administered to patients with advanced gastric cancer after failure of a previous line of therapy even in patients with impaired performance status.

If the lack of difference between chemotherapy agents was reported by other studies, we suggest a lack of difference between chemotherapy and ramucirumab. Further studies are urgently required to better understand the clinical or molecular characteristic for patient' selection.

Supporting Information

Checklist S1 Preferred Reporting Items for Systematic review and Meta-Analysis (PRISMA) statement.

Author Contributions

Analyzed the data: RI AF FP MDB AP. Contributed to the writing of the manuscript: RI FP MDB AP CM FR FDB.

References

1. Globocan 2012: Estimated cancer incidence, mortality and prevalence worldwide in 2012. Available: http://globocan.iarc.fr/Pages/fact_sheets_cancer.aspx. Accessed 2014 February 12.
2. Wagner AD, Unverzagt S, Grothe W, Kleber G, Grothey A, et al. (2010) Chemotherapy for advanced gastric cancer. Cochrane Database Syst Rev 3: CD004064.
3. National Comprehensive Cancer Network Guidelines for gastric cancer. Available: http://www.nccn.org/professionals/physician_gls/f_guidelines.asp. Accessed 2014 February 12.
4. European Society of Medical Oncology (ESMO) guidelines for gastric cancer. Available: http://www.esmo.org/Guidelines-Practice/Clinical-Practice-Guidelines/Gastrointestinal-Cancers/Gastric-Cancer. Accessed 2014 February 12.
5. Cunningham D, Starling N, Rao S, Iveson T, Nicolson M, et al. (2008) Capecitabine and oxaliplatin for advanced esophagogastric cancer. N Engl J Med 358: 36–46.
6. Koizumi W, Narahara H, Hara T, Takagane A, Akiya T, et al. (2008) S-1 plus cisplatin versus S-1 alone for first-line treatment of advanced gastric cancer (SPIRITS trial): a phase III trial. Lancet Oncol 9: 215–21.
7. Monti M, Foca F, Casadei Gardini A, Valgiusti M, Frassineti GL, et al. (2013) Retrospective analysis on the management of metastatic gastric cancer patients. A mono-institutional experience. What happens in clinical practice? Tumori 99: 583–8.
8. Elsing C, Herrmann C, Hannig CV, Stremmel W, Jäger D, et al. (2013) Sequential chemotherapies for advanced gastric cancer: a retrospective analysis of 111 patients. Oncology 85: 262–8.
9. Han FH, Li HM, Zheng DH, He YL, Zhan WH. (2010) The effect of the expression of vascular endothelial growth factor (VEGF)-C and VEGF receptor-3 on the clinical outcome in patients with gastric carcinoma. Eur J Surg Oncol 36: 1172–9.
10. Li J, Qin S, Xu J, Guo W, Xiong J, et al. (2013) Apatinib for chemotherapy-refractory advanced metastatic gastric cancer: results from a randomized, placebo-controlled, parallelarm, phase II trial. J Clin Oncol 31: 3219–25.
11. Fuchs CS, Tomasek J, Yong CJ, Dumitru F, Passalacqua R, et al. (2014) Ramucirumab monotherapy for previously treated advanced gastric or gastro-oesophageal junction adenocarcinoma (REGARD): an international, randomised, multicentre, placebo-controlled, phase 3 trial. Lancet 383: 31–9.
12. Jadad AR, Moore RA, Carroll D, Jenkinson C, Reynolds DJ, et al. (1996) Assessing the quality of reports of randomized clinical trials: Is blinding necessary? Control Clin Trials 17: 1–12.
13. Moher D, Liberati A, Tetzlaff J, Altman DG. (2009) Preferred reporting items for systematic reviews and meta-analyses: the PRISMA statement. BMJ 339: b2535.
14. Higgins JP, Thompson SG, Spiegelhalter DJ. (2009) A re-evaluation of random effects meta-analysis. Journal of the Royal Statistical Society (Series A) 172: 137–59.
15. Parmar MKB, Torri V, Stewart L (1998) Extracting summary statistics to perform meta-analyses of the published literature for survival endpoints. Statistics in Medicine 17: 2815–34.
16. R Core Team (2013) R: A language and environment for statistical computing. R Foundation for Statistical Computing, Vienna, Austria. Available: http://www.R-project.org/.
17. Ohtsu A, Ajani JA, Bai YX, Bang YJ, Chung HC, et al. (2013) Everolimus for previously treated advanced gastric cancer: results of the randomized, double-blind, phase III GRANITE-1 study. J Clin Oncol 31: 3935–43.
18. Ford HE, Marshall A, Bridgewater JA, Janowitz T, Coxon FY, et al. (2014) Docetaxel versus active symptom control for refractory oesophagogastric adenocarcinoma (COUGAR-02): an open-label, phase 3 randomised controlled trial. Lancet Oncol 15: 78–86.
19. Kang JH, Lee SI, Lim do H, Park KW, Oh SY, et al. (2012) Salvage chemotherapy for pretreated gastric cancer: a randomized phase III trial comparing chemotherapy plus best supportive care with best supportive care alone. J Clin Oncol 30: 1513–8.
20. Thuss-Patience PC, Kretzschmar A, Bichev D, Deist T, Hinke A, et al. (2011) Survival advantage for irinotecan versus best supportive care as second-line chemotherapy in gastric cancer–a randomised phase III study of the Arbeitsgemeinschaft Internistische Onkologie (AIO). Eur J Cancer 47: 2306–14.
21. Bang YJ, Van Cutsem E, Feyereislova A, Chung HC, Shen L, et al. (2010) Trastuzumab in combination with chemotherapy versus chemotherapy alone for treatment of HER2-positive advanced gastric or gastro-oesophageal junction cancer (ToGA): a phase 3, open-label, randomised controlled trial. Lancet 376: 687–97.
22. Kim HS, Kim HJ, Kim SY, Kim TY, Lee KW, et al. (2013) Second-line chemotherapy versus supportive cancer treatment in advanced gastric cancer: a meta-analysis. Ann Oncol. 24: 2850–4.
23. Ueda S, Hironaka S, Yasui H, Nishina T, Tsuda M, et al. (2012) Randomized phase III study of irinotecan (CPT-11) versus weekly paclitaxel (wPTX) for advanced gastric cancer (AGC) refractory to combination chemotherapy (CT) of fluoropyrimidine plus platinum (FP): WJOG4007 trial. J Clin Oncol 30: abstract 4002.
24. Roy AC, Park SR, Cunningham D, Kang YK, Chao Y, et al. (2013) A randomized phase II study of PEP02 (MM-398), irinotecan or docetaxel as a second-line therapy in patients with locally advanced or metastatic gastric or gastro-oesophageal junction adenocarcinoma. Ann Oncol 24: 1567–1573.
25. Sym SJ, Hong J, Park J, Cho EK, Lee JH, et al. (2013) A randomized phase II study of biweekly irinotecan monotherapy or a combination of irinotecan plus 5-fluorouracil/leucovorin (mFOLFIRI) in patients with metastatic gastric adenocarcinoma refractory to or progressive after first-line chemotherapy. Cancer Chemother Pharmacol. 71: 481–8.
26. Maruta F, Ishizone S, Hiraguri M, Fujimori Y, Shimizu F, et al. (2007) A clinical study of docetaxel with or without 5'DFUR as a second-line chemotherapy for advanced gastric cancer. Med Oncol 24: 71–5.
27. Wilke H, Van Cutsem E, Oh SC, Bodoky G, Shimada Y, et al. (2014) RAINBOW: A global, phase III, randomized, double-blind study of ramucir-

umab plus paclitaxel versus placebo plus paclitaxel in the treatment of metastatic gastroesophageal junction (GEJ) and gastric adenocarcinoma following disease progression on first-line platinum- and fluoropyrimidine-containing combination therapy rainbow IMCL CP12-0922 (I4T-IE-JVBE). J Clin Oncol 32: abstract LBA7.

28. Kasper S, Schuler M. (2014) Targeted therapies in gastroesophageal cancer. Eur J Cancer 50: 1247–58.

6

γδ T Cells Are Involved in Acute HIV Infection and Associated with AIDS Progression

Zhen Li[1☯], Wei Li[2☯], Ning Li[3], Yanmei Jiao[2], Dexi Chen[2], Lianxian Cui[1], Yu Hu[1], Hao Wu[2*], Wei He[1*]

1 Department of Immunology, School of Basic Medicine, Peking Union Medical College; National Key Laboratory of Medical Molecular Biology, Institute of Basic Medical Sciences, Chinese Academy of Medical Sciences, Beijing, China, **2** Center for Infectious Diseases, Beijing You'an Hospital, Capital Medical University, Beijing, China, **3** Beijing You'an Hospital, Capital Medical University, Beijing, China

Abstract

Background: Early diagnosis is vital to HIV control. γδ T cells play critical roles in viral infections, but their activation in acute HIV infected patients and follow up to 18 months has not been described.

Methods: Changes in γδ T cells, including subsets, function and activation, in treated and untreated acutely HIV-infected patients (n = 79) were compared by cytotoxicity assay and flow cytometry with healthy controls (n = 21) at month 0, 6, 12 and 18.

Results: In acutely HIV-infected patients, Vδ1 cell proportion was elevated ($P = 0.027$) with Vδ2 population reduced ($P = 0.002$). Effector and central memory γδ T cell factions were decreased ($P = 0.006$ and $P = 0.001$, respectively), while proportion of terminal γδ T cells increased ($P = 0.002$). γδ T cell cytotoxicity was compromised over time. Fraction of IL-17-producing cells increased ($P = 0.008$), and IFN-γ-producing cells were unaffected ($P = 0.115$). Elevation of a microbial translocation marker, sCD14, was associated with γδ T cell activation ($P = 0.001$), which increased in a time-dependent manner, correlating with CD4/CD8 T cell activation set-points and CD4 counts. Antiretroviral therapy did not affect these changes.

Conclusions: γδ T cell subpopulation and functions change significantly in acute HIV infection and over time. Early γδ T cell activation was associated with CD4/CD8 T cell activation set-points, which predict AIDS progression. Therefore, γδ T cell activation represents a potential surrogate marker of AIDS progression.

Editor: R. Keith Reeves, Beth Israel Deaconess Medical Center, Harvard Medical School, United States of America

Funding: This work was supported by Twelve-five Project for Control and Prevention of Major Infectious Diseases [2012ZX10001-006], National Natural Science Foundation of China [81371332], and Beijing Key Laboratory [BZ0089]. The funders had no role in study design, data collection and analysis, decision to publish, or preparation of the manuscript.

Competing Interests: The authors have declared that no competing interests exist.

* Email: wuhao90000@gmail.com (HW); suprahe@126.com (WH)

☯ These authors contributed equally to this work.

Introduction

Human γδ T cells are distinct from αβ T cells, and generally lack CD4 or CD8 antigen expression [1]. These cells represent approximately 3–4% of peripheral T cells in the blood, and are composed predominantly of two subsets, depending on expression of the Vδ1 or Vδ2 gene [2–5].Vδ1 T cells respond to antigens from bacterial pathogens [6]. Vδ2 T cells are believed to be activated by several chemical compounds, some of which are produced by activated host cells in response to pathogenic infections [7]. Vδ2 T cells can be further divided into four subtypes: naïve (Tnaïve), central memory (TCM), effector memory (TEM) and terminally differentiated (TEMRA), based on differential surface expression of CD45RA and CD27 [8,9]. Specific Tnaïve and TCM cells are located in the lymph nodes and possess distinct functions [10–12]. TEM and TEMRA are located preferentially at inflammatory sites and perform effector functions [10].

Functionally, γδ T cells produce interleukin 17 (IL-17), interferon gamma (IFN-γ), and other secreted protein factors after activation [13,14]. IL-17 is thought to be important for maintaining intestinal mucosal integrity and controlling microbial translocation [15]. IFN-γ levels are elevated in many proinflammatory events, and microbial translocation [16]. Furthermore, γδ T cells are involved in the recruitment of neutrophils in bacterial infections [17]. Reports indicate that, the cytotoxic activity of γδ T lymphocytes against human neuroblastoma cells, Burkitt's lymphoma cells and various cancer cell lines (human colon carcinoma, erytholeukemia, human neuroepithelioma and renal adenocarcinoma) is augmented upon activation [18,19].

It has been reported that, in all stages of HIV infections, gut functions are compromised, leading to enteropathy, which is commonly observed in HIV-infected patients. The permeability of the mucosal barrier is significantly elevated [20,21], which may result in translocation of microbial products into the systemic circulation, leading to immune stimulation. A number of studies have described the use of soluble CD14 (sCD14) as a marker of microbial translocation, and even as a predictive marker of disease progression in HIV infection [22–24].

The γδ T cell responses to HIV infection are complicated and not well-elucidated [25,26]. Previous reports have indicated increased levels of Vδ1 cells and decreased levels and function of Vδ2 cells in chronically HIV-infected patients [26,27]. Furthermore, changes in the Vδ2 T cell population correlated positively with CD4 T cell counts and negatively with viral loads [26].

The therapeutic efficacy of antiretroviral therapy (ART) on Vδ2 T cell recovery in HIV patients is controversial and inconclusive. A report in 2004 indicated partial recovery of the Vδ2 T cell population following long-term HIV-suppressive therapy [28]. However, a subsequent study demonstrated that ART failed to restore the Vδ2 repertoire in HIV-infected men [29].

In this study, we aimed to characterize the changes in γδ T cells, including subsets, function and activation status, in acutely HIV-infected patients who were followed from acute infection up to 18 months, and to define their possible correlation with other serum parameters, microbial translocation and AIDS progression.

Methods

Study subjects

The protocol for this study was approved by the Beijing You'an Hospital Ethics Committee and the Guidelines for Human Experimentation (PR. China) were followed throughout. Upon admission, all patients provided written informed consent for their information and clinical samples (blood and plasma) to be stored and used for research.

Seventy-nine homosexual men with acute HIV infection were enrolled in a prospective study (co-inclusion in the Beijing PRIMO cohort) conducted in Beijing You'an Hospital, Beijing, PR. China. Acute HIV infection was defined by a negative enzyme-linked immunosorbent assay (ELISA), and at least one of the following criteria: less than three bands on HIV Western blot (HIV BLOT2.2, 11039, Diagnostic Biotechnology Pte, Singapore), a positive p24 test or detectable plasma HIV-RNA. The estimated date of infection was calculated as follows: 2 weeks before onset of symptoms or 4 weeks before the first positive Western blot. All 79 enrolled HIV-infected patients were diagnosed early during the acute stage of HIV infection (median of estimated time post-infection: 41 days) and transmitted by homosexual activity. At baseline (day 0 of enrollment), all patients were treatment-naïve. During the follow-up times, some of the patients (11/79) started combination antiretroviral treatment (ART), based on CD4 cell counts (e.g. $<350/\mu L$ according to WHO recommendations) with or without clinical symptoms, and the consensus of both physicians and patients. Patients who were treated during the study received a combination of AZT, 3TC, and EFV or LPV/r. The courses of their treatment were not influenced by the authors for the purpose of this study. Most of them (68/79) were still untreated [CD4 counts: $>350/\mu L$ (n = 48), $<350/\mu L$ (n = 20)]. Twenty-one male homosexual healthy subjects were included as controls (HC). The ages of all groups were matched.

Blood samples were collected from patients at baseline (day 0 of enrollment), month 6, month 12 and month 18. Plasma samples were also collected from all healthy volunteers. All tests were performed at baseline. In addition, CD38/HLA-DR antigen expression and the cytotoxicity activity of γδ T cells were determined at baseline, month 6, month 12 and month 18.

Cell culture

Peripheral blood mononuclear cells (PBMCs) were isolated by Ficoll-Hypaque (TBD, Tianjin, China) centrifugation. The expansion of γδ T cells was performed as previously described [30]. Briefly, 24-well plates were coated with 500 μL purified anti-γδTCR antibody (IMMU510, 1 mg/mL; Immunotech, Beckman Coulter, Fullerton, CA, USA) at 37°C for 2 h. PBMCs were then added to the antibody-coated wells and cultured in RPMI 1640 medium supplemented with 10% FCS and 200 IU/mL recombinant human IL-2 (Beijing Read United Cross Pharmaceutical, Beijing, China). The culture medium was replaced every other day. The trypan blue exclusion method was employed to assess cell viability.

Antibodies and flow cytometry

Antibodies (Abs) used in this study were as follows: mouse phycoerythin (PE)-conjugated anti-human CD38 (HIT2) Ab, mouse allophycocyanin (APC)-conjugated anti-human HLA-DR (L243) Ab, mouse phycoerythin-cyanin7(PE-cy7)-conjugated anti-human CD3 (HIT3a) Ab, mouse peridinin-chlorophyll-protein complex cyanine 5.5 (PerCP-cy5.5) conjugated anti-human CD27 (O323) Ab, mouse APC-conjugated anti-human CD45RA (HI100) Ab were purchased from Biolegend (San Diego, CA, USA). Mouse Fluorescein isothiocyanate (FITC)-conjugated anti-human pan TCRγδ Ab, mouse PE-conjugated anti-human pan TCRγδ (IMMU510) Ab and mouse FITC-conjugated anti-human pan $V\delta_2$TCR (IMMU389) Ab were purchased from Immunotech (Beckman Coulter, Fullerton, CA, USA). Mouse FITC-conjugated anti-human pan $V\delta_1$TCR (TS8.2) Ab was purchased from Pierce (USA). Mouse PE-conjugated anti-human IL-17A (SCPL1362) Ab and mouse PE-conjugated anti-human IFN-γ (B27) Ab were purchased from BD Pharmingen (San Diego, CA, USA). The isotype control antibodies were purchased from the corresponding company, respectively.

Cells were stained with appropriate target Abs and isotype Abs using conventional surface- and/or intracellular staining methods. When both surface and intracellular staining was required, cells were first fixed and permeabilized using BD Cytofix/Cytoperm Fixation and Permeabilization Solution (BD Pharmingen), followed by staining for intracellular proteins. Cells were then washed extensively with PBS to remove excess Ab, stained for extracellular targets, and fixed with 2% formaldehyde. Fluorescence was evaluated with a FACSCalibur flow cytometer, and data were analyzed using FlowJo software (Tree Star, Ashland, OR, USA).

Cytotoxicity

The CytoTox 96 Non-Radioactive Cytotoxicity Assay (Promega, Madison, WI, USA), based on the colorimetric detection of released lactate dehydrogenase (LDH), was used according to the manufacturer's instructions. Tumor cells used as target cells were seeded into a 96-round well plate (5×10^4 cells/well). Expanded γδ T cells (effector cells) were added directly to individual wells at different effector/target (E/T) ratios. The plate was incubated at 37°C for 4 h before supernatants were collected and LDH activity was determined. Controls for spontaneous LDH release from effector and target cells, the target maximum release, as well as the culture medium background were assayed simultaneously. The cytotoxicity was calculated as follows: %Cytotoxicity = (Experimental-Effector spontaneous-Target spontaneous)/(Target maximum-Target spontaneous) ×100%.

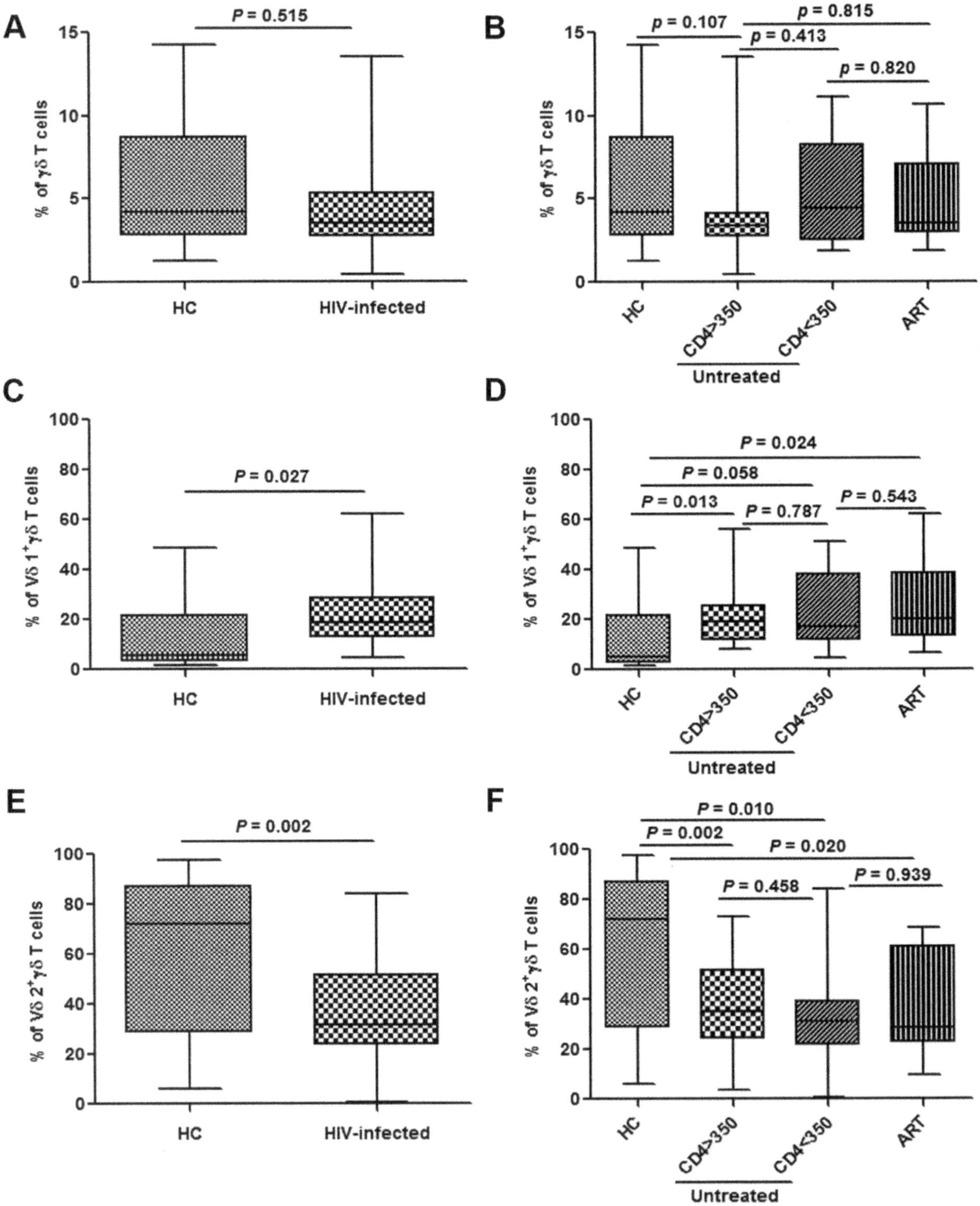

Figure 1. Changes in different subpopulations of γδ T cells in acutely HIV-infected patients. Frequencies of total γδ T cells, Vδ1 γδ T and Vδ2 γδ T cells were analyzed in healthy controls (HC) and acutely HIV-infected subjects at baseline (**A, C, E**). The patients were then subdivided based on the administration of antiretroviral therapy (ART) and CD4 levels (< or >than 350/μL). The frequencies were re-analyzed and compared with HC (**B, D, F**).

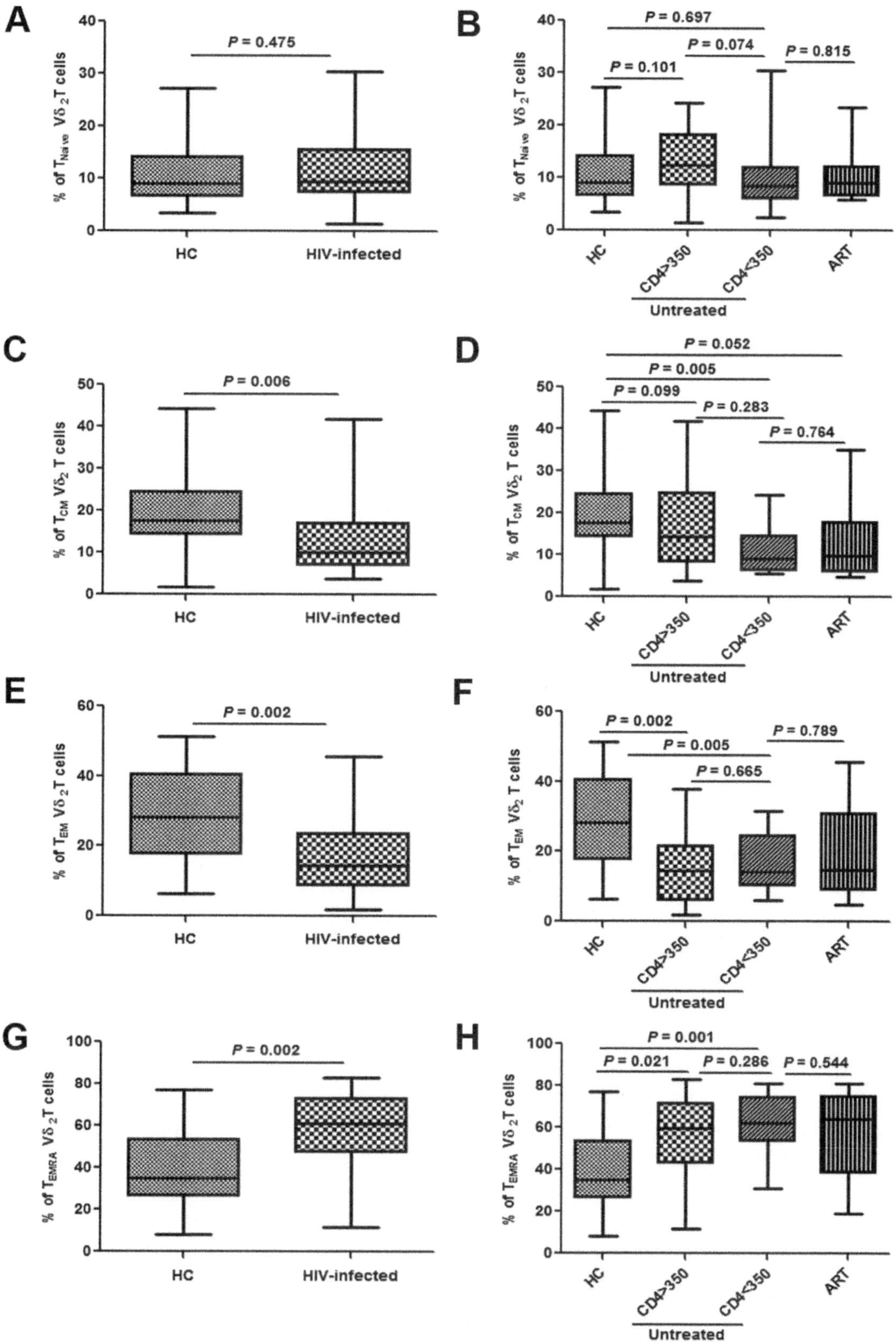

Figure 2. Changes in subpopulations of Vδ2γδ T cells in acutely HIV-infected patients. Frequencies of Tnaïve, TCM, TEM and TEMRA cells were analyzed in healthy controls (HC) and acutely HIV-infected subjects at baseline (**A, C, E, G**). The patients were then subdivided based on the administration of antiretroviral therapy (ART) and CD4 levels (< or >than 350/μL). The frequencies were re-analyzed and compared with HC (**B, D, F, H**).

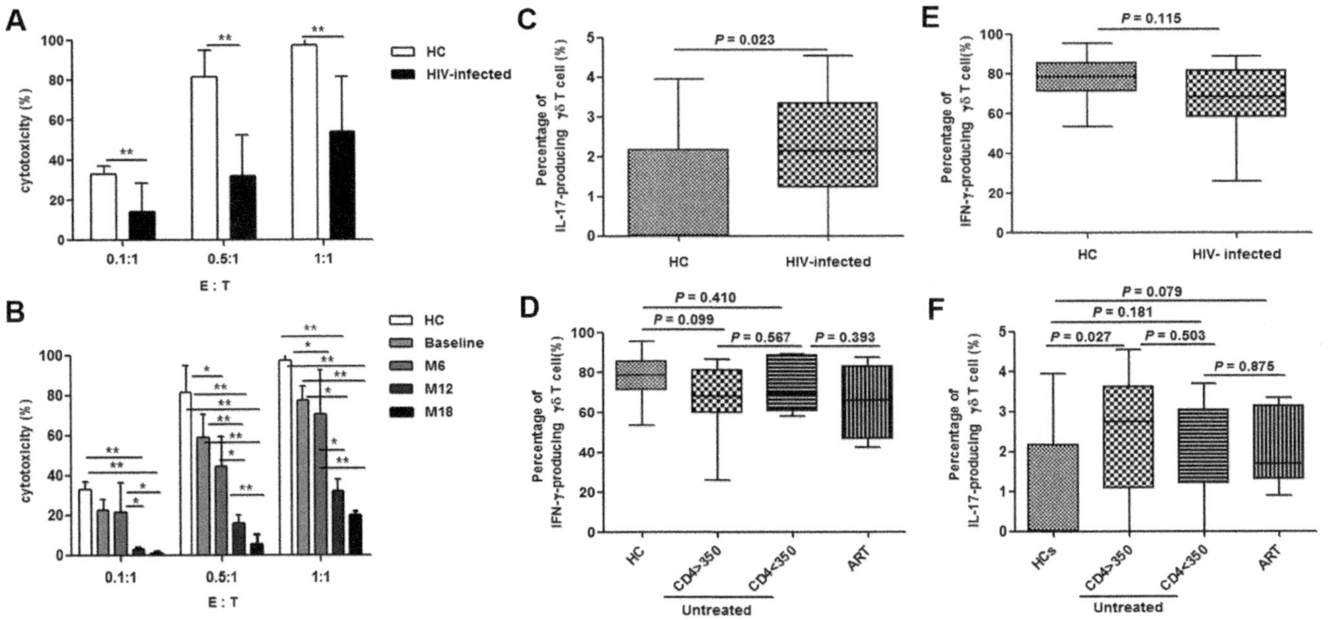

Figure 3. Altered functions of γδ T cells in acute HIV infection. Cytotoxicity (%) was assessed at different effector (E, γδ T cells) to target (T, Daudi cells) ratios (0.1:1, 0.5μ1 and 1:1) at baseline (**A**). Subsequently, longitudinal follow-up analysis of cytotoxicity was conducted at different time-points [baseline, month 6 (M6), month 12 (M12) and month 18 (M18)] (**B**). Frequencies of IFN-γ- and IL-17-producing γδ T cells were analyzed in healthy controls (HC) and acutely HIV-infected subjects at baseline (**C, E**). The patients were then subdivided based on the administration of antiretroviral therapy (ART) and CD4 levels (< or >than 350/μL). The frequencies were re-analyzed and compared with HC (**D, F**). *: $P<0.05$, **: $P<0.01$.

Quantification of plasma lipopolysaccharide (LPS), LPS-binding protein (LBP) and soluble CD14 (sCD14)

Commercially available ELISA kits were used in this study following manufacturers' instructions for measuring plasma concentrations of LBP (Cell Sciences, MA, USA) and sCD14 (soluble CD14) (R&D Systems, MN, USA). Plasma LPS levels were quantified using a *Limulus amebocyte* lysate (LAL) assay (Hycult Biotech, UDEN, Netherlands).

Statistical analysis

Non-parametric tests were used to avoid the impact of potential outlier values in a small study. Comparisons between groups were performed using the Mann-Whitney test. The Wilcox on matched pairs test was used to estimate the changes in the different variables throughout the follow-up. The Spearman's non-parametric correlation was used to estimate the association of two continuous variables of interest. *P*-values below 0.05 were considered statistically significant. Data of patients on ART were compared only with data of untreated patients with CD4 counts lower than 350/μL.

Results

Changes in γδ T cell subpopulations in acutely HIV-infected patients

Compared with healthy subjects, there was no marked difference in the fraction of γδ T cells in acutely HIV-infected patients (**Fig. 1A**). Nor were there any differences between patients with different CD4 counts, or those who did or did not receive ART (**Fig. 1B**). To characterize the changes in γδ T cells subpopulations in acutely HIV-infected patients, we first analyzed changes in the Vδ1 and Vδ2 subtypes. The proportion of Vδ1 cells among γδ T cells was elevated ($P=0.027$), while the Vδ2 population was significantly reduced ($P=0.002$) (**Fig. 1C and 1E**). However, there were no significant differences in both the proportions of Vδ1 and Vδ2 cells between patients with different CD4 counts (both $P>0.05$). Furthermore, Initiation of ART failed to bring about Vδ2 subtype recover, and had no effect on the Vδ1 population (**Fig. 1D and 1F**).

Changes in the levels of Vδ2 γδ T subgroups in acutely HIV-infected patients were investigated by analysis of the expression of surface CD27 and CD45RA antigens [8,9]. There was no difference in the proportion of naïve Vδ2γδT cells (CD27+/CD45RA+) observed ($P=0.475$), on which CD4 levels and ART showed no impact (**Fig. 2A and 2B**). The fractions of the TEM (effector memory Vδ2 γδ T cells, CD27−/CD45RA−) and TCM (central memory Vδ2 γδ T cells, CD27+/CD45RA−) populations were significantly decreased in acutely HIV-infected patients ($P=0.002$ and $P=0.006$, respectively), while the proportion of TEMRA (terminal Vδ2 γδ T cells, CD27−/CD45RA+) was increased ($P=0.002$) (**Fig. 2C, 2E, and 2G**). CD4 levels and the initiation of ART showed no correlation with these changes (**Fig. 2D, 2F, and 2H**).

Compromised γδ T cell functions in acutely HIV-infected patients

The cytotoxic activity of γδ T cells in the acute stage of HIV infection was analyzed at baseline, month 6, 12 and 18. Compared with data from healthy controls, γδ T cell cytotoxicity was gradually compromised in acutely HIV-infected patients over time (**Fig. 3A and 3B**). Furthermore, the fraction of IL-17-producing γδ T cells was elevated ($P=0.023$), while the fraction of IFN-γ-producing γδ T cells was unchanged ($P=0.115$) (**Fig. 3C and 3E**). These changes were not affected by CD4 counts or the initiation of ART (**Fig. 3D and 3F**).

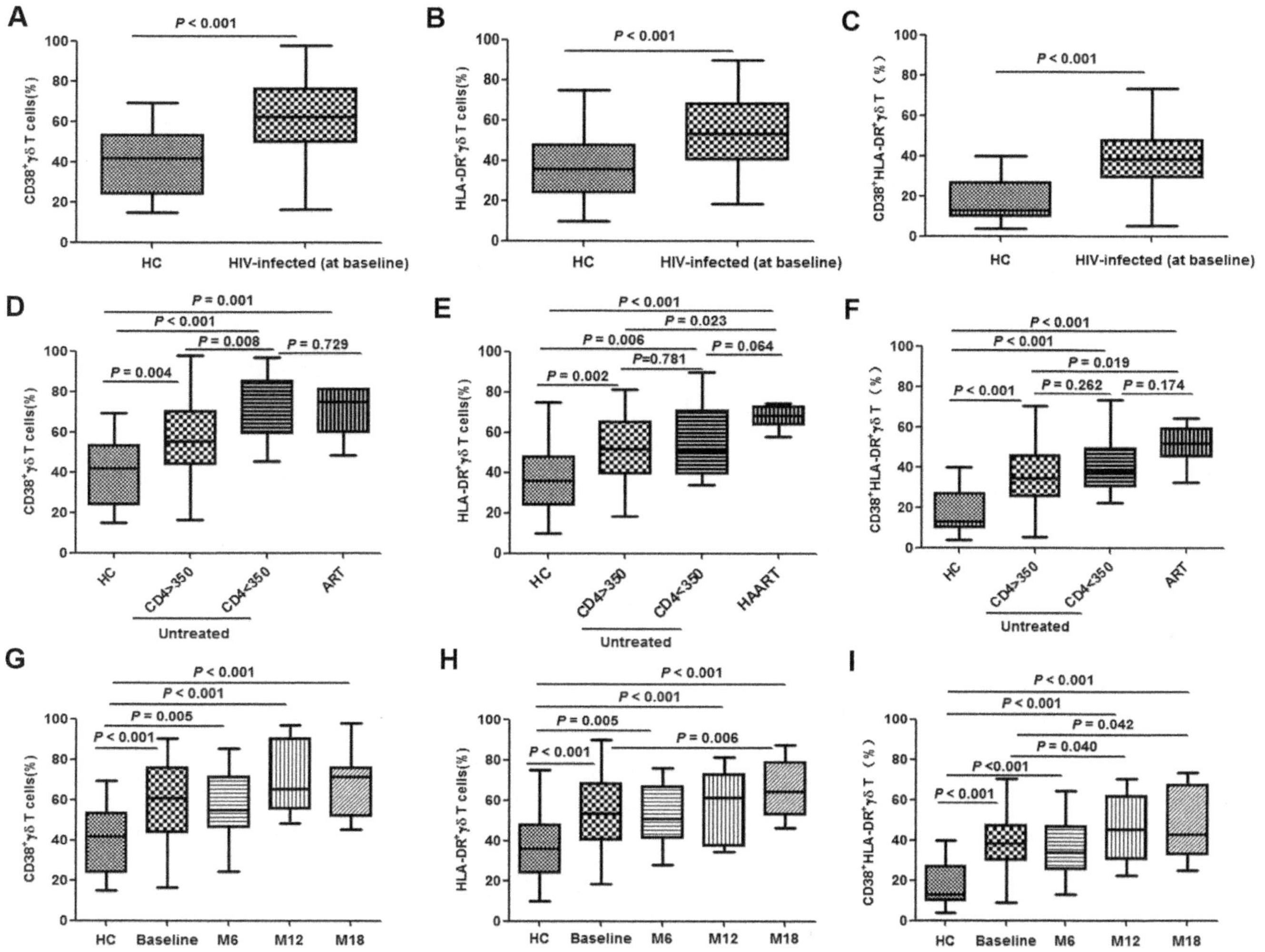

Figure 4. Activation of γδ T cells in acute HIV infection. Activation of γδ T cells was determined by analysis of CD38 and HLA-DR expression. Frequencies of CD38+, HLA-DR+ and CD38+/HLA-DR+ γδ T cells were tested in healthy controls (HC) and acutely HIV-infected subjects at baseline (**A–C**). The patients were then subdivided based on the administration of antiretroviral therapy (ART) and CD4 levels (< or >than 350/μL). The frequencies were re-analyzed and compared with HC (**D–F**). Longitudinal follow-up of γδ T cell activation in patients with acute HIV infection was conducted at baseline, month 6 (M6), month 12 (M12) and month 18 (M18). Frequencies of CD38+, HLA-DR+ and CD38+/HLA-DR+ γδ T cells were tested in healthy controls (HC) and acutely HIV-infected subjects at baseline (**G–I**).

Activation of γδ T cells

CD38 and HLA-DR are used as surrogate markers of γδ T cell activation [31,32]. Compared with healthy controls, the proportions of CD38+ or HLA-DR+ γδ T cells (single-positive and double-positive) were significantly elevated in acutely HIV-infected patients (all $P<0.001$) (**Fig. 4A–C**). For patients with a CD4 count of less than 350/μL, the fraction of CD38+ γδ T cells was higher than for those with a CD4 count of greater than 350/μL($P = 0.008$) (**Fig. 4D**); however, this CD4-dependent change was not evident in HLA-DR+ or double-positive cells ($P = 0.781$ and 0.262, respectively) (**Fig. 4E and 4F**). No significant differences were observed among patients on ART compared with untreated patients with a CD4 count of less than 350/μL (**Fig. 4D–F**).

To further characterize the activation of γδ T cells, CD38 and HLA-DR expression was determined at different time-points (baseline and months 6, 12, and 18). Steady increases in the fractions of all three groups of γδ T cells (CD38+, HLA-DR+ and CD38+/HLA-DR+) were observed over time, except for CD38+ cells, which showed a marginal reduction at month 6 (**Fig. 4G–I**).No correlation was detected between γδ T cell activation and baseline viral load (data not shown).

Correlations between early γδ T cell activation and CD4/CD8 T cell activation set-point

Activation of CD4/CD8 cells was determined on the basis of CD38 and HLA-DR surface marker expression at month 6 after baseline, when the T cell activation set-point of most patients should have been established [33]. Correlation analysis was performed by comparing CD4/CD8 activation at month 6 and γδ T cell activation at baseline. CD4 and CD8 T cell activation set-points were both positively correlated to γδ T cell activation (**Fig. 5A–B and 5E–F**). If HLA-DR alone was used, the CD8 T cell set-point was correlated ($P = 0.006$), but not the CD4 T cell activation set-point ($P = 0.490$) (**Fig. 5C and 5D**). These data demonstrated that early activation of γδ T cells was positively correlated withCD4 and CD8 T cell activation set-points.

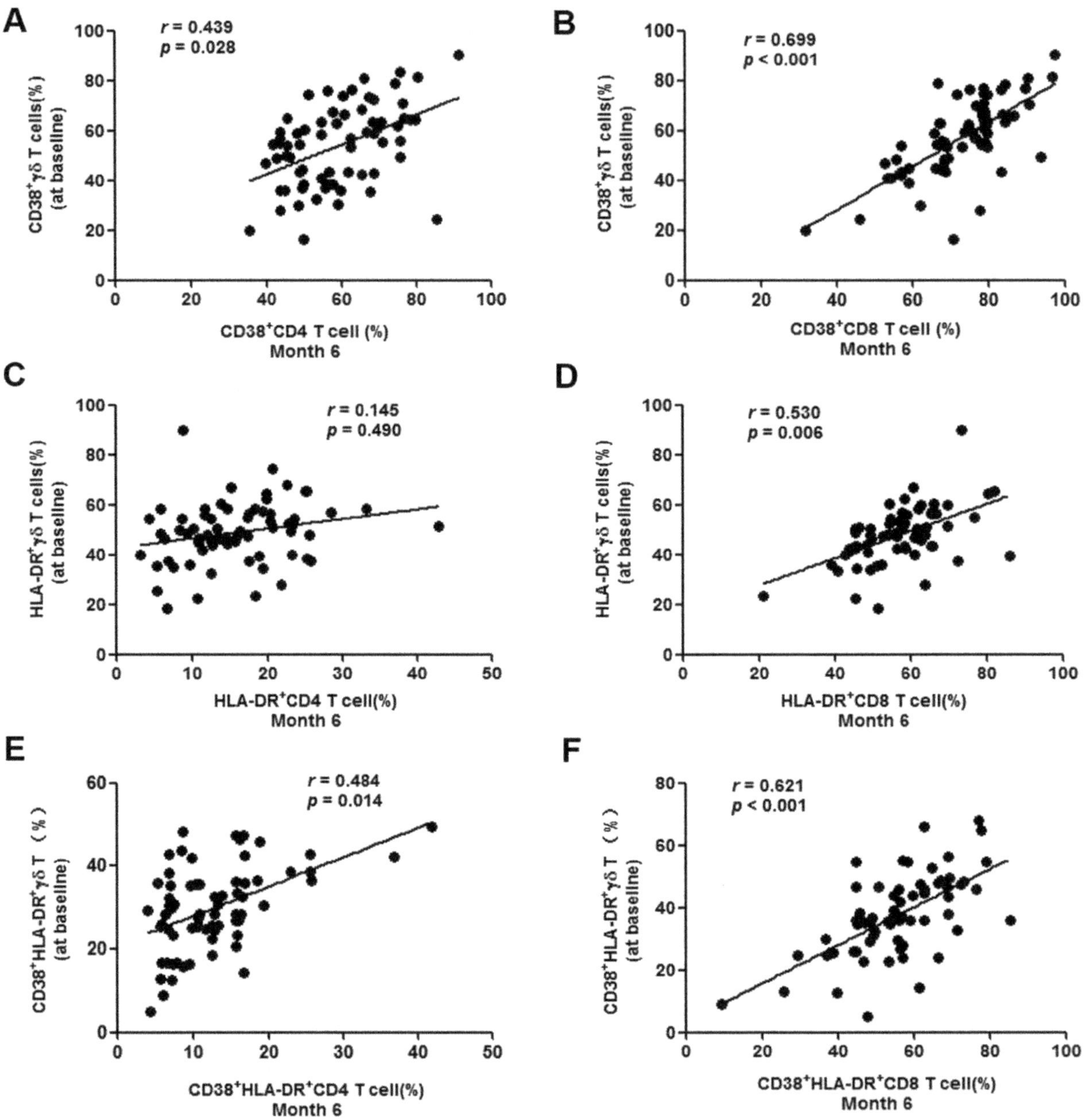

Figure 5. Correlation between early γδ T cell activation and CD4/CD8 T cell activation set-point. Correlation analysis of the proportion of CD38+/CD4+ T cells (**A**) or CD38+/CD8+ T cells (**B**) at month 6 after baseline and CD38+ γδ T cells frequency at baseline was performed by Spearman's rank correlation, where coefficients 'r' and corresponding *P*-values are indicated on each panel. Similar correlation analysis of the proportion of HLA-DR+ (**C, D**) and CD38+/HLA-DR+ cells (**E, F**) and CD4+ or CD8+ T cells at month 6 and γδ T cells at baseline was performed.

Microbial translocation versus γδ T cells

Microbial translocation in acute HIV infection was determined by analysis of the serum levels of LPS (lipopolysaccharides), LBP (LPS-binding protein) and sCD14 (soluble CD14). Only sCD14 levels were found to be elevated in acutely HIV-infected patients compared with those of healthy subjects ($P = 0.001$) (**Fig. 6A–C**). The levels of LPS and LBP did not differ between ART-treated and untreated patients, or between patients with different CD4 count levels (**Fig. 6D–E**). However, the sCD14 level in ART-treated patients was significant elevated compared with untreated patients with CD4>350 cells/µl ($P = 0.003$). Correlation analysis also showed no association between sCD14 levels and the total γδ T cell population ($P = 0.062$) (data not shown); however, further investigation demonstrated positive correlations between sCD14 levels and different γδ T subgroups. Statistical analysis indicated that sCD14 levels were positively correlated withCD38+/HLA-DR+ ($P = 0.002$) and HLA-DR+ ($P = 0.011$), but not with CD38+ ($P = 0.105$) (**Fig. 6G**). Fractions of Tnaïve, TCM and TEM were not correlated with sCD14 levels (all $P > 0.05$) (**Fig. 6J**). Proportions of Vδ1, Vδ2 and IL-17 levels were not associated with sCD14 levels (all $P > 0.05$) (**Fig. 6H, I, K**).The fraction of IFN-γ-producing γδ T cells was negatively correlated with sCD14 levels ($P = 0.046$) (**Fig. 6L**). Furthermore, sCD14 was associated with LPS levels ($P = 0.034$) (data not shown).

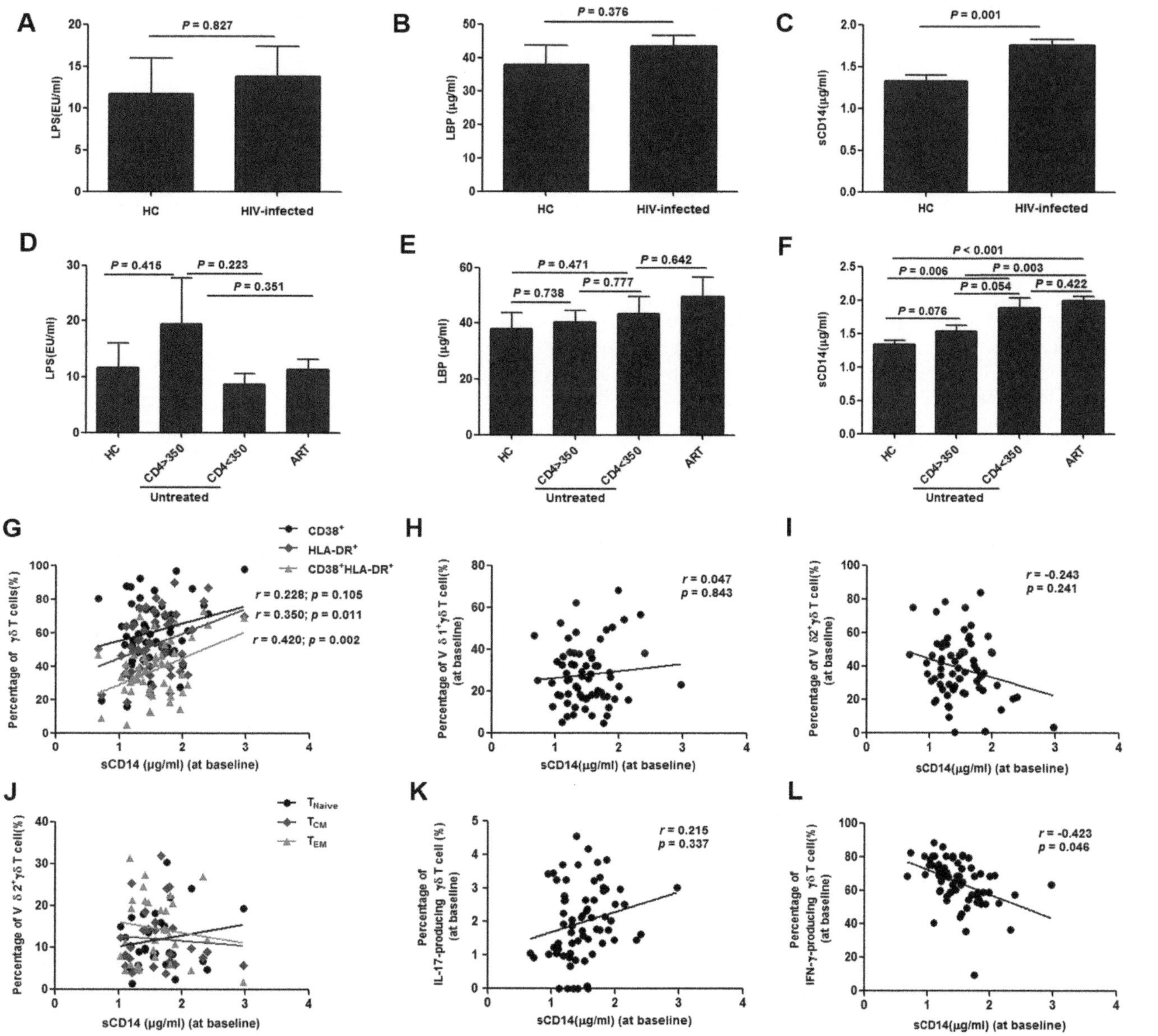

Figure 6. Correlation between microbial translocation and γδ T cells. Levels of LPS, LBP and sCD14 were analyzed in healthy controls (HC) and acutely HIV-infected subjects at baseline (**A–C**). The patients were then subdivided based on the administration of antiretroviral therapy (ART) and CD4 levels (< or >than 350/μL). The levels of LPS, LBP and sCD14 were re-analyzed and compared with HC (**D–F**). Correlation analysis between the proportion of activated γδ T cells determined by CD38+, HLA-DR+ or CD38+/HLA-DR and sCD14 levels at baseline was performed by Spearman's rank correlation, where coefficients 'r' and corresponding *P*-values are indicated on panel G. Similar correlation analysis was performed between sCD14 levels and different subsets of γδ T cells [Vδ1 (**H**) and Vδ2 (**I**)], subgroups of Vδ2 cells (Tnaïve, TCM and TEM) (**J**), IL-17-producing γδ T cells (**K**) or IFN-γ-producing γδ T cells (**L**).

Discussion

There is an increasing body of evidence indicating that γδ T cells play critical roles in HIV infections [25]. In this study conducted in a male homosexual population, we show that the proportion of Vδ1 cells among γδ T cells was elevated in acutely HIV-infected patients, while the Vδ2 population was significantly reduced. Among theVδ2γδT cells, the fractions of TCM and TEM cells were significantly decreased, while the proportion of TEMRA was increased. In term of γδ T cell functions, cytotoxicity was compromised significantly in acutely HIV-infected patients over time. The fraction of IL-17-producing cells was elevated, but not that of the IFN-γ-producing cells. Activation of γδ T cells increased over time and correlated withCD4 counts and CD4/CD8 T cell activation set-points. In acute HIV infection, levels of the microbial translocation marker, sCD14, were significantly elevated, regardless ofCD4 counts. However, sCD14 levels exhibited a positive association withCD38/HLA-DR and HLA-DR expression and a negative association with the fraction of IFN-γ-producing γδ T cells. Initiation of ART had no impact on any of these changes.

It has been shown that the Vδ2 population is depleted in chronically HIV-infected patients [25]. Our data indicated that this is an early event, occurring before the remarkable reduction in the total γδ T cell count, since baseline data showed no significant change in the fraction of γδ T cells among the total T cell population (**Fig. 1A, 1C and 1E**). Our results also demonstrated that there were no significant changes in the proportion of Tnaïve cells in the Vδ2 cell population. The fractions of TEM and TCM cells were decreased, while the TEMRA fraction was elevated, which indicated increased aging of Vδ2 cells in acute HIV infection (**Fig. 2**).

Activated γδ T cells are the subtype of this population of cells that function in response to HIV infections, and therefore, were the primary target cell population of this study. Our data showed a gradual increase in γδ T cell activation, along with compromised γδ T cell cytotoxicity over time, in addition to an elevation in the fraction of IL-17-producing γδ T cells in acute HIV infection (**Fig. 3, 4, 5**).These observations indicated that early initiation of ART failed to bring about recovery of the γδ T cell populations and their functions, which is consistent with a previous report [29], although additional studies are required to confirm this. More importantly, we found a positive correlation between baseline γδ T cell activation and CD4/CD8 T cell activation set-points (**Fig. 5**), which were used to predict subsequent CD4 T cell loss, disease progression and outcome [32,34]. It can therefore be speculated that activation of γδ cells represents a surrogate marker for disease progression and outcome.

Our results indicated potential associations between γδ T cell activation and levels of sCD14, a monocyte activation marker, which is also commonly used as an indicator of microbial translocation (**Fig. 6**). The level of sCD14 is used to predict mortality and disease progression in chronic HIV patients [24,33], in addition to other disorders, such as different cancers and hemodialysis [35,36]. Microbial translocation results in the release of bacterial products, such as LPS and LBP, which in turn activate monocytes in various clinical situations [37,38]. LPS and LBP levels have been shown to be increased in chronic HIV infection, but not in the acute stage. Furthermore, it has been reported that microbial translocation is not observed in primary HIV infection, although sCD14 levels are elevated [33,39]. Therefore, even though sCD14 levels were found to be elevated in acute HIV infection in our study, an association of γδ T cell activation with microbial translocation remains to be established.

No differences were observed between patients on early ART and untreated patients in any aspect investigated in this study. Although there are reports indicating that ART does not bring about recovery of the γδ T cell depletion, this conclusion is controversial. Our study was conducted in only 11 patients on ART, and two treatment plans including two different protease inhibitors; therefore, a larger study with a more stringent design is required to confirm this conclusion.

In conclusion, in acute HIV infections, a significant shift in the proportions and functions of the various γδ T cell subgroups was observed. Elevation of the microbial translocation marker, sCD14, was associated with γδ T cell activation. More importantly, early γδ T cell activation was associated with CD4 and CD8 T cell activation set-points, which predict AIDS progression. Therefore, γδ T cell activation represents a potential surrogate marker of AIDS progression. This conclusion requires confirmation in larger prospective cohort studies.

Author Contributions

Conceived and designed the experiments: WL WH HW. Performed the experiments: ZL YJ LC YH. Analyzed the data: WL NL ZL YJ HW LC. Contributed reagents/materials/analysis tools: NL DC. Contributed to the writing of the manuscript: WL WH.

References

1. Constant P, Davodeau F, Peyrat MA, Poquet Y, Puzo G, et al. (1994) Stimulation of human gamma delta T cells by nonpeptidic mycobacterial ligands. Science 264: 267–270.
2. Dik WA, Pike-Overzet K, Weerkamp F, de Ridder D, de Haas EF, et al. (2005) New insights on human T cell development by quantitative T cell receptor gene rearrangement studies and gene expression profiling. J Exp Med 201: 1715–1723.
3. Hayday AC (2000) [gamma][delta] cells: a right time and a right place for a conserved third way of protection. Annu Rev Immunol 18: 975–1026.
4. Xi X, Guo Y, Chen H, Xu C, Zhang H, et al. (2009) Antigen specificity of gammadelta T cells depends primarily on the flanking sequences of CDR3delta. J Biol Chem 284: 27449–27455.
5. Bukowski JF, Morita CT, Brenner MB (1999) Human gamma delta T cells recognize alkylamines derived from microbes, edible plants, and tea: implications for innate immunity. Immunity 11: 57–65.
6. Vincent MS, Roessner K, Sellati T, Huston CD, Sigal LH, et al. (1998) Lyme arthritis synovial gamma delta T cells respond to Borrelia burgdorferi lipoproteins and lipidated hexapeptides. J Immunol 161: 5762–5771.
7. Sanders JM, Ghosh S, Chan JM, Meints G, Wang H, et al. (2004) Quantitative structure-activity relationships for gammadelta T cell activation by bisphosphonates. J Med Chem 47: 375–384.
8. Battistini L, Caccamo N, Borsellino G, Meraviglia S, Angelini DF, et al. (2005) Homing and memory patterns of human gammadelta T cells in physiopathological situations. Microbes Infect 7: 510–517.
9. Qin G, Liu Y, Zheng J, Xiang Z, Ng IH, et al. (2012) Phenotypic and functional characterization of human gammadelta T-cell subsets in response to influenza A viruses. J Infect Dis 205: 1646–1653.
10. Dieli F, Poccia F, Lipp M, Sireci G, Caccamo N, et al. (2003) Differentiation of effector/memory Vdelta2 T cells and migratory routes in lymph nodes or inflammatory sites. J Exp Med 198: 391–397.
11. Eberl M, Engel R, Beck E, Jomaa H (2002) Differentiation of human gamma-delta T cells towards distinct memory phenotypes. Cell Immunol 218: 1–6.
12. Ribot JC, deBarros A, Pang DJ, Neves JF, Peperzak V, et al. (2009) CD27 is a thymic determinant of the balance between interferon-gamma- and interleukin 17-producing gammadelta T cell subsets. Nat Immunol 10: 427–436.
13. Fenoglio D, Poggi A, Catellani S, Battaglia F, Ferrera A, et al. (2009) Vdelta1 T lymphocytes producing IFN-gamma and IL-17 are expanded in HIV-1-infected patients and respond to Candida albicans. Blood 113: 6611–6618.
14. Battistini L, Borsellino G, Sawicki G, Poccia F, Salvetti M, et al. (1997) Phenotypic and cytokine analysis of human peripheral blood gamma delta T cells expressing NK cell receptors. J Immunol 159: 3723–3730.
15. Brenchley JM, Douek DC (2012) Microbial translocation across the GI tract. Annu Rev Immunol 30: 149–173.
16. Santos-Oliveira JR, Regis EG, Giacoia-Gripp CB, Valverde JG, Alexandrino-de-Oliveira P, et al. (2013) Microbial translocation induces an intense proinflammatory response in patients with visceral leishmaniasis and HIV type 1 coinfection. J Infect Dis 208: 57–66.
17. Shibata K, Yamada H, Hara H, Kishihara K, Yoshikai Y (2007) Resident Vdelta1+ gammadelta T cells control early infiltration of neutrophils after Escherichia coli infection via IL-17 production. J Immunol 178: 4466–4472.
18. Schilbach K, Geiselhart A, Handgretinger R (2001) Induction of proliferation and augmented cytotoxicity of gammadelta T lymphocytes by bisphosphonate clodronate. Blood 97: 2917–2918.
19. Thedrez A, Harly C, Morice A, Salot S, Bonneville M, et al. (2009) IL-21-mediated potentiation of antitumor cytolytic and proinflammatory responses of human V gamma 9V delta 2 T cells for adoptive immunotherapy. J Immunol 182: 3423–3431.
20. Kotler DP, Gaetz HP, Lange M, Klein EB, Holt PR (1984) Enteropathy associated with the acquired immunodeficiency syndrome. Ann Intern Med 101: 421–428.
21. Epple HJ, Schneider T, Troeger H, Kunkel D, Allers K, et al. (2009) Impairment of the intestinal barrier is evident in untreated but absent in suppressively treated HIV-infected patients. Gut 58: 220–227.
22. Romero-Sanchez M, Gonzalez-Serna A, Pacheco YM, Ferrando-Martinez S, Machmach K, et al. (2012) Different biological significance of sCD14 and LPS in HIV-infection: importance of the immunovirology stage and association with HIV-disease progression markers. J Infect 65: 431–438.
23. Wittkop L, Bitard J, Lazaro E, Neau D, Bonnet F, et al. (2013) Effect of cytomegalovirus-induced immune response, self antigen-induced immune response, and microbial translocation on chronic immune activation in successfully treated HIV type 1-infected patients: the ANRS CO3 Aquitaine Cohort. J Infect Dis 207: 622–627.
24. Sandler NG, Wand H, Roque A, Law M, Nason MC, et al. (2011) Plasma levels of soluble CD14 independently predict mortality in HIV infection. J Infect Dis 203: 780–790.

25. Li H, Chaudhry S, Poonia B, Shao Y, Pauza CD (2013) Depletion and dysfunction of Vgamma2Vdelta2 T cells in HIV disease: mechanisms, impacts and therapeutic implications. Cell Mol Immunol 10: 42–49.
26. Li H, Peng H, Ma P, Ruan Y, Su B, et al. (2008) Association between Vgamma2Vdelta2 T cells and disease progression after infection with closely related strains of HIV in China. Clin Infect Dis 46: 1466–1472.
27. Zheng NN, McElrath MJ, Sow PS, Mesher A, Hawes SE, et al. (2011) Association between peripheral gammadelta T-cell profile and disease progression in individuals infected with HIV-1 or HIV-2 in West Africa. J Acquir Immune Defic Syndr 57: 92–100.
28. Bordon J, Evans PS, Propp N, Davis CE Jr, Redfield RR, et al. (2004) Association between longer duration of HIV-suppressive therapy and partial recovery of the V gamma 2 T cell receptor repertoire. J Infect Dis 189: 1482–1486.
29. Hebbeler AM, Propp N, Cairo C, Li H, Cummings JS, et al. (2008) Failure to restore the Vgamma2-Jgamma1.2 repertoire in HIV-infected men receiving highly active antiretroviral therapy (HAART). Clin Immunol 128: 349–357.
30. Zhou J, Kang N, Cui L, Ba D, He W (2012) Anti-gammadelta TCR antibody-expanded gammadelta T cells: a better choice for the adoptive immunotherapy of lymphoid malignancies. Cell Mol Immunol 9: 34–44.
31. Sodora DL, Silvestri G (2008) Immune activation and AIDS pathogenesis. AIDS 22: 439–446.
32. Deeks SG, Kitchen CM, Liu L, Guo H, Gascon R, et al. (2004) Immune activation set point during early HIV infection predicts subsequent CD4+ T-cell changes independent of viral load. Blood 104: 942–947.
33. Chevalier MF, Petitjean G, Dunyach-Remy C, Didier C, Girard PM, et al. (2013) The Th17/Treg ratio, IL-1RA and sCD14 levels in primary HIV infection predict the T-cell activation set point in the absence of systemic microbial translocation. PLoS Pathog 9: e1003453.
34. Karim R, Mack WJ, Stiller T, Operskalski E, Frederick T, et al. (2013) Association of HIV clinical disease progression with profiles of early immune activation: results from a cluster analysis approach. AIDS 27: 1473–1481.
35. Alunno A, Bartoloni E, Bistoni O, Nocentini G, Ronchetti S, et al. (2012) Balance between regulatory T and Th17 cells in systemic lupus erythematosus: the old and the new. Clin Dev Immunol 2012: 823085.
36. Braga WM, Atanackovic D, Colleoni GW (2012) The role of regulatory T cells and TH17 cells in multiple myeloma. Clin Dev Immunol 2012: 293479.
37. Adib-Conquy M, Cavaillon JM (2007) Stress molecules in sepsis and systemic inflammatory response syndrome. FEBS Lett 581: 3723–3733.
38. Munoz C, Carlet J, Fitting C, Misset B, Bleriot JP, et al. (1991) Dysregulation of in vitro cytokine production by monocytes during sepsis. J Clin Invest 88: 1747–1754.
39. Brenchley JM, Price DA, Schacker TW, Asher TE, Silvestri G, et al. (2006) Microbial translocation is a cause of systemic immune activation in chronic HIV infection. Nat Med 12: 1365–1371.

THDP17 Decreases Ammonia Production through Glutaminase Inhibition. A New Drug for Hepatic Encephalopathy Therapy

M. Mar Díaz-Herrero[1], José A. del Campo[2]**, Pilar Carbonero-Aguilar[1], José M. Vega-Pérez[3], Fernando Iglesias-Guerra[3], Ignacio Periñán[3], Francisco J. Miñano[4], Juan Bautista[1]*, Manuel Romero-Gómez[2]***

1 Departamento de Bioquímica y Biología Molecular, Facultad de Farmacia, Universidad de Sevilla, Sevilla, Spain, **2** Unidad de Gestión Clínica de Enfermedades Digestivas & Centro de Investigación Biomédica en Red de Enfermedades Hepáticas y Digestivas (CIBERehd), Hospital Universitario de Valme, Sevilla, Spain, **3** Departamento de Química Orgánica y Química Farmacéutica, Facultad de Farmacia, Universidad de Sevilla, Sevilla, Spain, **4** Unidad de Farmacología Experimental y Clínica (UFEC), Hospital Universitario de Valme, Universidad de Sevilla, Sevilla, Spain

Abstract

Ammonia production is implicated in the pathogenesis of hepatic encephalopathy (HE), being intestinal glutaminase activity the main source for ammonia. Management of ammonia formation can be effective in HE treatment by lowering intestinal ammonia production. The use of glutaminase inhibitors represents one way to achieve this goal. In this work, we have performed a search for specific inhibitors that could decrease glutaminase activity by screening two different groups of compounds: i) a group integrated by a diverse, highly pure small molecule compounds derived from thiourea ranging from 200 to 800 Daltons; and ii) a group integrated by commonly use compounds in the treatment of HE. Results shown that THDP-17 (10 μM), a thiourea derivate product, could inhibit the intestinal glutaminase activity (57.4±6.7%). Inhibitory effect was tissue dependent, ranging from 40±5.5% to 80±7.8% in an uncompetitive manner, showing V_{max} and K_m values of 384.62 μmol min^{-1}, 13.62 mM with THDP-17 10 μM, respectively. This compound also decreased the glutaminase activity in Caco-2 cell cultures, showing a reduction of ammonia and glutamate production, compared to control cultures. Therefore, the THDP-17 compound could be a good candidate for HE management, by lowering ammonia production.

Editor: Yuan-Soon Ho, Taipei Medical University, Taiwan

Funding: Supported by research grants by Junta de Andalucía (Spain) P06-CTS-01887 and by Ministerio de Sanidad y Consumo of Spain PI070425. The funders had no role in study design, data collection and analysis, decision to publish, or preparation of the manuscript.

Competing Interests: The authors have declared that no competing interests exist.

* Email: jdbaut@us.es (JDB); mromerogomez@us.es (MRG)

These authors contributed equally to this work.

Introduction

Ammonia plays a major role in the pathogenesis of hepatic encephalopathy (HE). Systemic hyperammonemia has been largely found in patients with cirrhosis and HE. Ammonia derived from the gut and kidneys must be detoxified in the liver and muscle. In cirrhotic patients with impaired capacity for detoxification of ammonia into urea, peripheral glutamine synthase (mainly in muscle tissue) serves as the main alternative for ammonia detoxification pathway. In the brain, hyperammonemia promotes astrocyte swelling and impairment of neurotransmission [1,2]. In muscle tissue, it has been shown that glutaminase activity is increased in an animal model undergoing bile duct ligation and glutamine synthase has been shown to have an important role in detoxification of plasma ammonia in acute liver failure [3]. Moreover, the development of portosystemic shunting and alteration of blood flow is thought to affect circulating ammonia levels in cirrhosis [4]. These findings suggest that approaches to target glutaminase (by inhibition) may provide targets for ammonia detoxification as a valid therapeutic strategy for HE.

Glutamine deamidation by intestinal GA has been proposed as one of the main sources of ammonia production in patients with cirrhosis [5]. However, hyperammonemia found after portacaval shunt in rats is similar in both germ-free and non-germ-free animals [6,7] supporting the hypothesis that hyperammonemia and encephalopathy can be developed without the participation of bacteria [8].

Another supporting evidence is that the greatest hyperammonemia has been found in portal-drained viscera and derived mainly from glutamine deamidation in cirrhotic patients by phosphate-activated glutaminase (PAG) [9,10]. Glutamine synthesis is an effective mechanism to reduce the levels of ammonia; however, much of the newly synthesized glutamine is subsequently metabolized in mitochondria by phosphate-activated glutaminase, yielding glutamate and ammonia. In this manner, glutamine (the Trojan horse) is transported in excess from the cytoplasm to mitochondria serving as a carrier of ammonia [11].

PAG catalyzes the deamidation of glutamine to yield glutamate and ammonia. There are two genes that encoding two main isoforms of PAG: the kidney-type PAG (K-PAG) and the hepatic-type PAG (L-PAG). The K-PAG is the ubiquitous form. It can be found in kidney, brain, stomach, pancreas, muscle, or villus enterocyte; while L-PAG was thought to be restricted to the liver but in the last years it has also been found in nuclei of mammalian brain [12], and granule in human neutrophils [13]. The presence of LGA and/or KGA, together with GAC (Glutaminase C) form has also been reported in different tumour cells [14–16].

In healthy people, the higher PAG activity along the gastrointestinal tract has been found in the small intestine, being higher than 80% of the activity [17]. Lower but still substantial activity was found in the large intestine, approximately the 15%. In rats, the distribution of the PAG activity was found to be similar to humans [18]. Duodenal PAG activity, measured in mucosal biopsies from the first portion of the small intestine, has been found to be nearly four-times higher in cirrhotic patients than in healthy controls [10].

As the main source of ammonia in cirrhotic patients derives from portal-drained viscera owing to glutamine deamidation, PAG activity in the intestine seems to be responsible, largely, for systemic hyperammonemia production in these patients [5]. Besides, in portacaval shunted rats, a model for minimal HE [19], PAG activity has been found to be increased in duodenum and ileum compared to sham-operated rats [20]. All these data suggest that PAG is implicated in the production of systemic hyperammonemia from the intestine and could play a major role in the pathogenesis of HE. Therefore, a decrease of ammonia production in the small intestine by PAG inhibition may contribute to lower ammonia concentration in the systemic circulation. The aim of this work was the search for glutaminase inhibitors in order to find new ammonia-lowering drugs that could be used in the management of HE.

Materials and Methods

All chemicals used in this work were from high analytical grade and were obtained mainly from Sigma-Aldrich.

Compounds

Two groups of compounds were used:

The first group was integrated by a collection of diverse, highly pure, small molecule compounds derived from thiourea ranging from 200 to 800 Daltons. These compounds were named as THDP-xx, being xx an assigned number to each product. Compounds were received as powder and then dissolved in dimethyl sulfoxide (DMSO) to constitute a 10 mM stock concentration and stored at $-20°C$. Maximum DMSO concentration in the reaction mixture was 0.5% (vol/vol) to avoid toxic effects. Most of the compounds in the library can be dissolved at 50 μg/ml. For those compounds that did not dissolve (~5%), concentration was decreased to 10–25 μg/ml. All compounds were synthesized at the Departamento de Química Orgánica y Farmacéutica of Universidad de Sevilla (Spain) [21–23] as the result for a new simple thiourea derivatives library with potential use as enzymatic inhibitors.

The second group was integrated by compounds used or tested for the treatment of HE: metformine, metronidazole, neomicine, rifaximin, lactulose and lactitol (all purchased from Sigma-Aldrich).

In vitro inhibition screening

PAG inhibition analysis by different thiourea derivatives (THDP) was assayed *in vitro*, using the method of Heini for glutaminase activity determination [24], adapted to a microtiter procedure. This assay is based on the use of o-phthaldialdehyde/mercaptoethanol as a reagent for estimating ammonia released from glutamine. Briefly, 25 μl samples were added to 35 μl of a mixture of 150 mM potassium phosphate, 171 mM glutamine, 1 mM ammonium chloride, pH 8.0 and then incubated at 37°C for 45 min. The reaction was stopped by adding 10 μl 7% trichloroacetic acid, kept on ice for 15 min and centrifuged 5 min at 1000 × g. Aliquots of 5 μl/well were then mixed with 150 μl o-phthalaldehyde/mercaptoethanol reagent (14.4 ml 200 mM potassium phosphate, pH 7.4, 0.8 ml 72 mM mercaptoethanol in ethanol and 0.8 ml 186 mM o-phthaldialdehyde in ethanol). The samples were kept at room temperature in the dark and absorbance measured at 410 nm with a UV-Visible spectrophotometer (BioRad) after 45 min.

Kinetic assays

K-PAG activity was assayed, in presence and absence of the inhibitor; using isolated mitochondria from rat intestine and kidney, solubilized with 1% Triton X-100 as a source of glutaminase [25]. Enzymatic activity was measured using a two-step assay as described by Lund [26]. Protein concentration was

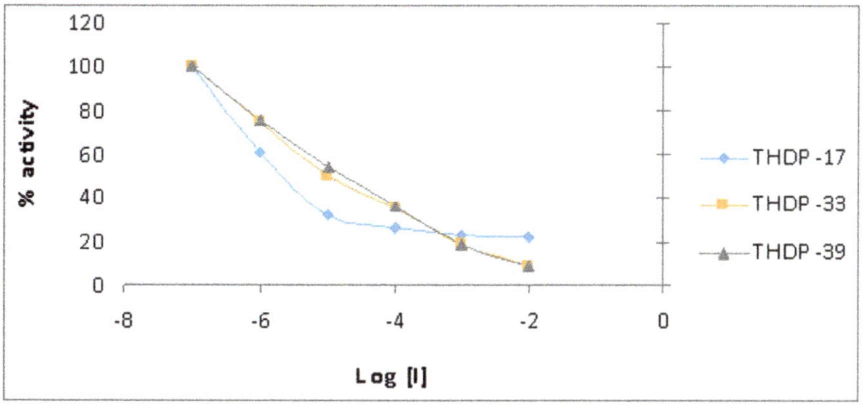

Figure 1. Inhibitory effect on glutaminase activity of several compounds tested in this study. Data are shown in log scale (inhibitor concentration), being 100% activity measured without inhibitor. There was not significant differences among compounds tested (Student t test; $p > 0.05$). THDP-17 was chosen on toxic effect base.

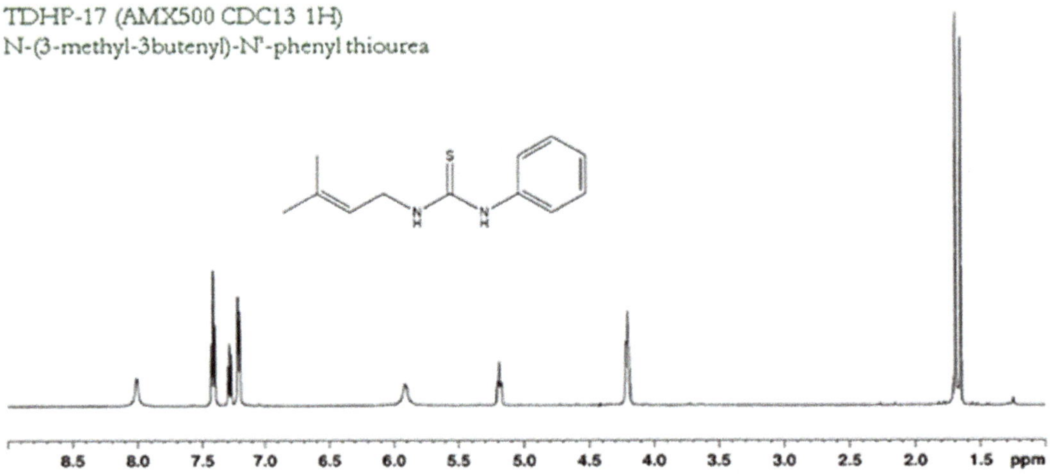

Figure 2. Molecular structure of N-(3-methyl 3-butenyl)-N′ phenyl thiourea (THDP17).

determined following Bradford method using serum albumin as standard [27].

Cell culture

Caco2 cells were purchased from Sigma-Aldrich (St. Louis, MO, USA) and grown with DMEM (Dulbecco's Modified Eagle Medium) medium supplemented with 2 mM L-Glutamine, 15% FBS, 1X antinbiotic/antimicotic solution, 1X non-essential amino acids (PAA Laboratories GmbH, Linz, Germany). Cells were incubated at 37°C and 5% CO_2.

Cytotoxicity assay

Cytotoxicity was evaluated in Caco-2 cells after 24 h-exposure to selected inhibitors using neutral red (NR) assay (Zhang et al. 1990) [28]. Regression analysis of dose-response curves and calculation of IC_{50} were done according to standard procedure using appropriated software.

Inhibition assay in cell culture

Inhibition assay was carried out by measuring the concentration of glutamate produced from glutamine by Caco-2 cells using HPLC, in presence and absence of the selected inhibitor (THDP-17) at different concentrations (0, 5, 20 and 100 μM) and time points (24 and 48 h). 6-Diazo-5-oxo-L-norleucine (DON) (0–100 μM) was used as positive control. The presence of the THDP-17 in the cells was checked by HPLC-MS.

Lethal Dose 50

Lethal dose was determined by oral administration of the compounds to C57BL/6 mice. Thirty mice were supplied by the Centre of Production and Experimental animal of Espartinas, University of Seville, Spain. The animals were housed ten per cage, and maintained at constant room temperature (22°C ±2°C), with artificial light-dark cycle of 12 h and a relative humidity of 65%−70%, and free access to food and water during 2 weeks for acclimation. All animal procedures were performed with approval from the Sevilla University Animal Ethics Committees, under the

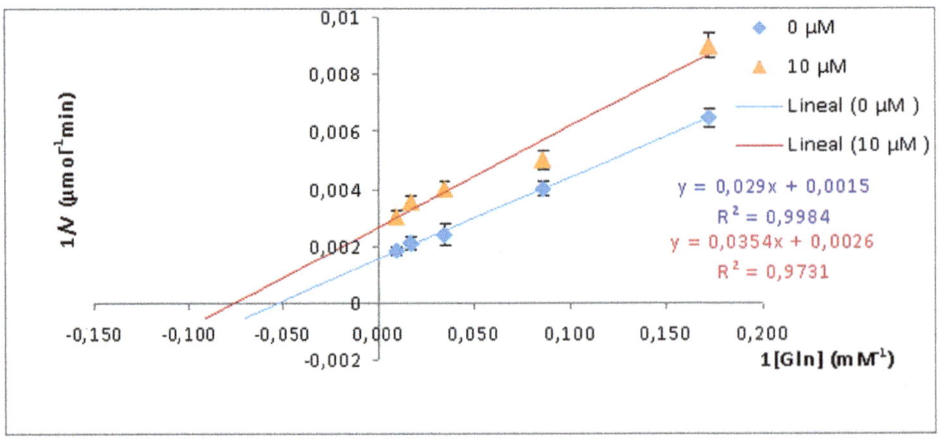

Figure 3. THDP17 inhibition kinetics. A lineal double reciprocal plot (1/v against 1/[Gln]) in absence (0) and presence (10 μM) of the inhibitor is shown. V_{max} and K_M values for PAG were 0.67 mmol min^{-1}, and 19.33 mM, respectively. In the presence of THDP-17 both parameters were modified: $V_{max\text{-}ap}$ = 0.38 mmol min^{-1}, and $K_{M\text{-}ap}$ = 13.62 mM, indicating an uncompetitive inhibition. All experiments were performed in triplicates; $p > 0.05$.

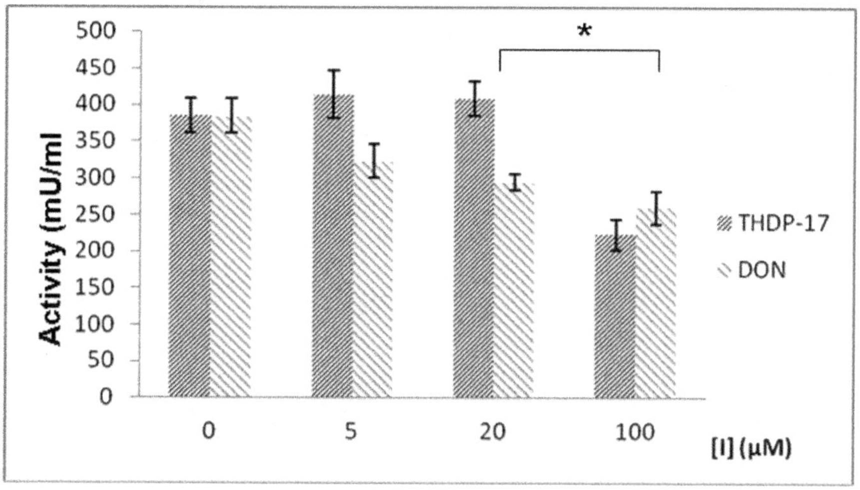

Figure 4. ***In vivo*** **inhibitory effect on glutaminase activity at different concentrations of THDP17 and DON (6-Diazo-5-oxo-L-norleucine) was tested in Caco-2 cell cultures.** PAG inhibition by THDP-17 in cell cultures is effective at concentration higher than 5 μM, showing an inhibition of 18±2.1% and 46±3.4% at 20 and 100 μM, respectively (* $p<0.05$).

guidelines of the directive 86/609/CEE of the European Union. Two assays in different conditions were performed. In the first assay fasted mice were divided in four groups of three animals and were administered 0, 50, 300 or 2000 mg THDP-17/kg, respectively. They were sacrificed 24 h latter and a macroscopic analysis of the tissues (lungs, heart, liver, stomach and intestine) was performed. In the second assay the mice were feed ad libitum, separated into three groups of four animals per group and administered them 0, 5 or 300 mg THDP-17/kg, respectively. In this case, mice were sacrificed 24 or 72 h latter and the same macroscopic analysis was performed.

Statistical analysis

The results were expressed as means and standard deviations (or standard errors) of three independent experiments. The results were analyzed with the Student t test. $P<0.001$, $P<0.01$, and $P<0.05$ were considered statistically significant.

Results and Discussion

Ammonia-lowering drugs have received great interest in the management of HE in the last few years. In order to develop adequate and selective inhibitors of intestinal PAG activity, we performed a screening of three different groups of compounds: i) a group integrated by a collection of diverse, highly pure, rationally selected small molecule compounds derived from thiourea ranging from 200 to 800 Daltons; ii) a group integrated by products showing some structural analogy with the compound THDP-17, and iii) a group integrated by compounds usually used in the treatment of HE.

i) Thiourea derived products

The first group analysed in this work is integrated by a small molecule compounds library derived from thiourea [21]. We found that some products derived from thiourea were able to reduce ammonia production *in vitro* from glutamine by an enriched preparation of glutaminase obtained from both kidney and intestine mitochondria (data not shown). For the screening, we have used the Heini's method to assay the glutaminase activity. Six synthetic inhibitors derived from thiourea were selected and extensively characterized. These products (THDP-17, THDP-33, THDP-39, THDP-41, THDP-42, and THDP-43) could inhibit K-PAG activity and were tested for toxicity and grade of inhibition on a cell culture based method, using the neutral red (NR) assay.

- Effects of selected inhibitors on cytotoxicity. Acute toxicity test in animals is typically the first step in the evaluation of the effects of a chemical compound, and its primary purpose is to provide information on potential health hazards that may result from a short-term exposure. Recent studies suggest that *in vitro* methods might be helpful in predicting acute toxicity and estimating toxic chemical concentrations in vivo [29]. In the

Table 1. THDP-17 was administered orally after 12 h fasted mice.

Group	N	Vehicle (DMSO)	THDP-17 (mg/Kg)	Deaths before		Sacrificed 24 h
				6 h	12 h	
I	3	+	0	1	1	1
II	3	+	50	1	1	1
III	3	+	300	1	1	1
IV	3	+	2000	2	1	-

Animals were divided in 4 groups and deaths after treatment were recorded.

Table 2. THDP-17 and/or DMSO were then administered orally to animals after free access to food and water overnight, and mice were sacrificed at 24 h and 72 h randomly.

Group	N	Vehicle (DMSO)	THDP-17 (mg/Kg)	Deaths before			Sacrificed	
				6 h	12 h	24 h	24	72
I	4	+	0	0	0	0	2	1
II	4	+	50	0	0	0	2	2
III	4	+	300	1*	0	0	1	2

*Animal without water.

present work, cytotoxicity of a 24 h-exposure to selected inhibitors was evaluated in Caco-2 cells using the MTT method developed by Mosmann [30] and modified by Denizot and Lang [31]. Based on cell viability the inhibitors THDP-41, THDP-42 and THDP-43 were rejected due to their high toxicity and the remaining compounds (THDP-17, THDP-33 and THDP-39) were selected for further inhibition studies.

- In vitro inhibition analysis. To evaluate the performance of the PAG-inhibitors THDP-17, THDP-33 and THDP-39 we used swine K-PAG (from kidney) in an *in vitro* optimised kinetic assay with and without the addition of the inhibitors. First, increasing concentration of inhibitors (ranging from 1 µM to 10 mM) were added to standard and optimized enzymatic assay conditions with 99 mmol/l glutamine, 150 mmol/l phosphate and 100 µg K-PAG (Figure 1). THDP-33 and THDP-39 were able to inhibit PAG completely at 10 mM. IC_{50} values were 19.1±1.3 µM for THPD-33, 31.6±2.7 µM for THDP-39 and 3.9±0.1 µM for THDP-17, indicating that THPD-33 and THPD-39 were less powerful inhibitors than THDP-17. THDP-33 and THDP-39 were able to completely inhibit PAG activity at low milimolar level (Figure 1) which can be harmful for the enterocyte, being discarded for future analysis.

THDP-17, N-(3-methyl-3-butenyl)-N′-phenylthiourea (Figure 2), was selected because it did not show cytotoxicity in cell culture, and because it inhibits PAG activity partially at low milimolar concentrations. In fact, there is a plateau (the enzyme activity is no longer affected by increasing the inhibitor concentration). This finding is highly relevant, since partial inhibition of PAG could be beneficial for the reduction of ammonia and the excess of glutamate production in the management of HE, without affecting the viability of the enterocyte.

PAG kinetic parameters were analyzed in absence and presence of THDP-17, at different concentrations (1 and 10 µM), under optimal analytical conditions, at different glutamine concentration (0 to 200 mM). A lineal double reciprocal plot (1/v against 1/[Gln]) in absence and presence of the inhibitor (Figure 3) shows that the V_{max} and K_M values for PAG were 0.67 mmol min^{-1}, and 19.33 mM, respectively, indicating that in presence of the inhibitor THDP-17 both parameters changes: $V_{max\text{-}ap} = 0.38$ m-mol min^{-1}, and $K_{M\text{-}ap} = 13.62$ mM). Data show an uncompetitive inhibition model indicating that THDP-17 binds PAG at a site created after enzyme-substrate complex formation. We propose that THDP-17 is a not-pure-uncompetitive inhibitor because the K_i value obtained from $V_{max\text{-}ap}$ and $K_{M\text{-}ap}$ (23.2 mM and 40.5 mM, respectively) is different.

- *in vivo* inhibition. Since *in vitro* inhibition by THDP-17 does not guarantee that the inhibitor also works *in vivo*, we evaluated the production of ammonia and glutamate from glutamine by enterocyte cells (Caco-2) in the presence and absence of THDP-17 at different concentrations (0, 5, 20 and 100 µM). Glutamate concentration was determined by HLPC in the supernatants and the presence of THDP-17 in the cells was analyzed by HPLC-MS. Results are shown on Figure 4. These data indicate that PAG inhibition by THDP-17 in cell cultures is effective at a concentration higher than 5 µM, showing an inhibition of 18±2.1% and 46±3.4% at 20 and 100 µM, respectively. This fact can be explained assuming that the inhibitor can diffuse or transported through the plasma and the outer mitochondrial membranes, reaching the PAG located at the inner mitochondrial membrane, interfering its function and lowering its activity. Since there is no direct evidence that this transport is taking place, we have studied the evolution of the concentration of THDP in the medium along the cultivation period, showing that

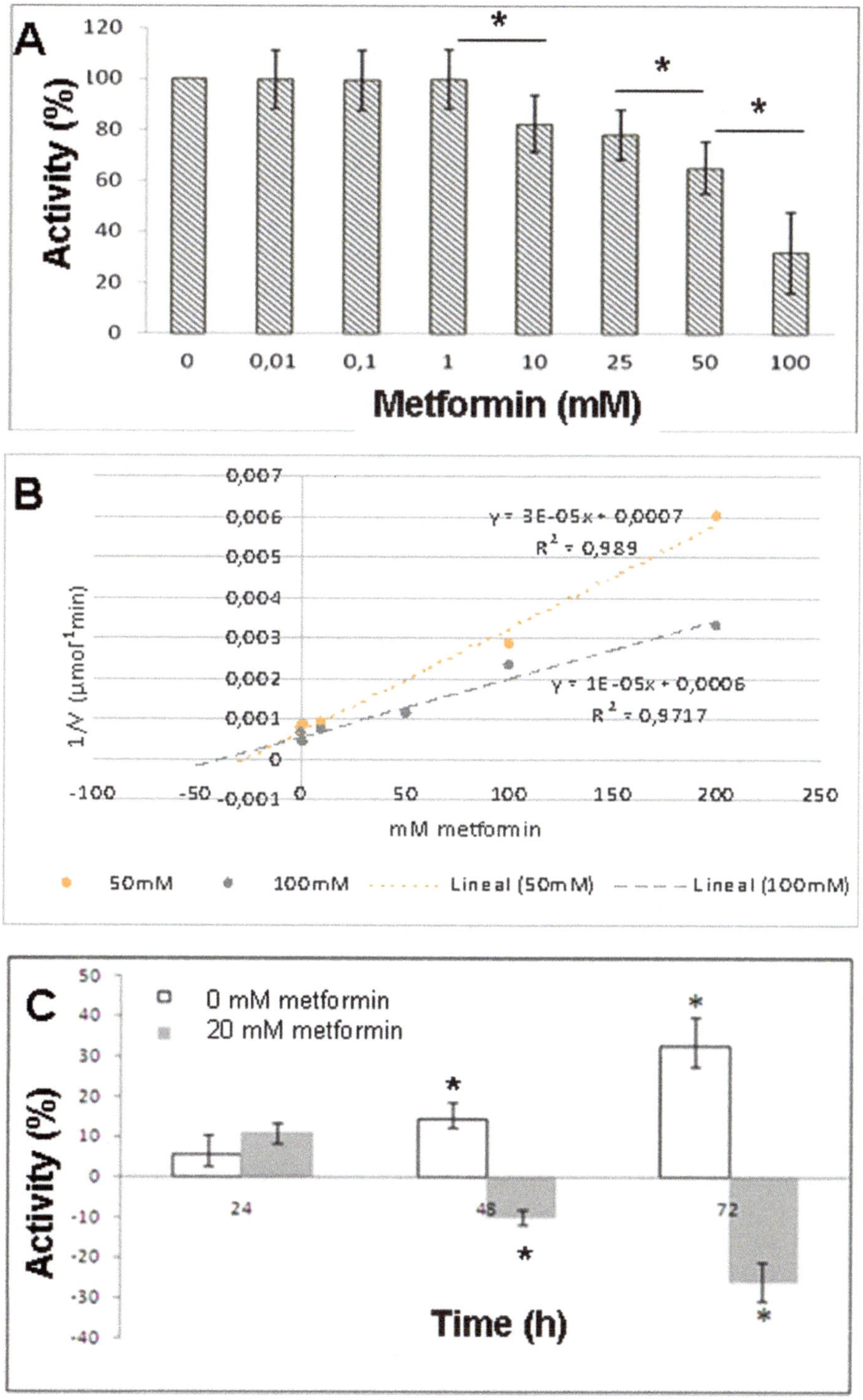

Figure 5. Metformin inhibits glutaminase activity in a dose dependent manner: from 17.5% at 10 mM up to 68% at 100 mM (A) (* p<0.05). This effect, analyzed by Dixon plot, showed competitive inhibition kinetics, with a K_i of 14.28 mM (B). In Caco2 cells, metformin 20 mM showed 24% inhibition of glutaminase activity at 72 hours compared to control cultures ($p<0.05$) (glutamate production was decreased from 26.85±0.74 µM to 19.9±2.05 µM; $p<0.05$) (C).

56% and 72% of the inhibitor is diffused or transported into the cells at 20 and 100 µM, respectively.

- Acute Toxicity of THDP17. Although THDP-17 did not show any cytotoxic effect in cell cultures, we have studied the effect of this compound ***in vivo*** using an animal model (C57BL mice) since thiourea-derived compounds can be toxic ***in vivo*** [32]. Two different assays were carried out: first, THDP-17 was administered orally after 12 h fasted. In this case deaths were observed at 6, 12 and 24 h in all groups (Table-1), including control group (without THDP-17) where only vehicle (DMSO) was administered.

However, lethality of DMSO-treated mice did not differ significantly in its lethality as compared to the groups treated with THDP-17. The undiluted compound has low systemic toxicity but a marked local necrotizing and inflammatory effect when it is administered orally. Animals showed gastrointestinal ulcers and duodenum bleeding (Figure S1). There were few fluctuations in histopathological values but no changes appear to be drug-related or of significance. No essential or histopathological damages were seen in mice treated with THDP-17 including lungs, heart or liver when compared to DMSO alone. We interpreted these results as DMSO and/or THDP-17 was toxic for the animals in fasted state. Therefore, we repeat the assay using feed animals. THDP-17 and/or DMSO were then administered orally to animals after free access to food and water overnight, and mice were sacrificed at 24 h and 72 h randomly. In this case, no deaths were observed during the experimental period (Table 2); probably due to protective effect of food. Macroscopic organs analysis from animals sacrificed at 24 and 72 h showed gastrointestinal ulcers and duodenum bleeding, although less pronounced than in the case of the first group (fasted group).

These results demonstrate that THDP-17 is no toxic *in vivo* and suggest that DMSO enhanced the toxicity of THDP in mice. Therefore, this compound is currently used as a primary group for the synthesis amphiphilic derivatives.

ii) Products currently used in HE therapy

A second group of compounds integrated by drugs habitually used in the treatment of HE was also tested for glutaminase inhibition: metformin, metronidazole, neomicine, rifaximin, lactulose and lactitol. PAG activity inhibition was analyzed using 10 mM concentration. PAG activity *in vitro* was no affected by neomycin, lactitol and lactulose, rather, an increased activity was observed.

Since metformin seems to be protective against hepatic encephalopathy in diabetic cirrhotic patients [33], we decided to analyze in vitro PAG inhibition by metformin. Results in Figure-5 show that PAG was inhibited in a dose dependent manner: from 17.5% at 10 mM up to 68% at 100 mM. This effect, analyzed by Dixon plot, showed competitive inhibition kinetics, with a K_i of 14.28 mM (Figure 5B). In Caco2 cells, metformin 20 mM showed 24% inhibition of glutaminase activity at 72 hours compared to control cultures ($p<0.05$) (glutamate production was decreased from 26.85 ± 0.74 µM to 19.9 ± 2.05 µM; $p<0.05$) (Figure-5C). Higher metformin concentrations (50, 100 and 200 mM) did not yield higher inhibition rates. These results demonstrated that metformin could serve as a new therapeutic option for HE management, although THDP17 was able to inhibit glutaminase (46 ± 3.4%) in a higher extent.

Conclusions

The major findings in this work are: i) although THDP-17 was reported first in 1960, this is the first time that its use as glutaminase activity inhibitor is described. THDP17 shows a partial uncompetitive inhibition *in vitro* and a significant reduction of ammonia and glutamate production *in vivo*. Therefore, THDP-17 could be a good candidate for HE management by lowering ammonia production. ii) Moreover, we have shown that metformin, an insulin sensitizer drug used for diabetes therapy, could also inhibit PAG activity, resulting in new therapeutic option for HE by reducing splenic ammonia production.

Supporting Information

Figure S1 Fasted C57BL/6 mice were treated with either DMSO (vehicle) or THDP-17. Both groups showed acute toxicity effects. A- Internal general appearance, showing gastrointestinal ulcers and bleeding in the duodenum; B- Bowels; C- Stomach.

Author Contributions

Conceived and designed the experiments: JAC JDB MDH IP FIG JMV. Performed the experiments: MDH JAC IP JMV. Analyzed the data: JDB JAC FIG MRG. Contributed reagents/materials/analysis tools: FIG JMV FJM MRG. Contributed to the writing of the manuscript: JAC MDH JDB FIG FJM MRG. Contributed dataset: PCA.

References

1. Zemtsova I, Görg B, Keitel V, Bidmon HJ, Schrör K, et al. (2011) Microglia activation in hepatic encephalopathy in rats and humans. Hepatology 54: 204–15.
2. Llansola M, Montoliu C, Cauli O, Hernández-Rabaza V, Agustí A, et al. (2013) Chronic hyperammonemia, glutamatergic neurotransmission and neurological alterations. Metab Brain Dis., 28: 151–154.
3. Jover-Cobos M, Noiret L, Lee K, Sharma V, Habtesion A, et al. (2014) Ornithine phenylacetate targets alterations in the expression and activity of glutamine synthase and glutaminase to reduce ammonia levels in bile duct ligated rats. J Hepatol. 60: 545–53.
4. Dam G, Keiding S, Munk OL, Ott P, Vilstrup H, et al. (2013) Hepatic encephalopathy is associated with decreased cerebral oxygen metabolism and blood flow, not increased ammonia uptake. Hepatology, 57: 258–265.
5. Romero-Gómez M, Bautista JD, Grande L, Ramos Guerrero RM, Sanchez Munoz D. (2004) New concepts in the physiopathology of hepatic encephalopathy and therapeutic prospects. Gastroenterol Hepatol. 27: 40–8.
6. Nance FC., Kline DG (1971). Eck's fistula encephalopathy in germ-free dogs. Ann Surg. 174: 856–61.
7. Warren KS, Newton WL (1959) Portal and peripheral blood ammonia concentrations in germ-free and conventional guinea pigs. Am J Pysiol. 197: 717–20.
8. Weber FJL, Veach GL (1979) The importance of the small intestine in gut ammonium production in the fasting dog. Gastroenterology 77: 235–40.
9. Olde-Damink SW, Jalan R, Redhead DN., Hayes PC, Deutz NE, et al. (2002) Inter-organ ammonia and amino acid metabolismo in metabolically stable patients with cirrhosis and a TIPSS. Hepatology 36; 1163–71.
10. Romero-Gómez M, Ramos-Guerrero R, Grande L, de Terán LC, Corpas R, et al. (2004). Intestinal glutaminase activity is increased in liver cirrhosis and correlates with minimal hepatic encephalopathy. J Hepatol, 41: 49–54.
11. Albrecht J, Norenberg MD (2006) Glutamine: a Trojan horse in ammonia neurotoxicity. Hepatology 44: 788–94.
12. Olalla L, Gutiérrez A, Campos JA, Khan ZU, Segura JA, et al. (2002) Nuclear localization of L-type glutaminase in mammalian brain. J Biol Chem 277: 28939–38944.
13. Castell L, Vance C, Abbott R, Márquez J, Eggleton P (2004) Granule localization of glutaminase in human neutrophils and the consequence of glutamine utilization for neutrophil activity. J Biol Chem 279: 13305–10.
14. Pérez-Gómez C, Campos-Sandoval JA., Alonso FJ, Segura JA., Manzanares E., et al. (2005) Co-expression of glutaminase K and L isoenzymes in human tumour cells. Biochem J 385: 535–542.
15. Roberg BA, Torgner IA, Kvamme E (2010) Kinetics of a novel isoform of phosphate activated glutaminasa (PAG) in SH.SY5Y neuroblastoma cells. Neurochem Res 35: 875–80.
16. Romero-Gómez M, Jover M, Del Campo JA, Royo JL, Hoyas E, et al. (2010) Variations in the promoter region of the glutaminase gene and the development of hepatic encephalopathy in patients with cirrhosis: a cohort study. Ann Intern Med.153: 281–288.
17. James LA, Lunn PG, Middleton S, Elia M (1998) Distribution of glutaminase and glutamine synthase activities in the human gastrointestinal tract. Clin Sci. 94: 313–19.
18. James LA, Lunn PG, Elia M (1988) Glutamine metabolism in the gastrointestinal tract of the rat assessed by the relative activity of glutaminase (EC 3.5.1.2) and glutamine synthetase (EC 6.3.1.2). Br J Nutrition. 79: 365–72.
19. Butterwort RF, Norenberg MD, Felipo V, Ferenci P, Albrecht J, et al. (2009) Experimental models of hepatic encephalopathy: ISHEN guidelines. Liver Int 29: 783–788.
20. Hawkins RA, Jessy J, Mans AM, Chedid A, DeJoseph MR (1994) Neomycin reduces the intestinal production of ammonia from glutamine. Adv Exp Med Biol. 368: 125–34.
21. Vega-Pérez JM, Periñán I, Argandoña M, Vega-Holm M, Palo-Nieto C, et al. (2012) Isoprenyl-thiourea and urea derivatives as new farnesyl diphosphate

analogues: synthesis and in vitro antimicrobial and cytotoxic activities. Eur J Med Chem. 58: 591–612.
22. Iliceto A, Fava A, Mazzucato U, Radici P (1960) Allyl and benzhydryl thiocyanates and isothiocyanates; thiocyanate-isothiocyanate equilibrium. Gazzetta Chimica Italiana, 90: 919–40.
23. Furdik M., Sutoris V (1961) Synergists of pyrethrum. VI. Synthesis of endo-cis-N-substituted 7-diphenylmethylenebicyclo[1.2.2]hept-5-ene-2,-3-dicarboximides. Chemicke Zvesti, 15: 173–80.
24. Heini HG, Gebhardt R, Brecht A, Mecke D (1987) Purification and characterization of rat liver glutaminasa. Eur J Biochem 162: 541–546.
25. Shapiro RA, Haser WG, Curthoys NP (1985) The orientation of phosphate-dependent glutaminasa on the inner membrane of rat renal mitochondria. Arch Biochem Bioph 243: 1–7.
26. Lund P (1988) L-glutamine and L-glutamate: UV-method with glutaminasa and glutamate dehydrogenase, in: H.U. Bermeyer (Eds.), Methods of enzymatic analysis, Weinheim, 57–363.
27. Bradford MM (1976) Rapid and sensitive method for the quantitation of microgram quantities of protein utilizing the principle of protein-dye binding. Anal Biochem 72: 248–254.
28. Zhang SZ, Lipsky MM, Trump BF, Hsu IC (1990) Neutral red (NR) assay for cell viability and xenobiotic-induced cytotoxicity in primary cultures of human and rat hepatocytes. Cell Biol Toxicol. 6: 219–34.
29. Spielmann H, Genschow E, Leibsch M, Halle W (1999) Determination of the starting dose for acute oral toxicity (LD50) testing in the up and down procedure (UDP) from cytotoxicity data. Alternatives to Laboratory Animals 27: 957–966.
30. Mossman T (1983) Rapid colorimetric assay for cellular growth and survival: application to proliferation and cytotoxicity assay. J. Immunol. Methods 65: 55–63.
31. Denizot F, Lang R (1986) Rapid colorimetric assay for cellular growth and survival. Modifications to the tetrazolium dye procedure giving improved sensitivity and reliability. J. Immunol. Methods 89: 271–277.
32. Smith PB, Crespi C (2002) Thiourea toxicity in mouse C3H/10T12 cells expressing human flavin-dependent monooxigenase 3. Biochemical Pharmacology, 63: 1941–1948.
33. Ampuero J, Ranchal I, Nuñez D, Díaz-Herrero MM, Maraver M, et al. (2012) Metformin Inhibits Glutaminase Activity and Protects against Hepatic Encephalopathy. PLoS One 7: e49279.

Risk Factors for Severe Neutropenia following Intra-Arterial Chemotherapy for Intra-Ocular Retinoblastoma

Ira J. Dunkel[1,2]*, Weiji Shi[3], Kim Salvaggio[4], Brian P. Marr[5,6], Scott E. Brodie[7], Y. Pierre Gobin[4], David H. Abramson[5,6]

1 Department of Pediatrics, Memorial Sloan-Kettering Cancer Center, New York, NY, United States of America, **2** Department of Pediatrics, New York Presbyterian Hospital Weill Cornell Medical College, New York, NY, United States of America, **3** Department of Epidemiology and Biostatistics, Memorial Sloan-Kettering Cancer Center, New York, NY, United States of America, **4** Department of Neurosurgery, New York Presbyterian Hospital Weill Cornell Medical College, New York, NY, United States of America, **5** Department of Surgery, Memorial Sloan-Kettering Cancer Center, New York, NY, United States of America, **6** Department of Ophthalmology, New York Presbyterian Hospital Weill Cornell Medical College, New York, NY, United States of America, **7** Department of Ophthalmology, Mount Sinai Medical Center, New York, NY, United States of America

Abstract

Purpose: Intra-arterial chemotherapy is a promising strategy for intra-ocular retinoblastoma. Neutropenia is the most commonly encountered systemic toxicity and in this study we aimed to determine the risk factors associated with the development of severe (≥grade 3) neutropenia.

Methods: Retrospective review of 187 evaluable cycles of melphalan-containing intra-arterial chemotherapy from the first three cycles administered to 106 patients with intra-ocular retinoblastoma from May 2006 to June 2011. Cycles were considered to be evaluable if (1) blood count results were available in the 7 to 14 days post-treatment interval and (2) concurrent intravenous chemotherapy was not administered. Toxicity was assessed via the Common Terminology Criteria for Adverse Events version 4.0.

Results: 54 cycles (29%) were associated with grade 3 (n = 43) or grade 4 (n = 11) neutropenia. Multivariate stepwise logistic regression revealed that a higher melphalan dose (>0.40 mg/kg) was significantly associated with severe neutropenia during all 3 cycles (odds ratio during cycle one 4.11, 95% confidence interval 1.33–12.73, p = 0.01), but the addition of topotecan and/or carboplatin were not. Prior treatment with systemic chemotherapy was not associated with severe neutropenia risk in any analysis.

Conclusions: Intra-arterial melphalan-based chemotherapy can cause severe neutropenia, especially when a dose of greater than 0.40 mg/kg is administered. Further study with a larger sample may be warranted.

Editor: John W. Glod, National Cancer Institute, United States of America

Funding: This work was supported by Perry's Promise Fund. The funders had no role in study design, data collection and analysis, decision to publish, or preparation of the manuscript.

Competing Interests: The authors have declared that no competing interests exist.

* Email: dunkeli@mskcc.org

Introduction

Retinoblastoma is the most common primary ocular tumor of childhood. In the United States and other socio-economically advantaged parts of the world, the vast majority of patients have intra-ocular disease at diagnosis. During the past 15 to 20 years advanced intra-ocular disease has most often been treated with enucleation, but super selective intra-arterial chemotherapy appears to be a promising option [1–2]. It is generally well tolerated, but grade 3 and 4 neutropenia may occur. We performed this analysis to try to determine risk factors associated with the development of severe neutropenia and hypothesized that factors associated with development of grade 3 or 4 neutropenia would be (1) the melphalan dose, (2) administration of topotecan and/or carboplatin in addition to melphalan, and (3) prior treatment with systemic chemotherapy.

Methods

Ethics statement

The Memorial Sloan-Kettering Cancer Center's Institutional Review Board/Privacy Board approved this retrospective review of existing data, granting a waiver for this to be done without obtaining consent from the subjects or their parents/legal guardians. All protected health information was handled in accordance with institutional policies that did not require the information to be anonymized and de-identified prior to analysis.

Patients

We retrospectively reviewed the first 106 consecutive patients with intra-ocular retinoblastoma treated with intra-arterial chemotherapy at our centers from May 2006 to June 2011.

Patients most frequently were treated with intra-arterial single-agent melphalan, but some cycles also included treatment with intra-arterial topotecan and/or carboplatin (Table 1). The number of agents to be used and the doses of chemotherapy administered were determined on a case by case basis. In general, patients received more than one agent if they had more severe disease or had been extensively pre-treated with intravenous chemotherapy and/or external beam radiation therapy, especially if we were treating the only remaining eye. In cycle 1, the dose was primarily determined by the patient's age. Patients 3 to 6 months of age generally received 2.5 mg of melphalan, 6 to 12 months of age, 3 mg of melphalan, 1 to 3 years of age, 4 mg of melphalan, and ≥3 years of age, 5 mg of melphalan [1]. In subsequent cycles we would consider increasing the dose(s) if the ophthalmic artery had large extra-ocular branches or if an inadequate response had been encountered without significant toxicity. We would consider decreasing the dose(s) if wedge flow was encountered or significant toxicity was encountered, such as an interval decrease in the eye's electroretinogram or an ocular inflammatory reaction. The median topotecan dose administered was 0.4 mg (range 0.2 to 2 mg) and the median carboplatin dose administered was 30 mg (range 25 to 80 mg).

We asked the parents to have a complete blood count performed 7 to 10 days after each dose of intra-arterial chemotherapy. However, many families did not reside in the New York area and returned home after the treatment, and so compliance was variable. Intra-arterial chemotherapy cycles were considered to be evaluable if (1) blood count results were available in the 7 to 14 days post-treatment interval, and (2) concurrent intravenous chemotherapy was not administered. However, if a blood count was not available within the 7 to 14 day post-treatment window, but grade 3 or 4 neutropenia was documented earlier or later in the cycle, the cycle was considered evaluable. Toxicity was assessed via the Common Terminology Criteria for Adverse Events (CTCAE) version 4.0.

Statistics

A binary melphalan dose is of interest in potential prediction. An optimal cut-off point of the dose was selected using Miller and Siegmund's minimum p-value approach and Altman, Lausen, Sauerbrei, and Schumacher's formula to adjust the minimum p-value selected from the systematic dependent multiple testing [3–4]. The highest and lowest 10% of the melphalan dose data were eliminated from the selection procedure. A univariate analysis of relationship between treatment factors and severe neutropenia (defined as grade 3 or 4 neutropenia) was performed using Chi-square, or Fisher's exact test. A multivariate analysis of the joint relationship was performed using a stepwise logistical regression method. Variables with p-value ≤0.20 on univariate analysis were candidates for the initial multivariate model. The statistical analysis was performed with the software SAS version 9.2 (SAS Institute Cary, NC) and r package ROCR (version 2.9.2). A p-value<0.05 was considered significant.

All analyses were based on evaluable cycles only. Grade 3 and 4 neutropenia were combined into the entity of severe neutropenia due to the small number of grade 4 neutropenia events (n = 5, 3, and 3 during cycles 1, 2 and 3, respectively).

Results

Melphalan dose and analysis of optimal cut-off point

A preliminary analysis showed that a higher melphalan dose (continuous variable) was significantly associated with severe neutropenia during cycles 2 and cycle 3 (p<0.0001, p<0.0001

Table 1. The number of patients treated per cycle with the various chemotherapy agents and the number of patients evaluable for analysis of neutropenia.

Cycle	Patient	M only	M + T	M+C	M+T+C	Inevaluable	Evaluable
1	106	74	21	1	10	33	73
2	100	67	25	1	7	36	64
3	86	53	24	1	8	36	50

M: melphalan; T: topotecan; C: carboplatin.

by t-test, respectively). The minimum p-value approach showed that 0.50 mg/kg was the best cut-off point during cycle 2 (>0.50 versus ≤50 mg/kg, $\chi^2 = 18.55$, adjusted p = 0.0007) and 0.40 mg/kg during cycle 3 (>0.40 versus ≤0.40 mg/kg, $\chi^2 = 26.26$, adjusted p<0.0001). Melphalan dose>0.40 mg/kg was also associated with severe neutropenia during cycle 2 ($\chi^2 = 12.34$, adjusted p = 0.01).

Analyses of treatment factors and severe neutropenia

Cycle 1. A univariate analysis regarding cycle 1 showed that a higher dose of melphalan (>0.40 versus ≤0.40 mg/kg) was significantly associated with severe neutropenia (p = 0.01, Table 2). The univariate analysis also showed that the severe neutropenia rate was significantly different among patients treated with different chemotherapy regimens (melphalan alone versus melphalan and either topotecan or carboplatin versus melphalan, topotecan and carboplatin, p = 0.04). Further pairwise comparison showed that patients treated with three chemotherapy agents more frequently experienced severe neutropenia compared to those treated with two agents (50% versus 9%, p = 0.02). However, a stepwise logistic regression revealed that while higher melphalan dose remained associated with severe neutropenia (odds ratio (OR) 4.11, 95% CI 1.33–12.73, p = 0.01), the number of chemotherapy agents administered did not.

Cycle 2. A univariate analysis revealed that a higher cycle 2 melphalan dose (>0.50 versus ≤0.50 mg/kg) was significantly associated with severe neutropenia (p<0.0001 by Chi-square test, adjusted p = 0.0007, Table 3). A stepwise logistic regression showed melphalan dose remained associated with severe neutropenia during cycle 2 (p = 0.0005, OR 10.86, 95% CI 2.84–41.57).

Cycle 3. A univariate analysis demonstrated that a higher cycle 3 melphalan dose (>0.40 versus ≤0.40 mg/kg) was once again significantly associated with severe neutropenia (p<0.0001, adjusted p<0.0001, Table 4). The univariate analysis and further pairwise comparison also showed that patients who had severe neutropenia during cycle 1 or 2 more frequently experienced severe neutropenia during cycle 3 compared to those who did not (yes versus unknown versus no, p = 0.02; yes versus no, p = 0.01). A stepwise multivariate logistic regression analysis revealed that while the higher melphalan dose remained associated with severe neutropenia (p = 0.0001, OR 72.0, 95% CI 7.93–653.4), severe neutropenia during cycle 1 or 2 did not.

Impact of inevaluable cycles

A greater proportion of later cycles were inevaluable (31%, 36%, and 42% in cycles 1, 2, and 3, respectively) and that could introduce bias on generalization. However, there was no significant difference in melphalan dose (continuous or binary> 0.40 versus ≤0.40 mg/kg), chemotherapy regimen, prior treatment, and severe neutropenia during cycle 1, 2 or 3 between the patients who had or had no record (at cycle 1, 2, or 3, p = 0.11 to 0.95), except that patients with record received a higher mean melphalan dose during cycle 1 (0.40±0.14 versus 0.34±0.14, p = 0.04). In addition, patients with record received a higher binary melphalan dose (>0.50 versus ≤0.50 mg/kg) during cycle 2 (27% versus vs 8%, p = 0.03).

Clinical consequences of severe neutropenia

Most patients who developed severe neutropenia were asymptomatic. Two patients were admitted to the hospital during three periods of severe neutropenia due to fever and/or mucositis. Both patients made complete recoveries.

Other severe hematological toxicities

Severe (grade 3 or 4) anemia or thrombocytopenia were encountered infrequently. Five patients suffered 1 episode each (5 episodes total) of grade 3 anemia and 2 patients suffered 1 episode each (2 episodes total) of grade 4 thrombocytopenia.

Discussion

Intra-arterial chemotherapy was first introduced into the treatment regimen for patients with intra-ocular retinoblastoma by Reese and colleagues in 1955 when they administered triethylenemelamine via the internal carotid artery [2]. It then fell out of favor due to toxicity concerns (that included death) and lack of clear benefit over other treatments until 1987 when Kaneko and colleagues began to administer selective ophthalmic arterial injection of melphalan using a balloon catheter. Their group recently reported a retrospective series of 343 patients (408

Table 2. Cycle 1: Univariate analysis of severe neutropenia.

Variable	Severe neutropenia (25%)		p-value
	No (n = 55)	Yes (n = 18)	
M dose, mg/kg (mean ± SD)	0.39±0.14	0.43±0.12	
Median (range)	0.35 (0.15–0.89)	0.46 (0.27–0.68)	
≤0.40	37 (67%)	6 (33%)	0.01
>0.40	18 (33%)	12 (67%)	
Chemotherapy agents			0.04
M only	30 (55%)	11 (61%)	
M+ T/C	20 (36%)	2 (11%)	
M+ T + C	5 (9%)	5 (28%)	
Prior treatment[a]			0.77
No	24 (45%)	7 (41%)	
Yes	29 (55%)	10 (59%)	

M: melphalan; T: topotecan; C: carboplatin.
[a]Data not available for 3 patients (2 had and 1 did not have severe neutropenia).

Table 3. Cycle 2: Univariate analysis of severe neutropenia[a].

Variable	Severe neutropenia (30%)		p-value
	No (n = 45)	Yes (n = 19)	
M dose, mg/kg (mean ± SD)	0.36±0.12	0.51±0.13	
Median (range)	0.34 (0.04–0.57)	0.52 (0.26–0.71)	
≤0.50	40 (89)	7 (37)	<0.001[c]
>0.50	5 (11)	12 (63)	
≤0.40	31 (69%)	4 (21%)	0.01[c]
>0.40	14 (31%)	15 (79%)	
Chemotherapy agents			0.26
M only	22 (49%)	9 (47%)	
M+ T/C	20 (44%)	6 (32%)	
M+ T + C	3 (7%)	4 (21%)	
Prior treatment[b]			0.13
No	22 (51%)	5 (29%)	
Yes	21 (49%)	12 (71%)	
Cycle 1 grade 3 or 4 neutropenia			0.66
No	27 (60%)	9 (47%)	
No record	10 (22%)	5 (26%)	
Yes	8 (18%)	5 (26%)	

M: melphalan; T: topotecan; C: carboplatin.
[a]Only one variable was associated with severe neutropenia at p≤0.05 on univariate analysis; multivariate analysis was not performed.
[b]Data not available for 4 patients (2 had and 2 did not have severe neutropenia).
[c]Cut-off point selection adjusted p-value.

Table 4. Cycle 3: Univariate analysis of severe neutropenia.

Variable	Severe neutropenia (34%)		p-value
	No (n = 33)	Yes (n = 17)	
M dose, mg/kg (mean ± SD)	0.34±0.11	0.50±0.08	
Median (range)	0.33 (0.13–0.70)	0.50 (0.33–0.69)	
≤0.40	27 (82%)	1 (6%)	<0.001[b]
>0.40	6 (18%)	16 (94%)	
Chemotherapy agents			0.15
M only	13 (39%)	4 (24%)	
M+ T/C	17 (52%)	8 (47%)	
M+ T + C	3 (9%)	5 (29%)	
Prior treatment[a]			0.83
No	16 (50%)	7 (47%)	
Yes	16 (50%)	8 (53%)	
Cycle 1 or 2, grade 3 or 4 neutropenia			0.02
No	14 (42%)	3 (18%)	
No record	11 (33%)	3 (18%)	
Yes	8 (24%)	11 (65%)	

M: melphalan; T: topotecan; C: carboplatin.
[a]Data not available for 3 patients (1 had and 2 did not have severe neutropenia).
[b]Cut-off point selection adjusted p-value.

eyes) treated with 1452 procedures from 1988 to 2007 [5]. They generally used single-agent melphalan at a dose of 5 to 7.5 mg/m^2 and did not encounter any> grade 1 decrease of white blood cells. Using a 30:1 conversion factor, these doses are approximately 0.17 to 0.25 mg/kg of melphalan.

In the patients reported in this series, we used a higher dose of melphalan (median dose of evaluable cycles 0.36 mg/kg, range 0.04 to 0.89 mg/kg) and have encountered severe neutropenia in 29% of the cycles. Fortunately, most of the episodes of severe neutropenia were grade 3 (n = 43) rather than grade 4 (n = 11) and were generally clinically insignificant. We counseled the parents about the risk of fever and neutropenia, but did not routinely prescribe filgrastim. We did not encounter any severe neutropenia during cycles in which the melphalan dose was ≤0.25 mg/kg (the approximate dose used in Japan). A limitation is that we may have overestimated the risk of neutropenia due to our decision to consider cycles with blood counts performed outside of the 7 to 14 day window without neutropenia to be inevaluable, but to consider cycles evaluable if grade 3 or 4 neutropenia was documented earlier or later in the cycle. We felt that potentially overestimating rather than underestimating the risk of neutropenia associated with the intra-arterial chemotherapy was the more conservative approach.

We hypothesized that (1) adding topotecan and/or carboplatin to the melphalan regimen and (2) a history of prior patient exposure to intravenous chemotherapy might be associated with increased risk for severe neutropenia, but that turned out not to be the case. The only factor that remained significant in analyses was melphalan dose, particularly when a dose of greater than 0.40 mg/kg was administered. Due to the relatively small number of events and sample size, we may not have enough power to detect some difference. We were also unable to select an optimal cut-off point of melphalan dose in a multivariate setting. Further study with a larger sample size may be warranted. This increased risk of neutropenia associated with a melphalan dose of greater than 0.40 mg/kg is comparable to the results in a small series reported by Argentine investigators. They noted that children receiving more than 0.48 mg/kg for bilateral tandem infusions had a significantly higher systemic area under the curve and a 50% probability of grade 3 or 4 neutropenia [6].

These data may help assist in the selection of safe intra-arterial chemotherapy doses for patients with intra-ocular retinoblastoma, but it is important to note that most patients who developed severe neutropenia were asymptomatic and did not require hospital admission for fever and neutropenia or an infection. To minimize the risk of severe neutropenia patients should be treated with a melphalan dose of 0.40 mg/kg or less, but should higher doses be required to adequately treat the retinoblastoma and cure the eye, the risk of severe neutropenia may be acceptable.

Acknowledgments

A preliminary version of this work was presented at the 2009 International Congress of Ocular Oncology meeting in Cambridge, UK.

Author Contributions

Conceived and designed the experiments: IJD BPM SEB YPG DHA. Analyzed the data: IJD WS. Contributed reagents/materials/analysis tools: IJD KS BPM SEB YPG DHA. Wrote the paper: IJD WS KS BPM SEB YPG DHA.

References

1. Gobin YP, Dunkel IJ, Marr BP, Brodie SE, Abramson DH. (2011) Intra-arterial chemotherapy in the management of retinoblastoma: Four-year experience. Arch Ophthalmol 129: 732–737.
2. Reese AB, Hyman GA, Merriam GR Jr, Forrest AW, Kligerman MM. (1954) Treatment of retinoblastoma by radiation and triethylenemelamine. AMA Arch Ophthalmol 53: 505–513.
3. Miller R, Siegmund D. (1982) Maximally selected chi square statistics. Biometrics 38: 1011–1016.
4. Altman DG, Lausen B, Sauerbrei W, Schumacher M. (1994) Dangers of using "optimal" cutpoints in the evaluation of prognostic factors. J Natl Cancer Inst 86: 829–835.
5. Suzuki S, Yamane T, Mohri M, Kaneko A. (2011) Selective ophthalmic arterial injection therapy for intraocular retinoblastoma: The long term prognosis. Ophthalmology 118: 2081–2087.
6. Schaiquevich P, Buitrago E, Taich P, Torbidoni A, Ceciliano A, et al. (2012) Pharmacokinetic analysis of melphalan after superselective ophthalmic artery infusion in preclinical models and retinoblastoma patients. Invest Ophthalmol Vis Sci 53: 4205–4212.

9

Aerosol Azacytidine Inhibits Orthotopic Lung Cancers in Mice through Its DNA Demethylation and Gene Reactivation Effects

Xuan Qiu[1], Yuanxin Liang[1], Rani S. Sellers[2], Roman Perez-Soler[1], Yiyu Zou[1]*

1 Departments of Medicine, Division of Medical Oncology, Albert Einstein College of Medicine, Bronx, NY, United States of America, 2 Department of Pathology, Albert Einstein College of Medicine, Bronx, NY, United States of America

Abstract

We devised an aerosol based demethylation therapy to achieve therapeutic efficacy in premalignant or in situ lesions of lung cancer, without systemic toxicity. Optimum regimens of aerosolized azacytidine (Aza) were designed and used in orthotopic human non-small cell lung cancer xenograft models. The therapeutic efficacy and toxicity of aerosol Aza were compared with intravenously administered Aza. We observed that 80% of the droplets of the aerosol Aza measured ~0.1–5 microns, which resulted in deposition in the lower bronchial airways. An animal model that phenocopies field carcinogeneisis in humans was developed by intratracheal inoculation of the human lung cancer cells in mice, thus resulting in their distribution throughout the entire airway space. Aerosolized Aza significantly prolonged the survival of mice bearing endo-bronchial lung tumors. The aerosol treatment did not cause any detectable lung toxicity or systemic toxicity. A pre-pharmacokinetic study in mice demonstrated that lung deposition of aerosolized Aza was significantly higher than the intravenous route. Lung tumors were resected after aerosol treatment and the methylation levels of 24 promoters of tumor-suppresser genes related to lung cancer were analyzed. Aerosol Aza significantly reduced the methylation level in 9 of these promoters and reexpressed several genes tested. In conclusion, aerosol Aza at non-cytotoxic doses appears to be effective and results in DNA demethylation and tumor suppressor gene re-expression. The therapeutic index of aerosol Aza is >100-fold higher than that of intravenous Aza. These results provide a preclinical rationale for a phase I clinical trial of aerosol Aza to be initiated at our Institution.

Editor: Jin Q. Cheng, H.Lee Moffitt Cancer Center & Research Institute, United States of America

Funding: The study is supported by a NIH grant 5R01CA154755. The funders had no role in study design, data collection and analysis, decision to publish, or preparation of the manuscript.

Competing Interests: The authors have declared that no competing interests exist.

* Email: yiyu.zou@einstein.yu.edu

Introduction

Lung cancer claims 1.4 million lives every year (http://www.who.int/mediacentre/factsheets/fs297/en/index.html) [1,2]. Lung cancer occurs as a result of cumulative damage in the bronchial epithelium caused by inhaled carcinogens. Because the cumulative damage affects the entire airway, damaged airway epithelium is prone to develop additional primary tumors during an individual's lifespan [3]. All non-small cell lung cancers (NSCLC) originate from the bronchial epithelium covering large or small airways, giving rise to central or peripheral tumors. For example, squamous cell lung carcinomas most often arise centrally in large bronchi, lung adenocarcinomas typically develop peripherally in the smaller airways, and large cell lung carcinomas may arise in either location. However, all tumors originate from transformed airway epithelial cells. Therefore, selective therapeutic intervention for tumors confined to the airways should effectively inhibit or delay their formation without causing systemic toxicity [4–7]. Unfortunately, no therapies specifically targeting the bronchial epithelium are currently available.

Increasing knowledge in the field of epigenetics and carcinogenesis has led to the conclusion that aberrant epigenetic changes play an important role during lung carcinogenesis; moreover, these changes are maintained through the entire process of disease progression [8,9]. For instance, silencing of tumor suppressor genes (TSGs) by aberrant hypermethylation has been found to play an important role in cancer initiation and development in multiple cancer types [10–19]. TSG promoter hypermethylation has also been shown to correlate with poor prognosis and resistance to chemotherapy [20–26]. In particular, all genetic lesions in lung cancers, including p53 and k-ras mutations, could be the consequence of aberrant epigenetic changes [10,27,28]. Epigenetic changes are reversible carcinogenic events [10,29]. The cancer-specificity of these epigenetic changes makes them unique targets for specific epigenetic therapies [30]. Therefore, we hypothesize that by targeting the airway epithelium with a demethylating agent by aerosol administration, we may affect favorably the natural history of lung cancer.

Previously we observed that hypermethylation in the promoter region of the Rassf1 gene in human NSCLC cell line H226 can be reversed by the demethylating agent azacytidine (Aza). We also demonstrated, using a clinically relevant animal model developed by our lab, that intratracheal injection (a similar form of local administration as aerosol) of Aza at sub-toxic doses can increase the survival of mice orthotopically implanted H358 and H460 lung cancer [31,32]. This clinically relevant animal model was proof of concept that airway targeted epigenetic therapy may have an advantage in the prevention and treatment of early NSCLC. Here we report on the efficacy of aerosolized Aza in the treatment of orthotopic human NSCLC xenograft models in mice as well as the efficacy of therapy on the demethylation of specific promoters of TSGs in tumor tissue. To elucidate the epigenetic therapeutic mechanisms, we resected tumor tissues from the xenografted animals treated with aerosol Aza, analyzed the methylation status of the promoters of 24 lung cancer-related TSGs, and determined the protein expression levels of those genes that showed significant promoter demethylation after aerosol Aza treatment.

Materials and Methods

Cell lines

Human NSCLC cell lines H226, H358, and H460 were purchased from American Type Culture Collection (ATCC) and cultured in RPMI-1640 medium supplemented with 10% fetal bovine serum (Invitrogen) and maintained in a 37°C incubator with 5% CO_2 and 95% air.

Animals

Male and female ICR and Athymic nude mice, 6–8 weeks old (Harlan) were housed in the animal facility at Albert Einstein College of Medicine. All animal studies were carried out in strict accordance with the recommendations in the Guide for the Care and Use of Laboratory Animals of the National Institutes of Health. The protocol (Protocol Number: 20130312) was approved by the Institutional Animal Care and Use Committee (IACUC) of Albert Einstein College of Medicine. In the survival study, the animals were observed daily. Humane endpoints were used during this study: moribund were humanely euthanized. We used two criteria to identify the moribund animals based on our animal use protocol: 1) mouse has difficulty breathing, eating, or drinking; 2) a mouse loses ≥15% body weight in 4 days. Mice were euthanized by CO2 in an inhalation chamber followed by exsanguination. Anesthesia was used to minimize discomfort during intratracheal injections.

Intratracheal injection

For intratracheal (IT) injection, a bolus of cells in suspension or drug was injected into the trachea. The procedure used for IT injection has been previously described by us [33]. Briefly, mice are anesthetized by with an Isoflurane vaporizer at 2.5% isoflurane delivered by 2.0 PSI of oxygen and immobilized using a restraint device. The mouth is opened with a forceps and a small tubular light device is inserted into the mouth to locate the trachea, if needed. A 22 G feeding needle attached to a 1 ml syringe containing the cell suspension or drug solution is inserted into the trachea. Approximately 3 µl solution/g body weight is then injected intratracheally. The mouse is released from the restraining device at the completion of the procedure and observed until full recovery from anesthesia.

Orthotopic endobronchial lung cancer xenograft models

Athymic nude mice were anesthetized as described previously and human NSCLC cancer cells in suspension were carefully inoculated into the trachea (approximately $2–4\times10^6$ cells/mouse) using the method described above. In these models, multiple tumors grow within the pulmonary airways. The lifespan of the animals correlates with the tumor burden, which is related to the size of the inoculum.

Aerosol formulation

Azacytidine (Sigma) was dissolved into sterile normal saline at 10 mg/ml right before dosing. The aerodynamic size of aerosol Aza was determined with extrusion-precipitation method using a 7-Stage Cascade Impactor linked to PARI's nebulizer system per manufacturer's instructions [34]. Aerosol administration. The aerosol was generated using a PARI's personal compressor and LC Star nebulizer system. The aerosol generation rate was about 0.25 ml/min. A nose-only exposure system (CH Technologies Inc. Westwood, NJ) linked to the PARI's aerosol system in a closed chemical hood was used for aerosol administration to mice. The aerosol administration time (min) (T) was strictly controlled. The aerosol dose deposited in the lungs (mg/m^2) (D) was calculated by multiplying the time of exposure (min) by the aerosol inhalation rate (ml/min) (R) and the drug concentration (mg/ml) (C) as described previously by us [34], i.e. $D = TRCI/W$, where I is body surface/weight index (mg/m^2 : mg/kg) (3 for mice) and W is animal body weight (kg). In a previous study, we measured aerosol inhalation rate in mice, defined as the volume of the aerosol liquid deposited in mouse lungs in 1 min was about 10^{-4} ml/min for a 25 g mouse [34]. The desired deposited dose is achieved by controlling the aerosol time as $T = DW/RCI$. For example, the aerosol administration time for a deposited dose of 2.5 mg/m^2 in a 25 gm mouse using a 10 mg/ml Aza concentration solution is $T = 2.5\times0.025/10^{-4}\times10\times3 = 20.83$ min.

Toxicity studies

ICR mice (6/group) were treated with aerosol Aza at 2.5 or 75 mg/m^2 daily×7 days. One group of mice was treated with IT Aza at the maximum tolerated dose of 270 mg/m^2 as positive control for lung toxicity. The lungs were resected 7 days after the last aerosol administration. The lung toxicity was determined by pathological evaluation (3 lungs/group) using a previously described method [34]. Additionally, the viability of the airway epithelial cells was also determined (3 lungs/group). Briefly the lung tissue was digested with liberase (Roche) and filtered to obtain a single cell suspension, which was labeled with anti-mouse Ep-CAM antibody and fluorescence-conjugated goat anti-rat secondary antibody (Biolegend, San Diego, CA), and sorted by flow cytometry. The percentage of viable cells was determined by the exclusion of 4′, 6-diamidino-2-phenylindole staining. The systemic toxicity was determined by measuring blood cell counts as an indicator of myelosuppression (the major toxicity of IV Aza) [35]. Blood samples were taken at the time of euthanasia by cardiac puncture 7 days after last treatment. The total number of white blood cells (WBC) was determined after removal of red blood cells as previously described [32].

Pre-pharmacokinetic study

Normal ICR mice were given aerosolized Aza at 2.5 mg/m^2 using the method described above. At each designed time point (5 min, 20 min, 2 h, 6 h and 24 h), three mice were euthanized and their lungs were resected after right ventricle perfusion with saline. Azacytidine in the lungs was extracted and quantitatively measured by a previously reported method using LC-MS system [36]. The quantitative detection was performed by Millis Scientific Inc. The sensitivity was 1 ng Aza/ml of sample. The Aza concentration in tissue vs. time data were analyzed and simulated with the best fitting (the highest R^2 value) under the Microsoft Excel program. The equation of the simulated curves $C = f(t)$ were used to calculated AUC from 0 to 24 hours as $AUC = \int_{0-24} f(t)\,dt$. Peaking time (T) and peak concentration were obtained directly from observation of the curves.

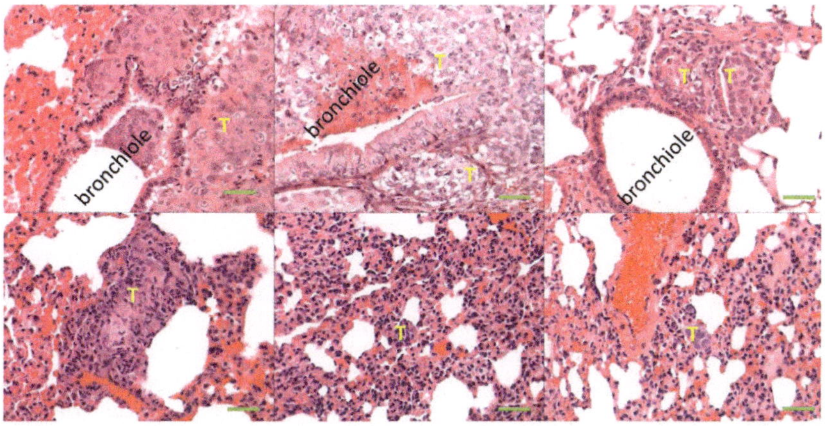

Figure 1. Orthotopic NSCLC model. The model was created by intratracheally injecting NSCLC H460 cells into nude mice. Top: H & E stained lung sections from IT inoculated mice on day 35. The tumors (T) arise within the airways and grow vicinally into the lung parenchyma. Bottom: H & E stained lung sections from the IV inoculated mice on day 35. Tumors arise within small vessels in the lung parenchyma. The objective magnification was 40X. The scale bars were 50 μm.

Antitumor efficacy

Ten days after the intratracheal inoculation of tumor cells, the nude mice were randomly divided into 3 groups (6 mice/group): 1) aerosolized Aza at 2.5 mg/m^2 daily for 7 days, 2) aerosolized normal saline (vehicle) daily for 7 days, 3) or IV Aza at 75 mg/m^2 daily for 7 days. In this survival study the animals were observed daily. Humane endpoints for euthanasia (previously described) were used during this study and moribund mice were humanely euthanized and mice were necropsied. The Increased lifespan (%ILS) [37] was used to assess the efficacy of each tested agent.

Histology of lung cancer xenografts

Mice from the anti-tumor efficacy study were necropsied after death and lung and tumor samples for histology were collected and placed into 10% neural buffered formalin. Samples were transferred to 70% ethanol and submitted to the Histology and Comparative Pathology Shared Resource at Einstein for routine processing to paraffin. Tissues were sectioned to 5 μm, stained with hematoxylin and eosin (H&E), and evaluated by a board certified veterinary pathologist.

Analysis of methylation status in lung tumors

Mice bearing orthotopic lung tumors were divided into 3 groups and treated with aerosol Aza, aerosol vehicle, or IV Aza. Two weeks after the final treatment, the mice were euthanized and visible lung tumors (1~3 mm) were resected, frozen on dry ice, and broken down into smaller (<0.5 mm) pieces by an eppendorf-tube homogenizer. Each tumor sample was divided into two aliquots: one for methylation assay and another for western blot. The methylation status of the promoter regions of 24 tumor suppressor genes involved in lung carcinogenesis was measured using a methylation qPCR array method by Qiagen (SA Bioscieces) as previously described [38].

Analysis of protein expression in lung tumors

The second aliquot of the lung tumor samples was used to determine protein expression of the TSGs. The TSGs showing significant promoter demethylation after aerosol Aza treatment as analyzed by the methylation qPCR array were selected for evaluation by western blot. The frozen tissues were homogenized in cold RIPA buffer with protease inhibitor cocktail (Sigma) using a hand homogenizer, followed by sonication. Protein samples were prepared for Western blot following a routine protocol from Life technologies (Grand Island, NY). Primary antibodies against human H-cadherin, OPCML, Rassf1a (Abcam), SFRP1 (Epitomics, Burlingame, CA), and beta-actin (Santa Cruz Biotechnology) were used to identify the targeted proteins. LI-COR C-DiGit Blot Scanner was used to scan the western blots; the bands were analyzed by Image Studio software (LI-COR). The density ratio of targeted protein vs. actin loading control of each sample was used to present the protein expression levels semi-quantitatively.

Statistical analysis

Differences among different groups were analyzed by two-side Log-rank (Mantel-Cox) test. A difference was considered statistically significant when $p<0.05$.

Results

Orthotopic NSCLC models

NSCLC is a disease of the airway epithelium as a result of chronic exposure of airborne carcinogens. In order to mimic this particular tumor growth in the airway in a xenograft model, we intra-tracheally inoculated human NSCLC cells directly into the airways, and then compared this mode of transplantation to IV injection of tumor cells into nude mice. Histological examination of the lung revealed that the tumor distribution and growth pattern was more similar to human lung cancer in the IT implantation model: the majority of the cancer cells seed on the surface of the airway epithelium at approximately one week after implantation. By 3 weeks, small mucosal tumor nodules were localized within the airway or the lung tissue adjacent to the airway, indicating primary tumor growth within the airway and invasion to the lung parenchyma (Figure 1 top row). The tumor growth and size are cell line- and inoculum size- dependent. In our experience, H460 (large cell carcinoma) tumors developed faster than H358 (bronchioalveolar carcinoma) and much faster than H226 (squamous cell carcinoma) when inoculated IT in mice. In the absence of any intervention, mice succumb to the lung tumors between days 40 and 102. There were no identifiable distant metastases. In contrast, almost all visible/detectable IV implanted cancer cells seeded in the capillary system of the lung and relatively far away from the airway (Figure 1 bottom row), more

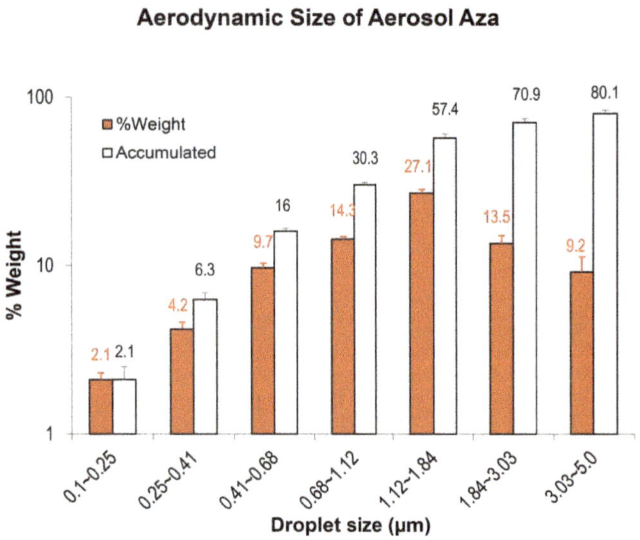

Figure 2. Aerodynamic size of Aerosol Aza. T determined with extrusion-precipitation method using a 7-Stage Cascade Impactor linked to PARI's personal compressor and LC star nebulizer system. Aza solution (4 ml) was aerosolized at an airflow rate of 5 L/min. The condensed aerosol samples were collected at 3 different intervals, from 1 to 1.5 min, from 3 to 3.5 min, and from 5 to 5.5 min. Aerodynamic size and fraction of aerosol with a particular size range were measured and calculated as per manufacturer's protocol. The data was mean ± standard deviation of the aerodynamic size based on weight (solid bars) and cumulative weight (empty bars) from 3 independent experiments.

closely resembling metastatic tumor cells migrating from other organs to the lung. The tumors also developed faster than after IT implantation, and the mice usually died on days 35 to 68 (data not shown). Our results suggest that IT implantation is a more appropriate orthotopic xenograft model for mimicking human NSCLC than the IV model or other ectopic models.

Aerosol formulation

We have developed an aerosol formulation for the prototypical demethylation agent, Azacytidine (Aza). The formulation is composed of Aza and injection grade sterile normal saline or water. The Aza power can be quickly dissolved (<30 sec) in the aqueous solution just prior to aerosol administration. It has been suggested that aerosol droplets <5 µm, particularly <3 µm, in diameter deposit most frequently in the lower airways and, therefore, are appropriate for pharmaceutical inhalation of aerosol preparations by humans [39] [40]. We measured the aerodynamic size of the Aza aerosol formulation, under our aerosol generation system with the extrusion-precipitation method [34], and found that about 80% of the droplets (weight%) were between 0.25–5 µm in diameter, and about 71% between 0.25~3 µm (Figure 2). Our results suggest that about 71~80% of the droplets should deposit in the distal airway in humans [39] [40].

Toxicity and dose selection

From our previous study we have learned that the effective demethylation concentration of Aza for cultured cells could be 3 logs lower than its IC_{50}. The antitumor effective dose (7.5 mg/m^2×3) by intratracheal injection of Aza in the mouse model did not cause lung toxicity or detectable systemic acute toxicity, and lung toxicity only occurred at the highest dose of 270 mg/m^2, which was the maximum tolerated dose by IV injection [32]. Based on these results, we selected a dose for aerosol Aza that we predicted would have demethylation function and therapeutic efficacy without significant toxicity. The results indicate that the selected aerosol dose of 2.5 mg/m^2 (about 21 min aerosol exposure to a 25 g mouse) daily for 7 days or a higher intratracheal bolus injection dose of 75 mg/m^2 for 5 days did not cause any detectable lung toxicity (evaluated histologically) or systemic toxicity (measured by myelosuppression) on day 7 after the final administration (Figure 3A, B). Pulmonary toxicity (pneumonitis) (Figure 3C) and myelosuppression (Figure 3D) only occurred when the mice received the highest dose of intravenous or intratracheal injection of Aza at 270 mg/m^2 (the MTD of intravenous Aza). Similarly, the aerosol Aza at the selected therapeutic dose did not cause any death of airway epithelial cells determined by flow cytometry, as compared to the positive control in mice treated with the highest IT doses of Aza at 270 mg/m^2 (Figure 3E).

Therapeutic efficacy

Systemic (IV) administration of Aza has led to only limited efficacy in NSCLC patients [41] [42], which has been recapitu-

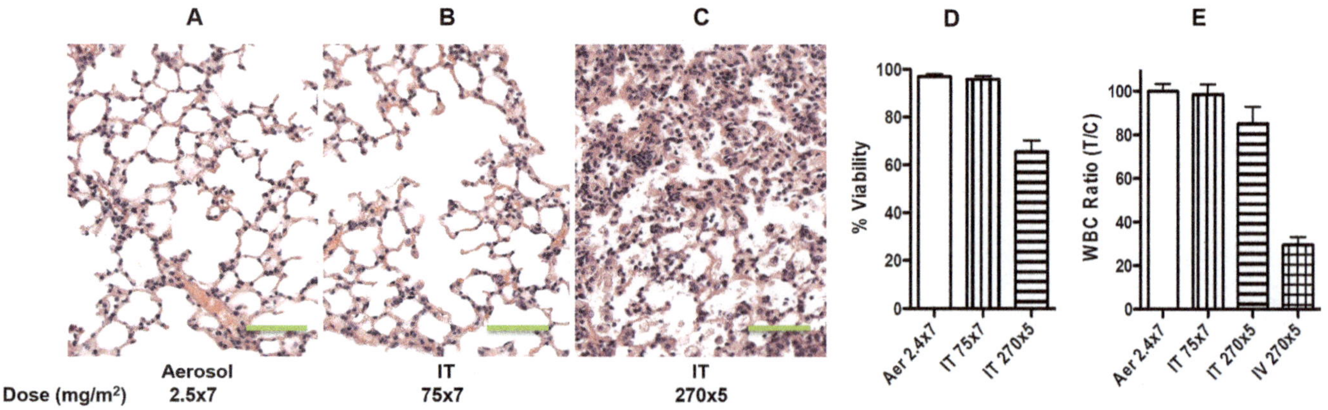

Figure 3. Toxicity of aerosol Aza. Lung sections of mice treated with aerosol Aza at 2.5 mg/m^2×7 (A), IT Aza at 75 mg/m^2×7 (B), and IT Aza at 270 mg/m^2×5 (C), respectively show no toxicity at 2.5 and 75 2.5 mg/m^2 but pneumonitis at 270 mg/m^2. The objective magnification was 40X. The scale bars were 50 µm. White blood cell count ratio (after treatment vs. before treatment, n = 6) (D) and percentage of viable cells of the sorted airway epithelial cells (n = 6) (E), respectively.

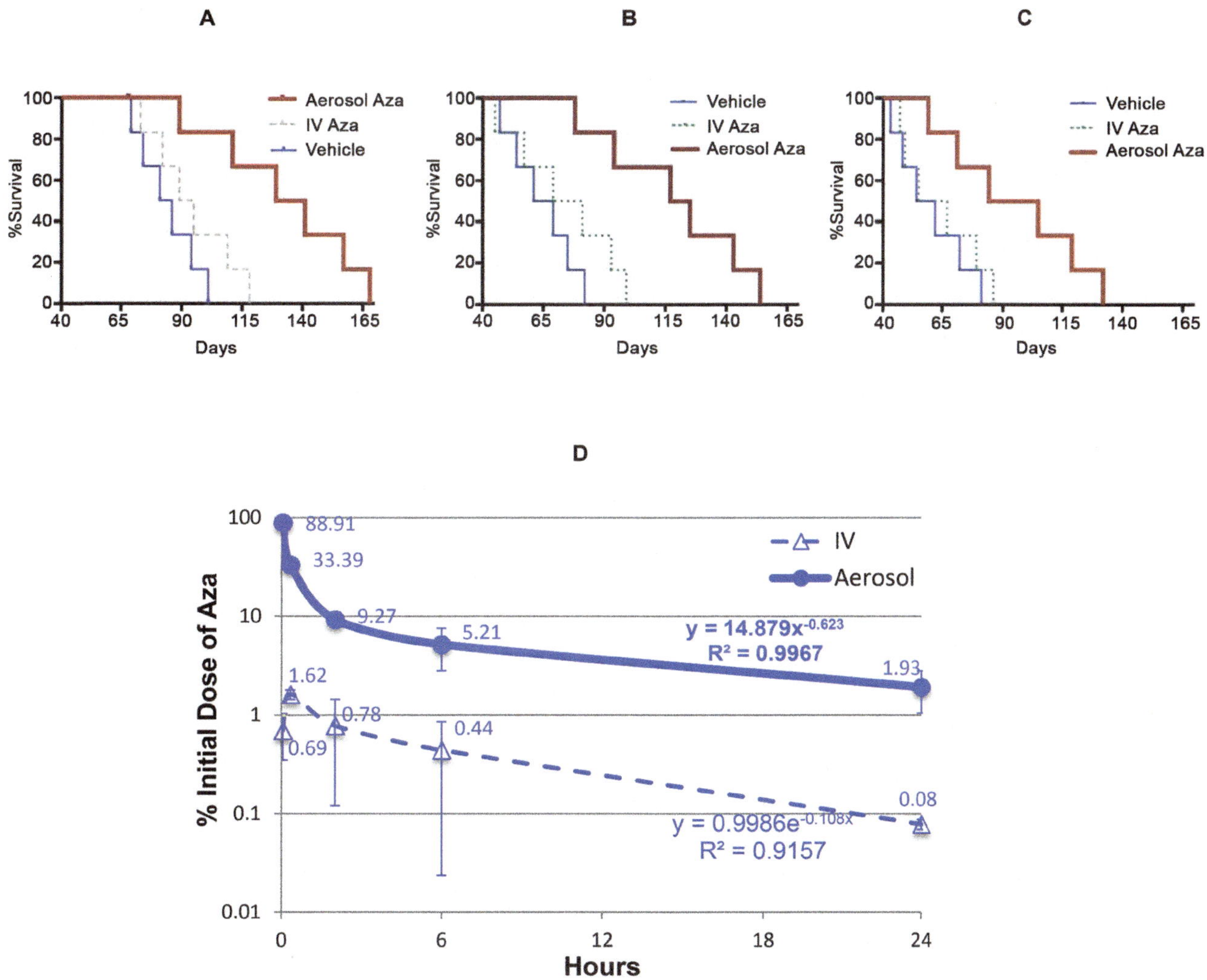

Figure 4. Aerosol administration of Aza significantly prolongs the survival of mice with orthotopic human NSCLC xenografts (A~C). Mice intratracheally inoculated with the cell lines H226 (A), H358 (B), or H460 (C) were treated with aerosol Aza (red thick line) at 2.5 mg/m^2 daily for 7 days or aerosol vehicle (blue thin line) with the same volume. Treatment with IV Aza (dash line) at the optimal dose of 75 mg/m^2 daily for 5 days was used as control. The statistical comparison of the survival curves was performed with Log-rank (Mantel-Cox) test. The p values for H266, H358, and H460 models were 0.0135, 0.0150, and 0.0719, respectively. Pre-pharmacokinetics of the aerosol Aza (D). ICR mice were treated with aerosol Aza at 2.5 mg/m^2. The data at each time point was the average ± standard deviation of the percent of given/initial dose from lungs of 3 mice. The equation for each curve was simulated curve with the best fitting (the highest R^2 value) under the Microsoft Excel program.

lated in our experimental mouse models of NSCLC treated with IV Aza, (Figure 4A–C). There is actually little experimental evidence supporting the use of a demethylation agent to effectively inhibit lung cancer in animal models [43] [32]. One of the possible reasons is that this very unstable drug was not efficiently delivered to the lung tumors. We were the first group to report a proof of concept study in which the administration of Aza directly into the airways effectively prolonged the survival of mice with orthotopic lung cancers [32]. In the current study, we are reporting the results using aerosol Aza for the treatment in human orthotopic NSCLC xenografts in mice. One of our goals was to demonstrate that aerosol delivery of Aza to the airway epithelium could effectively inhibit the growth of clinically relevant NSCLC models. In addition, these studies were further designed to investigate whether aerosol delivery of the demethylating agent could inhibit the growth of tumors within the airways at low demethylation doses rather than cytotoxic doses (Figure 4A–C).

Our results indicate that a low dose aerosol Aza significantly prolonged the lifespan of the mice bearing endobronchial tumors. The %ILS and median survival were prolonged by 5.1- and 1.5-, (p = 0.0135), 6.4- and 1.9- (p = 0.0150), and 8.9- and 1.6- (p = 0.0719) fold in the groups treated with aerosol Aza using a dose more than 20-fold lower than the dose used in the systemic treatment control group, in the H226, H358, and H460 model, respectively. These results clearly demonstrate that aerosol administration of Aza at demethylating, non-cytotoxic doses is superior to systemic administration of Aza in terms of tumor growth inhibitory efficacy.

Pre-pharmacokinetics

To determine the deposition of Aza in the lungs we conducted pre-pharmacokinetics studies in vivo. Intravenously injected Aza with the same dose was used a comparison. At the first time point (3 min, the earliest possible time point), we could recover ~89% of

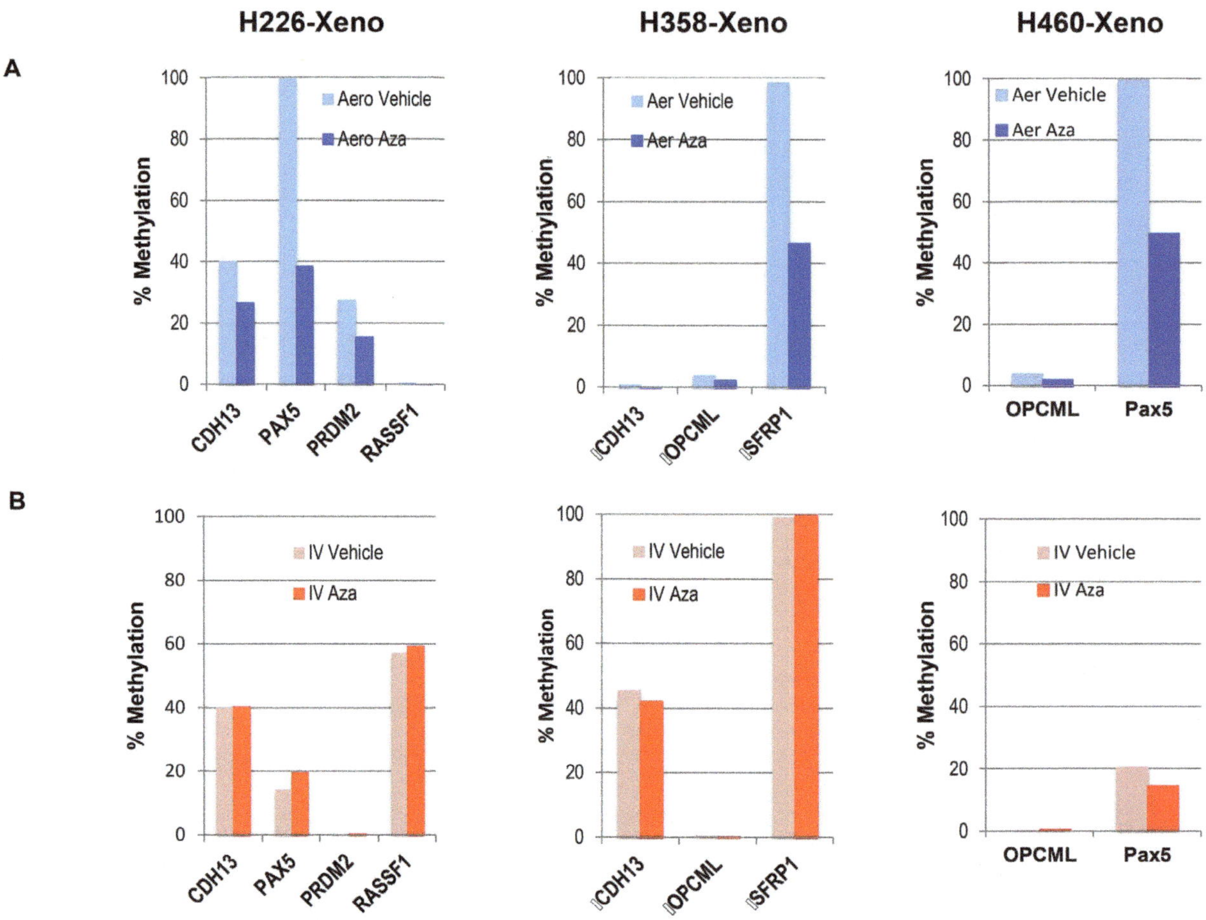

Figure 5. Methylation changes in the lung tumors. Tumor-bearing mice from 3 orthotopic lung cancer models (H226-xeno, H358-xeno, and H460-xeno) were treated with aerosol Aza (A) or IV Aza (B) at 2.5 mg/m^2 daily×7. The lungs were resected 14 days after the final treatment and tumor nodules larger than 1.5 mm were isolated for methylation detection by qPCR array technology at Qiagen (SA Biosciences). The data is %methylation of the total CG cytosine of the particular promoter region.

the initial aerosol dose from the lung. However, when the same dose was given by systemically injection, the peak concentration in the lungs was only 1.62% of the initial dose. The area under the concentration vs. time curve ($AUC_{0\sim24\ h}$) of aerosol Aza was 15-fold larger than that of IV Aza (Table 1). Aza given via both routes showed a similar decline pattern: a rapid decline within the first 2 h after the administration followed up by a slower decline. Aerosolized Aza showed a slower rate of decline in the slower phase while IV Aza showed a delayed peaking concentration in the lung (Figure 4D). These data demonstrate that the aerosolized Aza has higher tissue concentration in the lungs with predictable lower systemic exposure than the IV Aza.

Aerosol Aza demethylation

To analyze whether Aza has a demethylating effect in the clinically relevant NSCLC models, we harvested the lung tumors from the xenograft models treated with aerosol Aza or IV Aza and determined the methylation status of the promoter regions of 24 tumor suppressor genes, whose promoter hypermethylation has been reported as relevant in the lung cancer literature. Normal bronchial epithelial cells from animals without Aza treatment were

Table 1. Pre-pharmacokinetic parameters.

	T (min)	C_{max} (%ID)	$AUC_{0\sim24\ h}$ (%ID·h)
IV	20	1.62	9.45
Aerosol	3	88.9	146
Aero vs. IV	0.15	55	15

T is peaking time in minute. C_{max} is peak concentration of azacitidine presented as percent of given/initial dose (%ID). AUC is the area under the concentration vs. time curve from 0 to 24 h.

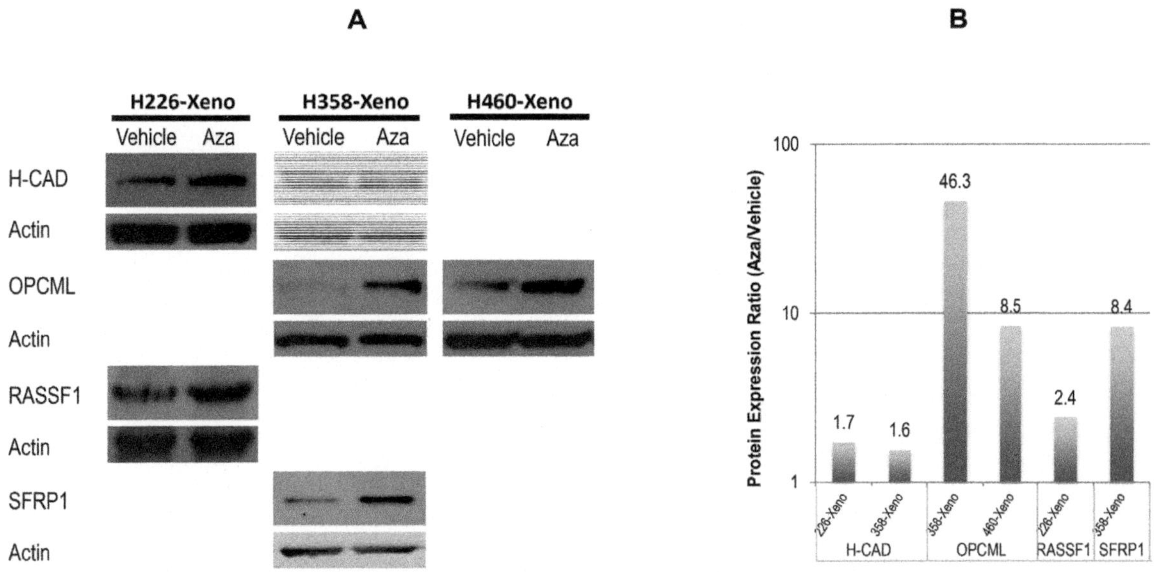

Figure 6. Protein expression of the TSGs in the lung tumors (A). Western blotting assay was used to determine the protein expression of the TSGs with significant promoter demethylation after aerosol Aza treatment identified by the methylation q-PCR array. The tumor samples were the same as used in Figure 5; Histogram of the Western blots (B). The western blot photo films were scanned and the density ratio of targeted protein vs. actin loading control of each sample was presented as the protein expression level.

used as methylation baseline control. The 24 TSGs selected are APBA1, APC, CADM1, CDH1, CDH13, CDKN1C, CDKN2A, CDKN2B, CXCL12, CYP1B1, DLC1, FHIT, MLH1, MTHFR, PAX5, PRDM2, RARB, RASSF1, RASSF2, SEMA3B, SFRP1, SLIT2, TCF21, and TGFB1. We found that 11 of the 24 promoters were hypermethylated in the tumor samples from 3 different xenograft models and that aerosol Aza could demethylate 5 of them (Figure 5A). In contract, IV Aza at the same dose did not show any demethylating effect (Figure 5B). The data shown in Figure 5 depict methylation level detected by a q-PCR methylation array and are presented as %methylation of the total CG cytosine of the particular promoter region.

TSG re-expression

In order to analyze whether the demethylation at the promoter region of TSGs can reactivate the TSGs, we resected the lung tumors and measured the protein expression of the TSGs whose promoters were significantly demethylated after the aerosol Aza treatment. Western blotting assay was used to detect the gene expression in the tumor samples from the same mice used in the demethylation study. We selected 5 TSGs whose promoters were significantly demethylated after the aerosol Aza treatment. We found that H-Cad and Rassf1 in the H266 tumors, H-cad, OPCML, and SFRP1 in H358 tumors, and OPCML in H460 tumors were reactivated at the protein level (Figure 6A and B) after the aerosol Aza treatment. Expression of these TSGs has been shown to inhibit or delay the cancer development in several types of cancers including lung cancer. For instance, we have reported before that loss of H-cadherin in human NSCLC is associated with tumorigenicity [44]. The promoter region of Rassf1a is also found heavily methylated in lung cancer [45]. OPCML, a broad tumor suppressor originally found in ovarian cancers, is found methylated in many lung cancer cell lines [46]. SFRP1 in the Wnt signaling pathway is transcriptionally silenced by promoter hypermethylation in NSCLC [47]. The results of this in vivo study imply that aerosol demethylating treatment may be effective in inhibiting endo-bronchial lung tumors and bronchial premalignant lesions.

Discussion

Lung cancer is viewed as a "genetic disease" [48,49]. Upon the demonstration that TSG deactivation and oncogene activation play a critical role in carcinogenesis, the therapeutic efforts have progressively shifted from optimizing the non-specific radiation and chemotherapy forms of therapy that mainly kill fast-dividing cells to develop agents that target the specific genetic changes. Recent evidence suggests a direct relationship between aberrant epigenetic changes and cancer development; it indicates that cancer is also in part an epigenetic disease. Unlike genetic changes, epigenetic changes occur earlier in cancer progression and are reversible. Therefore, a therapeutic strategy aiming at reversing aberrant epigenetic changes should be a more effective strategy than those focusing on the use of non-specific cytotoxic agents or on targeting irreversible genetic defects.

According to the field cancerization model [3] [50], all bronchial epithelial cells accumulate epigenetic and genetic lesions until one single cell acquires a malignant phenotype and gives rise to an invasive tumor, while the remaining of the field continues accumulating epigenetic and genetic damage and is at risk of developing other primary lung cancers (second primaries) [51] [52]. During this process, the aberrant epigenetic changes, such as hypermethylation of the promoters of TSGs, occur prior to and may cause the subsequent genetic changes, and they persist in tumor cells through the entire process of carcinogenesis.

On this basis, a strategy based on targeting the reversible aberrant epigenetic changes of the epithelial cells in the whole airway epithelium should be very effective in delaying or preventing primary and subsequent lung tumors. To test his hypothesis, we inoculated human lung cancer cell lines to mice by intratracheal injection. Tumor implantation within the airspace leads to tumor growth from the site where the cells are arrested, either large and/or small airways, and are thus more clinically relevant than other xenograft models implanted subcutaneously, injected intravascularly, or injected directly into the lung parenchyma. Therefore, it mimics more closely the whole airway field cancerization model and is also ideally suited to test the efficacy of aerosol treatments. The aerosolized Aza tested in our studies was effective in prolonging the lifespan of animals bearing tumors within the airways, thus strongly suggesting that this method has the potential to inhibit or prevent lung cancer tumors that arise in the bronchial epithelium of individuals exposed chronically to airborne carcinogens.

Our results also demonstrate that direct delivery of demethylating agents to the bronchial epithelium is a rational molecular strategy to reverse the aberrant hypermethylations associated with the lung carcinogenesis process. Using our orthotopic lung cancer mouse model we have showed that aerosol administration of Aza at non cytotoxic doses results in reactivation of several key TSGs through demethylating their hypomethylated promoters in lung tumor cells that are accessible by administration by aerosol. This constitutes experimental proof of principle that a single demethylation agent delivered by aerosol can inhibit orthotopic lung cancer at non-cytotoxic dose, and the results of pre-pharmacokinetic study demonstrated that the aerosolized drug deposits more efficiently in the lungs, and this is potentially one of the reasons for the higher efficacy in reactivating TSGs and antitumor activity with lower systemic toxicity. In addition, these effects were observed at a dose of Aza that is much lower dose than that used in the clinic and therefore without detectable local lung or systemic toxicity.

The lung carcinogenesis process involves very complicated epigenetic and genetic alterations affecting the entire genome. Aberrant methylation is one of the important epigenetic alterations that can lead to carcinogenesis. Based on current evidence, the majority of cancers have aberrant hypermethylation in the promoters of TSGs, although the whole genome may appear to be hypomethylated [53]. Our study was limited to 24 promoters, and the results could only identify methylation changes in a small number of the potential targets relevant to lung cancer. Ongoing efforts in our laboratory are focused on whole genome methylation screening using additional animal models to look at spontaneously occurring lung tumors as well as inhaled carcinogen-induced lung cancer. These studies are expected to yield much more valuable information on the biology of lung carcinogenesis, which should guide the development of improved therapeutic strategies.

Reed et al. have recently reported a similar study using aerosol Aza for the treatment of an orthotopic lung cancer model in nude rats [54]. They concluded that their aerosol and systemic deliveries are equally effective in terms of antitumor efficacy as well as gene demethylation [54]. Although the methodologies, tumor models, and endpoints used in Reed study and our study are significantly different, the fact that both studies lead to some similar conclusions using aerosol Aza represents strong proof of concept of the potential of aerosol demethylation therapy as a novel strategy for the management of lung cancer and bronchial premalignancy.

In our study, we found that the effective dose was much lower than the optimal IV dose, indicating that targeted aerosol delivery of the minimal effective dose rather than the maximum tolerated dose of an epigenetic agent is a potentially effective therapeutic strategy. In addition, our aerosol formulation showed a very good aerodynamic size range that makes it suitable for use in humans. We have recently completed preclinical toxicity studies of aerosol Vidaza in mice and FDA approved us to initiate a Phase I clinical study with aerosol Vidaza in patients with advanced lung cancer with disease mostly confined to the lung parenchyma that are not candidates for therapies of a higher priority. In this study, in addition to determining the toxicity profile of aerosol Vidaza, we will investigate by sequential bronchoscopy whether aerosol Vidaza can lead to TSG reexpression in the bronchial epithelium and or/tumor tissue. We expect that the results of this study will determine whether inhaled demethylation therapy should be further developed both as an early intervention or even prevention strategy. If that is the case, significant effort will have to be devoted to explore the use of other candidate agents as Aza is potentially carcinogenic [55] and therefore not an ideal candidate as a preventive agent.

Author Contributions

Conceived and designed the experiments: YZ RPS. Performed the experiments: XQ YL RS. Analyzed the data: XQ YZ. Contributed reagents/materials/analysis tools: XQ YZ. Wrote the paper: XQ YZ RPS.

References

1. Jemal A, Bray F, Center MM, Ferlay J, Ward E, et al. (2011) Global cancer statistics. CA Cancer J Clin 61: 69–90.
2. Siegel R, Naishadham D, Jemal A (2013) Cancer statistics, 2013. CA Cancer J Clin 63: 11–30.
3. Benner SE, Lippman SM, Hong WK (1995) Chemoprevention of lung cancer. Chest 107: 316S–321S.
4. Lippman SM, Lee JJ, Karp DD, Vokes EE, Benner SE, et al. (2001) Randomized phase III intergroup trial of isotretinoin to prevent second primary tumors in stage I non-small-cell lung cancer. J Natl Cancer Inst 93: 605–618.
5. Khuri FR, Cohen V (2004) Molecularly targeted approaches to the chemoprevention of lung cancer. Clinical Cancer Research 10: 4249s–4253s.
6. Wakelee H, Chhatwani L (2008) Adjuvant chemotherapy for resected non-small cell lung cancer. Seminars in Thoracic & Cardiovascular Surgery 20: 198–203.
7. Perez-Soler R (2009) Individualized therapy in non-small-cell lung cancer: future versus current clinical practice. Oncogene 28 Suppl 1: S38–45.
8. Baylin SB, Ohm JE (2006) Epigenetic gene silencing in cancer - a mechanism for early oncogenic pathway addiction? Nature Reviews Cancer. 6: 107–116.
9. Belinsky SA (2004) Gene-promoter hypermethylation as a biomarker in lung cancer. Nature Reviews Cancer. 4: 707–717.
10. Sawan C, Vaissiere T, Murr R, Herceg Z (2008) Epigenetic drivers and genetic passengers on the road to cancer. Mutation Research 642: 1–13.
11. Cecener G, Tunca B, Egeli U, Karadag M, Vatan O, et al. (2008) Mutation analysis of the FHIT gene in bronchoscopic specimens from patients with suspected lung cancer. Tumori 94: 845–848.
12. Seng TJ, Currey N, Cooper WA, Lee CS, Chan C, et al. (2008) DLEC1 and MLH1 promoter methylation are associated with poor prognosis in non-small cell lung carcinoma. British Journal of Cancer 99: 375–382.
13. Zochbauer-Muller S, Minna JD, Gazdar AF (2002) Aberrant DNA methylation in lung cancer: biological and clinical implications. Oncologist 7: 451–457.
14. Katayama H, Hiraki A, Fujiwara K, Matsuo K, Maeda T, et al. (2007) Aberrant promoter methylation profile in pleural fluid DNA and clinicopathological factors in patients with non-small cell lung cancer. Asian Pacific Journal of Cancer Prevention: Apjcp 8: 221–224.
15. Georgiou E, Valeri R, Tzimagiorgis G, Anzel J, Krikelis D, et al. (2007) Aberrant p16 promoter methylation among Greek lung cancer patients and smokers: correlation with smoking. European Journal of Cancer Prevention 16: 396–402.
16. Kaye FJ (2002) RB and cyclin dependent kinase pathways: defining a distinction between RB and p16 loss in lung cancer. Oncogene 21: 6908–6914.
17. Miyamoto K, Ushijima T (2005) Diagnostic and therapeutic applications of epigenetics. Japanese Journal of Clinical Oncology 35: 293–301.
18. Safar AM, Spencer H, Su X, Cooney CA, Shwaiki A, et al. (2007) Promoter hypermethylation for molecular nodal staging in non-small cell lung cancer. Archives of Pathology & Laboratory Medicine 131: 936–941.
19. Wang G, Hu X, Lu C, Su C, Luo S, et al. (2008) Promoter-hypermethylation associated defective expression of E-cadherin in primary non-small cell lung cancer. Lung Cancer 62: 162–172.
20. Fischer JR, Ohnmacht U, Rieger N, Zemaitis M, Stoffregen C, et al. (2006) Promoter methylation of RASSF1A, RARbeta and DAPK predict poor prognosis of patients with malignant mesothelioma. Lung Cancer 54: 109–116.
21. Kim JS, Kim JW, Han J, Shim YM, Park J, et al. (2006) Cohypermethylation of p16 and FHIT promoters as a prognostic factor of recurrence in surgically resected stage I non-small cell lung cancer. Cancer Res 66: 4049–4054.

22. Koga T, Takeshita M, Yano T, Maehara Y, Sueishi K (2011) CHFR hypermethylation and EGFR mutation are mutually exclusive and exhibit contrastive clinical backgrounds and outcomes in non-small cell lung cancer. International Journal of Cancer 128: 1009–1017.
23. Kontic M, Stojsic J, Jovanovic D, Bunjevacki V, Ognjanovic S, et al. (2012) Aberrant promoter methylation of CDH13 and MGMT genes is associated with clinicopathologic characteristics of primary non-small-cell lung carcinoma. Clin Lung Cancer 13: 297–303.
24. Lee SM, Lee WK, Kim DS, Park JY (2012) Quantitative promoter hypermethylation analysis of RASSF1A in lung cancer: comparison with methylation-specific PCR technique and clinical significance. Mol Med Report 5: 239–244.
25. Lin RK, Hsu HS, Chang JW, Chen CY, Chen JT, et al. (2007) Alteration of DNA methyltransferases contributes to 5′CpG methylation and poor prognosis in lung cancer. Lung Cancer 55: 205–213.
26. Bai X, Song Z, Fu Y, Yu Z, Zhao L, et al. (2012) Clinicopathological significance and prognostic value of DNA methyltransferase 1, 3a, and 3b expressions in sporadic epithelial ovarian cancer. PLoS One 7: e40024.
27. Lichter P (2008) The role of epigenetics in carcinogenesis. International Journal of Cancer 123: ix.
28. Wu J-Y, Wang J, Lai J-C, Cheng Y-W, Yeh K-T, et al. (2008) Association of O6-methylguanine-DNA methyltransferase (MGMT) promoter methylation with p53 mutation occurrence in non-small cell lung cancer with different histology, gender, and smoking status. Annals of Surgical Oncology 15: 3272–3277.
29. Ben-Avraham D, Muzumdar RH, Atzmon G (2012) Epigenetic genome-wide association methylation in aging and longevity. Epigenomics 4: 503–509.
30. Mufti G, List AF, Gore SD, Ho AY (2003) Myelodysplastic syndrome. Hematology Am Soc Hematol Educ Program: 176–199.
31. Mahesh SLY, Perez-Soler R, Zou Y (2008) Favorable toxicity profile of 5-azacytidine (5-aza) administered intratracheally (IT). Proceedings of the 99th Annual Meeting of the American Association for Cancer Research 99: 666.
32. Mahesh S, Saxena A, Qiu X, Perez-Soler R, Zou Y (2010) Intratracheally administered 5-azacytidine is effective against orthotopic human lung cancer xenograft models and devoid of important systemic toxicity. Clin Lung Cancer 11: 405–411.
33. Zou Y, Zong G, Ling YH, Hao MM, Lozano G, et al. (1998) Effective treatment of early endobronchial cancer with regional administration of liposome-p53 complexes. J Natl Cancer Inst 90: 1130–1137.
34. Zou Y, Tornos C, Qiu X, Lia M, Perez-Soler R (2007) p53 aerosol formulation with low toxicity and high efficiency for early lung cancer treatment. Clin Cancer Res 13: 4900–4908.
35. Keating GM (2012) Azacitidine: a review of its use in the management of myelodysplastic syndromes/acute myeloid leukaemia. Drugs 72: 1111–1136.
36. Zhao M, Rudek MA, He P, Hartke C, Gore S, et al. (2004) Quantification of 5-azacytidine in plasma by electrospray tandem mass spectrometry coupled with high-performance liquid chromatography. J Chromatogr B Analyt Technol Biomed Life Sci 813: 81–88.
37. Zou Y, Ling YH, Van NT, Priebe W, Perez-Soler R (1994) Antitumor activity of free and liposome-entrapped annamycin, a lipophilic anthracycline antibiotic with non-cross-resistance properties. Cancer Research 54: 1479–1484.
38. Jaspers I, Horvath KM, Zhang W, Brighton LE, Carson JL, et al. (2010) Reduced expression of IRF7 in nasal epithelial cells from smokers after infection with influenza. Am J Respir Cell Mol Biol 43: 368–375.
39. Bates DV, Fish BR, Hatch TF, Mercer TT, Morrow PE (1966) Deposition and retention models for internal dosimetry of the human respiratory tract. Task group on lung dynamics. Health Phys 12: 173–207.
40. Davies DS (1979) Pharmacokinetics of inhaled substances. Scand J Respir Dis Suppl 103: 44–49.
41. Digel W, Lubbert M (2005) DNA methylation disturbances as novel therapeutic target in lung cancer: preclinical and clinical results. Crit Rev Oncol Hematol 55: 1–11.
42. Bauman J, Verschraegen C, Belinsky S, Muller C, Rutledge T, et al. (2012) A phase I study of 5-azacytidine and erlotinib in advanced solid tumor malignancies. Cancer Chemother Pharmacol 69: 547–554.
43. Belinsky SA, Grimes MJ, Picchi MA, Mitchell HD, Stidley CA, et al. (2011) Combination therapy with vidaza and entinostat suppresses tumor growth and reprograms the epigenome in an orthotopic lung cancer model.
44. Zhong Y, Delgado Y, Gomez J, Lee SW, Perez-Soler R (2001) Loss of H-cadherin protein expression in human non-small cell lung cancer is associated with tumorigenicity. Clin Cancer Res 7: 1683–1687.
45. Donninger H, Vos MD, Clark GJ (2007) The RASSF1A tumor suppressor. J Cell Sci 120: 3163–3172.
46. Cui Y, Ying Y, van Hasselt A, Ng KM, Yu J, et al. (2008) OPCML is a broad tumor suppressor for multiple carcinomas and lymphomas with frequently epigenetic inactivation. PLoS One 3: e2990.
47. Fukui T, Kondo M, Ito G, Maeda O, Sato N, et al. (2005) Transcriptional silencing of secreted frizzled related protein 1 (SFRP 1) by promoter hypermethylation in non-small-cell lung cancer. Oncogene 24: 6323–6327.
48. Vogelstein B, Kinzler KW (2004) Cancer genes and the pathways they control. Nat Med 10: 789–799.
49. Sharma S, Kelly TK, Jones PA (2010) Epigenetics in cancer. Carcinogenesis 31: 27–36.
50. Spira A, Beane J, Shah V, Liu G, Schembri F, et al. (2004) Effects of cigarette smoke on the human airway epithelial cell transcriptome. Proc Natl Acad Sci U S A 101: 10143–10148.
51. Guo M, House MG, Hooker C, Han Y, Heath E, et al. (2004) Promoter hypermethylation of resected bronchial margins: a field defect of changes? Clin Cancer Res 10: 5131–5136.
52. Miyazu YM, Miyazawa T, Hiyama K, Kurimoto N, Iwamoto Y, et al. (2005) Telomerase expression in noncancerous bronchial epithelia is a possible marker of early development of lung cancer. Cancer Res 65: 9623–9627.
53. Daskalos A, Logotheti S, Markopoulou S, Xinarianos G, Gosney JR, et al. (2011) Global DNA hypomethylation-induced DeltaNp73 transcriptional activation in non-small cell lung cancer. Cancer Lett 300: 79–86.
54. Reed MD, Tellez CS, Grimes MJ, Picchi MA, Tessema M, et al. (2013) Aerosolised 5-azacytidine suppresses tumour growth and reprogrammes the epigenome in an orthotopic lung cancer model. Br J Cancer 109: 1775–1781.
55. Cavaliere A, Bufalari A, Vitali R (1987) 5-Azacytidine carcinogenesis in BALB/c mice. Cancer Letters 37: 51–58.

10

Compound A398, a Novel Podophyllotoxin Analogue: Cytotoxicity and Induction of Apoptosis in Human Leukemia Cells

Alethéia L. Silveira[1], Glaúcia V. Faheina-Martins[1,2], Raquel C. Maia[3], Demetrius A. M. Araújo[1]*

1 Departamento de Biotecnologia, Universidade Federal da Paraíba, João Pessoa, Paraíba, Brasil, **2** Departamento de Biologia Molecular, Universidade Federal da Paraíba, João Pessoa, Paraíba, Brasil, **3** Laboratório de Hemato-Oncologia Celular e Molecular, Instituto Nacional de Câncer (INCA), Rio de Janeiro, Rio de Janeiro, Brasil

Abstract

Despite advances in oncology research, cancer is one of the leading causes of death worldwide. Thus, there is a demand for the development of more selective and effective antitumor agents. This study showed that A398, a novel podophyllotoxin analogue, was cytotoxic to the HT-29, MCF-7, MOLT-4 and HL-60 tumor cell lines, being less active in human peripheral blood mononuclear cells and normal cell lines FGH and IEC-6. Tests using the HepG2 lineage indicated that its metabolites do not contribute to its cytotoxicity. In the HL-60 cells, A398 induced apoptosis in a time and concentration-dependent manner, promoting mitochondrial depolarization, inhibition of Bcl-2, phosphatidylserine exposure, activation of caspases -8, -9 and -3, and DNA fragmentation. The production of reactive oxygen species does not seem to be a crucial event for the apoptotic process. Pretreatment with specific inhibitors of kinases ERK1/2, JNK and p38 resulted in an increased percentage of death induced by A398. These results indicate that the compound induced apoptosis through activation of intrinsic and extrinsic death pathways with the mechanism involving the inhibition of the MAPKs and Bcl-2. Taken together, our findings suggest that A398 has an anticancer potential, proving itself to be a candidate for preclinical studies.

Editor: Arun Rishi, Wayne State University, United States of America

Funding: The authors thank Conselho Nacional de Desenvolvimento Científico e Tecnológico (CNPq) and Coordenação de Aperfeiçoamento de Pessoal de Nível Superior (CAPES) for their financial support in the form of grants and fellowship awards. DAMA, ALS, GVFM, and RCM are CNPq fellows. The funders had no role in study design, data collection and analysis, decision to publish, or preparation of the manuscript.

Competing Interests: The authors have declared that no competing interests exist.

* Email: demetrius@cbiotec.ufpb.br

Introduction

Existing schemes of therapeutic intervention for cancer treatment depend on the ability to induce tumor cell death by apoptosis [1,2]. This consists of a highly regulated, energy-dependent cell program, which plays an important role in many biological processes such as growth, proliferation, differentiation, immunity, removal of unwanted cells and the maintenance of homeostasis of the tissues [3,4,5].

The cascade of the apoptosis process may be mediated through two main pathways: the extrinsic and the intrinsic [6].The extrinsic pathway is initiated by the interaction of death ligands to their respective receptors present on the surface of the plasma membrane and results in activation of caspases -8 and -10 [7,8]. The intrinsic or mitochondrial pathway is triggered by a variety of stimuli such as ultraviolet light, reactive oxygen species, the action of cytotoxic agents and DNA damage [9,10]. The changes that occur in the inner mitochondrial membrane promote the release of several pro-apoptotic proteins into the cytosol, such as cytochrome C, Smac/DIABLO (Second mitochondrial activator of caspases/direct IAP binding protein with low PI) and endonuclease G. After being released from the mitochondria, cytochrome C binds to the adapter protein Apaf-1 (Apoptotic protease activating factor) to form the apoptosome, with activation of caspase -9 [11,12,13].

The extrinsic and intrinsic pathways converge at the phase of apoptosis execution, which is mediated mainly by effector caspases -3, -6 and -7 [14,15]. The cleavage of specific cellular substrates results in biochemical and morphological changes, such as: cell shrinkage, membrane blebbing, phosphatidylserine exposure, chromatin condensation and nuclear fragmentation [16,17]. Due to the physiological and pathological importance of cell death by apoptosis, this process is strictly regulated by several proteins, such as p53, members of the Bcl-2 family (B-cell lymphoma protein 2) and MAPKs (Mitogen-activated protein kinases) [18,19,20].

Natural products have been a rich source of agents of value to medicine. More than half of currently available drugs are natural compounds or are related to them, and in the case of cancer this proportion surpasses 60% [21,22]. Podophyllotoxin is a natural cyclolignan isolated from various plant species within the *Podophyllum* family. The cytotoxic effects of podophyllotoxin in different types of tumors have been attributed to its ability to bind to tubulin during mitosis and thus inhibit microtubule assembly. Unfortunately, its high toxicity has limited its application as a chemotherapeutic [23,24]. Continuous efforts towards the synthesis of its analogues led to the discovery of new anticancer drugs. For example its semisynthetic derivatives, etoposide and teniposide are currently used in the clinic for the treatment of a variety of malignancies. Due to the drug resistance developed by cancer cells as well as side effects associated with the use of these agents in clinic, the search for new effective anticancer analogues of podophyllotoxin remains an intense area of research [25,26,27]. In this study we evaluated the anticancer potential of a new podophyllotoxin analogue called (2S,6S,7R)-4-(4-fluoro-3-nitrophenyl)-7-methyl-12,14-dioxo-4-azatetracycle [7.7.0.02,6.011,15] hexadec-1(16),9,11(15)-triene-2,5-dione, which was coded as A398 (patent pending number PI1102759-2, deposited at the National Institute of Industrial Property, Brazil).

Materials and Methods

Ethics Statement

This study was approved by the Institutional Ethical Committee of Lauro Wanderley Hospital from Federal University of Paraíba, number 05878712.7.0000.5183. We used Human peripheral blood mononuclear cells (PBMC), which were isolated from samples of peripheral blood from healthy non-smoking donors who had not taken any medication for at least 15 days before donation, aged 18–30 years. As described in such approved document, these blood samples were provided by the Blood Center of Paraíba (Hemocentro, João Pessoa, Paraíba, Brazil) with the permission of blood donors and we also mention that we just used the blood samples that would be discarded by the Hemocentro.

Chemicals and antibodies

Propidium iodide (PI), 3-(4,5-dimethyl-thiazol-2-yl)-2,5-diphenyl-terazolium bromide (MTT), SB 203580, SP 600125, PD 98059, etoposide, tetramethyl rhodamine methyl ester (TMRM), phytohemagglutinine and carbonyl cyanide m-chlorophenyl hydrazone (CCCP) were purchased from Sigma (St. Louis, MO, USA). Alexa Fluor 488 and 2′,7′-dichlorodihydrofluorescein diacetate (H_2DCFH-DA) were obtained from Invitrogen (Carlsbad, CA, USA). Anti-Bcl-2 and β-actin were purchased from Santa Cruz Biotechology (Santa Cruz, CA, USA). All organic solvents and other chemicals were of analytical grade or complied with the standards needed for cell culture experiments.

Cell cultures

HL-60 (human acute promyelocytic leukemia), MOLT-4 (human acute lymphoblastic leukemia), MCF-7 (human breast cancer), HT-29 (human colon cancer), HepG2 (human hepatoma), FGH (human gingival fibroblasts) and IEC-6 (rat intestinal epithelial) cell lines were obtained from the Rio de Janeiro Cell Bank (Brazil). Cells were culture in RPMI 1640 medium (for the HL-60, MOLT-4, MCF-7 and IEC-6 cells) or DMEM medium (for the HT-29, HepG2 and FGH lineages), supplemented with 10% FBS and antibiotics (100 UI/mL of penicillin and 100 μg/mL of streptomycin). Alls cells were maintained in a 5% CO_2 humidified incubator at 37°C.

PBMC cells were isolated from heparin-anticoagulated blood of healthy volunteers by centrifugation with Ficoll-Paque Plus (Amersham Biosciences). PBMCs were also stimulated with phytohemagglutinine (PHA, 2 μg/ml) for 24 h before experimental treatment and were culture in RPMI 1640 medium.

Cytotoxicity assay

Cell viability was evaluated by measuring the mitochondrial-dependent reduction of MTT to a colored blue formazan. Briefly, cells was added to 96-well plates and incubated with increasing concentrations of compound A398 and etoposide for 24 hours. After the time exposure, MTT (5 mg/mL) was added and the plates were incubated at 37°C for 4 h. Formazan crystals were solubilized by the addition of 10% SDS in 0.01 N HCl overnight and the absorbance was measured at 570 nm on a microplate reader (Biotek Instruments EL800, USA). The cytotoxicity was expressed as an IC_{50} (concentration that inhibits 50% of the growth of cells), which was determined from the concentration-response curve.

Evaluation of apoptosis

The extent of apoptosis was quantified using annexinV/PI dual staining. HL-60 cells (5×10^5 cells/well) were exposed to compound A398 (4, 6 or 8 μM) or etoposide (5 μM) for 1 h, 3 h and 6 h. After the treatment periods, the cells were centrifuged (1.500 rpm, 5 minutes) and stained with 1 μl of annexin V-alexa Fluor 488 and 2 μl of 50 μg/ml PI in 100 μl binding buffer [10 mM HEPES (pH 7.4), 140 mM NaOH and 2.5 mM $CaCl_2$] for 15 min at room temperature in the dark. Quantification of apoptotic cells was performed by flow cytometry using a FASC Calibur (BD Biosciences, USA).

Determination of mitochondrial membrane potential ($\Delta\Psi m$)

Mitochondrial membrane potential was measured using TMRM, a cationic fluorochrome that rapidly accumulates in the mitochondria of living cells. HL-60 cells (5×10^5 cells/well) were treated with A398 (4, 6 or 8 μM) or with 5 μM of etoposide. As a positive control, cells were treated with 50 μM of the protonophore CCCP for 15 min. After 1 h, 3 h, and 6 h of exposure, the samples were centrifuged (1.500 rpm, 5 minutes) and stained with 150 nM of TMRM for 20 minutes in the dark at 37°C. The cells were then washed with PBS and analyzed by flow cytometry.

Estimation of caspases -8, -9, and -3 activity

HL-60 cells (2×10^6 cells/well) were incubated with A398 (6 μM) or etoposide (5 μM) for 1 h, 3 h, 6 h and 12 h. The activities of the caspases were carried out using colorimetric protease assay (Invitrogen, USA) following the manufacturer's protocol. Each kit contains a specific substrate: IETD (Ile-Glu-Thr-Asp), LEHD (Leu-Glu-His-Asp) and Ac-DEVD (acetyl-Asp-Glu-Val-Asp) for caspases -8, -9 and -3, respectively. Such substrates are labeled to the chromophore *p*-nitroanilide (*p*NA), which is released when they are cleaved by activated caspases and measured at 405 nm in a spectrophotometer (Biotek Instruments EL800, USA).

Cell cycle analysis and DNA content

Propidium iodide is a commonly used dye to quantitatively assess DNA content in cell cycle analysis. HL-60 cells (5×10^5 cells/well) were incubated with A398 (2, 4 or 6 μM) or etoposide (5 μM) for 1 h, 3 h and 6 h. The samples were centrifuged (1.500 rpm, 5 minutes) and resuspended in a hypotonic solution containing 0.1% sodium citrate, 0.1% triton X-100, RNase (10 μg/mL) and PI (50 μg/mL). The mixture was incubated in the dark at 4°C for 30 min and was then acquired in a FACS Calibur (BD Biosciences, USA).

Analysis of DNA fragmentation

DNA fragmentation was qualitatively analyzed by gel electrophoresis. HL-60 cells (2×10^6 cells/well) were treated with A398 (6 μM) or etoposide (5 μM). After 3 h, 6 h and 12 h, the cells were centrifuged (1,500 rpm, 5 minutes), washed with PBS and lysed with 250 μL of lysis buffer (10 mM EDTA, 50 mM Tris-HCL, 0.5% SDS) for 15 min at 55°C. Then proteinase K was added (500 μg/mL) and the samples were incubated for 1 h at 55°C. RNAse (1 μg/μL); was then added and the incubation continued for 90 more minutes at 55°C. The DNA was extracted with 250 μL of a solution containing phenol, chloroform and isoamyl alcohol (25:24:1) and precipitated with 0.1 volume of NaCl (2 M) and 2.5 volume of absolute ethanol. The precipitate was centrifuged at 12.000 rpm for 10 minutes and dissolved in a solution of Tris (10 mM) and EDTA (1 mM). The samples were separated on agarose gel (1.5%), stained with ethidium bromide (0.5 μg/mL), visualized under ultraviolet light and photographed.

Participation of MAPKs

To verify the involvement of MAPKs in the apoptosis process, HL-60 cells (5×10^5 cells/well) were pretreated with p38 inhibitor-SB 203580 (5 μM), JNK inhibitor-SP 600125 (10 μM) or ERK inhibitor-PD 98059 (10 μM) for 1 h and then incubated with derivative A398 (6 μM) for 6 h. The induction of apoptosis was determined by analyzing the sub-G_1 DNA damage using a FASC Calibur.

Measurement of intracellular oxygen species (ROS)

The intracellular ROS was estimated by fluorescent probe H_2-DCF-DA. HL-60 cells (5×10^5 cells/well) were treated with A398 (4, 6 or 8 μM) or with 5 μM of etoposide. Hydrogen peroxide (50 μM) was used as a positive control. After 1 h, 3 h and 6 h of exposure, the samples were centrifuged (1.500 rpm, 5 min), washed with PBS at 37°C and labeled with 10 μM of H_2-DCFH-DA for 30 min., in the dark, at 37°C. The cells were then washed with PBS and analyzed by flow cytometry.

Western blot analysis

HL-60 cells (2×10^6 cells/mL) were treated with 6 μM of compound A398 for 1 h, 3 h, 6 h and 12 h. Protein concentration was determined using Lowry assay and samples containing 50 μg of proteins were loaded and separated by sodium dodecyl sulphate-polyacrylaminde gel electrophoresis (10%). Protein bands were transferred to nitrocellulose membranes (Hybond-ECL - Amersham Biosciences, USA), blocked with 5% fat free dry milk and then incubated with specified primary antibodies (anti-Bcl2). After washing and incubation with an appropriate horseradish peroxidase-conjugated secondary antibody, the antigen-antibody complexes were visualized by enhanced chemiluminescence (ECL, Amersham Biosciences) using the manufacturer's protocol.

Statistical analysis

Data from the experiments were obtained from at least three independent tests performed in triplicate. Data were expressed as mean ± standard error of the mean (SE) and analyzed using Analysis of Variance (ANOVA) followed by the Newman-Keuls test, and were considered significant at $p<0.05$. The statistical software used was GraphPad Prism (GraphPad Software Incorporated, USA). In the cytometry experiments, 10.000 events were collected for each sample and data were analyzed using Summit software (Dako, USA).

Results

A398 is cytotoxic in human tumor lineages

The cytotoxic potential of A398 (Figure 1) was evaluated on several human tumor cell lines as well as normal FGH and IEC-6 cell lines and PBMC cells. As shown in table 1, A398 displayed broad and potent cytotoxicity on the tested human cancer cell lines but relatively minor cytotoxicity to normal cells. HL-60 cells were relatively more sensitive to compound A398 and were therefore selected for further analysis.

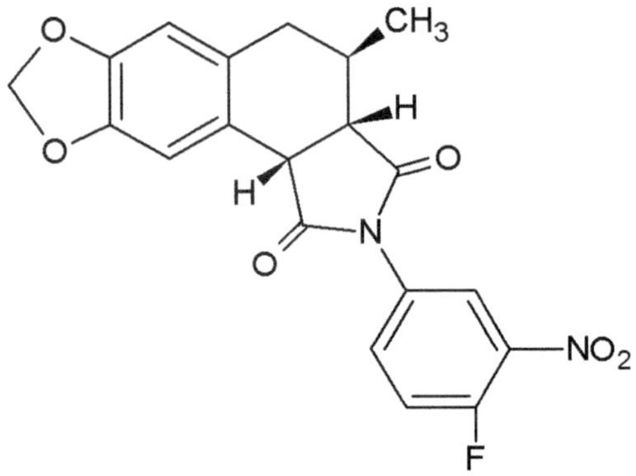

Figure 1. Chemical structure of A398.

A398 metabolites do not seem to contribute to its cytotoxic effect

To examine whether derivative A398 produces more toxic metabolites and to determine which enzymes are involved in their metabolism, the cytotoxicity of this compound was evaluated on the HepG2 hepatocytes lineage, using the MTT assay, with the absence and presence of dexamethasone (CYP3A inductor), ketoconazole (CYP3A inhibitor) or borneol (an inhibitor of UDP-glucuronosyltransferases - UGTs). Co-incubation with ketoconazole or with borneol promoted an increase in the cytotoxic effect of A398 while pretreatment with dexamethasone decreased the cytotoxicity of this compound. These results suggest that reactions involving the enzymes CYP3A and UGTs favor the detoxification of the compound (Table 2).

Apoptosis induced by A398 involves changes in the mitochondrial transmembrane potential (ΔΨm)

In order to determine whether A398 has an apoptosis-inducing activity on HL-60 cell line, dual staining with annexin V/PI was performed. Death promoted by A398 was time and concentration-dependent. The percentage of apoptotic cells significantly increased with 3 h of treatment and approximately 49.10% of the population was undergoing apoptosis after 6 h. Etoposide did not induce apoptosis until 6 h after the treatment (Figure 2A and Figure S1 in File S1).

To examine whether there is mitochondrial pathway involvement in apoptosis induced by A398, changes in ΔΨm were evaluated using the fluorochrome TMRM. The results showed that A398 promoted an increase in mitochondrial membrane depolarization in a time and concentration-dependent manner. A significant increase in the percentage of depolarized cells was detected with 1 h of treatment and reached 57.11% after 6 h. Alterations in ΔΨm induced by etoposide were detected at 6 h of incubation (Figure 2B and Figure S2 in File S1). Given that Bcl-2 is an anti-apoptotic protein that regulates mitochondrial outer membrane permeabilization, the effect of A398 on the expression of this protein was investigated using western blot. As shown in Figures 2C and 2D, the compound almost completely inhibited the expression of Bcl-2 at 1 h of incubation, which suggests that this protein is involved in regulating the intrinsic pathway of death.

Cell death promoted by A398 is mediated by activation of caspases

Because of the importance of caspases as executors of the apoptotic process, the participation of caspases -8, -9 and -3 was evaluated using specific fluorogenic substrates. As shown in

Table 1. Cytotoxicity of A398 and etoposide against several tumor cell lines (MCF-7, HT-29, MOLT-4 and HL-60) and normal cells (FGH, IEC-6 and PBMC).

Compound	IC_{50} (µM)						
	MCF-7	HT-29	MOLT-4	HL-60	FGH	PBMC	IEC-6
A398	41.4±1.6	37.9±2.3	7.3±1.8	6.3±0.8	241.2±15.8	75.3±2.6	127.4±3.8
Etoposide	42.3±8.1	72.9±8.6	0.8±0.3	2.2±1.4	183.1±18.5	102.6±2.4	-

Cells were treated with increasing concentrations of the substances for 24 h and cell viability was evaluated by MTT assay. Each IC_{50} value represented mean ± SE of three independent experiments.

Figure 3A, A398 promoted the significant activation of caspases -8, -9 and -3 after 3 hours of treatment and the maximum activity was reached at 6 hours of exposure. Etoposide induced a significant activation of caspases -8 and -3 after 6 h of treatment, while an increase in the activity of caspase -9 was only observed at 12 h.

Effects of A398 on the cell cycle and DNA content

To assess whether the induction of apoptosis by A398 is associated with changes in the distribution profile of cells in some stages of the cell cycle, we used the dye propidium iodide. In all times tested, A398 did not promote cell cycle arrest. However, the appearance of cells with hypodiploid content (sub-G_1), a characteristic of apoptotic cells with condensed nuclei and fragmented DNA, was time and concentration-dependent. The results show that the cytotoxic effect of A398 is not phase-specific, unlike etoposide, which promoted an accumulation of cells in S phase after 3 h and 6 h of treatment (Figure 3B and Figure S3 in File S1).

A398 induces damage to DNA

The degradation of DNA into fragments of 180–200 base pairs is a striking characteristic of death by apoptosis. Figure 3C shows that A398 promoted typical internucleosomal DNA fragmentation resulting from the induction of apoptosis or "ladder" formation all time points tested. The fluorescence of the bands was weak in the period of three hours, where maximum intensity was reached at a period of 6 h, with a reduction after 12 hours of exposure. Etoposide, when used in a concentration similar to that of compound A398, only induced DNA fragmentation in the period of 12 hours of treatment.

Effect of MAPKs inhibitors on apoptosis promoted by A398

MAPK pathways can mediate signals that either promote or suppress the growth of malignant hematopoietic cells. In view of evidence that p38, ERK, and JNK play a critical role in cell fate, the effect of A398 on these kinases was examined. Pretreatment with the inhibitors SB203580, PD98059 and SP600125 resulted in an increase in the percentage of death induced by A398 of 10.5, 14.4 and 22.8%, respectively (Figure 4A). This result suggests that A398 inhibits kinases p38, ERK1/2 and JNK and this effect certainly contributes at least in part, to the sensitivity of these leukemia cells to follow the apoptotic pathway.

Table 2. Effect of inhibitors and inducers of cytochrome P450 isoenzymes on the cytotoxicity of A398.

Compound	IC_{50} (µM)
A398	226.6±11.8
A398 + Dexamethasone	>250
A398 + ketoconazole	133.3±13.5
A398 + Borneol	170.1±7.3

Cell viability was assessed by the MTT method after 24 h of exposure. The treatment of HepG2 cells with A398 was done in the absence or presence of dexamethasone (inductor CYP3A), ketoconazole (inhibitor CYP3A) or borneol (inhibitor UGTs). In the latter condition, cells were pretreated with 10 µM of dexamethasone (for 24 h) or co-incubated with 15 µM ketoconazole or with 500 µM borneol. Each IC50 value represented mean ± SE of three independent experiments.

Reactive oxygen species are not crucial for A398 to induce death

Given that oxidative stress may participate in the process of apoptosis, we assessed whether derivative A398 promotes the generation of ROS using the probe H_2-DCFH-DA. Derivative A398 elevated ROS production in a time and concentration-dependent manner and the percentage of cells generating ROS reached 35.57% after 6 h of treatment (Figure 4B and Figure S4 in File S1). To determine whether reactive oxygen species may be involved in the process of apoptosis initiation, cells were pretreated with the antioxidant N-acetyl-cysteine (NAC) for 1 hour and subsequently exposed to A398 (6 µM) for 6 h. The induction of death was assessed by percentage of the sub-G_1 fraction. Figure 4C shows that pretreatment with NAC did not alter the percentage of cells in sub-G_1, suggesting that the generation of ROS appears not to be crucial for the induction of apoptosis by A398.

Discussion

Within the sphere of cancer, an important number of drugs currently commercialized were obtained from natural sources or their derivatives. Some of the compounds with chemotherapeutic activity, such as podophyllotoxin, serve as prototypes for the synthesis of new molecules with better pharmacological activity [28,23,22]. The present study showed that A398, a new podophyllotoxin analogue, was cytotoxic on different types of tumor cells and least active against normal cells.

In the early stages of cytotoxicity evaluation of a new substance, the use of hepatic cell lineages serves to detect the presence of more toxic metabolites and to assess which enzymes are involved

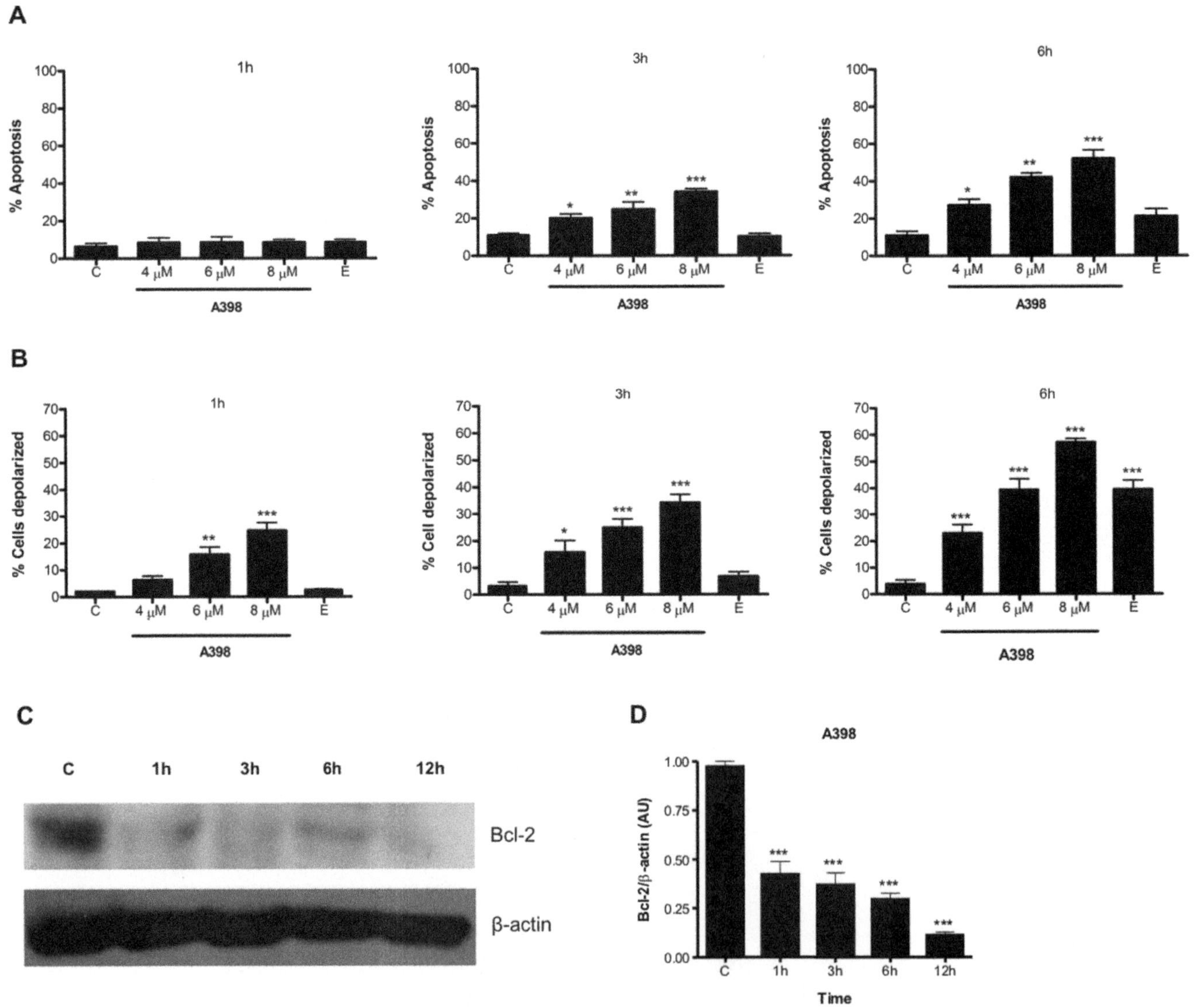

Figure 2. A398 induces apoptosis in HL-60 cell line. Cells were treated with A398 (4, 6 and 8 µM) or etoposide (5 µM) for 1 h, 3 h or 6 h and stained with annexin V-alexa fluor 488 and PI or with TMRM. The rate of apoptosis (A) and the percentage of depolarized cells (B) were quantified by flow cytometry. (C) Decreased expression of Bcl-2 in HL- 60 lineage. Cells were treated with 6 µM of A398 for 1 h, 3 h, 6 h and 12 h. The bands were revealed on nitrocellulose membrane after staining with secondary antibody labeled with peroxidase. β-actin was used as a loading control. C: control. (D) Bar plot showing arbitrary unit (AU) obtained from densitometry measure displayed in (C). Data are expressed as mean ± SE of three independent experiments. *$p<0.05$, **$p<0.01$ and ***$p<0.001$ compared with the control by ANOVA.

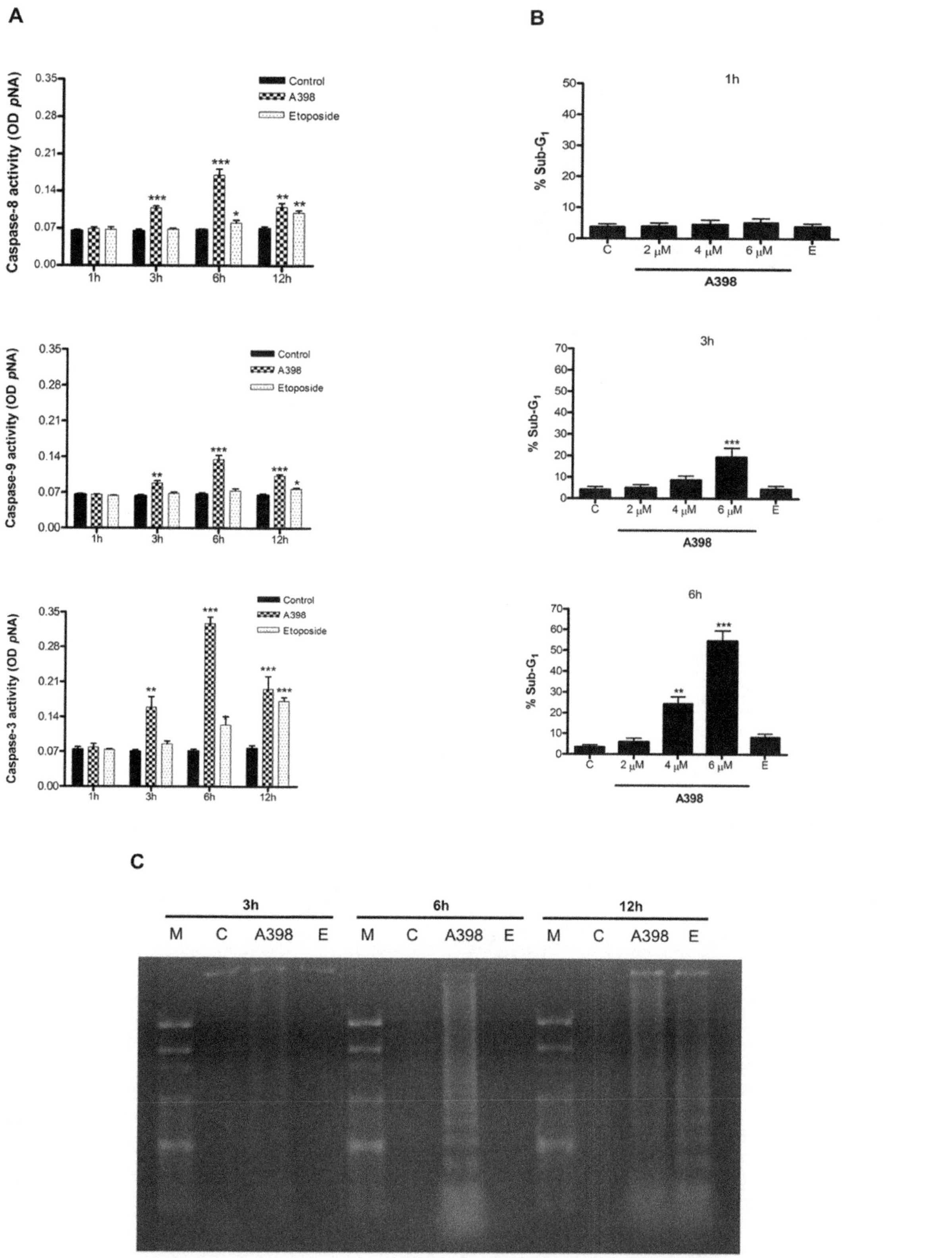

Figure 3. Activation of caspases and induction of DNA fragmentation by A98 in HL-60 cells. (A) Analysis of caspase activation by A398 in HL-60 lineage. Cells were treated with 6 μM of A398 or with 5 μM of etoposide for 1 h, 3 h, 6 h and 12 h. The level of activity of each caspase was expressed in relation to the optical density (OD) of the chromophore *p*NA. (B) A398 induces an increase in the sub-G_1 fraction in HL-60 lineage. Cells were treated with A398 (2, 4 and 6 μM) or etoposide (5 μM) for 1 h, 3 h or 6 h. After incubation, cells were stained with PI and analyzed by flow cytometry. (C) A398 promoted internucleosomal DNA fragmentation in HL-60 lineage. Cells were treated with 6 μM of A398 or 5 μM of etoposide for 3 h, 6 h and 12 h. DNA was extracted, separated on agarose gel and visualized with UV light. C: control; M: marker; E: etoposide. Data are expressed as mean ± SE of three independent experiments. *$p<0.05$, **$p<0.01$ and ***$p<0.001$ compared with the control by ANOVA.

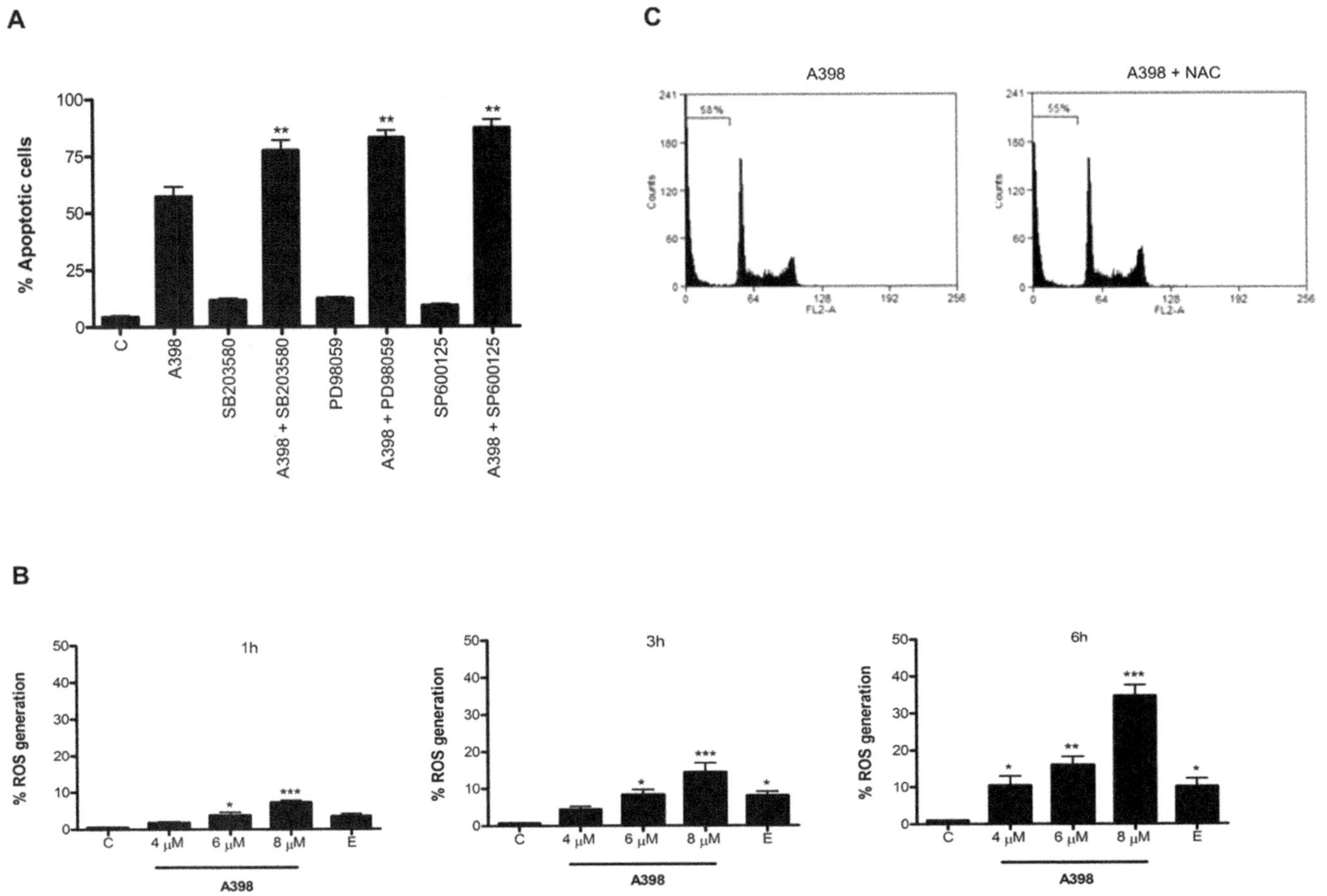

Figure 4. A398 inhibits the activation of MAPk and induces ROS generation in the HL-60 cells. (A) Effect of MAPKs inhibitors on apoptosis promoted by A398 in HL-60 lineage. Cells were pretreated with ERK inhibitor-PD 98059 (10 μM), JNK inhibitor-SP 600125 (10 μM), or p38 inhibitor-SB 203580 (5 μM) for 1 h and incubated with A398 (6 μM) during 6 h. Cells were stained with PI and analyzed by flow cytometry for sub-G_1 DNA damage. Data are expressed as mean ± SE of three independent experiments. **$p<0.01$, significantly different from derivative A398 treatment alone. (B) Evaluation of A398-induced ROS generation in HL-60 lineage. Cells were treated with A398 (4, 6 and 8 μM) or etoposide (5 μM) for 1 h, 3 h or 6 h, labeled with H_2- DCFH-DA and ROS production was quantified by flow cytometry. (D) Effect of antioxidant NAC on the induction of death of A398 in HL-60 lineage. Cells were pretreated with NAC and exposed to the compound (6 μM) for 6 h. After incubation, cells were stained with PI and analyzed by flow cytometry for sub-G_1 DNA damage. Data are expressed as mean ± SE of three independent experiments. *$p<0.05$, **$p<0.01$ and ***$p<0.001$ compared with the control by ANOVA.

in the metabolism of the substance being studied [29,30]. In this context, the cytotoxicity of A398 was assessed in the HepG2 lineage. The results suggest that the metabolites produced by A398 do not contribute to its cytotoxic activity and that metabolic reactions involving the enzymes CYP3A (phase I) and UGTs (phase II) favor the detoxification of the compound.

The suppression of apoptotic mechanisms is involved both in the process of tumorigenesis and maintenance of the malignant phenotype as in the development of drug resistance [31,12]. Therefore, the development of compounds that induce apoptosis is one of the therapeutic strategies in the field of oncology [6,32,4]. Our results showed that A398 promoted concentration- and time-dependent apoptosis in leukemia cell line HL-60.

The mitochondria integrates numerous signals of cell survival and death, thus exerting a decisive control over the various biochemical cascades that lead to apoptosis [33,34,13]. The BCL-2 family of proteins constitutes a critical control point in apoptosis residing immediately upstream of irreversible cellular damage, where family members control the release of apoptogenic factors from mitochondria [35,19]. Caspases are a family of cystine proteases which serve critical roles in initiation and execution of apoptosis [15,36]. Our studies clearly demonstrated that A398 promoted dissipation of mitochondrial transmembrane, inhibited the expression of anti-apoptotic Bcl-2 protein and induced activation of caspases -8, -9 and -3 in HL-60 cells, confirming the involvement of both intrinsic and extrinsic apoptotic pathways in apoptosis induced by A398.

In response to several types of cellular stress, cells may promote cell cycle arrest so as to prevent the progression of mutated and damaged cells [37]. Compound A398 exerted its cytotoxic effect on the HL-60 lineage without causing cell cycle arrest, with an increase in the sub-G_1 fraction, which shows induction of apoptosis. It is possible that the damage caused by this drug is so intense that there is no time for cell cycle arrest in an attempt to repair DNA, occurring rapidly an activation of apoptotic machinery [38,39]. Indeed, the results showed that A398 exerted its apoptotic effects early, much faster than etoposide in concentrations similar to this. Although cells are killed by

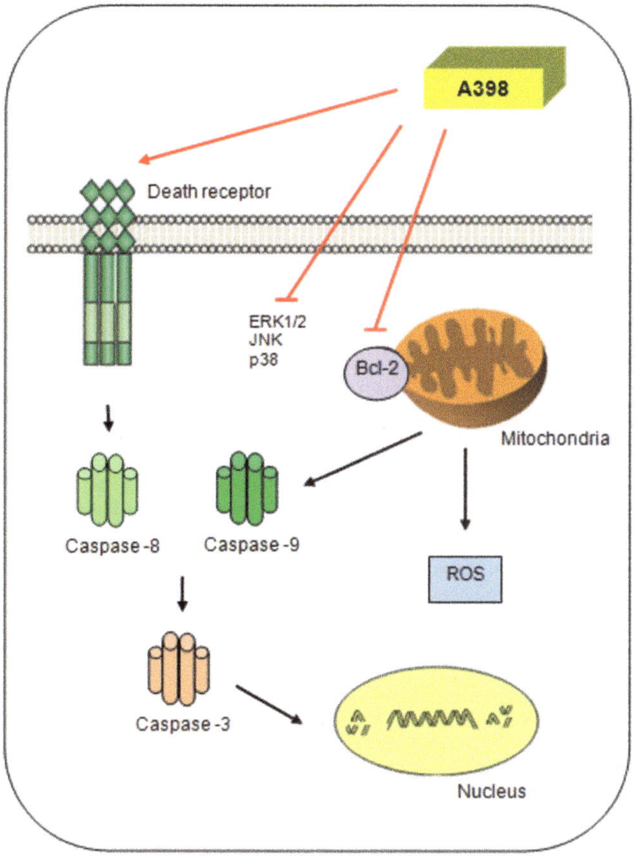

Figure 5. Hypothesized scheme of A398 events. For how compound A398 induced apoptosis in the leukemic HL-60 cells through the activation of the extrinsic and intrinsic pathways of death and its action on molecular targets, such as caspases, the anti-apoptotic protein Bcl-2 and MAPK proteins (ERK1/2, JNK and p38).

etoposide in any moment of the cell cycle, several data indicate a higher sensitivity from mid-S to G_2 phase [58]. Our studies demonstrated that etoposide promote arrest in the S-phase in HL-60, showing that A398 had a distinct mechanism of action compared to etoposide.

The degradation of DNA in a specific pattern of fragmentation is a characteristic of apoptosis, occurring due to activation of caspases -3 and endonucleases [40,41]. The electrophoresis in agarose gel showed that A398 promoted internucleosomal DNA fragmentation. Together with the other results, we found that the effector phase of the apoptotic process triggered by A398 starts at a time period of 3 hours, reaching a maximum induction of death in 6 hours of treatment, with a diminished effect from 12 hours treatment.

The three major mitogen-activated protein kinases (MAPKs) p38, JNK (c-Jun NH2-terminal kinases), and ERK (extracellular signal-regulated kinases) are signal transducers involved in a broad range of cell functions including survival, apoptosis, and cell differentiation [42,43]. Mutations leading to constitutive activation of the pathway Ras/Raf/MEK/ERK are highly prevalent in hematological cancers and mediates mitogenic and anti-apoptotic signals in leukemia cells [44,45]. Several studies suggest that p38 and JNK play pivotal roles in the leukemogenesis and may contribute to increased resistance to chemotherapeutics [46,47]. A398 appear to inhibit the activation of ERK1/2, p38 and JNK in HL-60 cells and this effect can contribute to the sensitivity of leukemia cells to the apoptotic effects of compound.

Reactive oxygen species (ROS) may play an important role in intracellular signaling cascades that trigger the initiation of apoptotic death [48,49]. The results showed that treatment with A398 promoted an increase in ROS. However, ROS does not seem to play a critical role in the initiation of apoptosis induced by A398, since the pretreatment of cells with the antioxidant NAC did not alter the percentage of death induced by this compound. Another evidence is that A398 did not promote the activation of the JNK and p38 pathways, which consist of the main mechanism by which ROS mediates the pro-apoptotic signaling [50,51]. Since the mitochondria becomes a source of generation of reactive oxygen species during apoptosis, we suggest that ROS production occurs after the dissipation of $\Delta\Psi m$, and may be involved in the amplification of death by apoptosis through the intrinsic pathway, since ROS promotes the depolarization of mitochondria [48,50].

The tumor suppressor protein p53 plays a critical role in regulating cell death, using various mechanisms to promote apoptosis [52,53]. Since the HL-60 cells are p53-deficient [54,55], A398 operates independently of the activation of this protein. So, this compound may also be effective in other tumor cells that have no functional p53, which is very important since the mutation and inactivation of this protein occurs in more than half of all types of cancer, and is correlated with the degree of aggressiveness of some tumors [56,57].

Studies indicate that an antitumor agent, in order to act effectively and more selectively in the treatment of cancer and to overcome the problem of drug resistance, should act in various molecular targets involved in tumorigenesis and the acquisition of the phenotype of multiple resistance to drugs [38,31]. In this scenario, A398 showed low cytotoxicity against normal cells and was cytotoxic to several human tumor cell lines, inducing apoptosis in the leukemic HL-60 cells through the activation of the extrinsic and intrinsic pathways of death and acting on molecular targets that are amenable to therapeutic intervention, such as caspases, the anti-apoptotic protein Bcl-2 and MAPKs (as indicated in the graphical abstract, Figure 5). Taken together, the current study suggested that A398 has an anticancer potential, proving to be a good candidate for preclinical research. Further studies *in vivo* are necessary to evaluate drug efficacy in animal models and clarify the detailed mechanism(s) involved in the observed antitumor effect.

Supporting Information

File S1 Combined file of supporting figures. Figure S1. Induction of apoptosis in HL-60 lineage. Cells were treated with derivative A398 (4, 6 and 8 μM) or with Etoposide (5 μM) for 1 h (A), 3 h (B) or 6 h (C). The cells were stained with annexin V-alexa fluor 488 and PI and evaluated by flow cytometry. The dual parametric dot plots show nonapoptotic live cells in the lower left quadrant (annexin V^-/PI^-), early apoptotic cells in the lower right quadrant (annexin V^+/PI^-), late apoptotic cells in upper right quadrant (annexin V^+/PI^+) or necrotic cells in the upper left (annexin V^-/PI^+). **Figure S2. Modification of mitochondrial transmembrane potential ($\Delta\Psi m$) in HL-60 lineage.** Cells were treated with derivative A398 (4, 6 and 8 μM) or with etoposide (5 μM) for 1 h (A), 3 h (B) or 6 h (C). The cells were stained with TMRM and the percentage of depolarized cells was quantified by flow cytometry. Right indicator: % cells with $\Delta\Psi m$ normal. Left indicator: % depolarized cells. CCCP (50 μM) was included as a positive control. **Figure S3. Analysis of cell cycle**

and DNA fragmentation in HL-60 lineage. Cells were treated with derivative A398 (2, 4 and 6 μM) or with etoposide (5 μM) for 1 h (A), 3 h (B) or 6 h (C). After treatment, cells were stained with PI and analyzed by flow cytometry. Indicator 1: % cells in sub-G_1. Indicator 2: % cells in G_1 phase. Indicator 3: % cells in S phase. Indicator 4: % cells in G_2/M phases. **Figure S4. Induction of ROS generation in HL-60 lineage.** Cells were treated with derivative A398 (4, 6 and 8 μM) or etoposide (5 μM) for 1 h (A), 3 h (B) or 6 h (C). They were then labeled with H_2-DCFH-DA and ROS production was quantified by flow cytometry. Indicator on the right: % cells with increased levels of ROS. H_2O_2 (50 μM) was used as a positive control.

Author Contributions

Conceived and designed the experiments: ALS DAMA RCM. Performed the experiments: ALS GVFM. Analyzed the data: ALS DAMA RCM. Contributed reagents/materials/analysis tools: ALS DAMA RCM. Wrote the paper: ALS DAMA.

References

1. Fulda S, Galluzzi L, Kroemer G (2010) Targeting mitochondria for cancer therapy. Nat Rev Drug Discov 9: 447–464.
2. Gyrd-Hansen M, Meier P (2010) IAPs: From caspase inhibitors to modulators of NF-κB, inflammation and cancer. Nat Rev Cancer 10: 561–574.
3. Elmore S (2007) Apoptosis: a review of programmed cell death. Toxicol Pathol 35: 495–516.
4. Ocker M, Höpfner M (2012) Apoptosis-modulating drugs for improved cancer therapy. Eur Surg Res 48: 111–120.
5. Plati J, Bucur O, Khosravi-Far R (2011) Apoptotic cell signaling in cancer progression and therapy. Integr Biol (Camb) 3: 279–296.
6. Bai L, Wang S (2014) Targeting apoptosis pathways for new cancer therapeutics. Annu Rev Med 65: 139–155.
7. Micheau O, Shirley S, Dufour F (2013) Death receptors as targets in cancer. Br J Pharmacol 169: 1723–1744.
8. Moffitt KL, Martin SL, Walker B (2010) From sentencing to execution - the processes of apoptosis. J Pharm Pharmacol 62: 547–562.
9. Fogg VC, Lanning NJ, Mackeigan JP (2011) Mitochondria in cancer: at the crossroads of life and death. Chin J Cancer 30: 526–539.
10. Wenner CE (2012) Targeting mitochondria as a therapeutic target in cancer. J Cell Physiol 227: 450–456.
11. Parsons MJ, Green DR (2010) Mitochondria in cell death. Essays Biochem 47: 99–114.
12. Portt L, Norman G, Clapp C, Greenwood M, Greenwood MT (2011) Anti-apoptosis and cell survival: a review. Biochim Biophys Acta 1813: 238–259.
13. Roy MJ, Vom A, Czabotar PE, Lessene G (2014) Cell death and the mitochondria: therapeutic targeting of the BCL-2 family-driven pathway. Br J Pharmacol 171: 1973–1987.
14. Alenzi FQ, Lotfy M, Wyse R (2010) Swords of cell death: caspase activation and regulation. Asian Pac J Cancer Prev 11: 271–280.
15. Fiandalo MV, Kyprianou N (2012) Caspase control: protagonists of cancer cell apoptosis. Exp Oncol 34: 165–175.
16. Hensley P, Mishra M, Kyprianou N (2013) Targeting caspases in cancer therapeutics. Biol Chem 394: 831–843.
17. Widlak P, Garrard WT (2009) Roles of the major apoptotic nuclease-DNA fragmentation factor-in biology and disease. Cell Mol Life Sci 66: 263–274.
18. Liu JJ, Lin M, Yu JY, Liu B, Bao JK (2011) Targeting apoptotic and autophagic pathways for cancer therapeutics. Cancer Lett 300: 105–114.
19. Ola MS, Nawaz M, Ahsan H (2011) Role of Bcl-2 family proteins and caspases in the regulation of apoptosis. Mol Cell Biochem 351: 41–58.
20. Santarpia L, Lippman SM, El-Naggar AK (2012) Targeting the MAPK-RAS-RAF signaling pathway in cancer therapy. Expert Opin Ther Targets 16: 103–119.
21. Karikas GA (2010) Anticancer and chemopreventing natural products: some biochemical and therapeutic aspects. J Buon 15: 627–638.
22. Mishra BB, Tiwari VK (2011) Natural products: An evolving role in future drug discovery. Eur J Med Chem 46: 4769–4807.
23. Gordaliza M (2007) Natural products as leads to anticancer drugs. Clin Transl Oncol 12: 767–776.
24. Shang H, Chen H, Zhao D, Tang X, Liu Y, et al. (2012) Synthesis and Biological Evaluation of 4α/4β-Imidazolyl Podophyllotoxin Analogues as Antitumor Agents. Arch Pharm Chem Life Sci 345: 43–48.
25. Kumar A, Kumar V, Alegria AE, Malhotra SV (2011) Synthetic and Application Perspectives of Azapodophyllotoxins: Alternative Scaffolds of Podophyllotoxin. Curr Med Chem 18: 3853–3870.
26. Sang CY, Liu JF, Qin WW, Zhao J, Hui L, et al. (2013) Synthesis and evaluation of the apoptosis inducing and CT DNA interaction properties of a series of 4β-carbamoyl 4′-O-demethylepipodophyllotoxins. Eur J Med Chem 70: 59–67.
27. You Y (2005) Podophyllotoxin Derivatives: Current Synthetic Approaches for New Anticancer Agents. Curr Pharm Des 11: 1695–1717.
28. Bailly C (2009) Ready for a comeback of natural products in oncology. Biochem Pharmacol 77: 1447–1457.
29. Guguen-Guillouzo C, Guillouzo A (2010) General review on in vitro hepatocyte models and their applications. Methods Mol Biol 640: 1–40.
30. Sahi J, Grepper S, Smith C (2010) Hepatocytes as a tool in drug metabolism, transport and safety evaluations in drug discovery. Curr Drug Discov Technol 7: 188–198.
31. Baguley BC (2011) The paradox of cancer cell apoptosis. Front Biosci 16: 1759–1767.
32. Ciavarella S, Milano A, Dammacco F, Silvestris F (2010) Targeted therapies in cancer. Biodrugs 24: 77–88.
33. Borutaite V (2010) Mitochondria as decision-makers in cell death. Environ Mol Mutagen 51: 406–416.
34. Galluzzi L, Kepp O, Kroemer G (2012) Mitochondria: master regulators of danger signalling. Nat Rev Mol Cell Biol 13: 780–788.
35. Leibowitz B, Yu J (2010) Mitochondrial signaling in cell death via the bcl-2 family. Cancer Biol Ther 9: 417–422.
36. Wyllie AH (2010) "Where, o death, is thy sting?" a brief review of apoptosis biology. Mol Neurobiol 42: 4–9.
37. Nowsheen S, Yang ES (2012) The intersection between DNA damage response and Cell death pathways. Exp Oncol 34: 243–254.
38. Bagnyukova TV, Serebriiskii IG, Zhou Y, Hopper-Borge EA, Golemis EA, et al. (2010) Chemotherapy and signaling: how can targeted therapies supercharge cytotoxic agents? Cancer Biol Ther 10: 839–853.
39. Malumbres M, Barbacid M (2010) Cell cycle, cdks and cancer: a changing paradigm. Nat Rev Cancer 9: 153–166.
40. Aleksandrushkina NI, Vanyushin BF (2012) Endonucleases and apoptosis in animals. Biochemistry (Mosc) 77: 1436–1451.
41. Kitazumi I, Tsukahara M (2011) Regulation of DNA fragmentation: the role of caspases and phosphorylation. FEBS J 278: 427–441.
42. Keshet Y, Seger R (2010) The Map Kinase signaling cascades: a system of hundreds of components regulates a diverse array of physiological functions. Methods Mol Biol 661: 3–38.
43. Plotnikov A, Zehorai E, Procaccia S, Seger R (2011) The Mapk Cascades: signaling components, nuclear roles and mechanisms of nuclear translocation. Biochim Biophys Acta 1813: 1619–1633.
44. Chung E, Kondo M (2011) Role of Ras/Raf/Mek/Erk signaling in physiological hematopoiesis and leukemia development. Immunol Res 49: 248–268.
45. Mccubrey JA, Steelman LS, Chappell WH, Abrams SL, Wong EW, et al. (2007) Roles of the Raf/Mek/Erk pathway in cell growth, malignant transformation and drug resistance. Biochim Biophys Acta 1773: 1263–1284.
46. Feng Y, Wen J, Chang CC (2009) p38 mitogen-activated protein kinase and hematologic malignancies. Arch Pathol Lab Med 133: 1850–1856.
47. Wagner EF, Nebreda AR (2009) Signal integration by JNK and p38 Mapk pathways in cancer development. Nat Rev Cancer 9: 537–549.
48. Burton GJ, Jauniaux E (2010) Oxidative stress. Best Pract Res Clin Obstet Gynaecol 25: 287–299.
49. Matés JM, Segura JA, Alonso FJ, Márquez J (2012) Oxidative stress in apoptosis and cancer: an update. Arch Toxicol 86: 1649–1665.
50. Sinha K, Das J, Pal PB, Sil PC (2013) Oxidative stress: the mitochondria-dependent and mitochondria-independent pathways of apoptosis. Arch Toxicol 87: 1157–1180.
51. Takeda K, Naguro I, Nishitoh H, Matsuzawa A, Ichijo H (2011) Apoptosis signaling kinases: from stress response to health outcomes. Antioxid Redox Signal 15: 719–761.
52. Golubovskaya VM, Cance WG (2013) Targeting the p53 pathway. Surg Oncol Clin N Am 22: 747–764.
53. Yoshida K, Miki Y (2010) The cell death machinery governed by the p53 tumor suppressor in response to DNA damage. Cancer Sci 101: 831–835.
54. Torres F, Quintana J, Díaz JG, Carmona AJ, Estévez F (2008) Trifolin acetate-induced cell death in human leukemia cells is dependent on caspase-6 and activates the Mapk pathway. Apoptosis 13: 716–728.
55. Kwak CS, Yeo EJ, Moon SC, Kim YW, Ahn HJ, et al. (2009) *Perilla* leaf, *Perilla frutescens*, induces apoptosis and g1 phase arrest in human leukemia hl-60 cells through the combinations of death receptor-mediated, mitochondrial, and endoplasmic reticulum stress-induced pathways. J Med Food 12: 508–517.
56. Bergholz J, Xiao ZX (2012) Role of p63 in Development, Tumorigenesis and Cancer Progression. Cancer Microenviron 5: 311–322.
57. Dai C, Gu W (2010) p53 post-translational modification: deregulated in tumorigenesis. Trends Mol Med 16: 528–536.
58. Montecucco A, Biamonti G (2007) Cellular response to etoposide treatment. Cancer Lett 252: 9–18.

The Susceptive Alendronate-Treatment Timing and Dosage for Osteogenesis Enhancement in Human Bone Marrow-Derived Stem Cells

Chih-Hsiang Chang[1], Chau-Zen Wang[1,2], Je-Ken Chang[1,3,4], Che-Yu Hsu[1], Mei-Ling Ho[1,2]*

1 Orthopaedic Research Center, Kaohsiung Medical University, Kaohsiung, Taiwan, 2 Department of Physiology, College of Medicine, Kaohsiung Medical University, Kaohsiung, Taiwan, 3 Department of Orthopaedics, College of Medicine, Kaohsiung Medical University Hospital, Kaohsiung Medical University, Kaohsiung, Taiwan, 4 Department of Orthopaedics, Kaohsiung Municipal Ta-Tung Hospital, Kaohsiung, Taiwan

Abstract

Recent studies indicated that alendronate enhanced osteogenesis in osteoblasts and human bone marrow-derived stem cells. However, the time- and dose-dependent effects of Aln on ostegenic differentiation and cytotoxicity of hBMSCs remain undefined. In present study, we investigated the effective dose range and timing of hBMSCs. hBMSCs were treated with various Aln doses (1, 5 and 10 μM) according to the following groups: group A was treated with Aln during the first five days of bone medium, groups B, C and D were treated during the first, second, and final five days of osteo-induction medium and group E was treated throughout the entire experiment. The mineralization level and cytotoxicity were measured by quantified Alizarin Red S staining and MTT assay. In addition, the reversal effects of farnesyl pyrophosphate and geranylgeranyl pyrophosphate replenishment in group B were also investigated. The results showed that Aln treatment in groups A, B and E enhanced hBMSC mineralization in a dose-dependent manner, and the most pronounced effects were observed in groups B and E. The higher dose of Aln simultaneously enhanced mineralization and caused cytotoxicity in groups B, C and E. Replenishment of FPP or GGPP resulted in partial or complete reverse of the Aln-induced mineralization respectively. Furthermore, the addition of FPP or GGPP also eliminated the Aln-induced cytotoxicity. We demonstrated that hBMSCs are susceptible to 5 μM Aln during the initiation stage of osteogenic differentiation and that a 10 μM dose is cytotoxic.

Editor: Pranela Rameshwar, Rutgers - New Jersey Medical School, United States of America

Funding: This study was supported by grants from the National Science Council: NSC 97-EC-17-A-17-S1-041, NSC 98-EC-17-A-17-S1-041, NSC 99-EC-17-A-17-S1-041, and NSC 102-2325-B-037-003. The funder's website is http://web1.most.gov.tw/. The funders had no role in study design, data collection and analysis, decision to publish, or preparation of the manuscript.

Competing Interests: The authors have declared that no competing interests exist.

* Email: homelin@kmu.edu.tw

Introduction

Bisphosphonates are clinically used to inhibit osteoclast activity and to treat osteoporosis. Alendronate (Aln), an aminobisphosphonate, is used as an antiosteoporotic agent and this agent down-regulates osteoclast activity by inhibiting farnesyl pyrophosphate synthase (FPP synthase), thus blocking prenylation of guanosine triphosphate (GTP)-binding proteins [1,2,3,4]. Moreover, Aln was also found to enhance osteogenic activities, including cell proliferation and osteo-differentation. The osteogenic effects of Aln were suggested due to increased alkaline phosphatase (ALP) activity, mineralization and osteogenic genes, such as bone morphogenetic protein-2 (BMP-2), type I collagen (Col I) and osteocalcin (OC) in osteoblasts [5,6,7,8], bone marrow-derived stem cells (BMSCs) [9,10] and human adipocyte-derived stem cells (human ADSCs) [11]. Previous studies suggested that Aln is uptaken by endocytosis of macrophages, osteoclasts and osteoblasts, the similar way may also happen in hBMSCs [12,13].

A wide range of effective osteo-inducing Aln doses has been reported for different cell types. In MG-63 osteoblastic cells, 1 nM to 100 μM Aln promoted cell proliferation and maturation [7,8]. In human BMSCs, 10 nM Aln enhanced osteo-differentiation [9,10]; however, the enhancement of osteo-differentiation required more than 14 days of treatment [9,10]. In our previous study, 5 μM Aln sufficiently enhanced osteo-differentiation in human ADSCs [11]. In contrast, aminobisphosphonate concentrations greater than 10 μM exerted strong inhibitory effects on both osteoblast proliferation and bone formation [6,14]. Moreover, the Aln treatment periods for stem cells are typically less than four days [5,7,8,9] or five days in osteo-induction medium [11]. The effective dose, treatment time and duration of Aln required for enhancing osteogenesis among different cell types were varied and those in mesenchymal stem cells have not been well characterized. Accordingly, in this study, we identified the appropriate treatment times and doses required for Aln-induced osteogenesis in human BMSCs. Simultaneously, the cell cytotoxic effects of Aln treatment were also investigated.

Materials and Methods

Reagents

Alendronate was obtained from Merck Pharmaceuticals (Whitehouse Station, NJ, USA). Alizarin Red S, thiazolyl blue tetrazolium bromide (MTT), Percoll, L-ascorbic acid-2-phosphate, beta-glycerolphosphate, sodium bicarbonate, dexamethasone, N-acetyl-L-cysteine (NAC), farnesyl pyrophosphate (FPP) and geranylgeranyl pyrophosphate (GGPP) were purchased from Sigma-Aldrich (Saint Louis, MO, USA). Dulbecco's Modified Eagle's Medium (DMEM), ascorbic acid, non-essential amino acids, penicillin/streptomycin, fetal bovine serum (FBS), keratinocyte-serum free medium (SFM), epidermal growth factor-bovine pituitary factor (EGF-BPE) and trypsin/EDTA were purchased from Gibco-BRL (Rockville, MD, USA). Dimethyl sulfoxide (DMSO) was purchased from Amresco LLC (Solon, OH, USA).

hBMSC isolation and culture

Human bone marrow mesenchymal stem cells (hBMSCs) were isolated from bone marrow aspirated from iliac crest of three male donors with informed consent and approval from the hospital ethics committee (IRB number: KMUK-IRB-970267). The donors were aged 18 to 22 years without any known bone disorders and underwent surgery due to hip trauma. The nucleated stromal cells were separated from bone marrow fluid using a 70% Percoll gradient. Then, these cells were selected by K-NAC medium containing keratinocyte-SFM supplemented with the EGF-BPE, 5% fetal bovine serum, 2 mM N-acetyl-L-cysteine, 0.1 mM L-ascorbic acid-2-phosphate, and 0.5% penicillin (10000 U/ml)/streptomycin (10000 μg/ml) in a humidified atmosphere of 5% CO_2 at 37°C. The hBMSCs were isolated in the K-NAC medium for 14 days after selection and the medium was changed every 3 days. The hBMSCs were used for experiments within 15 passages and the average doubling time of hBMSCs was 36–38 h. The hBMSCs were identified with CD markers as CD34−, CD45−, CD29+, CD49d+, CD90+ and CD166+ population [15,16], and their multi-potent differentiation properties were confirmed by inducing to osteogenic, chondrogenic and adipogenic differentiation, which are shown in Figure S1.

CD markers detection and multi-potent differentiation of isolated hBMSCs

Flow cytometry analysis was used to examine the presence of CD marker on the hBMSCs cell surface. Following the method of hBMSCs isolation, cells were harvested from the culture dishes by treating them with 0.25% trypsin/EDTA in PBS. One million hBMSCs were suspended in 500 μl of phosphate buffered saline (PBS) containing 20 μg/ml of an antibody. Phycoerythrin (PE)–conjugated antibodies (Becton Dickinson, Franklin Lakes, NJ) targeted against CD29, CD34, CD45, CD49d, CD90, CD166 were obtained from Becton Dickinson (BD; Becton Dickinson, Franklin Lakes, NJ). As an isotype control, PE-conjugated nonspecific mouse IgG was used as control. After incubation for 20 minutes at 4°C, the cells were washed with PBS 3 times and then suspended in 1 ml of PBS for analysis. Cell fluorescence was detected using a flow cytometer (FACS Vantage SE, Becton Dickinson) with a 570 nm filter for PE fluorescence, and the data were analyzed using WINMDI software.

We further analyzed the phenotypes of hBMSCs after induction into osteogenesis, chondrogenesis and adipogenesis. To induce osteogenic differentiation, cells were treated with OIM for 7 days and stained with ARS as described in methods. To induce chondrogenesis, cells were treated with chondro-induction medium (CIM) containing low glucose DMEM with 2.2 mg/ml sodium bicarbonate, 10% fetal bovine serum, 1% non-essential amino acid, 6.25 ug/mL insulin (Sigma-Aldrich, St. Louis, MO), 10 ng/mL TGF-β1 (Sigma-Aldrich, St. Louis, MO) 50 μmol/L L-ascorbate-2 -phosphate and 0.5% penicillin (10000 U/ml)/streptomycin (10000 μg/ml). After chondrogenic induction for 7 days, cells were fixed with 3.7% formaldehyde in phosphate buffered saline (PBS) for 30 minutes, washed once with PBS, and rinsed with distilled water; then they were processed for Alcian blue staining. Specimens were incubated with 0.05% Alcian blue solution overnight. Excess stain was removed by washing in PBS, rinsing with 5% acetic acid to remove nonspecific staining, and washing with PBS again. To induce adipogenesis, cells were treated with adipo-induction medium (AIM) containing Dulbecco's Modified Eagle's Medium supplemented with 0.5 mM 3-isobutyl-1-methylxanthine (Sigma-Aldrich), 1 μM dexamethasone (Sigma-Aldrich), and 10 μg/ml insulin. After 7 days, cells were washed twice with PBS and fixed with10% formalin in PBS for 1 hour; then washed three times with distilled water and dried. Cells were stained with oil red O for 15 minutes. Excess stain was removed by washing with 70% ethanol and distilled water.

Osteogenic differentiation of hBMSCs and Aln treatment

Isolated hBMSCs were seeded in 24-well plates at a density of 104 cells/well and cultured in bone medium (BM) containing low glucose DMEM with 2.2 mg/ml sodium bicarbonate, 10% fetal bovine serum, 1% non-essential amino acid solution, 1% ascorbic acid and 0.5% penicillin (10000 U/ml)/streptomycin (10000 μg/ml) in a humidified atmosphere of 5% CO2 at 37°C for 5 days. Osteogenesis was induced in the control group by culturing the cells for 20 days in the osteo-induction medium (OIM), which consisted of low glucose DMEM, 2.2 mg/ml sodium bicarbonate, 10% fetal bovine serum, 100 nM dexamethasone, 10 mM β-glycerophosphate, 0.1 mM L-ascorbic acid-2-phosphate, and 0.5% penicillin (10000 U/ml)/streptomycin (10000 μg/ml) after 5 days culture in BM. During the experimental period, the OIM was changed every 3 days. The hBMSCs were treated with various concentrations of Aln (1, 5 or 10 μM) at different culture stages, and the treatment groups are shown (Figure 1). In group A, cells were treated with Aln during the first five days (day 1–5) in BM. In group B, cells were treated with Aln during day 6–10 in OIM. In group C, cells were treated with Aln during day 11–15 in OIM. In group D, cells were treated with Aln during day 16–20 in OIM. In group E, cells were treated through entire experiment. In group B, 10 μM FPP or GGPP was replenished while 5 μM Aln treatment to test whether FPP or GGPP reverses Aln effect on mineralization.

Alizarin Red S staining and quantitative analysis

Mineralization or extracellular matrix calcification in Aln-treated hBMSCs was estimated by Alizarin Red S (ARS) staining. On each indicated day, cultured cells were fixed with ice-cold 10% formalin for 15 minutes. After washing with distilled water, the cells were stained with 1% Alizarin Red S for 15 minutes. The stained cells were washed twice with distilled water and then quantified. The degree of mineralization was quantified by adding 400 μl 10% acetic acid per well at 60°C for 10 minutes to extract the Alizarin Red S staining. Then, 100 μl of supernatant was transferred to a 96-well plate, and the absorbance at 415 nm (OD415) was measured by spectrometry.

MTT assay

The cell viabilities of Aln-treated hBMSCs were measured using MTT assay. On each indicated day, the cells were washed with PBS, and a one-tenth volume of MTT reagent (5 mg/ml stock)

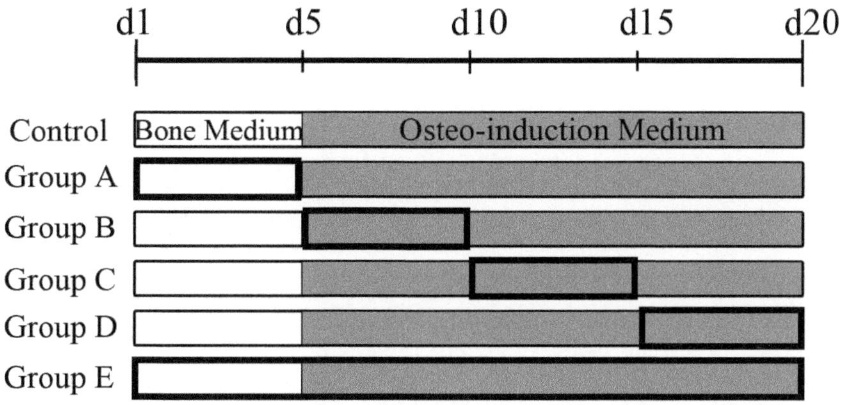

Figure 1. A schematic graph of Aln treatment strategies. All undifferentiated hBMSCs were seeded in 24-well plates and cultured for 5 days with bone medium (BM). Then, osteogenesis was induced using osteo-induction medium (OIM) for 15 days. The 5-day intervals of Aln treatments (indicated by bold frames) were performed during the first 5 days of culture (group A), day 6–10 (group B), 11–15 (group C), 16–20 (group D) or through entire experiment (group E). The control group was cultured with BM and then OIM without Aln treatment.

was added to each well, with a final volume of 500 µl in 24-well plate. The supernatant was carefully removed after 4 hours, and the formazan crystals were dissolved by adding 400 µl of DMSO per well at 37°C. The reactive product was transferred to an ELISA plate and the absorbance at 570 nm (with a reference absorbance in 655 nm) was measured by an ELISA reader (Tecan Sunrise, TECAN Deutschland GmbH, Crailsheim, Germany).

RNA isolation and Real-time polymerase chain reaction (real-time PCR)

At indicated days, cells were collected from wells and extracted total RNA by TRIzol (Gibco BRL, Rockville, MD). Briefly, 0.5–1 µg of total RNA per 20 µl of reaction volume were reverse transcribed into cDNA using the SuperScript First-Strand Synthesis System (Invitrogen). Real-time PCR reactions were performed and monitored using the iQ SYBR Green Supermix (Bio-Rad Laboratories Inc, Hercules, CA) and quantitative real-time PCR detection system (iQ5, Bio-Rad Laboratories Inc, Hercules, CA). The cDNA samples (2 µl for total volume of 25 µl per reaction) were analyzed for genes of interest and the reference gene glyceraldehyde-3-phosphate-dehydrogenase (GAPDH). The relative mRNA expression levels were calculated from the threshold cycle (C_t) value of each PCR product and normalized with GAPDH using the comparative C_t method [17]. All real-time PCR experiments were performed in triplicate and repeated at least three times. The primer sequences used were shown as followed: BMP-2: forward 5′-GGA ATG ACT GGA TTG TGG CT-3′; reverse 5′-TGA GTT CTG TCG GGA CAC AG-3′; Runx-2: forward 5′-AGA TGG GAC TGT GGT TAC TG-3′; reverse 5′-GTA GCT ACT TGG GGA GGA TT-3′; Osteocalcin: forward 5′-GTG CAG AGT CCA GCA AAG GT-3′; reverse 5′-CGA TAG GCC TCC TGA AAG C-3′; GAPDH: forward 5′-GAA GGT GAA GGT CGG AGT C-3′; reverse 5′-GAA GAT GGT GAT GGG ATT TC-3′.

Statistical methods

All experiments were performed in triplicate. The data presented as means±standard deviation (SD) at a significance level of $p<0.05$. Statistical differences between groups were calculated using a one-way ANOVA, and multiple comparisons were performed by Scheffe's method.

Results

Alendronate enhanced hBMSC osteogenesis during the initiation stage of osteogenic differentiation

The mineralization of induced hBMSCs were measured by ARS staining and quantified at the indicated days. As shown in Figure 2A, the ARS staining densities of group A were elevated with time. The quantified results showed that treatment with 10 µM Aln significantly increased mineralization after 15 days of treatment ($p<0.01$). The 1 µM or 5 µM Aln treatments did not significantly affect mineralization at day 15. At day 20, all doses of Aln significantly elevated mineralization, and 10 µM Aln showed the highest level of mineralization (Figure 2A, $p<0.01$). In group B, the mineralization levels were significantly elevated after Aln treatments in dose-dependent manner (Figure 2B). 1 µM and 5 µM Aln significantly elevated mineralization when compared to the control (Figure 2B, 1 and 5 µM, $p<0.01$), and 10 µM Aln treatment enhanced mineralization to a great degree than other doses at days 15 and 20 (Figure 2B, $p<0.01$). However, the mineralization levels of hBMSCs treated with Aln were not significantly different in group C (Figure 2E) and only weakly elevated in group D when compared to the control (Figure 2F). In group E, which consisted of hBMSCs treated with Aln during entire experiment, the mineralization effects were similar to those of group B (Figure 2G). The mineralization of hBMSCs treated with 5 µM Aln was significantly elevated at day 15 and 20 (Figure 2G, $p<0.01$), and hBMSCs treated with 10 µM Aln showed the highest degree of mineralization compared to the other doses (Figure 2G, $p<0.01$). In group A, gene expressions of BMP-2 and OC were significantly elevated in Aln treated hBMSCs than those in control group at day 5. Furthermore, the gene expressions of BMP-2, Runx-2 and OC in group B were significantly higher than those in group A and control group at day 7 (Figure S2A). We also measured the alkaline phosphatase (ALP) activity of control, group A and B at day 15. The result showed ALP activity in group B was the higher than other groups (Figure S2B).

We further confirmed the osteogenic effects observed in groups A and B on the hBMSCs obtained from donor II and III; the mineralization results were similar to those of donor I (Figure 2C and 2D). In group A of donor II and III, the osteogenic effects were significantly elevated after 10 µM Aln treatment at days 15 and 20 (Figure 2C, $p<0.01$). In group B, the elevated levels of

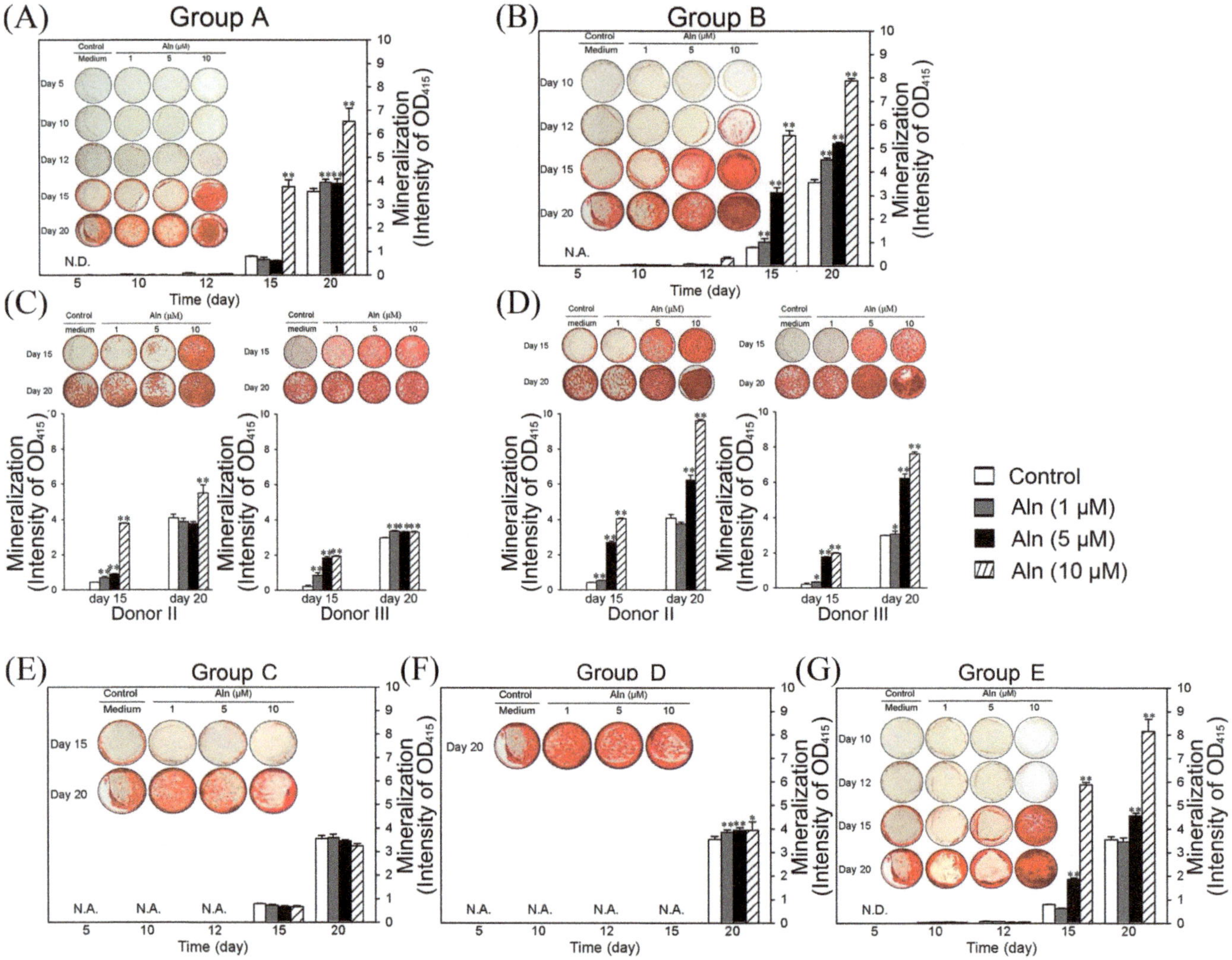

Figure 2. The mineralization effects of Aln on hBMSCs after different treatment strategies. Undifferentiated hBMSCs were treated with 1, 5 or 10 μM Aln during the culture time indicated in Figure 1. The mineralization levels were measured by ARS staining (shown as photographs), which was further quantified (shown as bar). The mineralization of group A (A) and group B (B) of donor I were measured at day 5, 10, 12, 15 and 20. The mineralization of donor II and donor III in group A (C) and group B (D) were measured at days 15 and 20. The mineralization of groups C (E), D (F) and E (G) of donor I were also determined using ARS staining (photographs) and quantified (bar). N.A: not available; N.D: not detectable; *$p<0.05$; **$p<0.01$ (n = 5).

mineralization were similar to those of donor I after treatment with 5 and 10 μM Aln, and the osteogenic effects were significantly increased at days 15 and 20 (Figure 2D, $p<0.01$). According to these results, hBMSCs were sensitive to the Aln-enhanced osteogenic effects during the initiation stage of incubation, regardless of the BM (group A) or OIM (group B) culture conditions. However, hBMSCs were insensitive to Aln in the later stage of OIM incubation, even after treatment with 10 μM Aln (groups C and D).

High doses of Aln caused cytotoxicity

The cell viabilities of Aln-treated hBMSCs in the five groups were also examined. As shown in Figure 3, cell viabilities were measured by MTT assay, and the cell morphology was observed by optical microscopy at the indicated times. Although MTT assay was associated to cell proliferation and mitochondrial activity, the reduction of MTT measurement indicated the loss of cell viability and also represented the cytotoxicity [18,19]. In group A, hBMSCs treated with 1 or 5 μM Aln showed no difference when compared to the control cells; only treatment with 10 μM Aln significantly reduced the cell viability after day 6 (Figure 3, group A, $p<0.01$). At day 15, hBMSCs treated with 1 or 5 μM Aln were as confluent as well as the control cells (Figure 3, a–c). In contrast, hBMSCs treated with 10 μM Aln were less confluent than the control cells (Figure 3, d). Treatment with 1 μM Aln reduced the cell viability slightly after day 7 in group B and day 14 in group C (Figure 3, group B and group C, $p<0.01$), and 5 μM Aln was slightly cytotoxic in both groups (Figure 3, group B and group C, $p<0.01$). Treatment with 10 μM Aln caused significant cytotoxic effects after day 7 in group B and day 14 in group C (Figure 3, group B, and group C, $p<0.01$). hBMSCs treated with 1 μM Aln in group B and 1 or 5 μM Aln in group C showed similar morphologies and confluence as the control cells at the indicated days (Figure 3, e, h and i). In group B, hBMSCs treated 5 μM Aln showed only some rounded cells (Figure 3, f), but cells treated with 10 μM Aln treated cells showed cytotoxic effect and were almost completely rounded at day 15 (Figure 3, g). In group C, treatment with 10 μM Aln also induced several rounded cells at day 20 (Figure 3, j). In group D, hBMSCs treated with any dose of Aln showed no signs of cytotoxic effects during culture (Figure 3, group D), and these cells grew and adhered well as the control (Figure 3, k, l and m). In group E, the cell viability and morphology were similar to group B, and treatment 1 μM Aln slightly reduced the cell viability after day 12 (Figure 3, group E, $p<0.01$). Cells treated with 5 μM Aln showed cytotoxic effects at day 8 (Figure 3, group E, $p<0.01$) and treatment with 10 μM Aln induced significant cytotoxic effects at day 6 (Figure 3, group E, $p<0.01$). hBMSCs treated with 10 μM Aln showed an almost completely rounded morphology at day 15 (Figure 3, p), while cells treated with 5 μM Aln showed only partial rounded cells (Figure 3, o). Cells treated with 1 μM Aln were not different than the control cells (Figure 3, n). The results indicated Aln treatment during the BM incubation was slightly cytotoxic only at 10 μM (Figure 3, group A, $p<0.01$). However, hBMSCs treated with 5 μM or 10 μM Aln during the OIM incubation showed significant cytotoxicity immediately (Figure 3, group B). These results indicated that higher doses of Aln are cytotoxic during the initiation stage of osteogenesis.

Replenishment of FPP or GGPP reversed the Aln-induced mineralization and cytotoxicity

We measured the mineralization of hBMSCs cultures that replenished with 5 to 20 μM of FPP and GGPP. The results showed no significant difference among treatment groups (Figure S4A and S4B). Based on this data, we used 10 μM FPP and GGPP to test the mineralization of group B hBMSCs. After FPP or GGPP replenishment, the effects of Aln treatments were partially reversed. Treatment with 10 μM FPP and 5 μM Aln significantly reduced the mineralization effects observed after treatment with 5 μM Aln in all three donors. The mineralization effects decreased by approximately 78%, 20% and 69% at day 15 (Figure 4A, $p<0.01$), and 14%, 37% and 20% at day 20 (Figure 4B, $p<0.01$). Moreover, treatment with 10 μM GGPP and 5 μM Aln nearly reversed the effects observed after treatment with Aln alone (Figure 4A and 4B, decreased approximately 97%, 77% and 86% at day 15, $p<0.01$, and 83%, 40% and 36% at day 20, $p<0.01$, respectively). The alkaline phosphatase (ALP) activity assay also showed similar results. The ALP activities of FPP or GGPP replenishment at day 15 were remarked reduced than Aln alone in group B (Figure S4C). According to these results, FPP replenishment partially reversed and GGPP replenishment almost completely reversed the Aln-induced mineralization.

The MTT assay results further indicated the Aln-induced cytotoxicity was partially reversed by FPP or GGPP replenishment (Figure 4C). The viability of cells treated with 5 μM Aln was only 38.8% of control hBMSCs (Figure 4C, $p<0.01$). After co-treatment with FPP and 5 μM Aln, the cell viability increased to 59.2% of control (Figure 4C, $p<0.01$). Co-treatment with GGPP and 5 μM Aln increased the cell viability to 69.2% of the control (Figure 4C, $p<0.01$). These results suggested that the mevalonate pathway may be involved in Aln-induced osteogenesis and cytotoxicity, and FPP or GGPP replenishment not only reduced osteogenesis but also reversed Aln-associated cytotoxicity.

Discussion

Aln was reported to enhance osteogenic differentiation in hBMSCs [9,10]; however, the timing of Aln treatment required for osteogenic effects has not been studied. In this report, we varied the dose and treatment timing of Aln during different stages of differentiation to examine the effects of Aln on hBMSCs.

We found that the most effective treatment was 5 μM Aln administered during the initiation stage of osteo-induction; however, the higher dose of Aln inhibited cell growth and even induced cytotoxic effect on hBMSCs during the earlier stage of osteo-induction. These findings not only demonstrated the significance of timing and the optimal effective dose on enhancing hBMSC osteogenesis in hBMSCs but also demonstrated the need to avoid simultaneous cytotoxic effect. By controlling the dose and time of Aln treatment, this combined use of Aln and hBMSCs may be locally applied as a material for enhancing bone regeneration.

Previous reports describe the osteogenesis of hBMSCs in different culture medium [20]. Elevated ALP activities were observed after approximately 7 days of culture in medium either with or without dexamethasome, and this effect was maintained for two weeks. In the absence of Aln, the hBMSC osteogenesis observed in this study showed similar pattern as the previous studies. The mineralization levels increased after culturing the cells in osteo-iduction medium (OIM) for 7 to 15 days (Figure 2, day 12 to 20 in control group). In regarding the effect of Aln on differentiation, the enhanced osteogenesis observed in group A and B was significantly higher than that observed in group C and D, and this effect was most pronounced in group B. In group A, Aln treatment prior to osteo-induction, required higher dose (10 μM) of Aln to enhance osteogenesis (4.74- and 1.84-folds increases when compared to the control at day 15 and day 20, respectively). In group B, treatment with 5 μM Aln during the initiation stage of osteo-induction effectively enhanced osteogenesis, and 10 μM Aln showed an even stronger effect. In comparison to the control cells, the mineralization levels observed in group B after treatment with 1, 5 or 10 μM Aln were 1.3-, 3.9- and 7- fold higher day 15, respectively, and 1.3-, 1.5- and 2.2- fold higher at day 20, respectively. Accordingly, we suggest that Aln treatment during the initiation stage of osteogenic differentiation enhanced the osteo-induction of hBMSCs. This effect was confirmed in BMSCs obtained from 3 different donors, all of which showed similar effects (Figure 2C and 2D). Only BMSCs obtained from donor III and treated in group A showed less sensitivity to the 10 μM Aln treatment.

In previous studies, BMP-2 expression increased after 3 to 4 days of Aln treatment in osteoblastic cells [7,8], ADSCs [11] and BMSCs [9]. However, the continuous supplement of Aln did not further increase BMP-2 expression in the later stage [10], which suggests that the initiation stage was more sensitive to the effects on Aln than the later stage. These results also suggest that Aln-induced osteogenesis in hBMSCs may act via BMP-2 expression and a subsequent increase in mineralization [10]. In this study, the continuous supplement of Aln in group E also confirmed that extending the duration of Aln treatment does not further enhance osteo-induction. The mineralization and cell viability pattern were similar to group B (Figure 2B and G; Figure 3 group B and E). Based on the findings from this study and previous studies, we suggest that the most effective timing of Aln on hBMSC osteogenesis is during the initiation stage of osteogenic differentiation.

In this study, we also investigated the cytotoxic effects of different Aln treatment strategies. In previous studies, Aln inhibited cellular proliferation at high dose in MG-63 osteoblastic cells (10^{-4} M) [7] and primary human osteoblasts ($<10^{-5}$ M) [5]. The toxic doses of Aln were usually greater than 1 μM in differentiated cells, such as chondrosacrcoma cell [21], epidermal cell [22], endothelial cell [23], fibroblast [24], intestinal epithelial cell [25], or macrophage [3,26,27]. In this study, the results showed that treatment with 10 μM Aln in group A inhibited the viability of hBMSCs (Figure 3, group A), but only 5 μM Aln

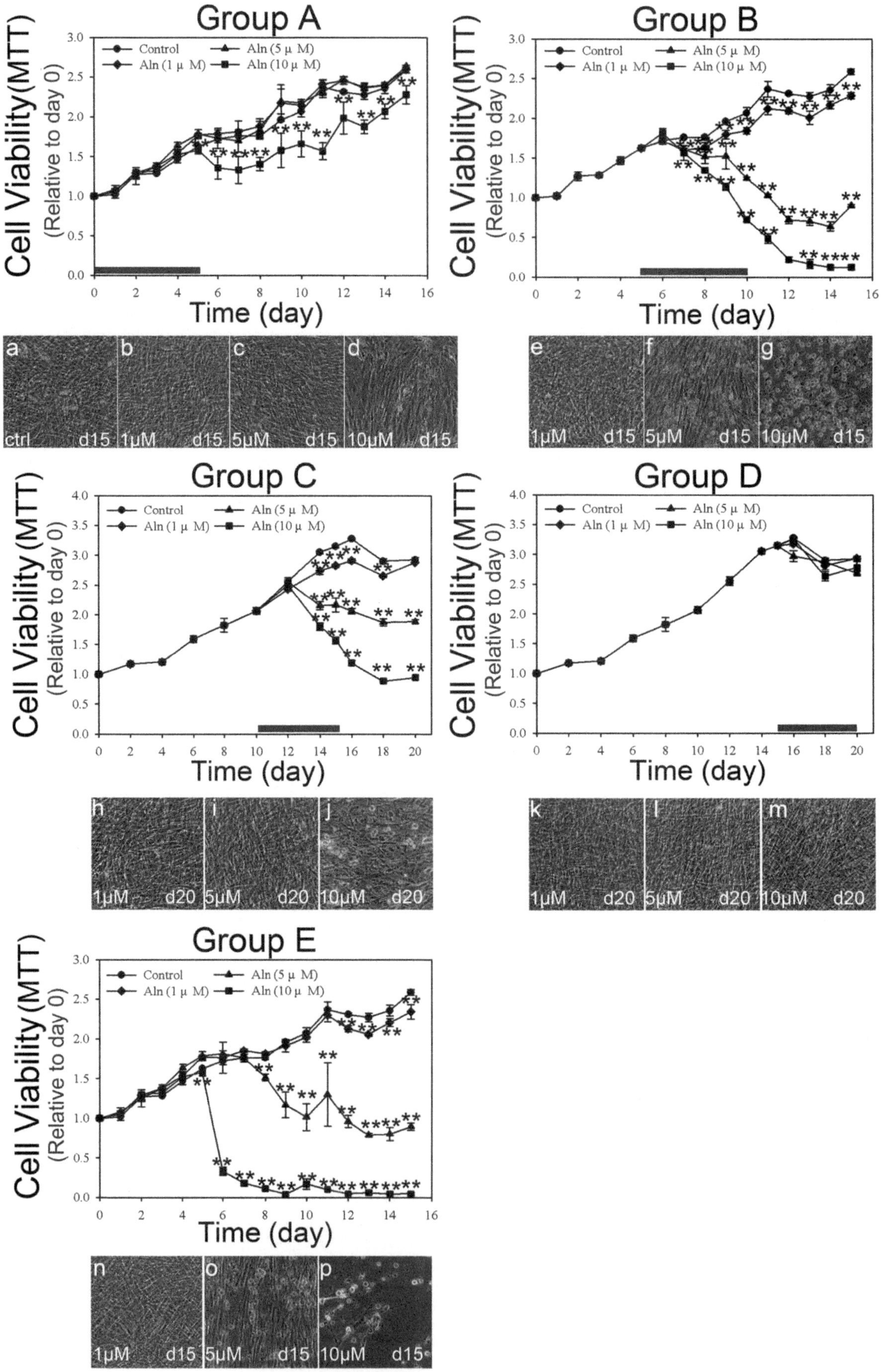
Group A
Cell Viability(MTT)
(Relative to day 0)
Control
Aln (1 μ M)
Aln (5 μ M)
Aln (10 μ M)
Time (day)
a ctrl d15
b 1μM d15
c 5μM d15
d 10μM d15
Group B
e 1μM d15
f 5μM d15
g 10μM d15
Group C
h 1μM d20
i 5μM d20
j 10μM d20
Group D
k 1μM d20
l 5μM d20
m 10μM d20
Group E
n 1μM d15
o 5μM d15
p 10μM d15

Figure 3. The cell viabilities of hBMSC after various Aln treatment strategies. The cell viabilities of cultures with various Aln treatment strategies were measured by MTT assays, and the cell morphology after various Aln doses for the indicated time are shown. The duration of Aln treatment is indicated by a bold bar. (Group A: a–d; Group B: e–g; Group C: h–j; Group D: k–m; Group E: n–p). $^{**}p<0.01$ (n = 4).

reduced its cell viability in groups B and C (Figure 3, group B and group C). These results revealed that the growth inhibitory or cytotoxic effect of Aln on hBMSCs might be more pronounced during the initiation stage of osteogenesis. Interestingly, the decreases of cell viability were consistent with the enhanced mineralization observed in hBMSCs treated with higher dose of Aln (Figure 2A and B; Figure 3, 10 μM in group A and 5–10 μM in group B). Accordingly, there may be some factors in Aln-induced osteo-induction that are also involved in cytotoxic effects of Aln treatment.

Recently, several studies indicated Aln suppresses bone resorption via inhibiting FPP synthase in osteoclasts, and thus suppressing osteoclastic differentiation and apoptosis [1,2,3,28]. Furthermore, Aln-induced apoptosis was partially reversed by FPP or GGPP replenishment [3,28]. However, the inhibition of isoprenylation was recently found to be involved in osteoblst osteogenesis [29,30,31]. The role of FPP or GGPP in the osteogenesis of osteoblast cells or human mesenchymal stem cells (hMSCs) was found to inhibit mineralization in osteoblast cultures [30,31,32,33]. Furthermore, the osteoblastogenesis of hBMSCs was increased by GGPP- or FPP-specific inhibitors in Aln-treated hMSCs but no effects on cell proliferation were observed [29]. In previous study, hMSCs treated with 40 μM FPP or 10 μM GGPP alone showed no significant effect on cell proliferation and apoptosis [34]. Five to 20 μM of FPP treatments were also used in the inhibition of zoledronic acid [5] induced mineralization in osteoblasts [31], which indicated FPP (lower than 40 μM) and GGPP (lower than 10 μM) did not affect cell proliferation individually, and GGPP (higher than 5 μM) inhibited ZA-induced mineralization. In the results of BMP-2 and Runx-2 expression, also showed no significant difference between control and 10 μM FPP or GGPP treated hBMSCs (Figure S3). In this study, we found that the Aln-induced osteogenesis of hBMSC was partially reversed by FPP replenishment and completely reversed by GGPP replenishment (Figure 4A and 4B). The Aln-induced cytotoxic effects were also reversed by FPP or GGPP replenishment (Figure 4C). The results indicate that Aln-reduced cell viability were at least partially due to the inhibition of FPP or GGPP in the mevalonate pathway.

Conclusions

We conclude that an Aln treatment strategy for hBMSC osteo-induction requires 5 μM Aln during the initiation stage of

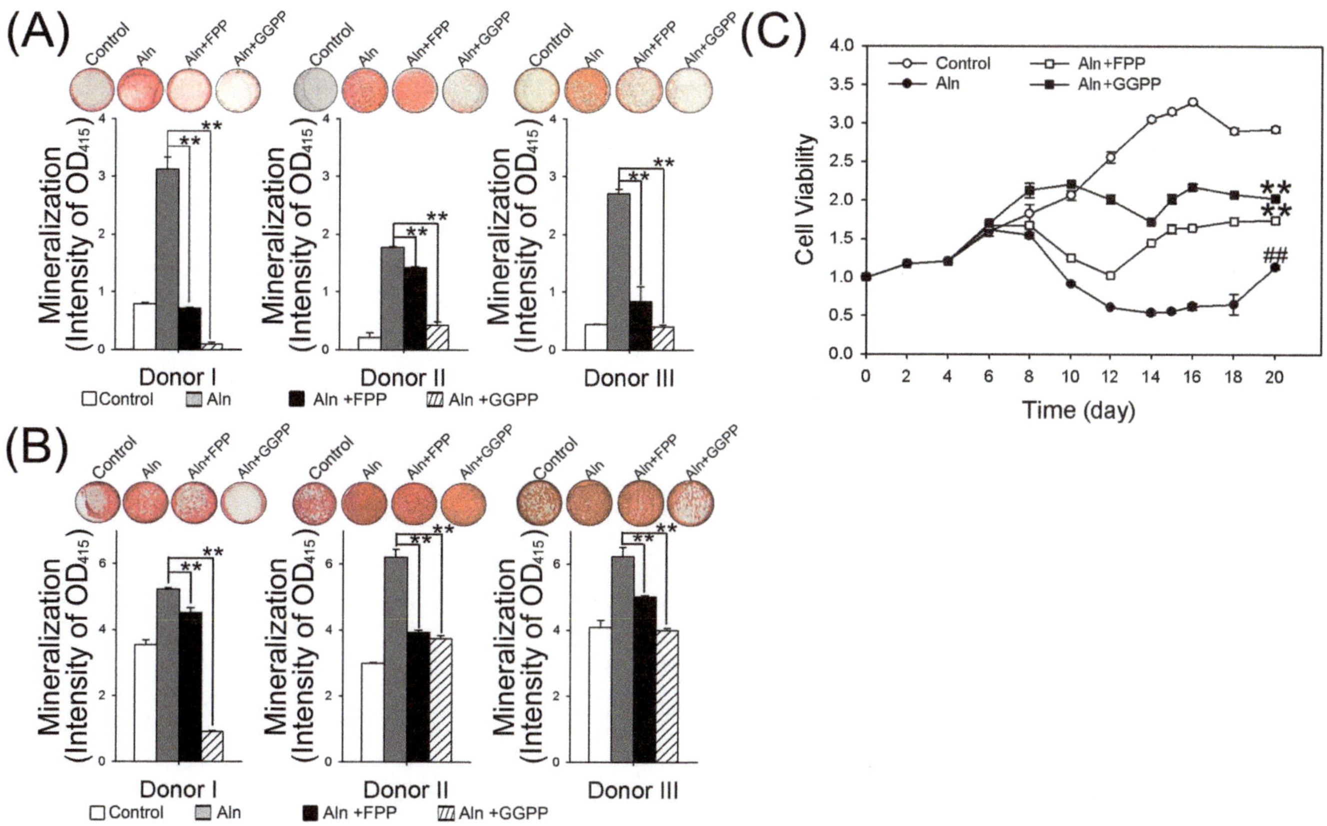

Figure 4. The reversal effects of FPP or GGPP on mineralization and cell viability of Aln-treated hBMSC with group B. The mineralization effects and cell viabilities of Aln-treated hBMSCs with group B were determined after additional treatment with FPP or GGPP. The ARS staining was performed at day15 (A) and day20 (B). The mineralization levels are shown as photographs and the quantified data shown as bar graph compared to 5 μM Aln-treated group. $^{**}p<0.01$, n = 5, compare with 5 μM Aln treatment. (C) The cell viabilities of hBMSCs treated with control, 5 μM Aln only, 5 μM Aln with 10 μM FPP or 5 μM Aln with 10 μM GGPP were measured by MTT assays and normalized to day 0. $^{**}p<0.01$, n = 4, compare with 5 μM Aln treatment; ## $p<0.01$, n = 4, compare to the control.

osteo-induction. A higher dose of Aln may be cytotoxic to hBMSCs at this stage. The results from this study suggest that using appropriate dose of Aln at specific time point may provide an application strategy for hBMSC-based material for tissue engineering on bone regeneration.

Supporting Information

Figure S1 Isolated hBMSCs were identified by surface markers and multi-potent differentiation potential. (A) Isolated hBMSCs were identified by CD markers that measured by flowcytometry using specific antibody against CD29, CD34, CD45, CD49d, CD90 and CD166. The white diagram indicated CD marker positive population and the red diagram indicated non-specific antibody control. (B) Isolated hBMSCs were induced to osteogenesis, chondrogenesis or adipogenesis by induction medium. Alcian blue, ARS or oil red staining was for sGAG, mineralization or oil droplet, respectively.

Figure S2 Aln increases osteogenic gene expressions and ALP activity in hBMSCs. (A) The mRNA expression of BMP-2, Runx-2 and osteocalcin in 5 μM Aln-treated hBMSCs were measured at day 5 and 7. The treatment strategy was shown in Figure 1. Cultured hBMSCs were untreated (control) or treated with 5 μM in the first 5 days in bone medium (grey bars; Group A) or the second five days in osteo-induction medium (black bars; Group B). $^{**}p<0.01$, n = 3; * $p<0.05$, n = 3, compared with control group. ##$p<0.01$, n = 3; #$p<0.05$, n = 3, compared with group A. (B) The ALP activities of Aln-treated hBMSCs were measured at day 15 with control group, group A, and group B. (TIF)

Figure S3 Treatment of FPP or GGPP did not affect the expression of BMP-2 and Runx-2 in hBMSCs. The mRNA expression of BMP-2 and Runx-2 in FPP or GGPP-treated hBMSCs were measured through two weeks. Cultured hBMSCs were untreated (control) or treated with 10 μM FPP (grey bars) or 10 μM GGPP (black bars) in bone medium for five days and then shift to osteo-induction medium without FPP or GGPP. Compare with the control, n = 3.

Figure S4 The reversal effects of different concentrations FPP or GGPP on mineralization of Aln-treated hBMSC with group B. The mineralization effects and cell viabilities of Aln-treated hBMSCs with group B were determined after additional treatment with 5 μM (light grey bars), 10 μM (deep grey bars) and 20 μM (black bars) FPP (A) or GGPP (B). The ARS staining was performed at day 15 and day 20. The mineralization levels are shown as photographs and the quantified data shown as bar graph compared to 5 μM Aln-treated group. $^{*}p<0.05$, n = 3, compare with 5 μM Aln treatment. (C) The ALP activities of 5 μM Aln-treated only or with additional 10 μM FPP or 10 μM GGPP were measured at day 15.

Methods S1

Author Contributions

Conceived and designed the experiments: C-HC C-ZW J-KC C-YH M-LH. Performed the experiments: C-HC C-YH. Analyzed the data: C-HC C-ZW C-YH. Contributed reagents/materials/analysis tools: J-KC. Contributed to the writing of the manuscript: C-HC C-ZW M-LH.

References

1. Dunford JE, Thompson K, Coxon FP, Luckman SP, Hahn FM, et al. (2001) Structure-activity relationships for inhibition of farnesyl diphosphate synthase in vitro and inhibition of bone resorption in vivo by nitrogen-containing bisphosphonates. J Pharmacol Exp Ther 296: 235–242.
2. Keller RK, Fliesler SJ (1999) Mechanism of aminobisphosphonate action: characterization of alendronate inhibition of the isoprenoid pathway. Biochem Biophys Res Commun 266: 560–563.
3. Luckman SP, Hughes DE, Coxon FP, Graham R, Russell G, et al. (1998) Nitrogen-containing bisphosphonates inhibit the mevalonate pathway and prevent post-translational prenylation of GTP-binding proteins, including Ras. J Bone Miner Res 13: 581–589.
4. Bergstrom JD, Bostedor RG, Masarachia PJ, Reszka AA, Rodan G (2000) Alendronate is a specific, nanomolar inhibitor of farnesyl diphosphate synthase. Arch Biochem Biophys 373: 231–241.
5. Enjuanes A, Ruiz-Gaspa S, Peris P, Ozalla D, Alvarez L, et al. (2010) The effect of the alendronate on OPG/RANKL system in differentiated primary human osteoblasts. Endocrine 37: 322–328.
6. Idris AI, Rojas J, Greig IR, Van't Hof RJ, Ralston SH (2008) Aminobisphosphonates cause osteoblast apoptosis and inhibit bone nodule formation in vitro. Calcif Tissue Int 82: 191–201.
7. Im GI, Qureshi SA, Kenney J, Rubash HE, Shanbhag AS (2004) Osteoblast proliferation and maturation by bisphosphonates. Biomaterials 25: 4105–4115.
8. Xiong Y, Yang HJ, Feng J, Shi ZL, Wu LD (2009) Effects of alendronate on the proliferation and osteogenic differentiation of MG-63 cells. J Int Med Res 37: 407–416.
9. Kim HK, Kim JH, Abbas AA, Yoon TR (2009) Alendronate enhances osteogenic differentiation of bone marrow stromal cells: a preliminary study. Clin Orthop Relat Res 467: 3121–3128.
10. von Knoch F, Jaquiery C, Kowalsky M, Schaeren S, Alabre C, et al. (2005) Effects of bisphosphonates on proliferation and osteoblast differentiation of human bone marrow stromal cells. Biomaterials 26: 6941–6949.
11. Wang CZ, Chen SM, Chen CH, Wang CK, Wang GJ, et al. (2010) The effect of the local delivery of alendronate on human adipose-derived stem cell-based bone regeneration. Biomaterials 31: 8674–8683.
12. Komatsu K, Shimada A, Shibata T, Wada S, Ideno H, et al. (2013) Alendronate promotes bone formation by inhibiting protein prenylation in osteoblasts in rat tooth replantation model. J Endocrinol 219: 145–158.
13. Thompson K, Rogers MJ, Coxon FP, Crockett JC (2006) Cytosolic entry of bisphosphonate drugs requires acidification of vesicles after fluid-phase endocytosis. Mol Pharmacol 69: 1624–1632.
14. Orriss IR, Key ML, Colston KW, Arnett TR (2009) Inhibition of osteoblast function in vitro by aminobisphosphonates. J Cell Biochem 106: 109–118.
15. Boxall SA, Jones E (2012) Markers for characterization of bone marrow multipotential stromal cells. Stem Cells Int 2012: 975871.
16. Jeong JA, Ko KM, Park HS, Lee J, Jang C, et al. (2007) Membrane proteomic analysis of human mesenchymal stromal cells during adipogenesis. Proteomics 7: 4181–4191.
17. Morrison TB, Weis JJ, Wittwer CT (1998) Quantification of low-copy transcripts by continuous SYBR Green I monitoring during amplification. Biotechniques 24: 954–958, 960, 962.
18. Ferrari M, Fornasiero MC, Isetta AM (1990) MTT colorimetric assay for testing macrophage cytotoxic activity in vitro. J Immunol Methods 131: 165–172.
19. Fotakis G, Timbrell JA (2006) In vitro cytotoxicity assays: comparison of LDH, neutral red, MTT and protein assay in hepatoma cell lines following exposure to cadmium chloride. Toxicol Lett 160: 171–177.
20. Hoemann CD, El-Gabalawy H, McKee MD (2009) In vitro osteogenesis assays: influence of the primary cell source on alkaline phosphatase activity and mineralization. Pathol Biol (Paris) 57: 318–323.
21. Susa M, Morii T, Yabe H, Horiuchi K, Toyama Y, et al. (2009) Alendronate inhibits growth of high-grade chondrosarcoma cells. Anticancer Res 29: 1879–1888.
22. Muller S, Migianu E, Lecouvey M, Kraemer M, Oudar O (2005) Alendronate inhibits proliferation and invasion of human epidermoid carcinoma cells in vitro. Anticancer Res 25: 2655–2660.
23. Moreira MS, Katayama E, Bombana AC, Marques MM (2005) Cytotoxicity analysis of alendronate on cultured endothelial cells and subcutaneous tissue. a pilot study. Dent Traumatol 21: 329–335.
24. Correia Vde F, Caldeira CL, Marques MM (2006) Cytotoxicity evaluation of sodium alendronate on cultured human periodontal ligament fibroblasts. Dent Traumatol 22: 312–317.
25. Suri S, Monkkonen J, Taskinen M, Pesonen J, Blank MA, et al. (2001) Nitrogen-containing bisphosphonates induce apoptosis of Caco-2 cells in vitro by inhibiting the mevalonate pathway: a model of bisphosphonate-induced gastrointestinal toxicity. Bone 29: 336–343.

26. Moreau MF, Guillet C, Massin P, Chevalier S, Gascan H, et al. (2007) Comparative effects of five bisphosphonates on apoptosis of macrophage cells in vitro. Biochem Pharmacol 73: 718–723.
27. Toyras A, Ollikainen J, Taskinen M, Monkkonen J (2003) Inhibition of mevalonate pathway is involved in alendronate-induced cell growth inhibition, but not in cytokine secretion from macrophages in vitro. Eur J Pharm Sci 19: 223–230.
28. Fisher JE, Rogers MJ, Halasy JM, Luckman SP, Hughes DE, et al. (1999) Alendronate mechanism of action: geranylgeraniol, an intermediate in the mevalonate pathway, prevents inhibition of osteoclast formation, bone resorption, and kinase activation in vitro. Proc Natl Acad Sci U S A 96: 133–138.
29. Duque G, Vidal C, Rivas D (2011) Protein isoprenylation regulates osteogenic differentiation of mesenchymal stem cells: effect of alendronate, and farnesyl and geranylgeranyl transferase inhibitors. Br J Pharmacol 162: 1109–1118.
30. Weivoda MM, Hohl RJ (2011) The effects of direct inhibition of geranylgeranyl pyrophosphate synthase on osteoblast differentiation. J Cell Biochem 112: 1506–1513.
31. Weivoda MM, Hohl RJ (2011) Effects of farnesyl pyrophosphate accumulation on calvarial osteoblast differentiation. Endocrinology 152: 3113–3122.
32. Yoshida T, Asanuma M, Grossmann L, Fuse M, Shibata T, et al. (2006) Geranylgeranyl-pyrophosphate (GGPP) synthase is down-regulated during differentiation of osteoblastic cell line MC3T3-E1. FEBS Lett 580: 5203–5207.
33. Weivoda MM, Hohl RJ (2012) Geranylgeranyl pyrophosphate stimulates PPARgamma expression and adipogenesis through the inhibition of osteoblast differentiation. Bone 50: 467–476.
34. Ebert R, Zeck S, Krug R, Meissner-Weigl J, Schneider D, et al. (2009) Pulse treatment with zoledronic acid causes sustained commitment of bone marrow derived mesenchymal stem cells for osteogenic differentiation. Bone 44: 858–864.

BL-7010 Demonstrates Specific Binding to Gliadin and Reduces Gluten-Associated Pathology in a Chronic Mouse Model of Gliadin Sensitivity

Justin L. McCarville[1], Yotam Nisemblat[2], Heather J. Galipeau[1], Jennifer Jury[1], Rinat Tabakman[2], Ad Cohen[2], Esmira Naftali[2], Bela Neiman[2], Efrat Halbfinger[2], Joseph A. Murray[3], Arivarasu N. Anbazhagan[4], Pradeep K. Dudeja[4], Alexander Varvak[5], Jean-Christophe Leroux[6], Elena F. Verdu[1]*

1 Farncombe Family Digestive Health Research Institute, McMaster University, Hamilton, Ontario, Canada, **2** BioLineRx, Ltd., Jerusalem, Israel, **3** Division of Gastroenterology and Hepatology, Mayo Clinic, Rochester, Minnesota, United States of America, **4** Division of Gastroenterology and Hepatology, Department of Medicine, University of Illinois and Chicago and Jesse Brown VA Medical Center, Chicago, Illinois, United States of America, **5** Chromatography Unit, Scientific Equipment Center, The Mina and Everard Goodman Faculty of Life Sciences, Bar-Ilan University, Ramat-Gan, Israel, **6** Institute of Pharmaceutical Sciences, Department of Chemistry and Applied Biosciences, ETH Zürich, Zürich, Switzerland

Abstract

Celiac disease (CD) is an autoimmune disorder in individuals that carry DQ2 or DQ8 MHC class II haplotypes, triggered by the ingestion of gluten. There is no current treatment other than a gluten-free diet (GFD). We have previously shown that the BL-7010 copolymer poly(hydroxyethyl methacrylate-*co*-styrene sulfonate) (P(HEMA-*co*-SS)) binds with higher efficiency to gliadin than to other proteins present in the small intestine, ameliorating gliadin-induced pathology in the HLA-HCD4/DQ8 model of gluten sensitivity. The aim of this study was to investigate the efficiency of two batches of BL-7010 to interact with gliadin, essential vitamins and digestive enzymes not previously tested, and to assess the ability of the copolymer to reduce gluten-associated pathology using the NOD-DQ8 mouse model, which exhibits more significant small intestinal damage when challenged with gluten than HCD4/DQ8 mice. In addition, the safety and systemic exposure of BL-7010 was evaluated *in vivo* (in rats) and *in vitro* (genetic toxicity studies). *In vitro* binding data showed that BL-7010 interacted with high affinity with gliadin and that BL-7010 had no interaction with the tested vitamins and digestive enzymes. BL-7010 was effective at preventing gluten-induced decreases in villus-to-crypt ratios, intraepithelial lymphocytosis and alterations in paracellular permeability and putative anion transporter-1 mRNA expression in the small intestine. In rats, BL-7010 was well-tolerated and safe following 14 days of daily repeated administration of 3000 mg/kg. BL-7010 did not exhibit any mutagenic effect in the genetic toxicity studies. Using complementary animal models and chronic gluten exposure the results demonstrate that administration of BL-7010 is effective and safe and that it is able to decrease pathology associated with gliadin sensitization warranting the progression to Phase I trials in humans.

Editor: Sabato D'Auria, CNR, Italy

Funding: This work was funded by a CIHR grant to EFV (MOP-123282) and by JA Campbell Young Investigator Awards from the Canadian Celiac Association to HJG and JLM, and a grant by BioLineRx. YN, RT, AC, EN, BN and EH (employees of BioLineRx) were involved in discussions about the dose ratios to be employed in the study and provided feedback to the manuscript's discussion. The funders had no role in data collection and analysis of results performed at McMaster University. EFV holds a Canada Research Chair. Studies in the laboratory of PKD were supported by NIH DK54016, DK92441 and Department of Veterans Affairs.

Competing Interests: EFV receives funding from BioLineRX and YN, RT, AC, EN, BN and EH are employed by BioLineRX.

* Email: verdue@mcmaster.ca

Introduction

Celiac disease (CD) is an autoimmune, chronic small intestinal enteropathy triggered by the consumption of gluten, composed of gliadins and glutenins, in genetically predisposed individuals expressing the HLA genes DQ2 or DQ8 [1,2]. Development of the disease may be accompanied by an unknown environmental risk factor [3,4]. Immune responses in these individuals mainly involve reaction with gliadin-derived peptides, such as the 33-mer [5], that have been modified by the native human enzyme, transglutaminase 2 (TG2), allowing greater binding affinity to HLA DQ2 and DQ8 haplotypes on gut immune cells [6]. About 1% of the population worldwide is affected by CD [7] and its prevalence is currently on the rise [8]. CD is characterized by a degree of small intestinal abnormalities including villus atrophy in the small intestine, an increase in intraepithelial lymphocytes (IELs) in villi tips and increased antibody titers to TG2 in serum [9,10].

Currently there is no effective pharmacological treatment for CD. Individuals with CD must adhere to a strict regimen of gluten-free foods [11,12]. Following a gluten-free diet (GFD) is difficult, and non-adherence can lead to ongoing symptoms, even when consuming as little as 50 mg/day [13], which is equivalent to a breadcrumb. Individuals with CD on a gluten containing diet have been shown to have nutritional deficiencies, neurological conditions, osteoporosis, increased mortality and higher risk for certain malignancies [7,12,14]. In addition, adult patients with CD that have adopted a GFD often experience residual symptoms or have evidence of persistent mucosal inflammation [15,16]. Therefore there is need for supportive therapy to a GFD. There are a number of therapeutic approaches under development for CD. Some of these include intestinal permeability modulators and endopeptidases that cleave gliadin peptides making them less immunogenic [17]. Some therapies are based on blocking or skewing the immune response to gliadin peptides (eg. HLA-DQ2 blockers) and vaccination with gluten components [18].

Our therapeutic approach involves the use of a high molecular weight, non-absorbable polymer, BL-7010, previously called P(HEMA-*co*-SS), which binds gliadin intraluminally, protecting it from enzymatic cleavage by digestive enzymes, therefore avoiding immunogenic effects of gliadin-derived peptides. Our group has demonstrated that BL-7010 was effective at binding gliadin and decreasing gliadin/gluten-associated pathology *in vitro* using cell culture [19] and, in a pilot study *in vivo*, using the acute transgenic HLA-HCD4/DQ8 mouse model of gluten sensitization [19,20]. We also provided preliminary evidence of activity upon prolonged exposure and, using tritium-labeled BL-7010 [20].

The purpose of the current study was to examine the effectiveness and safety of two batches of BL-7010, the previously described prototype polymer (hereby called Polymer A), that was found to be effective *in vitro* and *in vivo* in HCD4-DQ8 mice, and a new industrial batch (Polymer B), under chronic conditions of gluten sensitization. These two polymers have been produced under different synthesis methods. The reference polymer (Polymer B) was created by radical polymerization, whereas Polymer A was produced by an initiator and catalyst. These two polymers share a similar structure as used by Pinier *et al.* (2012) [20], and described by Liang *et al.* (2009) [21]. This comparison is key prior to the translation from preclinical to Phase I clinical trials using BL-7010. We chose to test this in a different model of gluten sensitivity, the NOD-DQ8 mouse that contains native mouse TCRs and develops moderate inflammation upon sensitization and chronic challenge with gliadin [22]. Additional aims were to determine the binding specificity and interactions of BL-7010 with gliadin and with nutrients that could be present in the small intestine, and to evaluate the genetic toxicity *in vitro* and the safety and systemic absorption of unlabeled BL-7010 as part of a 14-day repeated toxicology study in rats.

Materials and Methods

BL-7010 synthesis

BL-7010 co-polymer was synthesized by using two polymerization methods to form Polymer A and Polymer B of BL-7010. In the synthesis method of Polymer A, 2-hydroxyethyl methacrylate (HEMA) and sodium styrene sulfonate (SSNa) were polymerized in the presence of initiator and catalyst and then purified by filtration and precipitation to obtain the BL-7010 copolymer. The copolymer was obtained as white solid.

In the synthesis method of Polymer B, HEMA and SSNa were polymerized in the presence of a radical initiator. The work-up for the purification of the polymer was performed by few precipitation steps and finalized by dissolving the precipitation in water and drying. BL-7010 Polymer B was obtained as white solid with almost a double yield of Polymer A.

Binding studies using surface plasmon resonance (SPR)

BL-7010 was synthesized by radical polymerization method. Binding of the BL-7010 to the Biacore chip was conducted via biotin-streptavidin interaction using biotinylated BL-7010. Biotin was attached to the BL-7010 chain *via* coupling reaction using N, N′-Diisopropylcarbobiimide (DIC) and 4-Dimethylaminopyridine (DMAP) as a coupling agent. The reaction was performed in DMSO under inert condition of argon for 48 h. The product was isolated by work-up with water, filtration, dialysis (cut-off 7,000 Da) and freeze-drying to obtain the biotinylated BL-7010 as a white solid. Streptavidin was immobilized to the chip (CM5, GE Healthcare) at 200 μg/mL in immobilization buffer (10 mM sodium acetate pH 5.0) to reach values of 6000RU using flow rate of 10 μL/min for 1.5 minutes. Subsequently, free-non-reacted active sites on the chip were blocked with 1 M ethanolamine hydrochloride pH 8.5 at 10 μL/min for 7 min [23]. The interaction of BL-7010 with vitamins and enzymes was evaluated with Polymer A while the interaction with gliadin, extracted and purified from whole wheat gliadin (Sigma), was assessed with both forms of BL-7010, Polymers A and B. The analysis was performed using the Biacore T100 instrument (GE). Studies were performed at 25°C, using HBS-EP (10 mM HEPES, pH 7.4, 150 mM NaCl, 3 mM EDTA, 0.05% Polysorbate 20) as running buffer and PBS as the dilution buffer. Regeneration of active surfaces was accomplished by passing a solution of 1 mM NaOH (20 μL/min, 30 sec).

Binding of BL-7010 to the chip was conducted by running a solution of 200 μg/mL at a flow rate of 5 μL/min, for 10 min on the same channel. BL-7010 was linked with a value of 500-1000RU. The kinetics and affinity of interactions with gliadin (10-640 ng/mL), pepsin (500 and 1000 nM), pancreatin (500 and 1000 nM), bovine serum albumin (BSA), and vitamins B1, B2, B3, B5, B6, B9, B12, D3, E, A, K1 and C (10 and 100 μM), all obtained from Sigma, were assessed. For the SPR analysis, evaluation was undertaken a kinetic evaluation of 1:1 binding model of Biacore Evaluation Software v. 2.0.3, where all relevant QC parameters, such as residuals, goodness of fit, and curvature, were met.

Animals

The McMaster University Animal Care Committee approved all animal experiments following the guidelines prepared by the Canadian Council on Animal Care. The efficacy experiments were conducted using a transgenic mouse strain expressing HLA-DQ8, deficient in MHC class II, on a NOD background (NOD-DQ8s) bred at McMaster University. Mice were kept on a low-fat (4.4%), gluten-free diet (Harlan Laboratories). Sensitization began between 6 to 12 weeks of age. For toxicity studies in rats, the care or use of animals were reviewed and approved by WuXi AppTec's Institutional Animal Care and Use Committee (IACUC).

Sensitization and treatment protocol

Pepsin, trypsin digested gliadin (PT-gliadin), for sensitization, was prepared as described previously [20]. Briefly, gliadin (Sigma-Aldrich) was dissolved in endotoxin-free 0.02 N HCl and subsequently digested with 1 g of pepsin (Sigma-Aldrich) in a 37°C water bath for 2 h. Following the incubation, the pH was adjusted to 7.4 using endotoxin-free 2 M NaOH and trypsin (Sigma-Aldrich) was added and boiled for 30 min as described previously [24]. The PT-gliadin was then stored at −20°C until use. Gluten (Sigma-Aldrich), for challenge, was prepared by dissolving in 0.02 M acetic acid and stored at room temperature.

Mice were split into 4 groups (8 male and 8 female per group): a) non-sensitized negative control (negative), b) sensitized positive control (gluten challenge), c) sensitized group receiving Polymer B and d) sensitized reference group receiving Polymer A. Prior to sensitization, mice were injected twice, intraperitoneally (IP), with 250 μg of anti-CD25 antibody (Leinco Technologies), one week apart to deplete $CD4^+CD25^+Foxp3^+$ cells. Sensitization began when mice were 6 to 12 weeks old. Sensitized mice received, by gavage, a mixture of PT-gliadin (500 μg) and cholera toxin (CT) (25 μg) (List Biological Laboratories Inc.) once a week for 3 weeks. Following sensitization, mice were administered 2 mg of gluten by gavage, 3 times a week for 4 weeks. Mice received a gluten challenge 24 h before sacrifice. Mice receiving either Polymers A or B were gavaged with gluten and the polymer at a weight ratio of 1:3 (6 mg of BL-7010) during the challenge period. BL-7010 was administered as an oral gavage (dissolved in saline) 5 min prior to the gluten challenge. BL-7010 solution was given at a constant dose volume of 60 mg/ml. Sensitization and analysis were performed in blinded fashion. Non-sensitized mice received only CT during sensitization (no gliadin) and 0.02 M acetic acid for challenge (no gluten).

Histological and immunohistological analysis of small intestinal tissue

Proximal sections of the small intestine were fixed in 10% formalin, embedded in paraffin and H&E stained. Sections were evaluated for evidence of inflammation and the measurement of villus-to-crypt (V/C) ratios. $CD3^+$ staining was carried out as previously described [25]. Briefly, paraffin embedded sections of the small intestine were stained with a $CD3^+$ (1:2000, Dako) overnight, followed by an incubation with HRP and visualization via the substrate 3-amino-9-ethylcarbazole. $CD3^+$ IELs were counted for every 20 epithelial cells and averaged over 5 villi tips. Histology (4x and 10x magnification) and immunohistochemistry (20x magnification) were viewed via light microscopy (Olympus). Images were analyzed in Image Pro Plus and the average V/C ratio was calculated for each group.

Intestinal permeability

Intestinal permeability was measured as described previously [20]. Briefly, a 5-cm section from the jejunum was removed during sacrifice, cut along the mesenteric border, rinsed and mounted in an Ussing chamber (exposed surface area of 0.6 cm^2). The section was bathed in oxygenated Krebs buffer containing 10 mM glucose (serosal side) or 10 mM mannitol (luminal side) at 37°C. In order to measure the net active transport across the epithelium, a short circuit current (Isc) was applied through the tissue under voltage clamp conditions. Conductance (mS/cm^2) was recorded 15 min afterwards. Once equilibrium was achieved, ^{51}Cr-EDTA (Perkin Elmer) was added to the luminal buffer and serosal buffer (500 μL) was collected every 30 min for 2 h and replaced with fresh buffer to maintain optimal volume. The ^{51}Cr-EDTA was measured via a scintillation counter (Beckman Coulter LS6500 Multi-Purpose Scintillation Counter).

RNA extraction and real-time PCR

To quantitate the PAT-1 mRNA, total RNA was extracted from jejunal tissues from all the four groups using RNeasy mini kit (Qiagen) according to the manufacturer's instruction. The quality and quantity of total RNA were determined by Beckman DU640 spectrophotometer. Extracted RNA was amplified by Brilliant SYBR Green qRT-PCR Master Mix kit (Agilent Technologies, Santa Clara; CA) utilizing gene specific primer for PAT-1 (sense primer, 5′-AAATGGAGCTGCAGAGGAGA-3′; antisense primer, 5′ GCTGGAGCAGAAGAGAATGG-3′) and GAPDH (sense primer, 5'-TGTGTCCGTCGTGGATCTGA-3'; antisense primer, 5'-CCTGCTTCACCACCTCTTGAT-3'). The relative mRNA levels of PAT-1 were expressed as % of control normalized to GAPDH used as internal control gene.

14-Day toxicity study

The potential toxicity, assessment of the reversibility, persistence, or delayed occurrence of toxic effects of BL-7010 was tested in Sprague-Dawley (SD) rats administered by oral gavage for 14 days followed by 14-day recovery period. In addition, the toxicokinetics (TK) of BL-7010 was determined in satellite animals. Eighty (N = 80, F&M) rats were randomly assigned to 4 groups to determine the toxicity of BL-7010 (Polymer B) when administered once daily for 14 days by oral gavage at doses of 1000, 2000 or 3000 mg/kg/day. The control group was administered vehicle (saline). Study animals were 10 or 15/group, and TK animals were 3 in the control group and 9/group in the treated groups. Five of the study rats/group in the control and high dose groups were allocated for a two-week recovery period. Parameters evaluated during the study included clinical observation, body weight, food consumption, ophthalmology examinations, clinical pathology (serum chemistry, hematology, coagulation, and urinalysis), gross pathology, organ weight histopathology and TK.

Genetic toxicity studies

The potential mutagenic activity of BL-7010 was evaluated in accordance with the Organisation for Economic Co-operation and Development (OECD) 471 and 476 guidelines using bacterial and mammalian cells.

The bacterial reverse mutation assay was conducted to evaluate whether BL-7010 would induce reverse mutations at the histidine locus of the *Salmonella typhimurium* tester strains TA98, TA100, TA1535, and TA1537 or at the tryptophan locus of Escherichia coli tester strain WP2uvrA. The assay was conducted in the presence and absence of metabolic activation. Parallel testing was conducted with the negative control (vehicle alone) and positive controls. The mean number of revertants for the test article plates was compared to the mean number of revertants of the negative control plates for each of the five tester strains.

For the mammalian cells, the mouse lymphoma L5178Y/TK+/− cell line, heterozygous at the thymidine kinase (TK) locus, was used. This testing was conducted by exposing the cells to the test article dilutions and negative control (saline) and positive controls. The treatment was performed for 4 hours in the presence and absence of metabolic activation and 24 h in the absence of metabolic activation.

Evaluation of systemic exposure and urine levels using LC-MS/MS methods

Systemic exposure and urine level of BL-7010 following single and repeated administration was evaluated as part of the 14 days toxicology study. Sixty (60, 30/sex) satellite rats were assigned to 1 control group (3/sex) and 3 test article treated groups (9/sex/group) for the assessment of TK. Animals in test article treated groups were administered with the solution of BL-7010 (Polymer B) in vehicle by oral gavage at 3000 mg/kg/day for 14 days. Rats in the control group were dosed for 14 days with the vehicle only (0.9% sodium chloride for injection (saline)). Blood samples were collected at predose (0), 0.5, 1, 2, 4, 6, 8, 12, 24 on Day 1 and Day 14 and additional 48, 72 and 96 h postdose on Day 14 from 3 animals/sex/time point/group via the jugular vein as an alternating sampling design from the test article treated animals

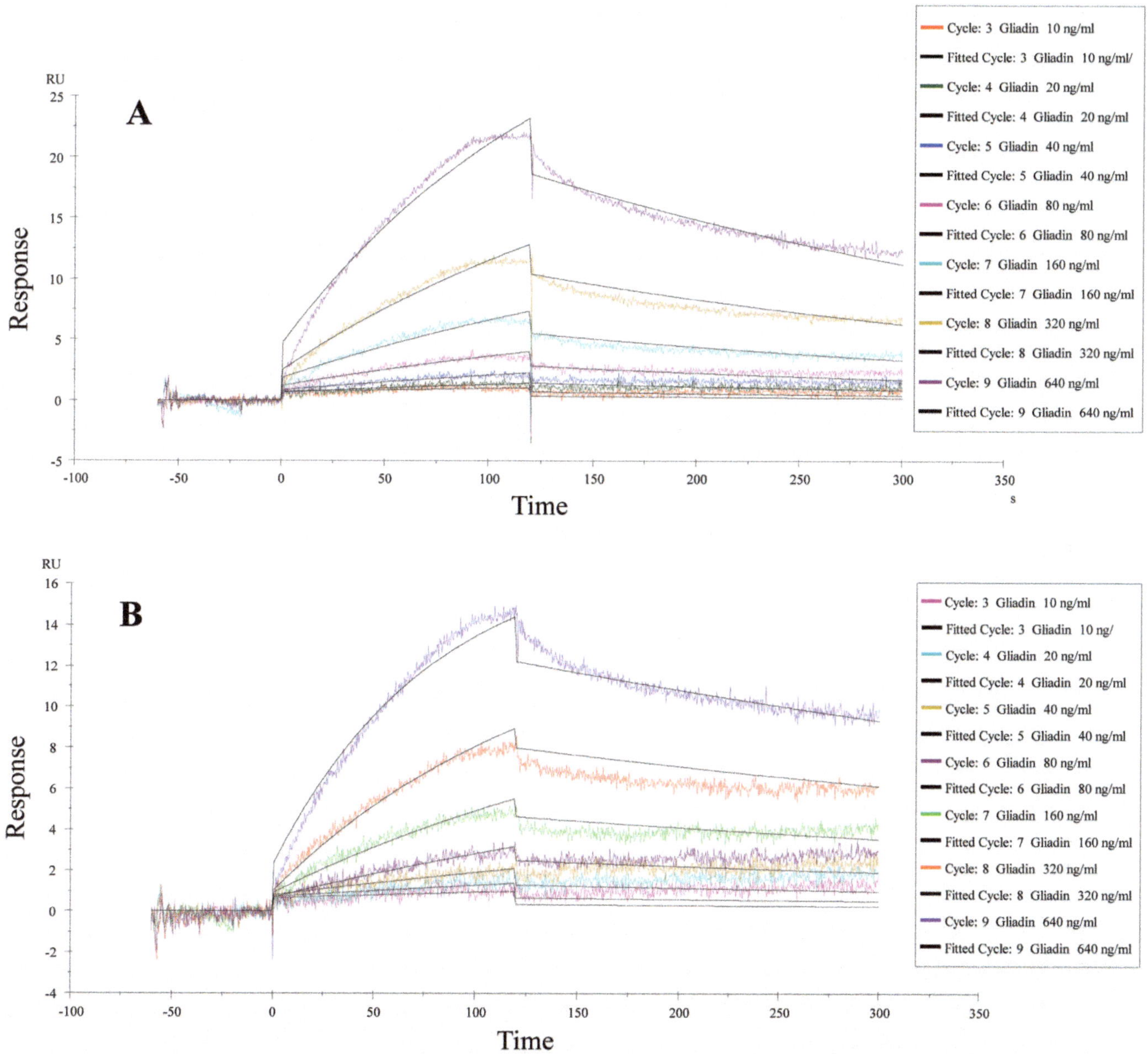

Figure 1. An overlay of sensograms showing Polymer A interaction with gliadin at a series of concentrations of 10-640 ng/mL. KD was determined to be 12.1 nM, with kon of 2.39×105 1/M·s and koff of 2.89×10-3 1/s. B. An overlay of sensograms showing Polymer B interacting with gliadin, with KD of 2.44 nM, kon of 6.11×105 1/M·s and koff of 1.49×10-3 1/s using Biacore Evaluation Software (see Methods). For both graphs, colored lines indicate experimental sensograms at various concentrations, while corresponding black lines denote fitting of experimental data using 1:1 binding model with global Rmax.

and only collected at 1, 4 and 24 h postdose from the control group animals. Samples were collected into appropriately labeled tubes containing K_2EDTA (Becton Dickinson). TK animals were placed in metabolic cages for urine collection up to 96 h post dosing. Urine samples were frozen at $\leq -60°C$ until analysis.

Plasma fraction was separated, aliquoted and stored at $-60°C$ pending analysis. For BL-7010 determination in plasma and urine, samples were analyzed by validated liquid chromatography - tandem mass spectrometry LC-MS/MS bioanalytical methods with Low Limit of Quantification of 5.00 μg/mL and 10.0 μg/mL for plasma and urine, respectively. Briefly, these methods utilize 30 μL aliquots of urine or 50 μL aliquots of plasma with dilution extraction for urine and protein digestion followed solid-phase extraction by Oasis HLB plates (30 μm, 10 mg) from Waters for plasma. Polyethylene glycol (PEG) was used as an internal standard. Waters Xbridge^{TM}, BEH300, 3.5 μm (2.1 mm I.D. x 50 mm) chromatographic column and isocratic elution of Mobile Phase A 0.1%NH4OH in 2 mM NH4Ac in water and Mobile Phase B 0.1%NH4OH in 20 mM NH4Ac in MeOH/AcN (v:v, 1:1) were used for separation. Analysis was done by negative ion Turbo Ion Spray in the multi-reaction monitoring mode (MRM). The LC-MS/MS technique employed very high declustering potentials at the orifice of the mass spectrometer before the mass filters. In this procedure, collision induced dissociation at the

Table 1. Interaction of BL-7010 (Polymer A) with gliadin, BSA, proteolytic enzymes and vitamins.

BL-7010 (Polymer B) – gliadin	**2.44E-9 M**
BL-7010 – gliadin	1.21E-8 M
BL-7010 – BSA	1.09E-8 M
BL-7010 – pepsin	No interaction[1]
BL-7010 – pancreatin	No interaction
BL-7010 – K1	No interaction
BL-7010 – D	No interaction
BL-7010 – E	No interaction
BL-7010 – A	No interaction
BL-7010 – C	No interaction
BL-7010 – B1	No interaction
BL-7010 – B2	No interaction
BL-7010 – B3	No interaction
BL-7010 – B5	No interaction
BL-7010 – B6	No interaction
BL-7010 – B12	No interaction

[1]"No Interaction" means that no interaction signal (no increase in resonance units) was detected while increasing the concentration of the tested material and therefore no kinetics parameters could be calculated.

orifice resulted in a partial decomposition of BL-7010 into lower molecular weight fragments. These fragments were detectable in the range of the mass spectrometer. These lower molecular weight ions were then utilized as parent ions that were further dissociated by MS/MS into characteristic product ions for detection.

Statistical analysis

Statistical analysis was completed via an ANOVA coupled with a Bonferroni post-hoc test.

Results

BL-7010 interacts specifically with gliadin

We first evaluated the interaction of biotinylated BL-7010 Polymers A and B, with gliadin using the SPR technique. Polymers A and B showed high affinities for gliadin with KD values of 1.21 10^{-8} M (Chi^2 of 0.263) (Figure 1A) and 2.44×10^{-9} M (Chi^2 of 0.385) (Figure 1B), respectively. These results demonstrate that BL-7010 interacted avidly with gliadin. The differences in KD could be attributed to the slight composition and molecular weight differences between the two batches, which is related to the fact that polymers are characterized on an average basis and are not single molecular entities. It could be due to the different position of binding of the polymers to the chip since the biotin groups in the two polymers are positioned differently (one at the chain-end of the polymer and the other at numerous places on the main chain).

The interaction of biotinylated Polymer A with vitamins, BSA and digestive enzymes was evaluated by SPR. The enzymes pepsin and pancreatin were tested at two concentrations, 500 and 1000 nM, and gliadin was injected as a positive control at a concentration of 62.5 nM. No interaction between BL-7010 and pepsin or pancreatin, even at a 16-fold higher concentration than gliadin, was observed. Polymer A was found to interact with BSA with a KD of 1.09×10^{-8} M (Chi^2 of 0.221). These results suggest that BL-7010 may also interact with human serum albumin but this is expected to have negligible effect on malnutrition and safety as discussed below. Due to their low molecular weights, the vitamins were evaluated at the high concentrations of 10 and 100 μM, while gliadin was injected as a positive control at a concentration of 100 nM. No binding of BL-7010 to the tested vitamins (K1, D, E, A, B5, C, B1, B2, B3, B6 and B12) was observed (Table 1).

BL-7010 attenuates gluten-induced pathology in a chronic mouse model

Villus-to-crypt (V/C) ratios were decreased by 66% in sensitized mice without treatment in comparison to the non-sensitized (control) mice ($p<0.0001$) due to villus shortening (Figure 2). Administration of BL-7010 Polymer B prevented the reduction in V/C ratios induced by gluten challenge in comparison to sensitized mice that did not receive BL-7010. However, V/C ratios were still slightly lower than in non-sensitized mice ($p<0.05$). On the other hand, BL-7010 Polymer A prevented reductions in V/C ratio ($p<0.0001$) in comparison to gluten challenged mice and led to V/C ratios similar to those found in non-sensitized controls (Figure 2).

Gliadin-sensitized non-treated mice also had increased $CD3^+$ IELs in villi tips of proximal small intestinal tissue in comparison to non-sensitized controls ($p<0.0001$) (Figure 3). Both Polymers B ($p<0.0001$) and A ($p<0.001$) were effective at lowering $CD3^+$ IELs in comparison to gliadin-sensitized control mice. The results show that the activities of both batches of BL-7010 are comparable and confirm the ability of BL-7010 to attenuate pathological changes associated with CD.

BL-7010 attenuates gluten-induced mucosal barrier dysfunction in a chronic mouse model

Sensitized non-treated mice had higher paracellular permeability in comparison to the non-sensitized mice ($p<0.0001$) (Figure 4A). Sensitized mice treated with Polymers A ($p<0.01$) or B ($p<0.001$) had lower paracellular permeability compared to the non-treated sensitized mice. BL-7010 treated mice had ^{51}Cr-EDTA flux levels that were comparable to non-sensitized control mice. Gliadin-sensitized non-treated mice also exhibited an increase in PAT-1 mRNA expression in comparison to non-sensitized control mice ($p<0.001$) (Figure 4B). Gliadin-sensitized mice receiving Polymer A or Polymer B had lower PAT-1 mRNA expression in comparison to sensitized non-treated mice ($p<0.05$), and comparable to that of the non-sensitized mice.

BL-7010 is not absorbed systemically

The presence of BL-7010 in rat plasma and urine was evaluated post single and 14-days once daily repeated administration of Polymer B at a dose of 3000 mg/kg/day. This dose is ~23 times higher than the expected effective human dose. BL-7010 was not found in any plasma sample tested following single or 14 days repeated administration of 3000 mg/kg, suggesting the lack of systemic exposure. BL-7010 was found at very low levels in the urine (0–<1.54% of dose) (data not shown). The trace amounts found in the urine are due to external contamination with feces in the metabolic cages.

BL-7010 safety and tolerability

Oral administration of BL-7010 once daily for 14 consecutive days was well tolerated at all dose levels. No mice reached endpoint to be sacrificed due to adverse effects. Treatment with BL-7010 at doses up to and including 3000 mg/kg/day did not cause adverse effects on body weight, food consumption, clinical pathology, ophthalmoscopy, organ weights, or gross pathology

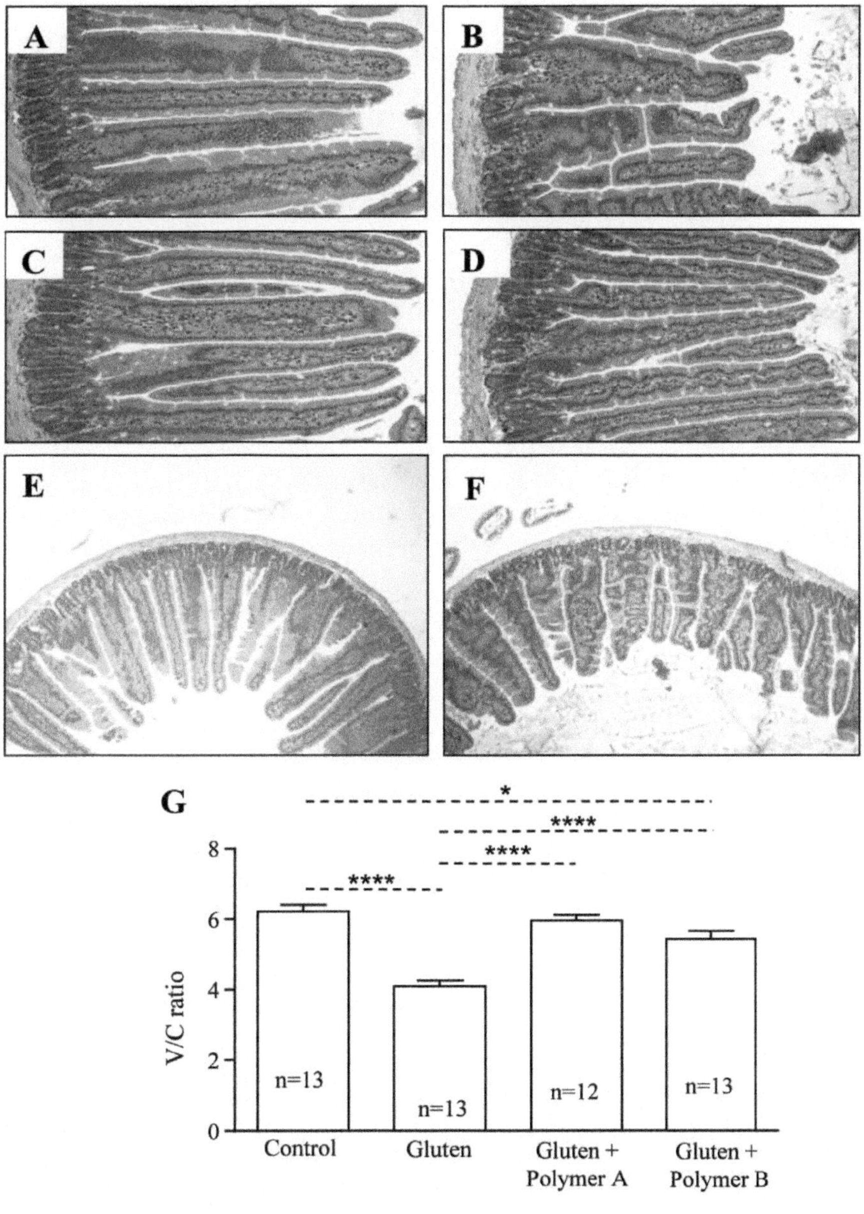

Figure 2. Administration of BL-7010 (Polymers A and B) at a gluten: BL-7010 ratio of 1:3 decreased gluten-associated histological pathology in the small intestine of sensitized mice. Small intestinal H&E stained sections from (A) non-sensitized mice (control) (B) gliadin-sensitized mice (gluten) (C) gliadin-sensitized mice, receiving Polymer A (D) gliadin-sensitized mice, receiving Polymer B. (E) non-sensitized mice (control) at different magnification (F) sensitized mice at different magnification (G) Quantification of V/C ratios from small intestinal villi. H&E sections were viewed via optical microscopy (4x and 10x magnification). Bars represent the mean +/− SEM, statistical analysis was completed via one-way ANOVA and Bonferroni's post-hoc test (*$p<0.05$, ****$p<0.0001$) ($n = 12$–13).

following the dosing or recovery phases. BL-7010 was clinically well tolerated at 1000, 2000, and 3000 mg/kg/day when administered by oral gavage to rats for 14 days. The no observed adverse effect level (NOAEL) was considered to be 3000 mg/kg/day for both males and females. In the bacterial reverse mutation assay BL-7010 was considered to be nonmutagenic to *S. typhimurium* tester strains TA98, TA100, TA1535, and TA1537, and to *E. coli* tester strain WP2uvrA. In the mammalian cells mouse lymphoma assay the test article did not cause a two-fold or greater increase in the mean mutant frequency over the negative control of the L5178Y/TK+/− cell line either in the presence or absence of metabolic activation and therefore is not consider mutagenic in this assay.

Discussion

There are several approaches for developing treatments for CD. One of them is the degradation of immunogenic gluten peptides to non-immunogenic fragments in the GI tract. This approach was found to be clinically effective [26] in patients receiving gluten degrading proteases proving that prevention of the formation of gliadin immunogenic peptides is clinically beneficial. In contrast to the enzymatic degradation approach, BL-7010 prevents the

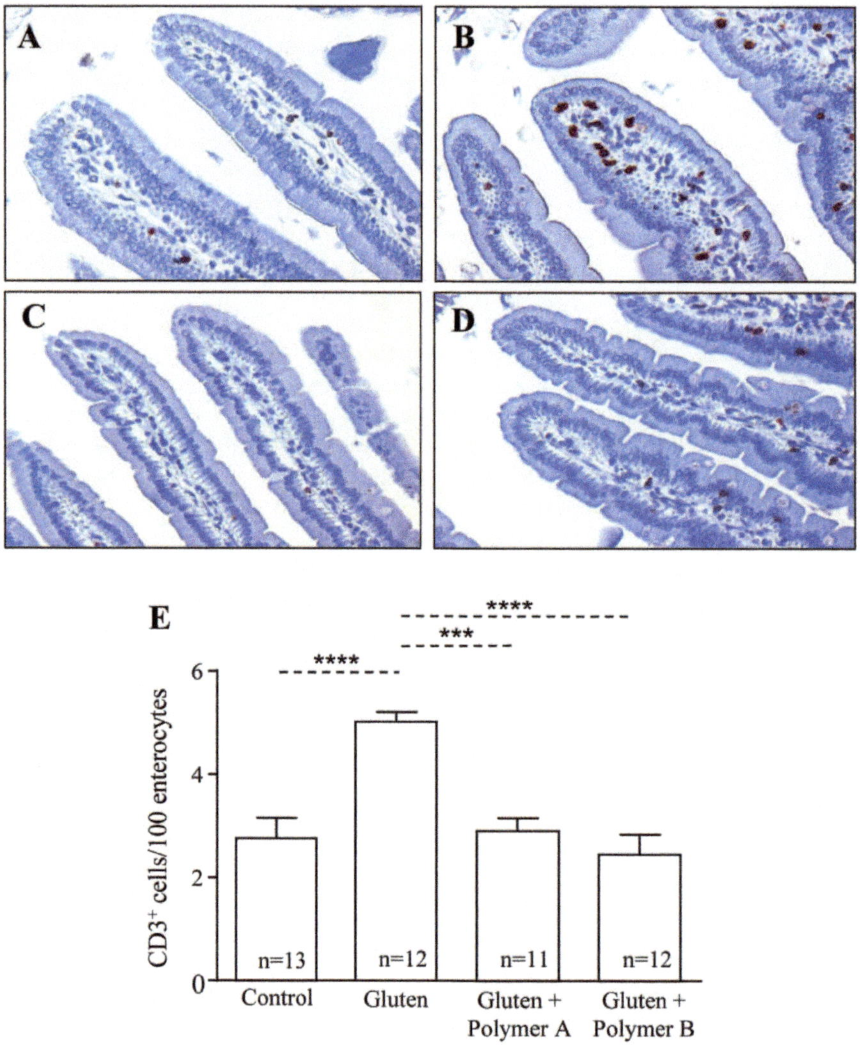

Figure 3. Administration of BL-7010 (Polymers A and B) decreased CD3$^+$ intraepithelial cells in villi tips of gliadin-sensitized mice. Small intestinal CD3$^+$ stained small intestinal tissues from (A) non-sensitized mice (control) (B) gliadin-sensitized mice (gluten) (C) gliadin-sensitized mice, receiving Polymer A (D) gliadin-sensitized mice, receiving Polymer B. (E) Quantification of CD3$^+$ cells/100 enterocytes in villi tips. Stained sections were viewed via optical microscopy (20X magnification). Bars represent the mean +/− SEM, statistical analysis was completed via one-way ANOVA and Bonferroni's post-hoc test (***$p<0.001$, ****$p<0.0001$) ($n = 11–13$).

formation of gliadin immunogenic peptides by sequestering gliadins in the GI tract [19,20].

The main objectives of this paper were to evaluate the efficacy of BL-7010, in NOD-DQ8 mice, a mouse model that develops moderate inflammation in the small intestine when sensitized and challenged with gluten [22]. We also tested the activity of 2 batches of polymer. Compared to classical low molecular weight drugs, polymers are not characterized by a precise molecular structure but are defined on an average fashion with respect to their composition and size. It is therefore important to validate the reproducibility of the scaled up manufacturing process suitable prior to initiating clinical trials in humans. In addition, the previous *in vivo* study [16] demonstrated the efficacy of Polymer A, which was synthesized by a preliminary synthesis method. The yield and cost effectiveness of this process was relatively poor. Therefore, we have developed a new synthesis method (Polymer B) which has improved cost effectiveness. Due to the fact that Polymer B has slightly different characteristics than Polymer A, the product of the preliminary synthesis method, we aimed to confirm its activity in an animal model of gluten sensitivity. Polymer A was used as a control in the *in vivo* and *in vitro* binding studies, as it was proven to be effective in the past [21]. In the current study we showed that the two batches of BL-7010, Polymers A and B, were effective at abrogating gluten-associated pathology in gliadin-sensitized NOD-DQ8 mice. Both batches of BL-7010 prevented the decrease in V/C ratio, decreased intraepithelial lymphocytosis and intestinal epithelial function in small intestinal tissues.

Recruitment of IELs and decreased V/C ratio, represent pathological features of CD [11,27]. Both batches of BL-7010 prevented changes in IELs and V/C ratios induced by gliadin, indicating amelioration of morphological changes as well as prevention of innate immune parameters in this model. Individuals with CD have been shown to have intestinal barrier dysfunction, which emerges early in the disease [28]. Our results demonstrated that BL-7010 prevented barrier dysfunction associated with gliadin sensitization. A leaky barrier may contribute to uptake of gliadin [29] or other proinflammatory luminal antigens of microbial origin [30], and promote progression of disease.

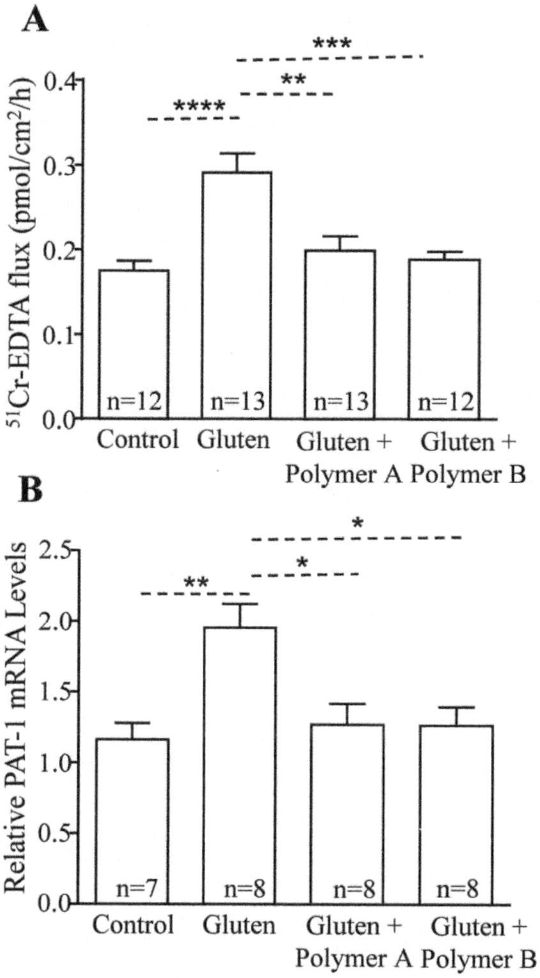

Figure 4. Administration of BL-7010 (Polymers A and B) decreased paracellular permeability and PAT-1 mRNA levels in gliadin-sensitized mice. (A) Gliadin-sensitized mice had higher ^{51}Cr-EDTA flux in comparison to non-sensitized control mice, and administration of Polymers A or B decreased it to the same level as the non-sensitized control mice. (B) Gliadin-sensitized mice had higher mRNA levels of PAT-1 in comparison to non-sensitized control mice, and administration of Polymers A or B to sensitized mice decreased expression to the same level as non-sensitized control mice. Bars represent the mean +/− SEM, statistical analysis was completed via one-way ANOVA and Bonferroni's post-hoc test (**$p<0.01$, ***$p<0.001$, ****$p<0.0001$) (n = 12–13).

Thus, drug development has targeted modulation of the intestinal barrier in CD, particularly using larazotide acetate [31]. This drug has demonstrated the ability to block some symptoms due to gluten sensitivity [31] via actin rearrangement and has also demonstrated the ability to decrease *in vitro* [32] and *in vivo* [33] permeability. BL-7010 normalized paracellular permeability most likely through an indirect mechanism, by reducing production of toxic peptides, limiting interaction of these peptides with the gut mucosa and through prevention of induction of immune responses. In addition to functional analysis of the intestinal epithelium, we measured ion transporter mRNA expression. PAT-1 mRNA expression, a small intestinal luminal membrane CL-/HCO_3- exchanger [34] primarily localized to the villi of the small intestine [35,36]. The expression of PAT-1 was increased in gliadin-sensitized mice and its high expression was restored to control levels in mice treated with both batches of BL-7010. The results suggest that ion transport mechanisms that are important for absorption and homeostatic mechanisms are affected by moderate gluten-induced inflammation and are normalized by BL-7010 administration, probably indirectly through prevention of the interaction of gliadin with the intestinal mucosa.

The mechanism of action of BL-7010 relies on electrostatic and hydrophobic interactions with gliadin and possibly on the formation of hydrogen bonds [21]. The relative specificity of BL-7010 to gliadin is a function of the amino acid sequence of gliadin and its resulting electrostatic charge and hydrophobicity. This kind of mechanism raises concerns about the specificity of BL-7010 to gliadin and possible side effects, such as malnutrition, which can result from interaction with nutrients, or an effect on the digestion process by interaction with digestive enzymes. Additionally, non-specific binding may reduce efficacy due to lower fraction of free BL-7010 that is available for neutralization of gliadin. However, the binding of BL-7010 to other wheat and barley fractions is of interest. We have previously shown that P(HEMA-*co*-SS) blocks the digestion of wheat gluten and barley hordein *in vivo* [20]. Incubation of P(HEMA-*co*-SS) with whole wheat gluten reduces the glutenin sequence SQQQQPPV by 60%, providing indirect evidence that BL-7010 may bind to glutenin. In this study, we demonstrated by SPR technique, that BL-7010 does not interact with the tested vitamins. No interaction was observed with the tested digestive enzymes, pepsin and pancreatin, implying that BL-7010 will not affect the digestion process. On the other hand, high affinity of BL-7010 to gliadin, for both polymeric batches, was observed. Some interaction of BL-7010 with BSA was noted but the fact that BL-7010 is not absorbed into the blood circulation and therefore is not expected to interact with human serum albumin, mitigate the risk of safety issues that can result from such interaction. The determination of KD constant by SPR to monitor the interactions between polymers and proteins should, however, be interpreted with caution since in solution the binder is not bound to a surface and therefore less conformationally hindered to interact with the protein. Our former data obtained by gel electrophoresis indicated that at pH 6.8, albumin interacted less than α-gliadin with BL-7010 [19]. A significant interaction of BL-7010 with nutrients seems therefore unlikely. In agreement with our previous study [30], we show here that a 14-day administration of a much higher dose was safe with no effects on body weight, food consumption, clinical pathology, ophthalmoscopy, organ weights, and gross pathology. This will nevertheless require confirmation in safety studies in humans.

Lack of systemic exposure is a key feature of BL-7010 that contributes to the safety and efficacy of this product. Due to its high molecular weight, BL-7010 is not anticipated to be absorbed systemically and it was found, using tritium-labeled BL-7010, that this compound is excreted almost completely in the feces, both in normal rats and in animals with intestinal barrier dysfunction, gliadin-challenged HLA-HCD4/DQ8 mice [20]. In the tritium-labeled study some radioactivity was detected in the urine (less than 2%), most likely due to proton exchange with water. Thus, in this study, we established a sensitive LC/MS/MS method that can specifically detect BL-7010 in plasma and urine at very high sensitivity and does not rely on radiolabeling, therefore excluding possible bias caused by proton exchange. We showed that BL-7010 is not found in the rat plasma of animals receiving single or 14 days repeated treatment at extremely high dose levels. The low levels of BL-7010 found in the urine could be explained by contamination of the urine samples with feces in the metabolic cage. Indeed, BL-7010 was most commonly found in the urine after more than 8 h, correlating to the time it was found in the feces in previous studies [20], and in accordance with the animals' circadian cycle, activity time and eating period. As there is no simple method to completely separate feces and urine samples from animals, the question of systemic exposure should be further evaluated in clinical studies.

The safety of BL-7010 (Polymer B) was evaluated in a repeated GLP toxicology study. Repeated administration of up to and including 3000 mg/kg/day doses was well tolerated and safe. Microscopic evaluation of the oral cavity, tongue, larynx, esophagus, stomach, small intestine (duodenum, jejunum, ileum) and large intestine (cecum, colon, rectum) revealed no evidence of irritation in these tissues. The no observed adverse effect level (NOAEL) was considered to be 3000 mg/kg/day for both males and females. *In vitro* genetic toxicity studies showed that BL-7010 has no mutagenic activity in both assays, indicate that BL-7010 is not mutagenic. Polymer A was also evaluated in a 14-day repeated toxicity study in rats up to dose of 2000 mg/kg/day and was found to be completely safe (data not shown).

In summary, this study demonstrated that BL-7010 is effective at abrogating gluten-associated pathology in a model of chronic gliadin-sensitization using NOD-DQ8 mice. Both batches of BL-7010 were effective in this model, suggesting that the synthesis method of BL-7010 is robust and produce a functional polymer. Furthermore, the results indicate that BL-7010 may be safe for long term use, as systemic absorption and interactions with vitamins and digestive enzymes were absent. BL-7010 safety was confirmed at doses that are far higher than the expected clinical effective dose. BL-7010 is currently undergoing clinical trials to further evaluate its efficacy, safety and systemic exposure in celiac disease patients.

Supporting Information

File S1 Supplementary information. Includes data for Figures 1, 2, 3, and 4.

Author Contributions

Conceived and designed the experiments: YN AV JCL EFV. Performed the experiments: HJG JJ RT AC EN BN EH ANA. Analyzed the data: JLM YN HJG JJ RT AC EN BN EH ANA PKD AV JCL EFV. Contributed reagents/materials/analysis tools: YN JAM PKD AV JCL EFV. Wrote the paper: JLM YN PKD JCL EFV.

References

1. Sollid LM (2002) Coeliac disease: dissecting a complex inflammatory disorder. Nat Rev Immunol 2: 647–655.
2. Abadie V, Sollid LM, Barreiro LB, Jabri B (2011) Integration of genetic and immunological insights into a model of celiac disease pathogenesis. Annu Rev Immunol 29: 493–525.
3. Tack GJ, Verbeek WH, Schreurs MW, Mulder CJ (2010) The spectrum of celiac disease: epidemiology, clinical aspects and treatment. Nat Rev Gastroenterol Hepatol 7: 204–213.
4. Sollid LM, Jabri B (2013) Triggers and drivers of autoimmunity: lessons from coeliac disease. Nat Rev Immunol 13: 294–302.
5. Shan L, Molberg O, Parrot I, Hausch F, Filiz F, et al. (2002) Structural basis for gluten intolerance in celiac sprue. Science 297: 2275–2279.
6. Molberg O, McAdam SN, Korner R, Quarsten H, Kristiansen C, et al. (1998) Tissue transglutaminase selectively modifies gliadin peptides that are recognized by gut-derived T cells in celiac disease. Nat Med 4: 713–717.
7. Fasano A, Berti I, Gerarduzzi T, Not T, Colletti RB, et al. (2003) Prevalence of celiac disease in at-risk and not-at-risk groups in the United States: a large multicenter study. Archives of internal medicine 163: 286.
8. Ludvigsson JF, Rubio-Tapia A, van Dyke CT, Melton LJ, 3rd, Zinsmeister AR, et al. (2013) Increasing incidence of celiac disease in a North American population. Am J Gastroenterol 108: 818–824.
9. Green PH, Cellier C (2007) Celiac disease. N Engl J Med 357: 1731–1743.
10. Barker JM, Liu E (2008) Celiac disease: pathophysiology, clinical manifestations, and associated autoimmune conditions. Adv Pediatr 55: 349–365.
11. Rubio-Tapia A, Hill ID, Kelly CP, Calderwood AH, Murray JA (2013) ACG clinical guidelines: diagnosis and management of celiac disease. Am J Gastroenterol 108: 656–676; quiz 677.
12. Fasano A, Catassi C (2001) Current approaches to diagnosis and treatment of celiac disease: an evolving spectrum. Gastroenterology 120: 636–651.
13. Catassi C, Fabiani E, Iacono G, D'Agate C, Francavilla R, et al. (2007) A prospective, double-blind, placebo-controlled trial to establish a safe gluten threshold for patients with celiac disease. Am J Clin Nutr 85: 160–166.
14. Fasano A, Catassi C (2012) Clinical practice. Celiac disease. N Engl J Med 367: 2419–2426.
15. Rubio-Tapia A, Ludvigsson JF, Brantner TL, Murray JA, Everhart JE (2012) The prevalence of celiac disease in the United States. The American journal of gastroenterology.
16. Lanzini A, Lanzarotto F, Villanacci V, Mora A, Bertolazzi S, et al. (2009) Complete recovery of intestinal mucosa occurs very rarely in adult coeliac patients despite adherence to gluten-free diet. Alimentary pharmacology & therapeutics 29: 1299–1308.
17. Crespo Perez L, Castillejo de Villasante G, Cano Ruiz A, Leon F (2012) Non-dietary therapeutic clinical trials in coeliac disease. Eur J Intern Med 23: 9–14.
18. Pinier M, Fuhrmann G, Verdu EF, Leroux JC (2010) Prevention measures and exploratory pharmacological treatments of celiac disease. Am J Gastroenterol 105: 2551–2561; quiz 2562.
19. Pinier M, Verdu EF, Nasser-Eddine M, David CS, Vézina A, et al. (2009) Polymeric binders suppress gliadin-induced toxicity in the intestinal epithelium. Gastroenterology 136: 288–298.
20. Pinier M, Fuhrmann G, Galipeau HJ, Rivard N, Murray JA, et al. (2012) The copolymer P(HEMA-co-SS) binds gluten and reduces immune response in gluten-sensitized mice and human tissues. Gastroenterology 142: 316–325 e311–312.
21. Liang L, Pinier M, Leroux JC, Subirade M (2009) Interaction of α-gliadin with poly (HEMA-co-SS): Structural characterization and biological implication. Biopolymers 91: 169–178.
22. Galipeau HJ, Rulli NE, Jury J, Huang X, Araya R, et al. (2011) Sensitization to gliadin induces moderate enteropathy and insulitis in nonobese diabetic-DQ8 mice. The Journal of Immunology 187: 4338–4346.
23. Correia JJ, Detrich HW (2008) Biophysical tools for biologists. London: Academic.
24. Thomas KE, Sapone A, Fasano A, Vogel SN (2006) Gliadin stimulation of murine macrophage inflammatory gene expression and intestinal permeability are MyD88-dependent: role of the innate immune response in Celiac disease. J Immunol 176: 2512–2521.
25. Verdu EF, Huang X, Natividad J, Lu J, Blennerhassett PA, et al. (2008) Gliadin-dependent neuromuscular and epithelial secretory responses in gluten-sensitive HLA-DQ8 transgenic mice. Am J Physiol Gastrointest Liver Physiol 294: G217–225.
26. Tye-Din JA, Anderson RP, Ffrench RA, Brown GJ, Hodsman P, et al. (2010) The effects of ALV003 pre-digestion of gluten on immune response and symptoms in celiac disease *in vivo*. Clinical Immunology 134: 289–295.
27. Marsh MN (1992) Gluten, major histocompatibility complex, and the small intestine. A molecular and immunobiologic approach to the spectrum of gluten sensitivity ('celiac sprue'). Gastroenterology 102: 330–354.
28. Camilleri M, Madsen K, Spiller R, Greenwood-Van Meerveld B, Verne GN (2012) Intestinal barrier function in health and gastrointestinal disease. Neurogastroenterol Motil 24: 503–512.
29. Clemente MG, De Virgiliis S, Kang JS, Macatagney R, Musu MP, et al. (2003) Early effects of gliadin on enterocyte intracellular signalling involved in intestinal barrier function. Gut 52: 218–223.
30. Natividad JM, Huang X, Slack E, Jury J, Sanz Y, et al. (2009) Host responses to intestinal microbial antigens in gluten-sensitive mice. PLoS One 4: e6472.
31. Leffler DA, Kelly CP, Abdallah HZ, Colatrella AM, Harris LA, et al. (2012) A randomized, double-blind study of larazotide acetate to prevent the activation of celiac disease during gluten challenge. Am J Gastroenterol 107: 1554–1562.
32. Gopalakrishnan S, Tripathi A, Tamiz AP, Alkan SS, Pandey NB (2012) Larazotide acetate promotes tight junction assembly in epithelial cells. Peptides 35: 95–101.
33. Gopalakrishnan S, Durai M, Kitchens K, Tamiz AP, Somerville R, et al. (2012) Larazotide acetate regulates epithelial tight junctions in vitro and in vivo. Peptides 35: 86–94.
34. Simpson JE, Walker NM, Supuran CT, Soleimani M, Clarke LL (2010) Putative anion transporter-1 (Pat-1, Slc26a6) contributes to intracellular pH regulation during H+-dipeptide transport in duodenal villous epithelium. American Journal of Physiology-Gastrointestinal and Liver Physiology 298: G683–G691.
35. Simpson JE, Schweinfest CW, Shull GE, Gawenis LR, Walker NM, et al. (2007) PAT-1 (Slc26a6) is the predominant apical membrane Cl−/HCO3− exchanger in the upper villous epithelium of the murine duodenum. American Journal of Physiology-Gastrointestinal and Liver Physiology 292: G1079–G1088.
36. Walker NM, Simpson JE, Brazill JM, Gill RK, Dudeja PK, et al. (2009) Role of Down-Regulated in Adenoma Anion Exchanger in HCO_3^- Secretion Across Murine Duodenum. Gastroenterology 136: 893–901. e892.

13

The Strong *In Vivo* Anti-Tumor Effect of the UIC2 Monoclonal Antibody Is the Combined Result of Pgp Inhibition and Antibody Dependent Cell-Mediated Cytotoxicity

Gábor Szalóki[1], Zoárd T. Krasznai[2], Ágnes Tóth[1], Laura Vízkeleti[3], Attila G. Szöllősi[4], György Trencsényi[5], Imre Lajtos[5], István Juhász[6,7], Zoltán Krasznai[1], Teréz Márián[5], Margit Balázs[3], Gábor Szabó[1¶], Katalin Goda[1*¶]

1 Department of Biophysics and Cell Biology, Faculty of Medicine, University of Debrecen, Debrecen, Hungary, 2 Department of Obstetrics and Gynecology, Faculty of Medicine, University of Debrecen, Debrecen, Hungary, 3 Department of Preventive Medicine, Faculty of Medicine, University of Debrecen, Debrecen, Hungary, 4 Department of Physiology, Faculty of Medicine, University of Debrecen, Debrecen, Hungary, 5 Department of Nuclear Medicine, Faculty of Medicine, University of Debrecen, Debrecen, Hungary, 6 Department of Dermatology, Faculty of Medicine, University of Debrecen, Debrecen, Hungary, 7 Department of Surgery and Operative Techniques, Faculty of Medicine, University of Debrecen, Debrecen, Hungary

Abstract

P-glycoprotein (Pgp) extrudes a large variety of chemotherapeutic drugs from the cells, causing multidrug resistance (MDR). The UIC2 monoclonal antibody recognizes human Pgp and inhibits its drug transport activity. However, this inhibition is partial, since UIC2 binds only to 10–40% of cell surface Pgps, while the rest becomes accessible to this antibody only in the presence of certain substrates or modulators (e.g. cyclosporine A (CsA)). The combined addition of UIC2 and 10 times lower concentrations of CsA than what is necessary for Pgp inhibition when the modulator is applied alone, decreased the EC_{50} of doxorubicin (DOX) in KB-V1 (Pgp^+) cells *in vitro* almost to the level of KB-3-1 (Pgp^-) cells. At the same time, UIC2 alone did not affect the EC_{50} value of DOX significantly. In xenotransplanted severe combined immunodeficient (SCID) mice co-treated with DOX, UIC2 and CsA, the average weight of Pgp^+ tumors was only ~10% of the untreated control and in 52% of these animals we could not detect tumors at all, while DOX treatment alone did not decrease the weight of Pgp^+ tumors. These data were confirmed by visualizing the tumors *in vivo* by positron emission tomography (PET) based on their increased ^{18}FDG accumulation. Unexpectedly, UIC2+DOX treatment also decreased the size of tumors compared to the DOX only treated animals, as opposed to the results of our *in vitro* cytotoxicity assays, suggesting that immunological factors are also involved in the antitumor effect of *in vivo* UIC2 treatment. Since UIC2 binding itself did not affect the viability of Pgp expressing cells, but it triggered *in vitro* cell killing by peripheral blood mononuclear cells (PBMCs), it is concluded that the impressive *in vivo* anti-tumor effect of the DOX-UIC2-CsA treatment is the combined result of Pgp inhibition and antibody dependent cell-mediated cytotoxicity (ADCC).

Editor: Mária A. Deli, Hungarian Academy of Sciences, Hungary

Funding: This work was supported by Szodoray Grant from the University of Debrecen (recipients are Katalin Goda and Zoárd T. Krasznai) and Astellas Pharma Kft and Libra Foundation (recipient is Gábor Szalóki). The work was also supported by Hungarian National Science and Research Foundation (OTKA) grants PD75994, K72762, NK101337 and by TÁMOP grants: TÁMOP 4.2.2-08/1-2008-0015, TÁMOP 4.2.1/B-09/1/KONV-2010-0007 and TÁMOP 4.2.2.A-11/1/KONV-2012-0023 "VÉD-ELEM" project. Gábor Szalóki's research was realized in the frames of TÁMOP 4.2.4. A/2-11-1-2012-0001 "National Excellence Program – Elaborating and operating an inland student and researcher personal support system convergence program." The funders had no role in study design, data collection and analysis, decision to publish, or preparation of the manuscript.

Competing Interests: The authors have declared that no competing interests exist.

* Email: goda@med.unideb.hu

¶ These authors are shared last authors on this work.

Introduction

One of the most common causes of cancer chemotherapy failure is the development of resistance against chemotherapeutic agents. In most cases the tumor cells are either intrinsically resistant, or become resistant in the course of chemotherapy, to a broad spectrum of chemotherapeutic agents, including compounds they have never met before [1]. This phenomenon is called multidrug resistance (MDR) and it is often associated with high-level expression of active transporter proteins belonging to the ATP Binding Cassette (ABC) super-family, such as ABCB1 (MDR1, P-glycoprotein, Pgp), ABCC1 (MRP1, multidrug resistance protein 1) or ABCG2 (BCRP, breast cancer resistance protein)[2,3]. Pgp was the first transporter described in connection with multidrug resistance, and it seems to have the most significant role in clinical cases [3].

The Pgp molecule consists of two almost identical halves connected by a 75 amino acid long intracellular linker region. Both halves comprise six membrane spanning α-helices forming a transmembrane domain (TMD) and a nucleotide binding domain

(NBD). The two TMDs define the substrate binding sites and the translocation pathway, allowing the protein to transport various hydrophobic compounds out of the cells [4]. The overall energy requirement of drug efflux is covered by ATP hydrolysis conducted by the two NBDs (for possible models, see e.g. Senior [5], Ambudkar et al. [6]).

Pgp is generally expressed in tissues having barrier functions (e.g., in endothelial cells of the blood-brain barrier, in hepatocytes, in epithelial cells of the kidney and the intestines) and it is suggested to have an important role in protection of the body from toxic substances [2,3,7]). However, the loss of the *abcb1a/b* genes in mice (homologues of the human *ABCB1* gene) is not accompanied by major physiological consequences [8,9]; hence, inhibition of Pgp molecules may be a plausible strategy of overcoming drug resistance without serious side effects. The classical pharmacological approach involves co-administration of the cytotoxic compounds that are substrates of Pgp with pump inhibitors, to increase the accumulation of the former into the tumor cells. Unfortunately, Pgp inhibitors often induce unpredictable and intolerable pharmacokinetic interactions and toxicity through inhibiting other drug transporters or cytochrome P450, by changing the clearance and metabolism of the co-administered chemotherapeutic agents [10–12]

Several monoclonal antibodies (mAb) recognizing extracellular epitopes have been developed against Pgp. A few of them (e.g., MRK16, MRK17, MC57, HYB-241, and UIC2) are thought to recognize discontinuous conformation sensitive epitopes. Upon binding, these antibodies can partially inhibit Pgp mediated drug transport *in vivo* and *in vitro* [13–16]. However, this inhibitory effect is often weak [13–18], its extent may depend on the transported substrate [15,17,18], and it is variable even in the case of the same substrate according to general experience.

UIC2 is an IgG2a isotype mouse monoclonal antibody raised against human Pgp. It recognizes a complex epitope involving at least the first [19] and the third extracellular loops of the protein [20]. In the absence of Pgp substrates and modulators UIC2 can bind only to 10–40% of cell surface Pgps, while the rest adopts the UIC2 binding conformation only in the presence of a distinct group of substrates or modulators (e.g., cyclosporine A (CsA), SDZ PSC 833, vinblastine and paclitaxel [21–23]. In previous studies we have demonstrated that the UIC2 antibody itself completely inhibits Pgp function when it is applied together with any of the above Pgp modulators added at low, sub-inhibitory concentrations [22]. The above phenomenon was also confirmed *in vitro*, by measuring the cellular accumulation of various fluorescent Pgp substrates including DNR, R123, calcein and a radioactive tracer ^{99m}Tc-Mibi [22]. In line with the above, combined treatment with either CSA + UIC2 or paclitaxel + UIC2 specifically decreased the rate of glucose metabolism in Pgp$^+$ cells, as measured in 2-[^{18}F]fluoro-2-deoxy-D-glucose (^{18}FDG) accumulation experiments, also suggesting that Pgp inhibition with concomitant decrease of energy consumption has occurred [23]. On the other hand, we also demonstrated in xenotransplanted severe combined immunodeficient (SCID) mice that UIC2 could readily penetrate into the compact solid tumors, intensively staining cell surface Pgps and increasing daunorubicin accumulation in the Pgp$^+$ tumors to the level of the Pgp$^-$ ones [22].

In the present study we tested whether the combined treatment with CsA and UIC2 mAb can potentiate the anti-tumor effect of doxorubicin (DOX) in Pgp$^+$ tumors to achieve clinically relevant reduction in tumor size, applying the above experimental model [22]. Tumor growth was followed by weighing the mass of the tumors in sacrificed animals and also *in vivo* on the basis of ^{18}FDG accumulation. In the latter case a small-animal Positron Emission Tomography (PET) camera was applied to visualize tumors on the basis of their increased rate of glucose metabolism [24–26]. Our data demonstrate that the combined application of a class of modulators (including CsA) used at sub-inhibitory concentrations and of the UIC2 antibody may serve as an effective tool for blocking the growth of Pgp expressing tumors.

Materials and Methods

Ethics Statement

The experiments using human blood were done with the approval of the Scientific and Research Ethics Committee of the Medical Research Council (ETT TUKEB, permission number: 25364-1/2012/EKU (449/P1/12.)). Written informed consent was obtained from donors prior to blood donation, and their data were processed and stored according to the principles expressed in the Declaration of Helsinki.

In animal experiments the *Principles of Laboratory Animal Care* (National Institute of Health) was strictly followed, and the experimental protocol was approved by the Laboratory Animal Care and Use Committee of the University of Debrecen (Permission Numbers: 26/2006/DE-MAB and 122/2009/DE-MAB).

Cell Lines

KB-3-1 human epidermoid carcinoma cell line and KB-V1, its Pgp positive counterpart were used in the experiments (obtained from Michael Gottesman's lab, NIH, Bethesda) [27,28]. The cells were grown as monolayer cultures at 37°C in Dulbecco's modified Eagle's medium (DMEM) containing 4.5 g/l glucose and supplemented with 10% heat-inactivated fetal bovine serum (FBS), 2 mM L-glutamine and 25 μM/ml gentamycin. The KB-V1 cells were cultured in the presence of 180 nM vinblastine until 3 days before their use. The viability of the cells in our experiments was always higher than 90%, as assessed by the trypan blue exclusion test. The cells were regularly checked for mycoplasma by the Plasmo Test mycoplasma detection kit (San Diego, CA) and found to be negative.

Chemicals

All the Pgp substrates, modulators, cell culture media and supplements were from Sigma–Aldrich (Budapest, Hungary). The UIC2, 15D3, 5D3 and QCRL-3 mAbs were purified from the supernatants of hybridoma cell lines using affinity chromatography. The hybridoma cell lines were purchased from the American Type Culture Collections, Manassas, VA, USA), except the 5D3 hybridoma cell line, which was a kind gift from Brain P. Sorrentino (Division of Experimental Hematology, Department of Hematology/Oncology, St. Jude Children's Research Hospital, Memphis, Tennessee). The mAb preparations were>97% pure by SDS/PAGE. The glucose analogue 2-[^{18}F]fluoro-2-deoxy-D-glucose (^{18}FDG) was synthesized and labeled with the positron-decaying isotope ^{18}F according to Hamacher et al. [29].

Indirect immunofluorescence

For detection of Pgp and ABCG2 living cells (10^6 cells/ml) were incubated in the presence of 30 μg/ml 15D3 anti- Pgp mAb or 2 μg/ml 5D3 anti-ABCG2 mAb for 30 min at 37°C. For measurement of MRP1 (ABCC1) expression cells were fixed and permeabilised with 1% para-formaldehyde and 0.1% TritonX-100 in PBS (15 min; 4°C) and then labeled with 2 μg/ml QCRL-3 mAb (30 min; 4°C). After two washes with ice-cold PBS containing 1% bovine serum albumin (BSA-PBS), cells were incubated with goat anti-mouse IgG (0.5 μg/ml CruzFluor 647

(CFL647-GaMIgG), Santa Cruz Biotechnology, Inc., Texas) for 30 min at 4°C. Fluorescence intensities were detected using a Becton Dickinson FACSAria III Cell Sorter (Becton Dickinson) measuring the 633 nm/660±20 nm fluorescence intensities. Fluorescence signals were collected in logarithmic mode and the cytofluorimetric data were analyzed by the BDIS CELLQUEST (Becton Dickinson) software.

To detect Pgp expression in tumors 5-µm-thick cryosections were prepared, dried at room temperature and fixed in pre-cooled acetone (-20°C) for 10 min. Sections were then washed with PBS and blocked with 1% BSA-PBS for 20 minutes to avoid non-specific labeling and further incubated at room temperature with the UIC2 (10 µg/ml) mouse mAb for 60 min. To visualize the binding of the primary antibody an Alexa-488 conjugated goat anti-mouse IgG (A488-GaMIgG, Invitrogen) was used at 1:1000 dilution. Negative controls were obtained by omitting the primary antibody.

A Zeiss LSM 510 confocal laser-scanning microscope was used for the measurements. Alexa-488 was excited at the 488 nm line of an argon-ion laser. Fluorescence was detected through a 505–550 nm band pass filter. Images of 512×512 pixels were obtained in extended focus mode, through a 63× (numerical aperture = 1.4) Plan-Apochromat oil immersion objective.

In vitro cytotoxicity tests

Cells were seeded in 96-well plates at a cell density of 5×10^3 cells/well. 24 hours later DOX was added at different concentrations with CsA and/or UIC2 mAb or the modulator alone, and the plates were further incubated for 72 h at 37°C. The cell viability was determined using the AlamarBlue assay (Serotec, UK) measuring the 530/590 nm fluorescence intensity of the dye in an automated microplate reader (BioTec Synergy HT, US). The fluorescence intensities of the samples were normalized to the fluorescence of the untreated (DOX, antibody and CsA free) control sample, and plotted as a function of DOX concentration. Dose-response curves were fitted, and EC_{50} values were calculated by SigmaPlot 12.0 programme (Systat Software, Inc., USA).

In vitro antibody-dependent cell-mediated cytotoxicity (ADCC) assay

ADCC experiments were performed as described previously [30] with minor modifications. Human peripheral blood mononuclear cells (PBMCs) were prepared from the blood of healthy donors by Ficoll (Histopaque-1077, Sigma-Aldrich, Budapest) density gradient centrifugation. The PBMC rich fraction (effector cells) was washed three times and re-suspended in DMEM containing 10% FCS. After trypsinization, KB-V1 and KB-3-1 (target cells) cells were labeled with 5(6)-carboxyfluorescein diacetate N-succinimidyl ester (CFDA-SE) at a concentration of 10 µM at 37°C for 10 min. Then, the cells were washed thrice with DMEM containing 10% FCS and 1% BSA to remove unbound CFDA-SE and finally re-suspended in DMEM containing 10% FCS. 1.5×10^5 target cells were mixed with effector cells at target/effector cell ratios of 1:5, 1:10, 1:50 and 1:100 in a final volume of 1 ml. Samples were incubated in the absence or presence of 0.1 µM CsA and/or 20 µg/ml UIC2 at 37°C for 8 hours in CO_2 incubator. After incubation the cells were washed and then re-suspended in ice cold phosphate-buffered saline (PBS), containing 8 mM glucose and 5 µg/ml propidium iodide (PI) and were analyzed using a Becton Dickinson FACScan flow cytometer (Mountain View, CA). The labeling distinguishes four populations of cells as it is demonstrated in Fig. S1: **1**. living target cells in green (CFDA-SE positive cells); **2**. dead target cells in green and red (CFDA-SE and PI double positive cells); **3**. dead effector cells in red (PI positive cells), and **4**. live effector cells, which remain unstained. The negative control sample did not contain PBMCs, while tumor cells killed by 4% para-formaldehyde served as the positive control. The percentage of killed target cells was calculated dividing the number of PI and CFDA-SE double positive cells (dead cells) by the number of the CFDA-SE positive cells.

In vitro complement-dependent cytotoxicity (CDC) assay

Samples containing 2.5×10^5 KB-V1 or KB-3-1 cells and 20 µg/ml UIC2 mAb in the presence or absence of 0.1 µM CsA were treated with freshly prepared human serum at different dilutions in DMEM medium. The samples were incubated for 4 hours in CO_2 incubator at 37°C and then stained with 5 µg/ml PI. The percentage of the PI positive dead cell was determined by a Becton Dickinson FACScan flow cytometer (Mountain View, CA).

The hemolytic complement activity of the human serum samples was determined applying sheep red blood cells sensitized with a rabbit stroma antibody (SSRBC) kindly provided by Attila Bácsi (Inst. of Immunology, University of Debrecen, Faculty of Medicine). 50 µl of 1% SSRBC was mixed with different dilutions of human serum or heat inactivated human serum (inactivated at 56°C for 30 min) and incubated for 30 min at 37°C. After centrifugation the absorbance of the supernatants was measured at 541 nm by BioTek Synergy HT plate reader. The HC_{50} value where 50% of the RBCs were lysed was determined for the human sera and found to be normal.

Animal model and study design

Adult (10 to 12 week-old), pathogen-free B-17 severe combined immunodeficiency (SCID) mice were used in this study [31]. Animals were housed under pathogen free circumstances at a temperature of 26±2°C, with 50±10% humidity and artificial lighting with a circadian cycle of 12 h. The food and drinking water (sterilized by autoclaving) were available ad-libitum to all animals.

SCID mice were injected subcutaneously with KB-3-1 (Pgp^-; 1.5×10^6 cells in 150 µl sterile phosphate-buffered saline (PBS)) on the left and KB-V1 (Pgp^+; 3×10^6 cells in 150 µl sterile PBS) cells on the right thighs. To obtain approximately similar tumor sizes in case of KB-V1 and KB-3-1 cells we had to inject double number of KB-V1 cells, because of their slower cell proliferation rate. In another experiment we grafted four tumors per animal in order to limit the number of animals and to maximize the number of tumors imaged. In this case each animal received two injections in the shoulders and two in the upper part of the thighs. Tumors were grown for 4 days and then the mice were treated with DOX alone (5 mg/kg, i.v.), or DOX combined with either UIC2 mAb (5 mg/kg, i.v.) or CsA (Sandimmun, Novartis, Basel, Switzerland; 10 mg/kg i.p.) or both. The animals were killed 8 days after treatment with chemotherapy by cervical dislocation and the tumors were removed to weigh them and then they were snap frozen in liquid nitrogen and stored at -70°C till mRNA expression analysis.

mRNA expression analysis

The frozen tumor sections were equilibrated in 10 volumes pre-chilled RNAlater-ICE solution (Applied Biosystem, CA) at -20°C overnight to protect RNA from degradation and then total RNA was isolated using the RNeasy Mini kit (Qiagen Inc., CA) according to the protocol. RNA quantity was determined by NanoDrop ND-1000 UV-Vis Spectrophotometer (NanoDrop Technologies, Wilmington, DE). RNA was then subjected to reverse transcription-real time quantitative polymerase chain reaction (RT-qPCR) using the Taqman assay with stocked primers and probes (Applied Biosystem, CA). Pgp mRNA expression was normalized to the human glyceraldehyde-3-phosphate dehydrogenase (GAPDH) expression.

Small-animal PET imaging using ^{18}FDG

After the implantation of tumor cells ^{18}FDG PET scans were repeated at different time points. Prior to PET measurements, mice were fasted overnight. On the day of PET imaging mice were pre-warmed to a body temperature of 37°C and this temperature was maintained throughout the uptake and scanning period to minimize the visualization of brown fat. Mice were injected via the tail vein with 5.5±0.2 MBq of ^{18}FDG. 40 min after tracer injection animal were anaesthetized by 3% isoflurane with a dedicated small animal anesthesia device and then 20 min long static single-frame PET scans were acquired using a small-animal PET scanner (MiniPET-II, Department of Nuclear Medicine, Faculty of Medicine, University of Debrecen) to visualize the tumors.

The MiniPET-II system is a dedicated small animal PET scanner developed with the help of a Hungarian project. The MiniPET-II scanner consists of 12 detector modules including LYSO (Cerium Doped Lutetium Yttrium Orthosilicate) scintillation crystal blocks and position sensitive photo multiplier tubes. Each crystal block comprises 35×35 pins of 1.27×1.27×12 mm size. Scanner normalization and random correction were applied on the data and the images were reconstructed with the standard maximum likelihood expectation maximization iterative algorithm. The pixel size was 0.27×0.27×1.35 mm and the spatial resolution varies between 1.4 to 2.1 mm from central to 25 mm radial distances. The system sensitivity is 11.4%.

^{18}FDG-PET data analysis

The ^{18}FDG uptake was expressed in terms of standardized uptake values (SUVs) and tumor to muscle (T/M) ratios. Ellipsoidal 3-dimensional regions of interest (ROI) were manually drawn around the edges of the tumor xenografts by visual inspection using BrainCad software (Institute of Nuclear Medicine, University of Debrecen). The standardized uptake value (SUV) was calculated as follows: SUV = [ROI activity (Bq/ml)]/[injected activity (Bq)/animal weight (g)], assuming a density of 1 g/cm^3. The T/M ratios were computed as the ratio of the mean activity in the tumor volume of interest (VOI) and the background (muscle) VOI.

Statistical analysis

The displayed data are the means ± SD of the results of at least three independent experiments. Data have been analyzed using SigmaPlot 12.0 programme (Systat Software, Inc., USA) and IBM SPSS Statistics 20 (IBM Corp., USA). Statistical significance was assessed using analysis of variance (ANOVA), applying Holm-Sidak method for post hoc pair-wise comparison of the different samples. In the case of unequal variances Dunnett T3 post hoc pair-wise comparison method was used. Differences were considered to be significant at $P<0.05$.

Results

In vitro cytotoxicity measurements

KB-V1 cells express Pgp at high level, while other drug transporting ABC proteins ABCG2 and ABCC1 were not detectable by means of indirect immunofluorescence and flow cytometry (Fig. S2). KB-3-1 cells do not express any of the above ABC transporters at measurable level (Fig. S2). In accordance with the high Pgp expression level of KB-V1 cells the EC_{50} value of DOX was 2.19±0.39 μM in these cells, while it was only 44±3 nM in KB-3-1 (Pgp$^-$) cells (Fig. 1). In KB-V1 cells CsA co-treatment decreased the EC_{50} value of DOX in a dose dependent manner, while UIC2 had only a weak, statistically not significant effect. Interestingly, the combined treatment of KB-V1 cells with 1 μM CsA and a saturating concentration of UIC2 mAb decreased the EC_{50} value to 33±19.7 nM, which could be achieved by 10 times higher CsA concentration when it was applied alone (Fig. 1). In contrast, administration of the UIC2 mAb and 1 μM CsA to cultures of KB-3-1 cells, simultaneously or alone, had no significant effects on their DOX sensitivity.

mRNA expression analysis of Pgp in the tumor xenografts and cells used for grafting the tumors

Based on the above *in vitro* results, we have designed *in vivo* experiments to test the effectiveness of the combined treatment with low dose of CsA, UIC2 and DOX. SCID mice were inoculated with KB-V1 and KB-3-1 cells, respectively. Palpable subcutaneous tumors developed in 10–12 days. Their Pgp expression was compared to that of the tumor cells used for grafting the tumors. Since the immunofluorescence intensities in frozen tissue sections and cell monolayers are not directly comparable, Pgp expression was examined at the mRNA level. In the KB-3-1 tumors a well detectable ~60-fold increment of Pgp mRNA levels occurred compared to the inoculated cells, while the Pgp mRNA level of the KB-V1 tumors did not change upon proliferation of the inoculated cells (Fig. 2). The Pgp mRNA levels proved to be at least three orders of magnitude higher in the KB-V1 tumors compared to the KB-3-1 xenografts (see Fig. 2). Thus, the KB-V1 tumor xenografts retained their MDR phenotype as it was also proved by indirect immunofluorescent labeling (see Fig. S3), while the KB-3-1 cells continued to express Pgp at very low levels (not detectable by immunofluorescence, Fig. S3) in the developed tumors on the time scale of the *in vivo* experiments.

^{18}FDG accumulation in xenotransplanted tumors

Grafted tumors were grown for 4 days, then the mice were treated with DOX alone (5 mg/kg, i.v.) or DOX combined with either the UIC2 mAb (5 mg/kg, i.v.) or CsA (10 mg/kg, i.p.), or

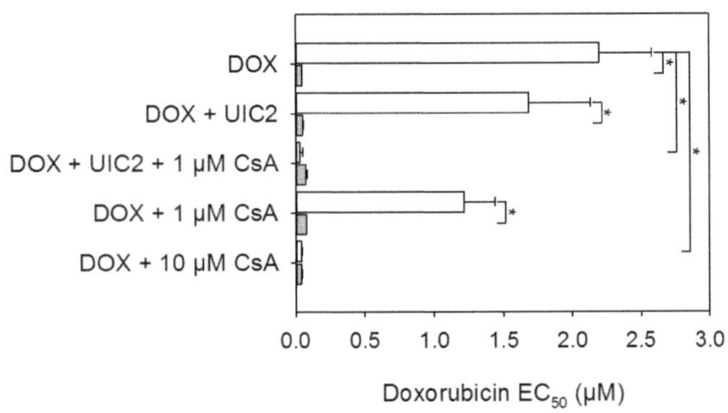

Figure 1. Cytotoxic effect of DOX in KB-3-1 (Pgp$^-$; grey bars) and KB-V1 (Pgp$^+$; empty bars) cells and its potentiation by treatments with CsA (used at the indicated concentrations) and/or UIC2 mAb (20 μg/ml). EC_{50} values were calculated by fitting the dose-response curves. Values are means (± SD) of 3 independent experiments. Statistically significant differences are shown by * ($P<0.05$).

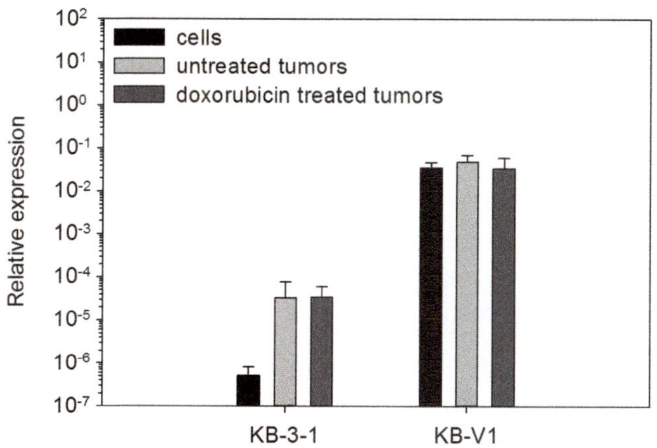

Figure 2. Relative Pgp mRNA expression levels of the KB-V1 and KB-3-1 tumors and the cells applied for grafting the tumors. Pgp expression levels are normalized to the expression level of GAPDH mRNA (mean ± SD, n = 5). The Pgp expression levels of inoculated KB-V1 cells and tumors were significantly different from those of the KB-3-1 cells and tumors (P<0.01), while the Pgp expression levels of tumors were not statistically different from those of the cells they originated from.

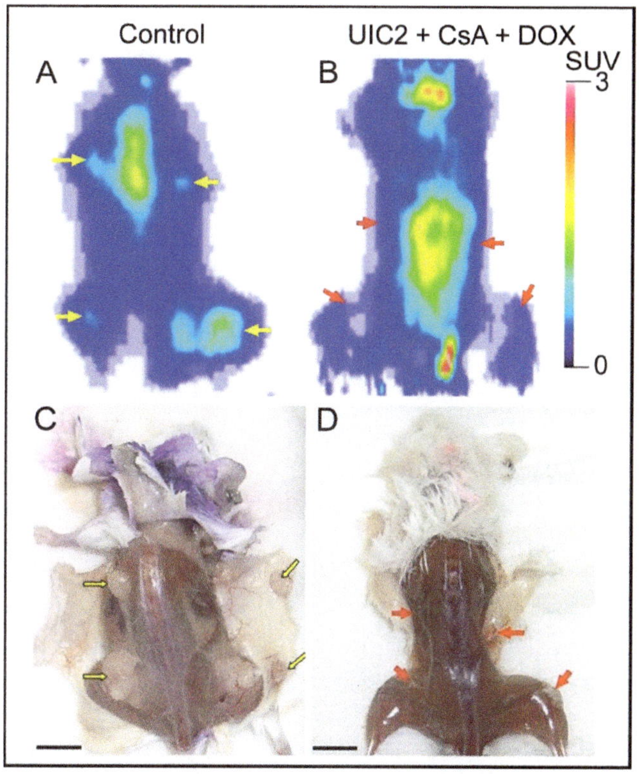

Figure 3. Effect of combined treatment with DOX+CsA+UIC2 on the KB-V1 and KB-3-1 tumors visualized *in vivo* by ^{18}FDG-miniPET. Coronal section of ^{18}FDG-miniPET image of a representative control (*A*) and a DOX-CsA-UIC2-treated tumor bearing mouse (*B*). Standardized uptake values (SUV) are calculated as described in Materials and Methods. The sites of tumor cell inoculation are shown by arrows (KB-V1 tumors, right side arrows; KB-3-1 tumors, left side arrows). Below: autopsies of the same control (*C*) and treated (*D*) animals. Bar: 10 mm.

both. For an *in vivo* visualization of their effects, miniPET-^{18}FDG accumulation measurements were performed, carrying out the scans 6–8 days after the above treatments. Fig. 3 shows representative ^{18}FDG accumulation scans and the anatomical pictures of untreated (Fig. 3*A and C*) and DOX-UIC2-CsA treated (Fig. 3*B and D*) SCID mice bearing Pgp$^+$ (KB-V1, right shoulder and thigh) and Pgp$^-$ (KB-3-1, left shoulder and thigh) tumor xenografts. At the time of the measurements the tumors of untreated mice were 5–8 mm in diameter (6.43±1.13 mm mean ± SD, KB-3-1; 6.0±1.15 mm mean ± SD, KB-V1), and well-detectable on the basis of their increased rate of glucose metabolism.

The tumor (T) to skeletal muscle (M) ^{18}FDG accumulation ratio (T/M) was 4.2±0.6 in the case of Pgp$^-$ and 4.8±0.7 for the Pgp$^+$ tissues (n = 7, ± SD), indicating significantly higher rate of glucose consumption in tumors compared to the muscle cells in both cases.

No visible or palpable tumors developed in 20% of the animals treated with the combination of DOX-UIC2-CsA. This observation was confirmed by the mini-PET scans, since no significant ^{18}FDG accumulation was observed at the sites of tumor cell inoculation in these animals, as reflected by the T/M ^{18}FDG accumulation ratios being close to 1 (T/M = 1.1±0.2 for Pgp$^+$ and T/M = 0.97±0.1 for Pgp$^-$ tumors; (n = 4)).

Effect of DOX treatment combined with UIC2 and/or a low dose of CsA on the weights of grafted tumors

Eight days after chemotherapy, the animals were sacrificed by cervical dislocation and the tumors were removed to weigh them. As it is shown in Fig. 4*A*, DOX treatment alone was almost ineffective in the case of KB-V1 tumors, while the weight of the KB-3-1 tumors decreased considerably. Co-administration of 10 mg/kg CsA decreased the size of the KB-V1 tumors only mildly and did not affect the KB-3-1 tumors. Combined treatment with DOX-CsA-UIC2 decreased the mean weight of the KB-V1 tumors 9 fold compared to the animals treated with DOX alone. The combined treatment also decreased the mean weight of the KB-3-1 tumors, but not in a statistically significant extent. Importantly, only 52% of the grafted Pgp$^+$ or Pgp$^-$ tumors developed into detectable tumors in the DOX-CSA-UIC2 treated animals (see Fig. 4*B*), and 20% of the animals remained completely tumor-free. In contrast, we always detected tumors in the other treatment groups. Co-administration of UIC2 and DOX also decreased tumor size significantly compared to DOX alone in the case of KB-V1 tumors. This finding, in view of the fact that UIC2 treatment alone does not affect the EC_{50} value of DOX in *in vitro* cytotoxicity tests (Fig. 1), suggested to us that the growth inhibitory effect of the antibody is not exclusively due to Pgp inhibition. UIC2 binding did not affect the *in vitro* cell viability significantly (Fig. 5); therefore, the contribution of the immune system, that is partly functional in the SCID mice, was tested.

In vitro ADCC and CDC measurements

Fig. 6 shows the effect of human peripheral blood mononuclear cells (PBMCs), and of human serum samples, on UIC2 mAb treated KB-V1 and KB-3-1 cells, *in vitro*. PBMCs killed about 70-80% of the UIC2 treated KB-V1 cells both in the presence and absence of CsA, at target to effector cell ratios of 1:50 and 1:100, respectively (Fig. 6*A*), in contrast with the UIC2 treated KB-3-1 cells that exhibited a survival rate similar to that of the untreated control (Fig. 6*B*). In the absence of UIC2, the percentages of dead target cells were low (see Fig. 6*A and B*).

In order to assess the possible role of CDC in UIC2 mediated *in vivo* tumor cell killing, the cytotoxicity of human serum samples

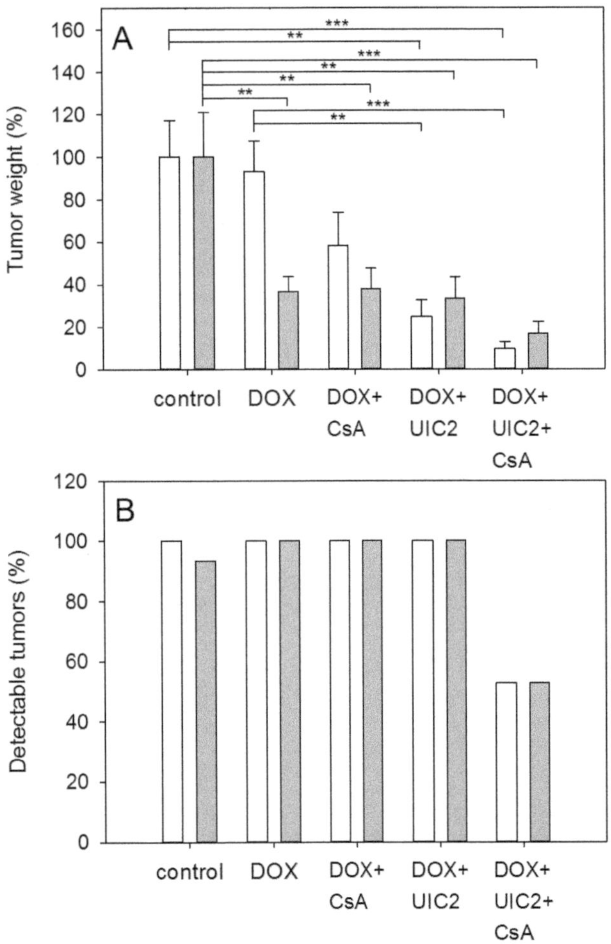

Figure 4. Effect of treatment with DOX combined with UIC2 and/or a low dose of CsA on the weights of the grafted tumors (*A*) and on the percentage of the detectable tumors in the different treatment groups (*B*). Tumor weights were expressed as a percentage of the average weight of the tumors of untreated animals measured at the time of the termination of the experiment (mean values ± SEM, n = 8). Statistically significant differences relative to the untreated control and the DOX-only treated groups are marked with *: $P<0.05$; **: $P<0.01$; ***: $P<0.001$). Grey bars: KB-3-1 (Pgp^-) and empty bars: KB-V1 (Pgp^+) tumors.

was measured *in vitro*. Cell killing did not increase in the UIC2 treated KB-V1 (Fig. 6*C*) and KB-3-1 (Fig. 6*D*) samples despite the strong hemolytic activity of the applied sera (Fig. 6*D, inset*).

Discussion

In the present experiments SCID mice xenotransplanted with Pgp^+ and Pgp^- tumors were used to study the efficacy of an antibody-based multidrug resistance reversal strategy. The KB-V1/KB-3-1 cell pair does not express ABCG2 and MRP1 (ABCC1) at detectable levels, while the KB-V1 cells have high Pgp expression level (see Fig. S2 and Fig. 2). They are growing fast and develop into subcutaneous tumors of ~1 cm diameter in about 10-12 days after inoculation of $1–3\times10^6$ cells into the animals. An advantage of the fast tumor growth in this model system is that the Pgp expression level of the tumor cells does not decline in the absence of Pgp substrates (see Fig. 2) on the time scale of the *in vivo* experiments.

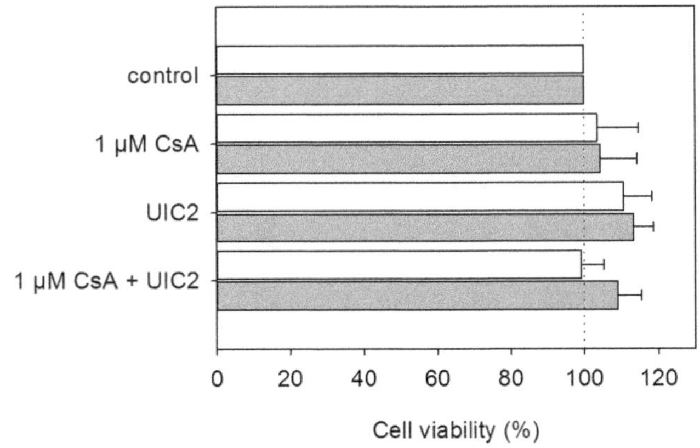

Figure 5. Effect of the UIC2 mAb treatment on the *in vitro* viability of KB-3-1 (Pgp^-; grey bars) and KB-V1 cells (Pgp^+; empty bars) in the presence or absence of 1 μM CsA. Cell viability was expressed as percentage of the untreated control. Values are means (± SD) of three independent experiments.

In the above model system, co-treatment with UIC2 + CsA potentiated the anti-tumor effect of DOX and inhibited or hindered the development of KB-V1 Pgp^+ tumors *in vivo* (Fig. 4). At the same time, DOX treatment alone did not have a significant effect on the size of the KB-V1 tumors. These data are in line with the conclusions of our previously published *in vitro* and *in vivo* drug accumulation studies [22] and with the results of the *in vitro* cytotoxicity measurements shown in Fig. 1. Although these observations may suggest that the dramatic antitumor effect of the combined treatment is the result of increased antibody binding with consequential Pgp inhibition and DOX accumulation, the mechanism proved more complex.

SCID mice have intact complement system as well as functioning macrophages, natural killer cells and polymorphonuclear cells [32]. Therefore, antibody binding to the tumor cells may elicit cytotoxicity directly through complement binding (complement-dependent cytotoxicity, CDC) or indirectly, via the recruitment of the above effector cells to the antibody covered tumor cells (antibody-dependent cell-mediated cytotoxicity, ADCC). The possibility, that immunological factors contribute to the cytotoxic effect of our treatment protocol is strongly supported by the difference between the outcomes of *in vitro* cytotoxicity measurements and of the *in vivo* experiments conducted in SCID mice (compare Fig.1 and Fig. 4). In the *in vitro* cytotoxicity assay, UIC2 alone only mildly aggravated DOX cytotoxicity (see Fig. 1) what could be explained by the trapping and inhibition of only a small fraction of cell surface Pgps by UIC2 in the absence of CsA [22]. In contrast, we experienced *in vivo* a marked decrease of tumor size in response to the DOX+UIC2-only treatment, as the average weight of the KB-V1 tumors was approx. 4 fold smaller compared to the animals treated with DOX alone (Fig. 4*A*). These data are in line with the positive results of our *in vitro* ADCC assays (Fig. 6), supporting the notion that the UIC2 mAb also induces ADCC *in vivo* in the SCID mice.

The involvement of ADCC in the *in vivo* anti-tumor effect of UIC2 treatment is an unexpected finding of our experiments, since IgG_2 antibodies are mostly inefficient at supporting effector functions and are chosen for antibody therapy when effector functions are unnecessary or undesirable [33]. However, there is an example [34] when the antitumor and antimetastatic effects of two IgG_2 isotye anti-Pgp antibodies (the mouse-human chimeric Ab (MH162) and its mouse counterpart (MRK16)) was attributed

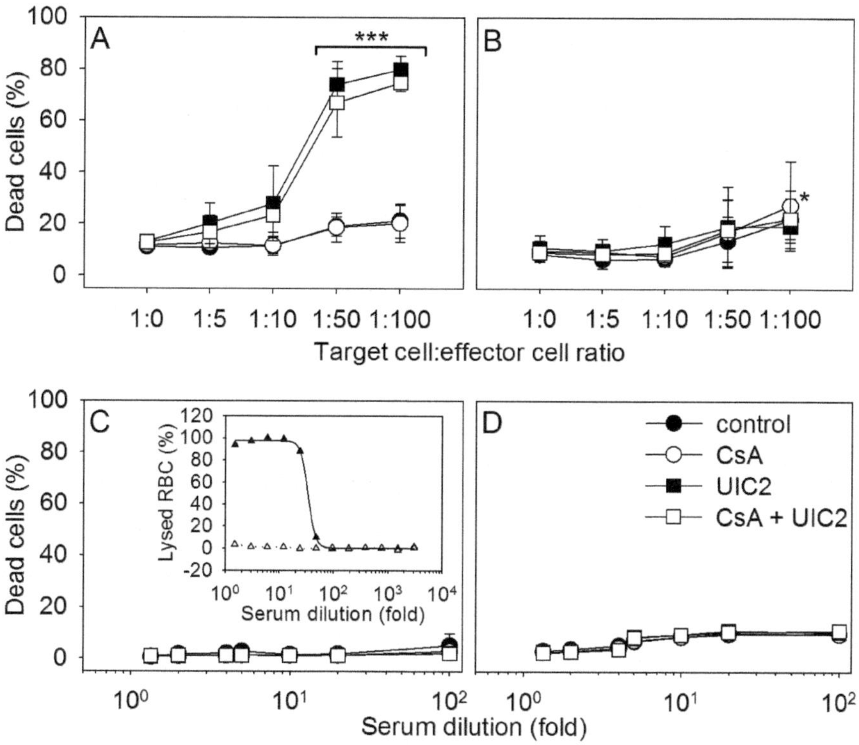

Figure 6. Antibody-dependent cell-mediated cytotoxicity (ADCC, *panels A* and *B*) and complement mediated lysis (CDC, *panels C* and *D*) induced by UIC2 mAb treatment *in vitro*. In the ADCC assay KB-V1 (*A*) and KB-3-1 (*B*) tumor cells were labeled with CFDA-SE then mixed with PBMCs freshly isolated from peripheral blood at different target to effector cell ratios. Samples were treated with 10 μM CsA (○), 20 μg/ml UIC2 mAb (■), 10 μM CsA and 20 μg/ml UIC2 mAb (□) or buffer (●). After 8 h incubation at 37°C, samples were stained with PI and analyzed by flow cytometry. In the CDC assay, cells were incubated with human serum at different dilutions for 4 h. *Inset of panel C*: Hemolytic effect of the serum (▲) and of heat inactivated serum (△) on sensitized sheep red blood cells served as positive and negative control, respectively. The percentages of killed cells were calculated as described in Materials and Methods. Values are means (± SD) of four independent experiments, ***P<0,001.

to ADCC. Similarly, it was proven in a recent study that a human IgG_2 isotype mAb specific for epidermal growth factor receptor effectively triggers ADCC by recruting monocytes and neutrophils [35] via FcγRIIa binding [36,37]. Since IgG_2 isotype antibodies do not trigger natural killer cell mediated ADCC [35], therefore in our *in vitro* ADCC experiments carried out with PBMCs cell killing was mediated by monocytes. Since SCID mice also have monocytes [32] the same mechanism is functional and probably explains our *in vivo* results.

The strong anti-tumor effect of the combined treatment might be attributed exclusively to ADCC triggered by the UIC2 mAb binding. However, the fact that the extents of the *in vitro* ADCC effects were indistinguishable in the presence of UIC2 or UIC2+ CsA suggests that binding of the antibody to a small fraction of the cell surface Pgps (20-40%) is sufficient to induce a maximal ADCC effect (Fig. 6*A*). Consequently, the differences in the size of the KB-V1 tumors between the UIC2 and UIC2+CsA treated animals and the lack of the KB-V1 tumors in 52% of these animals (Fig 4*A and B*) argue against the above assumption and suggests that the stronger Pgp inhibitory effect of the UIC2+CsA combination mediates the anti-tumor effect at least in part. Pgp inhibition by the antibody requires saturating antibody concentrations that seems to be reached in our experiments, since strong UIC2 staining and a two fold increase in the accumulation of a Pgp substrate daunorubicin was measured in the tumor sections prepared from the tumors 8 hours after the injection of UIC2 and CsA [22] added at similar conditions. Taken together, our *in vitro* and *in vivo* data suggest that strong anti-tumor effect can be reached by the combinative treatment studied, as a joint result of Pgp inhibition and ADCC. However, the relative contribution of these mechanisms to the anti-tumor effect of the UIC2 mAb is not known.

ADCC can be triggered at relatively low receptor occupancy by the antibody or at low receptor abundance [35,38]. Thus, the 60 fold increased Pgp expression level of the KB-3-1 tumors compared to the KB-3-1 cells (see Fig. 2.) is probably sufficient to trigger ADCC effect, when the Pgp molecules are saturated by the antibody in the presence of CsA. In line with this assumption in 52% of the DOX-UIC2-CsA treated animals we could not detect KB-3-1 tumors, while they appeared in all of the DOX or DOX+UIC2 treated animals (see Fig. 4*B*).

It could be a limitation of our study that only one cell line pair was used in the experiments. However, the observed effects have been proven to be Pgp specific, since decreased tumor size was detected exclusively in those animal groups that received UIC2 mAb treatment. In addition, in our previous studies we compared the KB-V1 cell line with mdr1 transfected NIH 3T3 murine fibroblast cells [22], and Pgp^+ $A2780^{AD}$ ovarian carcinoma cells [23] and found them equivalent in every aspect of their multidrug resistant phenotype including the inhibition of Pgp-mediated drug transport by UIC2 mAb. On the other hand, ADCC effect is not dependent on the tissue origin of the target cells; rather it is determined by the interaction between the Fc part of the antibody and the Fc receptor of the effector cells.

Doubts about the possible clinical application of an anti-Pgp mAb based tumor therapy are related to the likely side effects that may arise as a result of either Pgp inhibition or ADCC, both exerted on cells expressing Pgp at physiological barriers of the body. For instance, inhibition of Pgp expressed in the blood-brain barrier may lead to increased accumulation of its substrates in the central nervous system leading to neurotoxicity, as it was experienced in mdr1a/b knock-out mice [8]. However, administration of Pgp modulators in clinical trials does not seem to cause toxicity to the central nervous system [39] probably because other ABC transporters (e.g. MRP1, BCRP1) may compensate for the loss off Pgp's gatekeeper function in the blood-brain barrier [40]. However, in contrast to Pgp inhibition, ADCC may damage the tissues at the physiological Pgp expression sites. Our *in vitro* assays using human PBMCs confirmed that UIC2 mAb effectively triggers ADCC. Since ADCC is mediated via the Fc portion of the antibody, to avoid this side-effect upon human applications the whole UIC2 antibody could be substituted for by its Fab fragments (that behave very similarly to whole antibodies *in vitro*; our unpublished data) or upon humanization of the antibody its effector functions may be fine-tuned by the design of the Fc part [33]. However, it remains to be studied if inhibition of Pgp function alone may augment DOX cytotoxicity sufficiently enough for a therapeutically significant anti-tumor effect.

The UIC2 mAb does not bind to mouse Pgp [14], therefore the SCID mouse model system is not applicable for studying the possible side effects brought about by antibody binding to physiological Pgp expression sites. Since the UIC2 mAb also recognizes primate [14] and sheep [41] Pgp, such animal models may be used for the evaluation of the feasibility of the strategy demonstrated herein. Direct injection of the antibody into the tumor tissue may also be tested for the purposes of reducing antibody dose and decrease systemic side effects.

In our model system, treatments were applied shortly (four days) after injection of the tumor cells, when the tumors were still rather small, a situation perhaps analogous to the clinical setting when systemic therapy is applied to prevent or hinder the development of multidrug resistant primary or metastatic tumors.

Each multidrug resistant tumor may have a unique signature of resistance mechanisms. Consequently, cancer therapy will need to be personalized, not only with respect to the mechanisms of malignant transformation but also regarding the mechanisms of resistance [42]. The strategy of Pgp inhibition demonstrated herein is offered to enrich the repertoir of possible protocols that can be considered for the treatment of multidrug resistant tumors, once humanized UIC2 becomes available.

Supporting Information

Figure S1 In the ADCC assay KB-V1 (panels *A*, *B*, *C*) and KB-3-1 (*D*, *E*, *F*) tumor cells were labeled with CFDA-SE, then mixed with PBMCs freshly isolated from peripheral blood at 1:5 (*B*, *E*) or 1:100 (*C*, *F*) target to effector cell ratios. Samples were treated with 20 µg/ml UIC2 mAb. After 8 h incubation at 37°C, samples were stained with PI and analyzed by flow cytometry. *Green gates* mark living target cells, while *red gates* show dead target cells.

Figure S2 Flow cytometric analysis of Pgp, ABCG2 and MRP1 expression in KB-V1 (*A*, *C*, *E*) and KB-3-1 (*B*, *D*, *F*) cell lines applying indirect immunofluorescence staining by CFL647-conjugated GaMIgG. Pgp was detected by 15D3 mAb, while ABCG2 and MRP1 (ABCC1) were labeled by 5D3 and QCRL-3 mAbs, respectively. Grey filled histograms show antibody untreated cells, while dashed lines represent samples treated with secondary antibody only. The ABCG2 positive MDCK ABCG2 cells (*G*) and the MRP1 (*I*) expressing GLC4-ADR cell line and their nonexpressing counterparts (*H* and *J*) were used as controls. The GLC4 human small cell lung carcinoma cell line pair was a kind gift from Pinedo HM (Department of Medical Oncology, Academic Hospital Vrije Universiteit, Amsterdam), while the MDCK (Madin-Darby canine kidney) cell line and its ABCG2 transfected counterpart was kindly provided by Sarkadi B (Biomembrane Institute of Molecular Pharmacology, Budapest)

Figure S3 Pgp expression of KB-V1 (*A*) and KB-3-1 (*C*) tumor xenografts visualized by indirect immunofluorescence. Panels (*B*) and (*D*) are phase contrast images of the same tumor slices. The 5 µm thick cryosections were fixed in acetone, labeled with UIC2 mAb followed by A488-GaMIgG at room temperature for 60 min.

Acknowledgments

We thank Anita Szabóné Jeney (Department of Biophysics and Cell Biology), Judit Pélyi and Tamás Nagy (Department of Nuclear Medicine) for their technical assistance.

Author Contributions

Conceived and designed the experiments: G. Szalóki KG G. Szabó ZK TM ZTK GT IL IJ. Performed the experiments: G. Szalóki ZTK LV AGS IL IJ GT ÁT ZK TM. Analyzed the data: G. Szalóki ZTK GT ZK TM MB. Wrote the paper: KG G. Szabó GT TM G. Szalóki.

References

1. Gottesman MM, Fojo T, Bates SE (2002) Multidrug resistance in cancer: role of ATP-dependent transporters. Nat Rev Cancer 2: 48–58.
2. Borst P, Elferink RO (2002) Mammalian ABC transporters in health and disease. Annu Rev Biochem 71: 537–592.
3. Glavinas H, Krajcsi P, Cserepes J, Sarkadi B (2004) The role of ABC transporters in drug resistance, metabolism and toxicity. Curr Drug Deliv 1: 27–42.
4. Rosenberg MF, Callaghan R, Ford RC, Higgins CF (1997) Structure of the multidrug resistance P-glycoprotein to 2.5 nm resolution determined by electron microscopy and image analysis. J Biol Chem 272: 10685–10694.
5. Senior AE (2012) Two ATPases. J Biol Chem 287: 30049–30062.
6. Ambudkar SV, Kim IW, Sauna ZE (2006) The power of the pump: mechanisms of action of P-glycoprotein (ABCB1). Eur J Pharm Sci 27: 392–400.
7. Leonard GD, Fojo T, Bates SE (2003) The role of ABC transporters in clinical practice. Oncologist 8: 411–424.
8. Schinkel AH, Smit JJ, van Tellingen O, Beijnen JH, Wagenaar E, et al. (1994) Disruption of the mouse mdr1a P-glycoprotein gene leads to a deficiency in the blood-brain barrier and to increased sensitivity to drugs. Cell 77: 491–502.
9. Schinkel AH, Mayer U, Wagenaar E, Mol CA, van Deemter L, et al. (1997) Normal viability and altered pharmacokinetics in mice lacking mdr1-type (drug-transporting) P-glycoproteins. Proc Natl Acad Sci U S A 94: 4028–4033.
10. Friedenberg WR, Rue M, Blood EA, Dalton WS, Shustik C, et al. (2006) Phase III study of PSC-833 (valspodar) in combination with vincristine, doxorubicin, and dexamethasone (valspodar/VAD) versus VAD alone in patients with recurring or refractory multiple myeloma (E1A95): a trial of the Eastern Cooperative Oncology Group. Cancer 106: 830–838.
11. Chico I, Kang MH, Bergan R, Abraham J, Bakke S, et al. (2001) Phase I study of infusional paclitaxel in combination with the P-glycoprotein antagonist PSC 833. J Clin Oncol 19: 832–842.
12. Szakacs G, Varadi A, Ozvegy-Laczka C, Sarkadi B (2008) The role of ABC transporters in drug absorption, distribution, metabolism, excretion and toxicity (ADME-Tox). Drug Discov Today 13: 379–393.
13. Chaudhary PM, Mechetner EB, Roninson IB (1992) Expression and activity of the multidrug resistance P-glycoprotein in human peripheral blood lymphocytes. Blood 80: 2735–2739.

14. Mechetner EB, Roninson IB (1992) Efficient inhibition of P-glycoprotein-mediated multidrug resistance with a monoclonal antibody. Proc Natl Acad Sci U S A 89: 5824–5828.
15. Rittmann-Grauer LS, Yong MA, Sanders V, Mackensen DG (1992) Reversal of Vinca alkaloid resistance by anti-P-glycoprotein monoclonal antibody HYB-241 in a human tumor xenograft. Cancer Res 52: 1810–1816.
16. Jachez B, Cianfriglia M, Loor F (1994) Modulation of human P-glycoprotein epitope expression by temperature and/or resistance-modulating agents. Anticancer Drugs 5: 655–665.
17. Watanabe T, Naito M, Kokubu N, Tsuruo T (1997) Regression of established tumors expressing P-glycoprotein by combinations of adriamycin, cyclosporin derivatives, and MRK-16 antibodies. J Natl Cancer Inst 89: 512–518.
18. Naito M, Watanabe T, Tsuge H, Koyama T, Oh-hara T, et al. (1996) Potentiation of the reversal activity of SDZ PSC833 on multi-drug resistance by an anti-P-glycoprotein monoclonal antibody MRK-16. Int J Cancer 67: 435–440.
19. Schinkel AH, Arceci RJ, Smit JJ, Wagenaar E, Baas F, et al. (1993) Binding properties of monoclonal antibodies recognizing external epitopes of the human MDR1 P-glycoprotein. Int J Cancer 55: 478–484.
20. Zhou Y, Gottesman MM, Pastan I (1999) The extracellular loop between TM5 and TM6 of P-glycoprotein is required for reactivity with monoclonal antibody UIC2. Arch Biochem Biophys 367: 74–80.
21. Nagy H, Goda K, Fenyvesi F, Bacsó Z, Szilasi M, et al. (2004) Distinct groups of multidrug resistance modulating agents are distinguished by competition of P-glycoprotein-specific antibodies. Biochem Biophys Res Commun 315: 942–949.
22. Goda K, Fenyvesi F, Bacsó Z, Nagy H, Márián T, et al. (2007) Complete inhibition of P-glycoprotein by simultaneous treatment with a distinct class of modulators and the UIC2 monoclonal antibody. J Pharmacol Exp Ther 320: 81–88.
23. Krasznai ZT, Tóth A, Mikecz P, Fodor Z, Szabó G, et al. (2010) Pgp inhibition by UIC2 antibody can be followed in vitro by using tumor-diagnostic radiotracers, 99mTc-MIBI and 18FDG. Eur J Pharm Sci 41: 665–669.
24. Pauwels EK, Ribeiro MJ, Stoot JH, McCready VR, Bourguignon M, et al. (1998) FDG accumulation and tumor biology. Nucl Med Biol 25: 317–322.
25. Wahl RL (1996) Targeting glucose transporters for tumor imaging: "sweet" idea, "sour" result. J Nucl Med 37: 1038–1041.
26. Mankoff DA, Bellon JR (2001) Positron-emission tomographic imaging of cancer: glucose metabolism and beyond. Semin Radiat Oncol 11: 16–27.
27. Shen DW, Cardarelli C, Hwang J, Cornwell M, Richert N, et al. (1986) Multiple drug-resistant human KB carcinoma cells independently selected for high-level resistance to colchicine, adriamycin, or vinblastine show changes in expression of specific proteins. J Biol Chem 261: 7762–7770.
28. Akiyama S, Fojo A, Hanover JA, Pastan I, Gottesman MM (1985) Isolation and genetic characterization of human KB cell lines resistant to multiple drugs. Somat Cell Mol Genet 11: 117–126.
29. Hamacher K, Coenen HH, Stocklin G (1986) Efficient stereospecific synthesis of no-carrier-added 2-[18F]-fluoro-2-deoxy-D-glucose using aminopolyether supported nucleophilic substitution. J Nucl Med 27: 235–238.
30. Barok M, Isola J, Pályi-Krekk Z, Nagy P, Juhász I, et al. (2007) Trastuzumab causes antibody-dependent cellular cytotoxicity-mediated growth inhibition of submacroscopic JIMT-1 breast cancer xenografts despite intrinsic drug resistance. Mol Cancer Ther 6: 2065–2072.
31. Márián T, Szabó G, Goda K, Nagy H, Szincsák N, et al. (2003) In vivo and in vitro multitracer analyses of P-glycoprotein expression-related multidrug resistance. Eur J Nucl Med Mol Imaging 30: 1147–1154.
32. Bosma MJ, Carroll AM (1991) The SCID mouse mutant: definition, characterization, and potential uses. Annu Rev Immunol 9: 323–350.
33. Carter PJ (2006) Potent antibody therapeutics by design. Nat Rev Immunol 6: 343–357.
34. Yano S, Hanibuchi M, Nishioka Y, Nokihara H, Nishimura N, et al. (1999) Combined therapy with anti-P-glycoprotein antibody and macrophage colony-stimulating factor gene transduction for multiorgan metastases of multidrug-resistant human small cell lung cancer in NK cell-depleted SCID mice. Int J Cancer 82: 105–111.
35. Schneider-Merck T, Lammerts van Bueren JJ, Berger S, Rossen K, van Berkel PH, et al. (2010) Human IgG2 antibodies against epidermal growth factor receptor effectively trigger antibody-dependent cellular cytotoxicity but, in contrast to IgG1, only by cells of myeloid lineage. J Immunol 184: 512–520.
36. van de Winkel JG, Anderson CL (1991) Biology of human immunoglobulin G Fc receptors. Journal of leukocyte biology 49: 511–524.
37. Ravetch JV, Bolland S (2001) IgG Fc receptors. Annu Rev Immunol 19: 275–290.
38. Bleeker WK, Lammerts van Bueren JJ, van Ojik HH, Gerritsen AF, Pluyter M, et al. (2004) Dual mode of action of a human anti-epidermal growth factor receptor monoclonal antibody for cancer therapy. J Immunol 173: 4699–4707.
39. Sikic BI, Fisher GA, Lum BL, Halsey J, Beketic-Oreskovic L, et al. (1997) Modulation and prevention of multidrug resistance by inhibitors of P-glycoprotein. Cancer Chemother Pharmacol 40 Suppl: S13–19.
40. Loscher W, Potschka H (2005) Blood-brain barrier active efflux transporters: ATP-binding cassette gene family. NeuroRx 2: 86–98.
41. Bougoin S, Lomet D, Kerboeuf D, Le Vern Y, Malpaux B, et al. (2008) Evidence that the choroids plexus in female sheep express P-glycoprotein. Neuro Endocrinol Lett 29: 438–442.
42. Gottesman MM, Ludwig J, Xia D, Szakacs G (2006) Defeating drug resistance in cancer. Discov Med 6: 18–23.

14

Levels of Alpha-Toxin Correlate with Distinct Phenotypic Response Profiles of Blood Mononuclear Cells and with agr Background of Community-Associated *Staphylococcus aureus* Isolates

Srikanth Mairpady Shambat[1], Axana Haggar[1], Francois Vandenesch[2], Gerard Lina[2], Willem J. B. van Wamel[3], Gayathri Arakere[4], Mattias Svensson[1], Anna Norrby-Teglund[1]*

1 Karolinska Institutet, Centre for Infectious Medicine, Stockholm, Sweden, **2** CIRI, International Center for Infectiology Research, LabEx Ecofect, Université Lyon 1, Inserm U1111, Ecole Normale Supérieure de Lyon, CNRS UMR5308, Centre National de Référence des Staphylocoques, Hospices civils de Lyon, Lyon, France, **3** Department of Medical Microbiology and Infectious Diseases, Erasmus Medical Centre, Rotterdam, Netherlands, **4** Society for Innovation and Development, Indian Institute of Science Campus, Bangalore, India

Abstract

Epidemiological studies of *Staphylococcus aureus* have shown a relation between certain clones and the presence of specific virulence genes, but how this translates into virulence-associated functional responses is not fully elucidated. Here we addressed this issue by analyses of community-acquired *S. aureus* strains characterized with respect to antibiotic resistance, ST types, *agr* types, and virulence gene profiles. Supernatants containing exotoxins were prepared from overnight bacterial cultures, and tested in proliferation assays using human peripheral blood mononuclear cells (PBMC). The strains displayed stable phenotypic response profiles, defined by either a proliferative or cytotoxic response. Although, virtually all strains elicited superantigen-mediated proliferative responses, the strains with a cytotoxic profile induced proliferation only in cultures with the most diluted supernatants. This indicated that the superantigen-response was masked by a cytotoxic effect which was also confirmed by flow cytometry analysis. The cytotoxic supernatants contained significantly higher levels of α-toxin than did the proliferative supernatants. Addition of α-toxin to supernatants characterized as proliferative switched the response into cytotoxic profiles. In contrast, no effect of Panton Valentine Leukocidin, δ-toxin or phenol soluble modulin α-3 was noted in the proliferative assay. Furthermore, a significant association between *agr* type and phenotypic profile was found, where *agr*II and *agr*III strains had predominantly a proliferative profile whereas *agr*I and IV strains had a predominantly cytotoxic profile. The differential response profiles associated with specific *S. aureus* strains with varying toxin production could possibly have an impact on disease manifestations, and as such may reflect specific pathotypes.

Editor: J. Ross Fitzgerald, University of Edinburgh, United Kingdom

Funding: This work was supported by grants from Karolinska Institutet (ANT), Swedish Research Council (521-2011-2540, ANT; 521-2009-3900 MS), The Swedish Research Links Programme (348-2009-6501; ANT & GA), Knut and Alice Wallenberg Foundation, European Union FP7 INFECT-project, NPP grant New INDIGO ERA Net, Hospices Civils de Lyon (GL, FV) and INSERM (GL, FV). The funders had no role in study design, data collection and analysis, decision to publish, or preparation of the manuscript.

Competing Interests: The authors have declared that no competing interests exist.

* Email: anna.norrby-teglund@ki.se

Introduction

Staphylococcus aureus (*S. aureus*) is a significant cause of human infections and an emerging health problem globally due to its increasing resistance to beta-lactams (methicillin-resistant *S. aureus*, MRSA). A special cause of concern is the rise in community-acquired (CA) *S. aureus* strains, and particularly concerning are reports of specific CA-MRSA clones associated with highly aggressive infections, including necrotizing fasciitis and pneumonia in otherwise healthy individuals [1–3]. Although there are large geographical differences, epidemiological studies have shown that more than 20 distinct CA-MRSA lineages are present globally [4].

The pathogenesis of invasive staphylococcal infections involves a variety of virulence factors. In severe invasive infections, several exotoxins have been implicated in disease pathogenesis, such as superantigens [5–7], as well as the cytotoxins Panton Valentine Leukocidin (PVL) [8–10], alpha-toxin (α-toxin) [8,10], and phenol soluble modulins (PSMs) [10,11]. There are to date 24 distinct superantigens identified in *S. aureus*, including the well characterized TSST-1 and the staphylococcal enterotoxins (SE) A–C. Superantigens have been attributed a central role in mediating the systemic toxicity and shock by virtue of their ability to induce hyper-inflammatory responses [6,7], and they have also been implicated in severe pulmonary infections [5]. Cytotoxins have been associated with cell and tissue injury as well as inflammation in severe invasive infections, including necrotising infections of the skin and lung [10]. Several of the toxins are encoded by mobile genetic elements resulting in great diversity in toxin gene content among *S. aureus* strains [12,13] Many of these virulence factors are regulated and controlled by a global regulator system called the accessory gene regulator (agr) system encoded by the *agr* locus [14–17]. The *agr* locus has diverged among different *S. aureus* strains with polymorphism in *agrBDC* region, resulting in four major allelic types of *agr*, i.e. *agr*I–IV [15,17,18]. A recent study demonstrated the impact of allelic variation on *agr* induction dynamics, which translated into significant differences in expression of several virulence factors [19].

S. aureus infected patients have been shown to develop antibodies against both superantigens and cytotoxins [20–24]; thus demonstrating that patients are exposed to a combination of exotoxins secreted by the strains during infection. Exactly how these different combinations of toxins affect virulence and disease outcome is, as of yet, not fully elucidated. One example of interactions between superantigens and cytotoxins was reported by Broshnan et al [25] who showed that cytolysins promoted increased penetrance of superantigens in mucosa. Here we set out to explore how diversity in exotoxin profiles among CA *S. aureus* strains translates into virulence-associated functional responses. To this end, we determined the effect on peripheral blood mononuclear cells (PBMC) elicited by CA *S. aureus* strains of different sequence types (ST) with distinct *agr* allelic types and toxin gene profiles. Our data revealed that the strains induced differential responses in PBMC, characterized by either cytotoxic or proliferative responses, which were linked to specific *agr* types and levels of α-toxin expression.

Materials and Methods

Ethics Statement

This study includes blood samples from buffy coats of blood provided by the blood bank at the Karolinska University Hospital. The buffy coats were provided anonymously; hence informed consent was not required. The ethical research committee at Huddinge University Hospital (Forskningskommitte Syd) approved the study.

Clinical isolates

Strains (n = 38) were selected from a heterogeneous cohort of CA *S. aureus* representing a diverse collection of isolates with different ST and toxin profiles collected from colonized individuals (nasal swabs) or patients with varying *S. aureus* infections in India (Table 1). The isolate collection has previously been characterized with respect to antibiotic resistance profile, molecular typing including ST, *agr*-types, and toxin profile as determined by either PCR for specific genes or by a DNA microarray covering 185 *S. aureus* genes [26]. The study also included a confirmatory cohort of 31 isolates collected from patients with CA *S. aureus* pneumonia included in a prospective cohort study in France [9] or from cases referred to the French national reference laboratory for staphylococci.

A clinical USA300 strain (LUG2012) from a patient from South-Ouest of France and its isogenic mutants deficient for either α-toxin (LUG2209) or PVL (LUG2040) were also included to confirm toxin-mediated effects. The deletion mutants were obtained by using pMAD [27] carrying the replacement cassette that was electroporated to RN4220 recipient strain and then to LUG2012. Growth at non-permissive temperature (44°C) was followed by several subcultures at 30°C and 37°C to favor double crossing over as previously described [28]. Validation of deletion of the *hla* (LUG2209) or *luk*SF-PV (LUG2040) genes were done by PCR and the production of α-toxin and PVL were determined with previously described toxin-specific ELISA (see below).

Preparation of bacterial culture supernatants

The strains were cultured overnight at 37°C in 25 ml casein hydrolysate and yeast extract (CCY) medium. Cell-free supernatants were prepared through centrifugation at 3350 g followed by filter sterilization.

Toxin determination

Bacterial supernatants were assessed for the presence of selected exotoxins, including α-toxin, staphylococcal enterotoxin B (SEB), and toxic shock syndrome toxin 1 (TSST-1), using a multiplex competition immunoassay based on Luminex technology [29], and the amounts of α-toxin and PVL in the bacterial supernatants were determined by specific ELISA as previously described [30,31] using specific antibodies kindly provided, respectively, by GSK Biologicals Inc. (USA) and bioMerieux R&D Immunodiagnostic (France).

PBMC proliferation assay

Human PBMC were isolated from healthy donors using Lymphoprep density centrifugation. The cells were cultured in RPMI-1640 medium supplemented with 10% FCS, 10 mM of L-glutamine, Penicillin (100 U/ml)/Streptomycin (100μg/ml) and 25 mmol/L HEPES (all from Thermo Scientific HyClone, USA). PBMC were seeded at 2×10^5 cells/well and stimulated at 37°C with serial dilutions of bacterial supernatants. After 72 hours, the cells were pulsed for 6 hours with 1 μCi/well of 3H-thymidine (Perkin-Elmer) after which 3H-uptake was measured in a beta-scintillation counter. Phytohemagglutinin-L (PHA) (1 μg/ml) (Sigma-Aldrich, St. Louis, USA) was used as a positive control for polyclonal T cell activation. The cytotoxic/inhibitory effect was tested by the addition of bacterial supernatants in combination with PHA in the proliferation assay. The bacterial culture medium CCY was included as a negative control, and was found to have negligible effect on proliferation (mean CPM 3088, 1785, 1719 for 1:50; 1:100 and 1:1000 dilution, resp.) and no inhibitory effect on PHA-induced proliferation. PBMC were also stimulated with bacterial supernatants or PHA in combination with purified α-toxin (Sigma-Aldrich, St. Louis, USA), recombinant PVL, recombinant δ-toxin and purified PSM-α3 (all from IBT Bioservices, Gaithersburg, USA) and proliferation assessed.

IVIG inhibition assay

PBMC were stimulated with bacterial supernatants, α-toxin or PHA in the presence or absence of different concentrations of IVIG (Gammagard S/D, Baxter). Proliferative responses were determined after 72 hours as described above.

Flow cytometry analysis

Stimulated PBMC were washed and incubated for 30 minutes on ice with directly conjugated antibodies (CD3: SK7, R&D Systems; CD45: T29/33, BD Biosciences; HLA-DR: TU36, Life technologies) in combination with a dead cell marker (Live/Dead Fixable near IR; Molecular Probes). Analyses were done using a Beckton Dickinson LSRII SORP flow cytometer and FlowJo 9.5.3.

Statistical evaluation

Data were analyzed by GraphPad Prism version 4.0 for Windows (GraphPad software). Two-sided Mann-Whitney *U* test

Table 1. Characteristics of *S. aureus* strains with respect to antibiotic resistance, ST and agr types, toxin genotype and production *in vitro*.

Strains	Clinical Diagnosis	MSSA/MRSA	*agr* type	SPA type	CC/ST type	*pvl*	*luk*D/E	*se/tsst-1/egc*	*Etd/edinB*	Response Profile#	Luminex*
Colonizing strains											
Sa559		MSSA	I	t005	ST22/CC22	+	ND	−/−/+	−	C (96.6)	α-toxin+++
Sa115		MRSA-IV	I	t852	ST22/CC22	+	ND	−/−/+	−	P (12.3)	α-toxin +
Sa165		MRSA-IV	I	t852	ST22/CC22	+	ND	−/−/+	−	C (95.7)	α-toxin +++
Sa95		MSSA	I	t3840	ST672	−	+	−/−/+	−	C (98.0)	α-toxin +++
Sa134		MRSA-V	I	t064	ST1208/CC8	+	+	*se*-A, B, K, Q/−/−	−	P (76.5)	α-toxin ++/SEB++
Sa180		MSSA	I	t4285	ST6	−	ND	*se*-L/*tsst-1*/−	−	C (97.5)	α-toxin +++/TSST+++
Sa168		MSSA	I	t937	ST291	−	−/+	−/−/−	+	C (96.7)	α-toxin +++
Sa337		MSSA	I	t3096	ST291	−	ND	−/−/−	+	C (96.1)	α-toxin +++
Sa18		MRSA-V	II	t657	ST772/CC1	+	−	*se*-A, C, L/−/+	−	P (27.0)	α-toxin ++
Sa289		MSSA	II	t1839	ST772/CC1	+	−	*se*-A/−/+	−	P (28.5)	α-toxin ++
Sa1437		MSSA	II	t345	ST772/CC1	+	ND	*se*-A, C, L/*tsst*-1/+	−	P (23.7)	α-toxin +
Sa1446		MRSA-V	II	t657	ST772/CC1	+	ND	*se* - A, E/−/+	−	P (25.4)	α-toxin +
Sa233		MRSA-V	II	t657	ST772/CC1	+	ND	*se*-A, C, E, L/−/+	−	P (30.0)	α-toxin ++
Sa159		MSSA	II	t774	ST199/CC15	−	+	−/−/−	−	Ambigious	α-toxin +
Sa160		MSSA	II	t774	ST199/CC15	−	+	−/−/−	−	C (69.6)	α-toxin +
Sa32		MRSA-IV	III	t021	CC30	+	−	−/−/+	−	P (21.2)	α-toxin +/−
Sa368		MSSA	IV	t1999	CC121	+	ND	−/−/+	−	C (93.7)	α-toxin +++
Sa14		MSSA	IV	t3204	CC121	+	−/+	*se*-B/−/+	−	C (95.0)	α-toxin +++/SEB+++
Disease strains											
Sa37	SSTI	MRSA-IV	I	t852	ST22/CC22	+	−	−/−/+	−	P (12.8)	α-toxin +
Sa08	SSTI	MRSA-IV	I	t852	ST22/CC22	+	−	−/−/+	−	C (98.0)	α-toxin +++
Sa113	Brain abscess	MRSA-IV	I	t852	ST22/CC22	+	−	−/−/+	−	C (95.0)	α-toxin +++
Sa114	Cerebral abscess	MRSA-IV	I	t852	ST22/CC22	+	−	−/−/+	−	C (95.8)	α-toxin +++
SaN08	Meningitis	MRSA-IV	I	t852	ST22/CC22	+	−	−/−/+	−	C (96.5)	α-toxin +++
Sa1	Invasive infection	MRSA-IV	I	t1309	ST672	−	+	−/−/+	−	C (94.6)	α-toxin +++
Sa754	Invasive infection	MRSA-IV	I	t852	ST22/CC22	+	−	−/−/+	−	C/P (87.3)	α-toxin +++
Sa755	Invasive infection	MSSA	I	ND	ST22/CC22	+	−	−/−/+	−	P (3.31)	α-toxin ++
Sa118	Pyomyositis	MRSA-V	II	t657	ST772/CC1	+	−	*se*-A, C, L/−/+	−	P (30.7)	α-toxin ++
Sa3957	Breast abscess	MRSA-V	II	t1387	ST772/CC1	+	ND	*se*-A, E/−/+	−	P (21.5)	α-toxin +/−
Sa3989	Pneumonia	MRSA-V	II	t3596	ST772/CC1	+	ND	*se*-A, E/−/+	−	P (24.8)	α-toxin ++
Sa120/1	Cerebral abscess (pus)	MRSA-V	II	t657	ST772/CC1	+	ND	ND	ND	P (24.7)	α-toxin ++
Sa2332	Pleural Empyema	MSSA	III	t021	CC30	+	−	−/−/+	−	P (36.2)	α-toxin +/−

Table 1. Cont.

Strains	Clinical Diagnosis	MSSA/MRSA	*agr* type	SPA type	CC/ST type	*pvl*	*lukD/E*	*se/tsst-1/egc*	*Etd/edinB*	Response Profile#	Luminex*
SaP1	Suture induced infiltrate	MSSA	III	t127	ST1	–	ND	*se*-D, E/–/–	–	P (80.5)	ND
SaP3	Keratitis	MSSA	III	t8078	ST1	–	ND	*se*-D/–/–	–	P (27.1)	ND
SaP6	Keratitis	MSSA	III	t127	ST1	–	ND	*se*-D, E/–/–	–	P (60.1)	ND
SaP7	Orbital abscess	MRSA-V	III	t2526	ST88	+	ND	*se*-D/–/+	–	C (97.3)	ND
Sa753	Necrotizing pneumonia	MSSA	IV	t159	ST121	+	+	–/–/+	–	C (98.3)	α-toxin +++
Sa796	Necrotizing pneumonia	MSSA	IV	t159	ST121	+	+	–/–/+	–	C (97.4)	α-toxin +++
Sa1059	Invasive inf.	MSSA	IV	t159	ST121	+	+	–/–/+	–	C (98.5)	α-toxin +++

MRSA, methicillin-resistant S. aureus; MSSA, methicillin-susceptible S. aureus; ND, not determined; *se*, staphylococcal enterotoxin;P, proliferative; C, cytotoxic; SSTI, skin and soft tissue infection.
#The response profile is determined by the pattern of proliferative responses elicited by different dilutions of bacterial supernatants. A proliferative profile was denoted if all dilutions elicited a proliferative response. A cytotoxic pattern was denoted if a) proliferation was only noted at the highest dilution of the supernatants, and b) the supernatants (1:50 dilution) resulted in inhibition of PHA-induced responses in co-stimulation experiments. Cells stimulated with supernatants (1:50 dilution) were also analyzed by flow cytometry and the value in parenthesis shows % cells stained positive for dead cell marker.
*Analyses of selected exotoxins including α-toxin (alpha-toxin), SEB (staphylococcal enterotoxin B) and TSST-1 (toxic shock syndrome toxin 1) are assessed in overnight bacterial culture supernatants by luminex. Only positive results are indicated and based on the dilution series response a semi quantitative measure is given as +/–, +, ++, +++.

or Fisher's exact test were used for comparison between two groups. Comparisons between multiple groups were done using ANOVA and Dunn's multiple comparison test. Differences were considered significant when $p<0.05$.

Results

Distinct functional phenotypic profiles of clinical *S. aureus* strains

A proliferation assay was employed in which PBMC were exposed to bacterial supernatants prepared from CA *S. aureus* strains. The strain collection included CA MRSA and MSSA strains of varying ST types, *agr* types and toxin gene profile isolated from patients or colonized individuals (Table 1). All strains, except two (strains Sa159 and Sa160), harbored superantigen encoding genes (Table 1). To confirm that the genes were expressed during *in vitro* culture, the bacterial supernatants were analyzed for toxin content using a customized luminex assay. Due to technical limitations only a selected set of exotoxins, i.e. α-toxin, SEB and TSST-1, were included for which the protein detection concurred with the toxin gene profile of respective strain (Table 1).

PBMC from four different donors were stimulated with serial dilutions of the bacterial supernatants and proliferative responses assessed (Figure 1A). This assay is commonly used to functionally assess superantigen-mediated proliferation, but the response will be influenced by the presence of cytotoxins targeting PBMC. Accordingly, supernatants prepared from the different *S. aureus* strains induced starkly different ($p<0.0001$) and highly reproducible response profiles (Figures 1A, 1B). While some strains induced consistently high proliferative responses at all dilutions (1:50, 1:100 and 1:1000) of the bacterial supernatants (denoted as a proliferative profile) (Figures 1A, 1B), other strains induced proliferation only at the highest dilution (1:1000) of bacterial supernatants (Figures 1A, 1B). Thus, demonstrating that the supernatants contain superantigens that trigger proliferation, but this activity is masked in more concentrated supernatants from certain strains, potentially through toxin-mediated cytotoxicity.

To investigate this further, proliferation was assessed following stimulations of PBMC with the polyclonal T cell activator PHA in combination with supernatants displaying a proliferative or cytotoxic profile (Figure 1C). Using the 1:50 dilutions of bacterial supernatants, the PHA-response was completely abolished by the cytotoxic supernatants, whereas augmented by the proliferative supernatants (Figure 1D). To directly assess whether bacterial supernatants were cytotoxic, flow cytometry analysis was applied on cells stimulated with supernatants and subsequently stained with a dead cell marker in combination with antibodies directed towards defined cell markers. PBMC stimulated with a proliferative supernatant showed a relative expansion of the T cell population similar to that seen for PHA (Figure 1E). Notably, stimulation with a cytotoxic supernatant resulted in extensive cell death as almost all cells stained positive for the dead cell marker (Figure 1E). As shown in Table 1, cell death was significantly more pronounced among cells exposed to cytotoxic, as compared to proliferative, supernatants (mean % of cells staining positive for dead cell marker: 96.3% and 33.2% for cytotoxic and proliferative supernatants, resp.; $p<0.001$).

The two functional response profiles were found in both colonizing as well as patient isolates and there was no significant association between functional profile and ST-type/clonal complex or MRSA/MSSA type (Table 1). In contrast, an association with *agr* type was evident, and *agr* II and *agr* III strains had predominantly a proliferative profile whereas *agr* I and IV strains were cytotoxic ($p<0.015$) (Figure 1F). As these analyses were

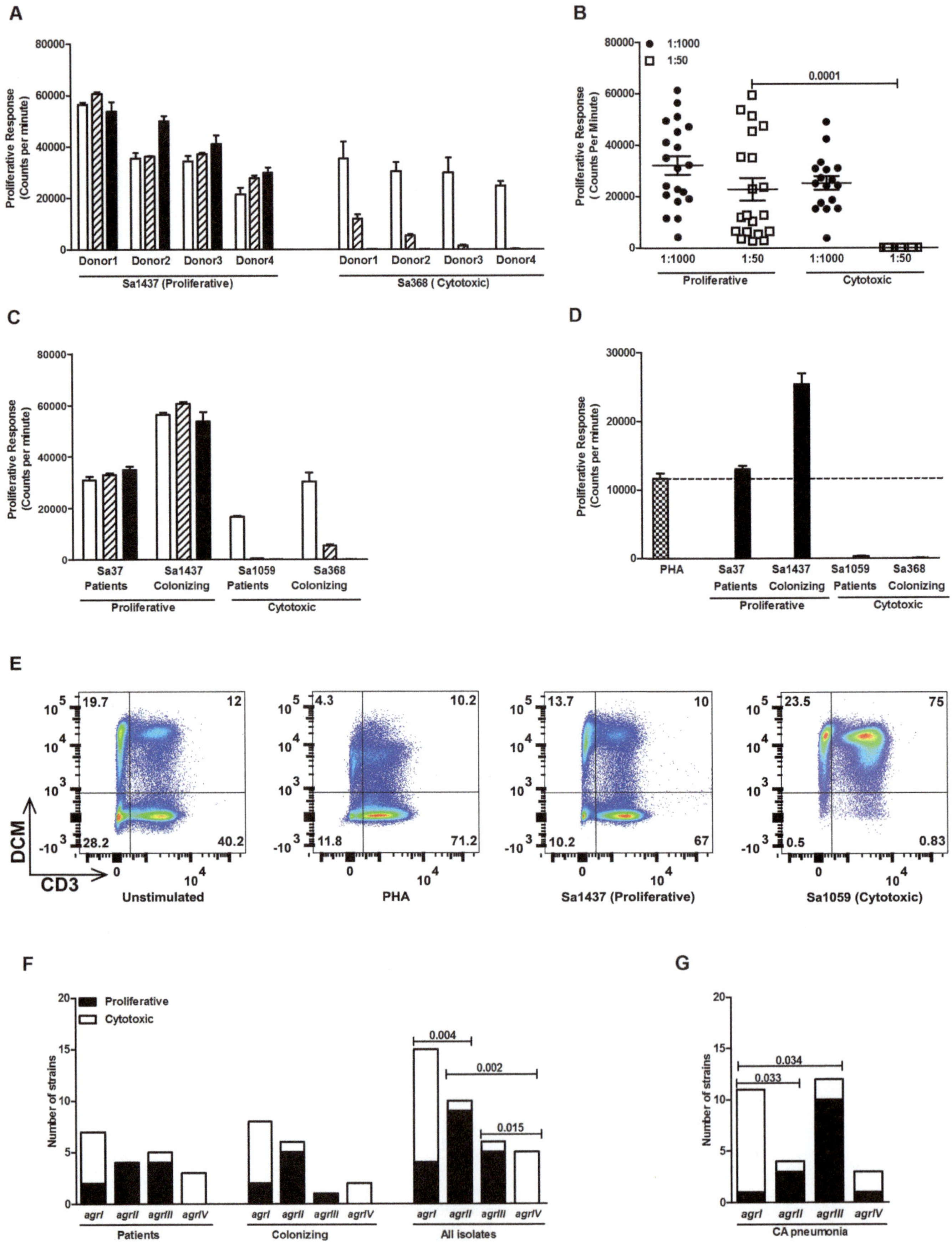
A
Proliferative Response (Counts per minute)
Donor1 Donor2 Donor3 Donor4
Sa1437 (Proliferative)
Sa368 (Cytotoxic)
B
1:1000
1:50
0.0001
Proliferative Response (Counts Per Minute)
Proliferative
Cytotoxic
C
Sa37 Patients
Sa1437 Colonizing
Sa1059 Patients
Sa368 Colonizing
Proliferative
Cytotoxic
D
PHA
E
DCM
CD3
Unstimulated
PHA
Sa1437 (Proliferative)
Sa1059 (Cytotoxic)
F
Proliferative
Cytotoxic
Number of strains
0.004
0.002
0.015
agrI agrII agrIII agrIV
Patients
Colonizing
All isolates
G
0.034
0.033
CA pneumonia

Figure 1. Proliferative or cytotoxic responses elicited by *S. aureus* strains. Human peripheral blood mononuclear cells (PBMC) isolated from healthy donors were stimulated with dilution series (1:1000 (open), 1:100 (stripped) and 1:50 (filled) bars) of bacterial supernatants prepared from overnight cultures of *S. aureus* strains. Proliferative responses were determined by 3H-thymidine uptake and are presented as mean counts per minute ± SD. The supernatants induced either a proliferative or cytotoxic response profile. A cytotoxic profile was assigned when strains did not elicit a proliferative response in the more concentrated supernatants but only in the most diluted samples. A) Shows the response of a representative proliferative and cytotoxic supernatant in separate experiments using cells from four healthy donors. B) Scatter plot of proliferative responses induced by bacterial supernatants with a proliferative or a cytotoxic profile. Mean values of four different experiments are shown for dilutions 1:1000 (filled symbols) and 1:50 (open symbols). C) Bacterial supernatants with proliferative or cytotoxic profiles were found among both patients (n = 20) and colonized individuals (n = 17). The figure shows proliferative responses (mean ± SD) of one representative, out of five, experiments using cells from different donors stimulated with dilution series (1:1000 (open), 1:100 (stripped) and 1:50 (filled) bars) of bacterial supernatants. D) To confirm a cytotoxic effect, proliferative responses were assessed in cells stimulated with PHA alone (the dashed line shows the average PHA response) or in combination with supernatants (proliferative and cytotoxic; dilution 1:50 (filled) bars). The figure shows one representative of three experiments using cells from different donors. E) Flow cytometry analysis on PBMC stimulated with PHA and bacterial supernatants (1:50 dilution). Total PBMC were gated based on CD45 expression and CD45 positive cells were further analysed for dead cell marker (DCM, Y axis) and CD3 (X axis) positivity. The figure shows one representative of three individual experiments using cells from different donors. F) Relation between *agr* types and proliferative (black bars) or cytotoxic (white bars) community *S. aureus* strains collected from colonizing individuals (n = 17) or patients with infections (n = 20) in India. G) Relation between *agr* type and proliferative or cytotoxic profiles elicited by community-acquired (CA) pneumonia patients (n = 31). Statistical significant differences were determined by use of the two-sided Mann Whitney test and Fisher's exact test with two sided p value and p values are indicated in the figure.

conducted on a highly heterogeneous strain cohort, we expanded the analyses to include a more homogenous strain cohort collected from patients with CA *S. aureus* pneumonia. Also in this cohort, a similar association with *agr* I and IV being significantly more cytotoxic than *agr* II or III was noted ($p<0.034$) (Figure 1G).

High α-toxin expression is linked to the cytotoxic profiles

Quantitation of the cytotoxins α-toxin and PVL revealed that the cytotoxic supernatants had significantly higher levels of α-toxin than the proliferative, regardless of whether the strains were collected from patients, colonized individuals, or CA pneumonia ($p<0.002$) (Figure 2A). In contrast, there was no correlation between the response profile and PVL expression (Figure 2B) which is in line with the reported lack of susceptibility of PBMC to PVL [32]. In agreement with the data in Figure 1F demonstrating that different *agr* types are linked to either proliferative or cytotoxic profiles, significantly higher amounts of α-toxin were found in strains belonging to the cytotoxic *agr* type I and IV, as compared to the proliferative *agr* type II and III strains ($p<0.05$) (Figure 2C). In this context, no association between PVL levels and specific *agr* types was seen (Figure 2D). Also, expression data on the *psm-α* gene revealed no difference between strains eliciting either response profile (data not shown).

These data implied that α-toxin is a main mediator of the noted cell death in PBMC. To further test this, PBMC were stimulated with PHA in combination with purified α-toxin revealing a clear dose response pattern where increasing amounts of α-toxin resulted in reduced proliferation (Figure 3A). Furthermore, supplementing a proliferative (Sa1437) or a cytotoxic (Sa368) *S. aureus* supernatant with purified α-toxin, resulted in a switch to a cytotoxic response both at the 1:1000 and 1:50 dilutions for the proliferative strain and an increased cytotoxic response at the 1:1000 dilution for the cytotoxic strain (Figure 3B). Further support for an α-toxin-mediated effect was provided by using a strain of the USA300 lineage (LUG2012), which belongs to the cytotoxic *agr* type I and produces high levels of α-toxin (α-toxin 83 μg/ml, PVL 14 μg/ml), and its isogenic mutants for PVL (α-toxin 54 μg/ml, PVL 0 μg/ml) and α-toxin (α-toxin 0 μg/ml, PVL 11 μg/ml). The USA300 and the PVL-deficient mutant both showed a cytotoxic profile, whereas the α-toxin mutant showed a proliferative response profile (Figure 3C). Also when PBMC were stimulated with PHA in combination with these supernatants a reduction of PHA-induced responses was noted in the presence of supernatants from USA300 and the PVL-deficient mutant (Figure 3D). In contrast, the α-toxin mutant supernatant resulted in an augmentation of the PHA-induced response (Figure 3D). Flow cytometry analysis further confirmed a potent cytotoxic effect of USA300 and the PVL-deficient mutant supernatants (97.2% and 85.8% dead cell marker positivity, resp.), whereas the α-toxin mutant supernatant had limited cytotoxicity (20.4% dead cell marker positivity). It should be noted that although the supernatant of the α-toxin mutant elicited a proliferative response at all dilutions tested, the proliferative response increased with more diluted supernatants (Figure 3C); thus, indicating that there are inhibitory factors, other than α-toxin, present in the USA300 supernatant. Other cytotoxins tested, including purified PVL, PSM-α3 and δ-toxin alone or in combination with PHA did not elicited any cytotoxic or proliferative responses in PBMC (Figure 3E).

IVIG inhibition of toxin mediated responses in PBMC

Studies have demonstrated the presence of antibodies against defined *S. aureus* virulence factors, including α-toxin, PVL and superantigens, in intravenous polyclonal immunoglobulin (IVIG) preparations [33–37]. Here we tested whether IVIG could inhibit the toxin-mediated functional response profiles, in particular the cytotoxic effect associated with high α-toxin levels. To this end, PBMC were stimulated with PHA and different concentrations of purified α-toxin in the presence or absence of IVIG. We found a clear dose response of IVIG-mediated inhibition of α-toxin cytotoxicity, evident by increased proliferative responses (Figure 4A). At the lowest concentration of α-toxin (60 ng/ml), a significant inhibition was achieved with 0.1 mg/ml IVIG ($p = 0.05$) while at the highest concentration of 480 ng/ml of α-toxin significant neutralization was achieved only at concentrations >1.0 mg/ml of IVIG ($p = 0.034$) (Figure 4A). Similarly we found that IVIG was able to inhibit both the proliferative as well as the cytotoxic responses elicited by superantigens and cytotoxins present in *S. aureus* culture supernatants (Figure 4B).

Discussion

In this study we demonstrate that CA *S. aureus* strains with distinct toxin profiles exhibit stable robust phenotypic profiles evident by their ability to elicit either a proliferative or cytotoxic response profile in human PBMC. All experiments were done using bacterial supernatants containing superantigens and cytotoxins secreted by the strains. The data shows that the vast majority of strains elicit superantigenic activity as demonstrated by the induction of proliferative responses in PBMC. However, this

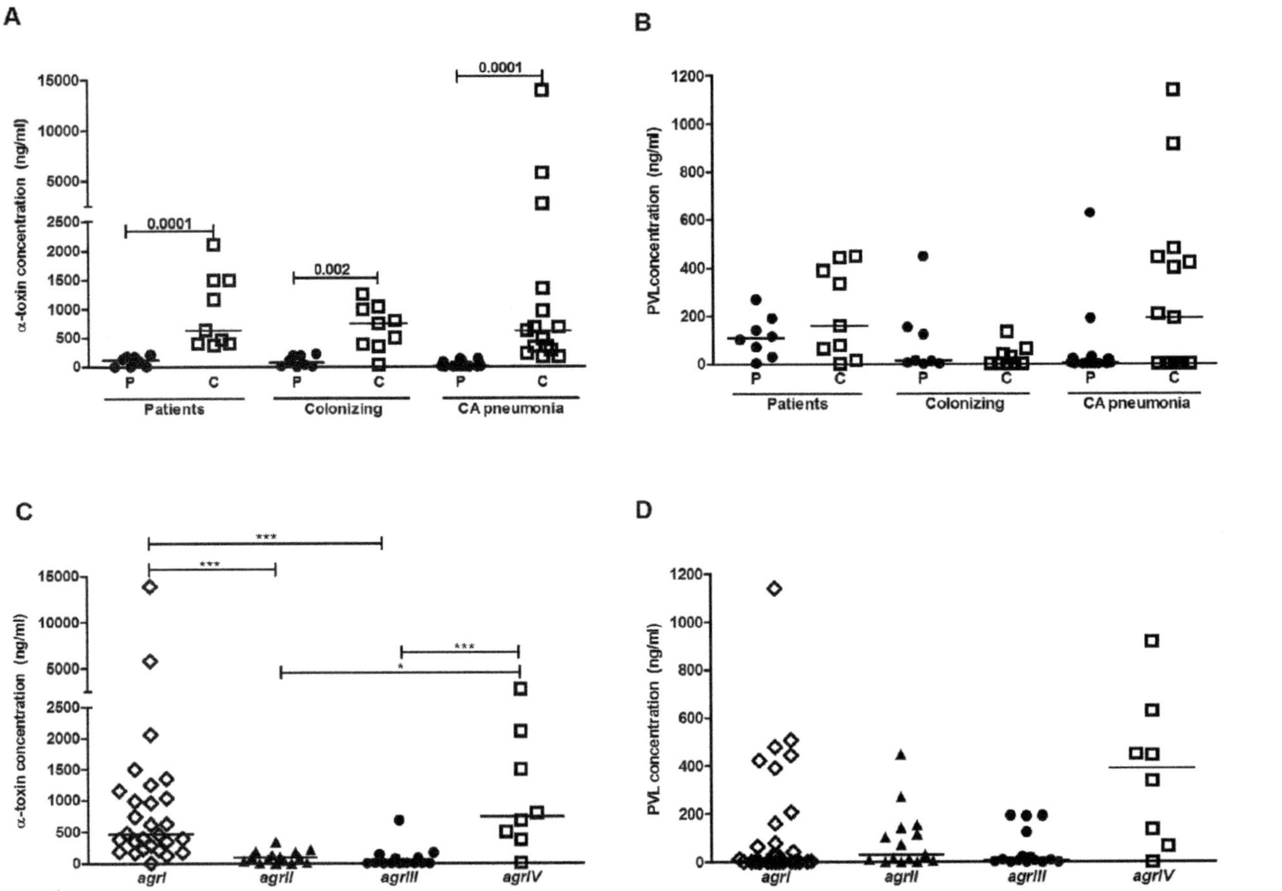

Figure 2. Levels of alpha-toxin expression correlate with cytotoxic profile and *agr* types. Amounts of alpha-toxin (α-toxin) (A) and Panton Valentine Leukocidin (PVL) (B) were determined by ELISA in the *S. aureus* bacterial supernatants (1:50 dilutions) having either a cytotoxic (C; open symbols) or a proliferative (P; filled symbols) response profile. Supernatants prepared from isolates collected from colonized individuals (n = 17), patients (n = 17), or a confirmatory cohort of CA pneumonia (n = 31) are shown. Association between amount of α-toxin (C) and PVL (D) in the *S. aureus* bacterial supernatants (1:50 dilutions; n = 65) with their respective *agr* types (*agr* I and IV, open symbol; *agr* II and III, filled symbol). Statistical significant differences were determined by use of the two-sided Mann Whitney test or with ANOVA and Dunn's multiple comparison test for comparisons of multiple groups and p values are indicated in the figure.

response was masked in cytotoxic strains due inhibitory/cytotoxic factors present in the supernatants. This illustrates the relevance of using such a mixture of secreted factors from clinical isolates, which allows simultaneous analyses of the combined activity of toxins and which more closely reflects what the patients are exposed to. The PBMC assay provided an efficient tool to assess both superantigenic and cytotoxic effects; thereby allowing for identification of distinct phenotypic response profiles among CA S. *aureus* isolates. However, it should be noted that this assay involves only PBMC and hence, excludes effects of pore-forming toxins, such as PVL, LukDE and LukAB, that targets cell populations which are either missing or represent a minor subset in PBMCs [32,38,39]. Therefore, future studies should include also other clinically relevant cells, such as neutrophils, that are highly susceptible to the above mentioned cytotoxins.

Our findings demonstrated a striking association between high α-toxin levels and a cytotoxic phenotypic profile of the *S. aureus* strains. Supernatants that displayed a cytotoxic profile had in average 20-fold higher α-toxin levels than the proliferative supernatants ($p<0.0001$). Furthermore addition of purified α-toxin to either bacterial supernatants or to PHA resulted in a dose-dependent transition from proliferative to cytotoxic response profiles. Also, a USA300 α-toxin deficient mutant revealed a proliferative profile in contrast to the wildtype USA300 strain that was distinctly cytotoxic. As a vast majority (29/33) of the cytotoxic supernatants contained ≥225 ng/ml α-toxin, whereas all proliferative supernatants had levels below 221 ng/ml (in the 1:50 dilutions), we speculate that there might be a critical α-toxin concentration determining cytotoxic responses in PBMC. This underscores the value of quantifying the levels of toxins produced and secreted by the strains as this seems to dramatically influence the cellular responses and potentially virulence.

Similarly, a significant association between response profiles, α-toxin levels and the *agr* type of the strain was noted. Significantly increased α-toxin levels were found in the cytotoxic *agr*I and IV types, as compared to the proliferative *agr*II and III strains. The complexity of agr regulation of *S. aureus* virulence factors was emphasized in the report by Geisinger et al [19] in which protein A, α-toxin, PVL and TSST-1 were tested in congenic strains each harboring a unique *agr* allele (I–IV). They found significant variation in the kinetics and degree of the *agr* signal resulting in differential induction of specific virulence factors. Of special interest, is their finding that *agr* I and IV are the earliest and strongest, followed by *agr* II and III, which is the same hierarchical order as we report here in regards to cytotoxicity. The molecular basis for the noted association between toxin

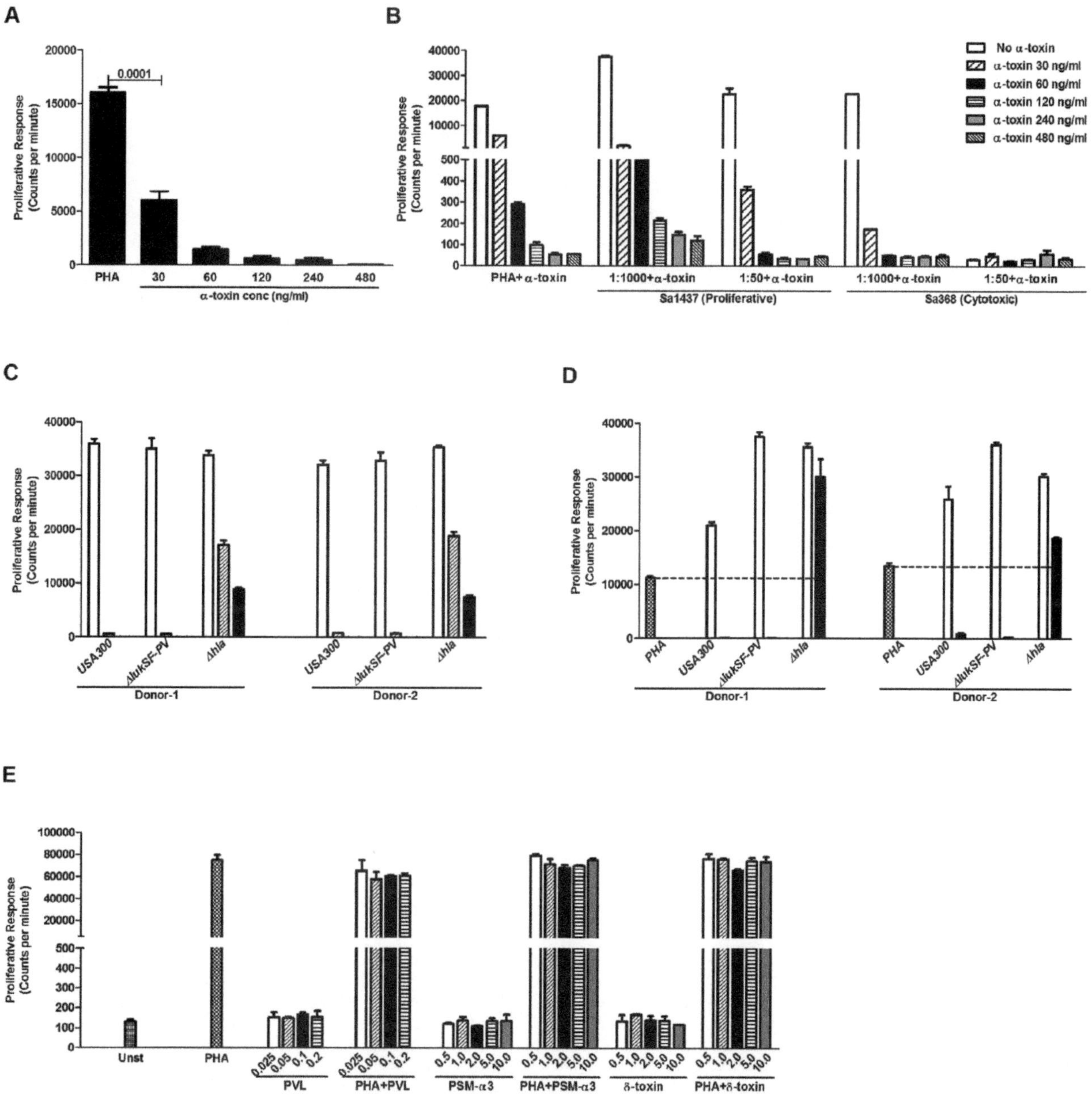

Figure 3. Alpha-toxin mediates cytotoxicity in PBMC. A) Proliferation assay using human PBMC stimulated with PHA alone or in the presence of different concentrations of α-toxin. Proliferative responses were determined by 3H-thymidine uptake and are presented as mean counts per minute ± SD. Mean values of four different experiments from different donors are shown. B) Inhibition of PHA-induced and bacterial supernatant-induced proliferation of PBMCs by addition of increasing concentrations of α-toxin (60 to 480 ng/ml). The figure shows one representative of two experiments using cells from different donors. C) Proliferation assay using cells from two donors stimulated with different dilutions (1:1000 (open), 1:100 (stripped) and 1:50 (filled) bars) of supernatants prepared from USA300 (LUG2012), and its isogenic mutants of PVL (Δ*lukSF-PV*) and α-toxin (Δ*hla*). D) Proliferative responses assessed in cells stimulated with PHA alone or in combination with indicated supernatants at dilutions 1:1000 (open) and 1:50 (filled) bars. The figure shows experiments using cells from two donors. E) PBMC were stimulated with increasing concentrations of PVL, PSM α3 and δ-toxin at indicated concentrations (μg/ml) alone and in combination with PHA. Statistical significant differences were determined by Fisher's exact test with two sided p value and p value is indicated in the figure.

production and *agr* types translating into distinct response profiles has yet to be determined but could possibly be linked to varying levels of secreted auto inducing peptides (AIP). Another possibility is that the AIP-AgrC interaction kinetics may vary between the proliferative and cytotoxic isolates, as previous studies [15,17,18,40] have shown that *agr* I and IV AIPs share almost identical primary sequence with only one amino acid difference whereas agr II and III AIPs show greater variations in their primary sequence.

Another interesting aspect is the potential link between genetic background, *agr* type and disease manifestation [18,41]. Although the link is far from exclusive, *agr* IV are often emphasized as a disease-related isolate [42,43]. Furthermore Jarraud et al [41] reported a predominance of *agr* II and III in TSST-1 mediated TSS and scarlet fever (90–94%, respectively). Here we report

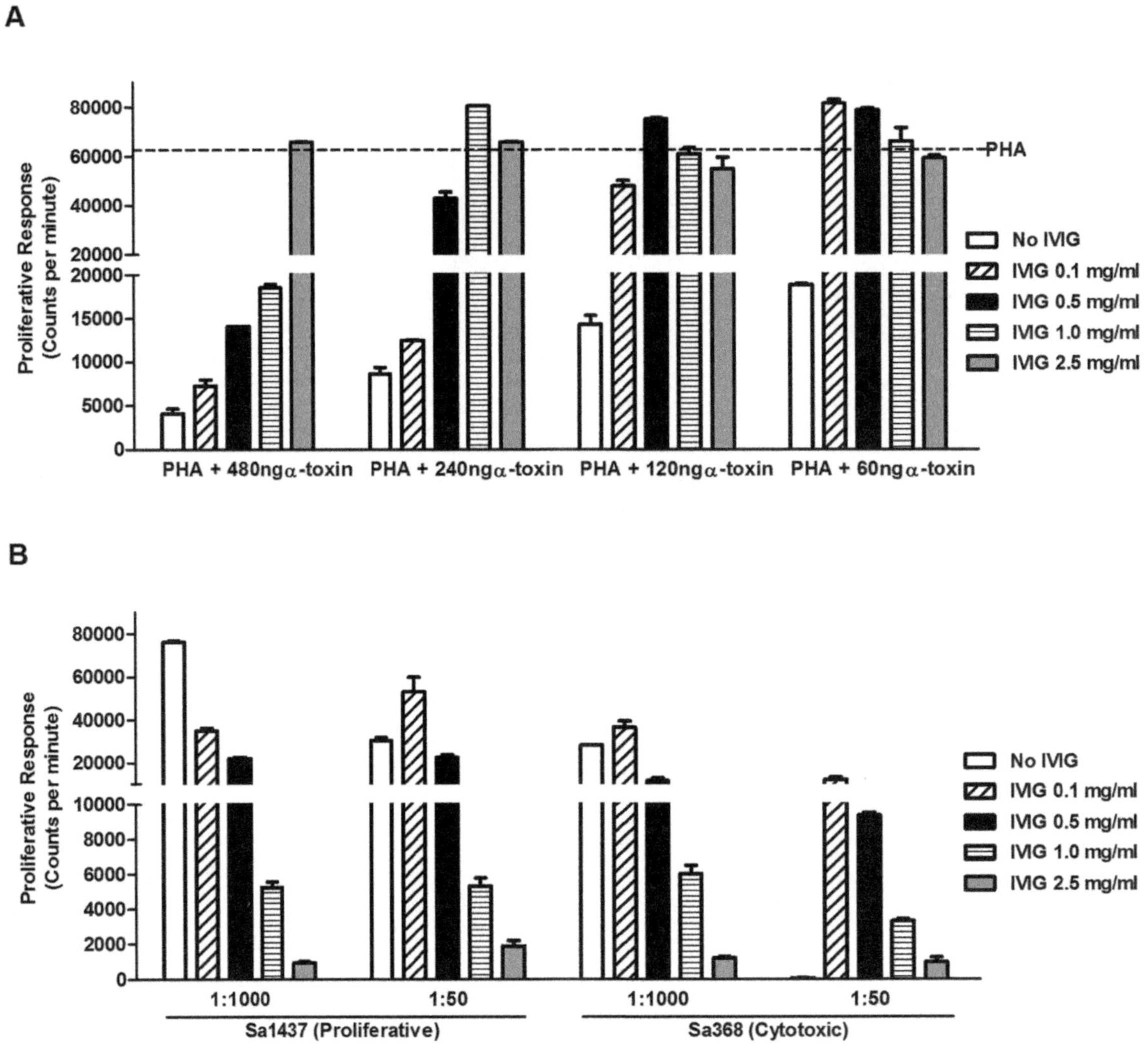

Figure 4. Inhibition of α-toxin and staphylococcal supernatant mediated cytotoxicity and proliferation of PBMC by IVIG. A) Neutralization of α-toxin induced cytotoxicity of PBMC by different concentrations of IVIG. PBMC were stimulated with PHA with increasing concentrations of α-toxin (60 to 480 ng/ml) in the presence or absence of IVIG (0.1 to 2.5 mg/ml, as indicated). The dashed line indicates the mean PHA-induced proliferative response. B) Proliferative responses induced by bacterial supernatants from both proliferative and cytotoxic strains (dilutions 1:1000 and 1:50) in the presence or absence of different concentrations of IVIG (0.1 to 2.5 mg/ml, as indicated). 3H-thymidine uptake after 72 hours of culture is presented as mean counts per minute ± SD. The figure shows one representative of two experiments using cells from different donors.

distinct phenotypic response profiles elicited by clinical CA *S. aureus* isolates, which are determined by the combined action of α-toxin and superantigens, and the data shows a striking association to specific *agr* types, such as the *agr* II and III that demonstrated a superantigen-mediated proliferative profile whereas *agr* I and IV were cytotoxic. Whether these phenotypic response profiles reflect specific pathotypes that contribute to distinct disease manifestations can at present time only be speculated upon and can only be addressed by the use of large well-defined clinical cohorts. For instance, the course of bacteremia, toxic shock syndrome and/or necrotizing infections of the skin and lung might be connected to a specific ratio of superantigenic activity versus cytotoxicity, which would determine the overall biological effect on patients' immune cells. However, we fully appreciate the complexity of identification of pathotypes dictated by toxin-mediated cellular responses. Clinical manifestations and outcome will depend on a variety of factors including among others varying host susceptibility at both the cellular and genetic level as well as site of infection. In addition, patients can present with both necrotizing infections and severe sepsis/septic shock, which raises the currently unexplored question of expression levels and effects of superantigens and cytotoxins systemically and locally. Nevertheless, identification of specific pathotypes would be of great clinical value as it enables recognition of risk patients and allows for epidemiologic surveillance.

Although clinical data are limited, IVIG has been proposed as adjunctive therapy in fulminant invasive *S. aureus* infections and the mechanistic action suggested to involve neutralizing antibodies against superantigens [33,35], PVL [37] and α-toxin [34]. Here we found that addition of IVIG to PBMC cultures efficiently inhibited both α-toxin-mediated cytotoxicity as well as superantigen-mediated proliferation at physiological concentrations. Thus, our data support a beneficial role of IVIG that through its broad spectrum of antibodies can prevent toxin-mediated damage, including both excessive proliferation as well as cytolytic events.

Acknowledgments

The skillful technical assistance provided by Sushma Prabhakara, Yvonne Benito, Michele Bes, Florence Couzon and Cedric Badiou is gratefully acknowledged. We are grateful to bioMérieux and GlaxoSmithKline Vaccines for providing the antibodies required for PVL ELISA and α-toxin ELISA, respectively.

Author Contributions

Conceived and designed the experiments: SMS ANT MS. Performed the experiments: SMS AH FV WvW. Analyzed the data: SMS AH FV WvW GA MS ANT. Contributed reagents/materials/analysis tools: FV GL GA. Contributed to the writing of the manuscript: SMS FV GA MS ANT.

References

1. Vandenesch F, Naimi T, Enright MC, Lina G, Nimmo GR, et al. (2003) Community-acquired methicillin-resistant Staphylococcus aureus carrying Panton-Valentine leukocidin genes: worldwide emergence. Emerg Infect Dis 9: 978–984.
2. (1999) Four pediatric deaths from community-acquired methicillin-resistant Staphylococcus aureus - Minnesota and North Dakota, 1997–1999. Morb Mortal Wkly Rep 48: 707–710.
3. Miller LG, Perdreau-Remington F, Rieg G, Mehdi S, Perlroth J, et al. (2005) Necrotizing fasciitis caused by community-associated methicillin-resistant Staphylococcus aureus in Los Angeles. N Engl J Med 352: 1445–1453.
4. Mediavilla JR, Chen L, Mathema B, Kreiswirth BN (2012) Global epidemiology of community-associated methicillin resistant Staphylococcus aureus (CA-MRSA). Curr Opin Microbiol 15: 588–595.
5. Strandberg KL, Rotschafer JH, Vetter SM, Buonpane RA, Kranz DM, et al. (2010) Staphylococcal superantigens cause lethal pulmonary disease in rabbits. J Infect Dis 202: 1690–1697.
6. Xu SX, McCormick JK (2012) Staphylococcal superantigens in colonization and disease. Front Cell Infect Microbiol 2: 52.
7. Spaulding AR, Salgado-Pabon W, Kohler PL, Horswill AR, Leung DY, et al. (2013) Staphylococcal and streptococcal superantigen exotoxins. Clin Microbiol Rev 26: 422–447.
8. Bubeck Wardenburg J, Bae T, Otto M, Deleo FR, Schneewind O (2007) Poring over pores: alpha-hemolysin and Panton-Valentine leukocidin in Staphylococcus aureus pneumonia. Nat Med 13: 1405–1406.
9. Gillet Y, Issartel B, Vanhems P, Fournet JC, Lina G, et al. (2002) Association between Staphylococcus aureus strains carrying gene for Panton-Valentine leukocidin and highly lethal necrotising pneumonia in young immunocompetent patients. Lancet 359: 753–759.
10. Vandenesch F, Lina G, Henry T (2012) Staphylococcus aureus hemolysins, bi-component leukocidins, and cytolytic peptides: a redundant arsenal of membrane-damaging virulence factors? Front Cell Infect Microbiol 2: 12.
11. Wang R, Braughton KR, Kretschmer D, Bach TH, Queck SY, et al. (2007) Identification of novel cytolytic peptides as key virulence determinants for community-associated MRSA. Nat Med 13: 1510–1514.
12. Deghorain M, Van Melderen L (2012) The Staphylococci phages family: an overview. Viruses 4: 3316–3335.
13. Malachowa N, DeLeo FR (2010) Mobile genetic elements of Staphylococcus aureus. Cell Mol Life Sci 67: 3057–3071.
14. Foster TJ (2005) Immune evasion by staphylococci. Nat Rev Microbiol 3: 948–958.
15. Ji G, Beavis R, Novick RP (1997) Bacterial interference caused by autoinducing peptide variants. Science 276: 2027–2030.
16. Lindsay JA, Holden MT (2004) Staphylococcus aureus: superbug, super genome? Trends Microbiol 12: 378–385.
17. Novick RP (2003) Autoinduction and signal transduction in the regulation of staphylococcal virulence. Mol Microbiol 48: 1429–1449.
18. Jarraud S, Lyon GJ, Figueiredo AM, Lina G, Vandenesch F, et al. (2000) Exfoliatin-producing strains define a fourth agr specificity group in Staphylococcus aureus. J Bacteriol 182: 6517–6522.
19. Geisinger E, Chen J, Novick RP (2012) Allele-dependent differences in quorum-sensing dynamics result in variant expression of virulence genes in Staphylococcus aureus. J Bacteriol 194: 2854–2864.
20. Croze M, Dauwalder O, Dumitrescu O, Badiou C, Gillet Y, et al. (2009) Serum antibodies against Panton-Valentine leukocidin in a normal population and during Staphylococcus aureus infection. Clin Microbiol Infect 15: 144–148.
21. Jacobsson G, Colque-Navarro P, Gustafsson E, Andersson R, Mollby R (2010) Antibody responses in patients with invasive Staphylococcus aureus infections. Eur J Clin Microbiol Infect Dis 29: 715–725.
22. Kanclerski K, Soderquist B, Kjellgren M, Holmberg H, Mollby R (1996) Serum antibody response to Staphylococcus aureus enterotoxins and TSST-1 in patients with septicaemia. J Med Microbiol 44: 171–177.
23. Ide F, Matsubara T, Kaneko M, Ichiyama T, Mukouyama T, et al. (2004) Staphylococcal enterotoxin-specific IgE antibodies in atopic dermatitis. Pediat Int 46: 337–341.
24. Verkaik NJ, Dauwalder O, Antri K, Boubekri I, de Vogel CP, et al. (2010) Immunogenicity of toxins during Staphylococcus aureus infection. Clin Infect Dis 50: 61–68.
25. Brosnahan AJ, Mantz MJ, Squier CA, Peterson ML, Schlievert PM (2009) Cytolysins augment superantigen penetration of stratified mucosa. J Immunol 182: 2364–2373.
26. Shambat S, Nadig S, Prabhakara S, Bes M, Etienne J, et al. (2012) Clonal complexes and virulence factors of Staphylococcus aureus from several cities in India. BMC Microbiol 12: 64.
27. Arnaud M, Chastanet A, Debarbouille M (2004) New vector for efficient allelic replacement in naturally nontransformable, low-GC-content, gram-positive bacteria. Appl Environ Microbiol 70: 6887–6891.
28. Boisset S, Geissmann T, Huntzinger E, Fechter P, Bendridi N, et al. (2007) Staphylococcus aureus RNAIII coordinately represses the synthesis of virulence factors and the transcription regulator Rot by an antisense mechanism. Genes Dev 21: 1353–1366.
29. Hansenova Manaskova S, Bikker FJ, Veerman EC, van Belkum A, van Wamel WJ (2013) Rapid detection and semi-quantification of IgG-accessible Staphylococcus aureus surface-associated antigens using a multiplex competitive Luminex assay. J Immunol Methods.
30. Badiou C, Dumitrescu O, Croze M, Gillet Y, Dohin B, et al. (2008) Panton-Valentine leukocidin is expressed at toxic levels in human skin abscesses. Clin Microbiol Infect 14: 1180–1183.
31. Rasigade JP, Trouillet-Assant S, Ferry T, Diep BA, Sapin A, et al. (2013) PSMs of hypervirulent Staphylococcus aureus act as intracellular toxins that kill infected osteoblasts. PLoS One 8: e63176.
32. Spaan AN, Henry T, van Rooijen WJ, Perret M, Badiou C, et al. (2013) The staphylococcal toxin Panton-Valentine Leukocidin targets human C5a receptors. Cell Host Microbe 13: 584–594.
33. Darenberg J, Soderquist B, Normark BH, Norrby-Teglund A (2004) Differences in potency of intravenous polyspecific immunoglobulin G against streptococcal and staphylococcal superantigens: implications for therapy of toxic shock syndrome. Clin Infect Dis 38: 836–842.
34. Farag N, Mahran L, Abou-Aisha K, El-Azizi M (2013) Assessment of the efficacy of polyclonal intravenous immunoglobulin G (IVIG) against the infectivity of clinical isolates of methicillin-resistant Staphylococcus aureus (MRSA) in vitro and in vivo. Eur J Clin Microbiol Infect Dis 32: 1149–1160.
35. Nishi JI, Kanekura S, Takei S, Kitajima I, Nakajima T, et al. (1997) B cell epitope mapping of the bacterial superantigen staphylococcal enterotoxin B: the dominant epitope region recognized by intravenous IgG. J Immunol 158: 247–254.
36. Takei S, Arora YK, Walker SM (1993) Intravenous immunoglobulin contains specific antibodies inhibitory to activation of T cells by staphylococcal toxin superantigens. J Clin Invest 91: 602–607.
37. Gauduchon V, Cozon G, Vandenesch F, Genestier AL, Eyssade N, et al. (2004) Neutralization of Staphylococcus aureus Panton Valentine leukocidin by intravenous immunoglobulin in vitro. J Infect Dis 189: 346–353.
38. Alonzo F 3rd, Kozhaya L, Rawlings SA, Reyes-Robles T, DuMont AL, et al. (2013) CCR5 is a receptor for Staphylococcus aureus leukotoxin ED. Nature 493: 51–55.
39. DuMont AL, Torres VJ (2014) Cell targeting by the Staphylococcus aureus pore-forming toxins: it's not just about lipids. Trends Microbiol 22: 21–27.
40. Novick RP, Geisinger E (2008) Quorum sensing in staphylococci. Annu Rev Genet 42: 541–564.
41. Jarraud S, Mougel C, Thioulouse J, Lina G, Meugnier H, et al. (2002) Relationships between Staphylococcus aureus genetic background, virulence factors, agr groups (alleles), and human disease. Infect Immun 70: 631–641.
42. Collery MM, Smyth DS, Twohig JM, Shore AC, Coleman DC, et al. (2008) Molecular typing of nasal carriage isolates of Staphylococcus aureus from an Irish university student population based on toxin gene PCR, agr locus types and multiple locus, variable number tandem repeat analysis. J Med Microbiol 57: 348–358.
43. Szkaradkiewicz A, Karpinski TM, Zeidler A, Szkaradkiewicz AK, Masiuk H, et al. (2012) Cytokine response in patients with chronic infections caused by Staphylococcus aureus strains and diversification of their Agr system classes. Eur J Clin Microbiol Infect Dis 31: 2809–2815.

15

Enhanced Cytotoxicity of Natural Killer Cells following the Acquisition of Chimeric Antigen Receptors through Trogocytosis

Fu-Nan Cho[1]☯, Tsung-Hsien Chang[2]☯, Chih-Wen Shu[2], Ming-Chin Ko[3], Shuen-Kuei Liao[4], Kang-Hsi Wu[5], Ming-Sun Yu[6], Shyh-Jer Lin[6], Ying-Chung Hong[6], Chien-Hsun Chen[7], Chien-Hui Hung[3], Yu-Hsiang Chang[3,8,9]*

1 Department of Obstetrics and Gynecology, Kaohsiung Veterans General Hospital, Kaohsiung, Taiwan, **2** Department of Medical Education and Research, Kaohsiung Veterans General Hospital, Kaohsiung, Taiwan, **3** Department of Pediatrics, Kaohsiung Veterans General Hospital, Kaohsiung, Taiwan, **4** Graduate Institute of Cancer Biology and Drug Discovery and Center of Excellence for Cancer Research, Taipei Medical University, Taipei, Taiwan, **5** Department of Pediatrics, Children's Hospital and School of Chinese Medicine, China Medical University Hospitals, Taichung, Taiwan, **6** Haematology-Oncology Section, Department of Medicine, Kaohsiung Veterans General Hospital, Kaohsiung, Taiwan, **7** Department of Radiation Oncology, Kaohsiung Veterans General Hospital, Kaohsiung, Taiwan, **8** Faculty of Medicine, National Yang-Ming University, Taipei, Taiwan, **9** Department of Nursing, Tajen University, Yanpu Township, Pingtung County, Taiwan

Abstract

Natural killer (NK) cells have the capacity to target tumors and are ideal candidates for immunotherapy. Viral vectors have been used to genetically modify *in vitro* expanded NK cells to express chimeric antigen receptors (CARs), which confer cytotoxicity against tumors. However, use of viral transduction methods raises the safety concern of viral integration into the NK cell genome. In this study, we used trogocytosis as a non-viral method to modify NK cells for immunotherapy. A K562 cell line expressing high levels of anti-CD19 CARs was generated as a donor cell to transfer the anti-CD19 CARs onto NK cells via trogocytosis. Anti-CD19 CAR expression was observed in expanded NK cells after these cells were co-cultured for one hour with freeze/thaw-treated donor cells expressing anti-CD19 CARs. Immunofluorescence analysis confirmed the localization of the anti-CD19 CARs on the NK cell surface. Acquisition of anti-CD19 CARs via trogocytosis enhanced NK cell-mediated cytotoxicity against the B-cell acute lymphoblastic leukemia (B-ALL) cell lines and primary B-ALL cells derived from patients. To our knowledge, this is the first report that describes the increased cytotoxicity of NK cells following the acquisition of CARs via trogocytosis. This novel strategy could be a potential valuable therapeutic approach for the treatment of B-cell tumors.

Editor: Jacques Zimmer, Centre de Recherche Public de la Santé (CRP-Santé), Luxembourg

Funding: This study was supported by grants from National Health Research Institutes (PS9808, http://english.nhri.org.tw/NHRI_WEB/nhriw001Action.do), and Kaohsiung Veterans General Hospital (VGHKS 102-104 and 103-123, http://www.vghks.gov.tw/English/) to YHC. This work is also supported in part by the Ministry of Health and Welfare of Taiwan (MOHW103-TD-B-111-01), http://www.mohw.gov.tw/EN/Ministry/Index.aspx) to SKL. The funders had no role in study design, data collection and analysis, decision to publish, or preparation of the manuscript.

Competing Interests: The authors have declared that no competing interests exist.

* Email: yhchang@vghks.gov.tw

☯ These authors contributed equally to this work.

Introduction

Natural killer (NK) cells have the ability to recognize and eliminate tumor cells, making them ideal candidates for tumor immunotherapy [1,2]. NK cell activity is regulated by the cumulative effects of multiple activating and inhibitory signals that are transmitted through the receptors on the NK cell surface. We have previously genetically modified *in vitro* expanded NK cells to express DAP10 and the chimeric NKG2D receptor containing the CD3ζ signal domain, which altered the balance between the activating and inhibitory signals of NK cells and enhanced the cytotoxicity against NKG2D ligand-bearing tumors [3]. Further, expression of anti-CD19 chimeric antigen receptors (CARs) containing 41BB and CD3ζ signal domains on NK cells enhanced the activating signals originating from CD19 antigen engagement, leading to cytotoxicity specifically against B-cell leukemia [4].

Trogocytosis is a process in which membrane patches are exchanged between target and immune cells [5–7]. When an NK cell interacts with a target cell, an immune synapse, which is strong enough to allow the transfer of small membrane patches from one cell to its partner cell, is formed [8,9]. Therefore, target cell surface molecules can be found on the surface of NK cells. The chemokine receptor CCR7 has been shown to be transferred from donor cells onto the surface of NK cells via trogocytosis, and this transfer stimulated NK cell migration, leading to enhanced lymph node homing [10,11]. Similarly, T cells captured NKG2D and NKp46 ligands on tumor cells through trogocytosis and promoted NK cell activity [12].

CD19 is an ideal target antigen for immunotherapy because it is expressed on nearly all leukemia cells in most patients with B-cell acute lymphoblastic leukemia (ALL) and chronic lymphoblastic leukemia (CLL) [13,14]. T cells expressing anti-CD19 CARs containing 41BB and CD3ζ signaling domains have shown remarkable antileukemic effects, leading to prolonged survival [15,16]. Autologous T cells transduced with anti-CD19 CARs have been reported to induce complete remission in patients with chronic lymphoblastic leukemia (CLL) and acute lymphoblastic leukemia (ALL) [17–20].

In this study, we sought to express anti-CD19 CARs on expanded NK cells to enhance their cytotoxicity against B-ALL cells. Viral vectors have been used to genetically modify expanded NK cells to express CARs [4,21]. Because of the safety concerns regarding viral integration into the NK cell genome, non-viral mRNA electroporation methods have been developed to modify NK cells and induce NK cell-mediated killing of leukemia cells [22,23]. Although viral gene transduction and mRNA electroporation are feasible methods, their application is limited because of the high costs and complexity. Therefore, we developed a fast, easy, and low-cost method to modify NK cells via trogocytosis. To the best of our knowledge, this is the first report that describes the use of trogocytosis as a tool to modify NK cells with chimeric antigen receptors to enhance their cytotoxicity against B-cell leukemia cells.

Materials and Methods

Cell lines and B-ALL cells from patients

The human B-lineage ALL cell line OP-1 [t(9;22) (q34;q11)/BCR-ABL] was a generous gift from Dario Campana (St. Jude Children's Research Hospital) [24]. The human B-ALL cell lines RS4;11 and SUP-B15 and the non-B leukemia cell line U937 were obtained from American Type Culture Collection (ATCC; Rockville, MD). The K562 cell line was purchased from Bioresource Collection and Research Center (BCRC; Hsinchu, Taiwan). RPMI-1640 (Invitrogen, Carlsbad, CA) supplemented with 10% fetal bovine serum (FBS; Gibco, Carlsbad, CA) and 100 mg/mL penicillin/streptomycin (Invitrogen) was used to maintain K562, OP-1, and RS4;11 cells. The SupB15 cells were maintained in Iscove's Modified Dulbecco's Medium (IMDM; Gibco, Carlsbad, CA).

Following the approval of the protocols and the written informed consent form by the Institutional Review Board of Kaohsiung Veterans General Hospital (Protocol number: VGHKS13-CT6-11), the patients' bone marrow samples were collected strictly adhering to the current ethical principles of the Declaration of Helsinki. Bone marrow samples were collected only after receiving written informed consents from all individuals. The acute B-cell lymphoblastic leukemia cells were separated by centrifugation on a Lymphoprep (Nycomed, Oslo, Norway) density step. The NK cells were thawed and cultured overnight in NK cell culture medium. The cytotoxicity of NK cells against the primary B-cell leukemia cells was assessed immediately after sample collection to avoid massive cell death after the freeze/thaw cycle.

Human NK cell expansion

Peripheral blood samples were obtained from healthy adult donors. Mononuclear cells collected from the samples by centrifugation on a Lymphoprep (Nycomed, Oslo, Norway) density step were washed twice using medium. To expand the CD56+CD3- NK cells, we co-cultured the peripheral blood mononuclear cells (PBMCs) and the genetically modified K562-mb15-41BBL cell line [4,25]. PBMCs (3×10^6) were co-cultured in a 6-well tissue culture plate with 2×10^6 irradiated or freeze/thaw-treated K562-mb15-41BBL cells in 5 mL of RPMI-1640 medium containing and 10% FBS and 10 IU/mL human IL-2 (eBioscience, San Diego, CA). Once every 2 days, fresh culture medium containing 10% FBS and 20 IU/mL human IL-2 was added to double the volume, and NK cells were split from one well into two wells. After 7 days of co-culture, the T cells were removed using CD3 Dynabeads (Invitrogen, Carlsbad, CA), which yielded cell populations containing >95% CD56+CD3- NK cells. The purified NK cells were stored in liquid nitrogen for further experiments.

For the freeze/thaw cycle, ethanol contained in a 50-mL tube was chilled in a −80°C freezer. Following this, a suspension of K562-mb15-41BBL cells in RPMI medium (5×10^6/mL) contained in a 1.5-mL tube was rapidly frozen using pre-chilled ethanol for 2 min. The K562-mb15-41BBL cells were then thawed quickly in a water bath at 37°C for co-culturing with PBMCs as described earlier.

Plasmids, virus production, and gene transduction

The pMSCV-IRES-GFP, pEQ-PAM3(-E), pRDF, and anti-CD19-BB-ζ were generous gifts from Dario Campana (St. Jude Children's Research Hospital) [15]. To generate the RD144-pseudotyped retrovirus, 2.5×10^6 293T cells maintained in 10-cm tissue culture dishes for 16 h were transfected with 3.5 μg of cDNA encoding anti-CD19-BB-ζ constructs, 3.5 μg of pEQ-PAM3(-E), and 3 μg of pRDF using fuGENE 6 (Roche, Indianapolis, IN) reagent [4,15]. The culture supernatant containing the retrovirus was harvested at 48, 72, and 96 h post-transfection. For gene transduction, the supernatant-containing virus particles were filtered and added to RetroNectin (TakaRa, Otsu, Japan)-coated polypropylene tubes. After centrifugation at 1,400×g for 10 min, the tubes were incubated at 37°C for 4 h. After additional centrifugation and removal of the supernatant, K562 cells (5×10^4) were added to the tubes, and the tubes were incubated at 37°C for 24 h. This procedure was repeated for 7 more days. Cells were then maintained in RPMI-1640 supplemented with FBS and antibiotics.

The expression of anti-CD19-BB-ζ on the K562 cell surface was analyzed by flow cytometry on a FACSCalibur instrument using CellQuest software (BD Biosciences, San Jose, CA). Biotin-SP-conjugated AffiniPure goat anti-mouse IgG,F(ab')2 fragment-specific antibody (Jackson ImmunoResearch 115-065-072) and PE-conjugated streptavidin (Jackson ImmunoResearch 016-110-084) were used for labeling the cells. Single K562-anti-CD19-BB-ζ cells with the highest expression of anti-CD19-BB-ζ were sorted with a FACSAria cell sorter (BD Biosciences, San Jose, CA).

Trogocytosis and staining

Immediately before trogocytosis, donor or control cells (5×10^6 cells/mL in culture medium) were rapidly frozen in chilled ethanol (−80°C) for 2 min and then thawed in a water bath maintained at 37°C. Trogocytosis of antiCD19BB-ζ was achieved by co-culturing the NK cells with freeze/thaw-treated K562-antiCD19BBζ cells (donor cells) or K562 cells (control cells) in a 24-well plate as described earlier at an acceptor-to-donor cell ratio of 1:1. The plate was centrifuged for 2 min to increase cell-cell contact and was then incubated at 37°C. After trogocytosis, the cells were gently pipetted to disrupt cell-cell interaction of the immune synapses. The NK cells were separated from the donor or control cells by density gradient centrifugation on Lymphoprep (Nycomed, Oslo, Norway) at 400×g for 20 min and were cultured in NK cell medium. For staining, the cells were fixed with 1% para

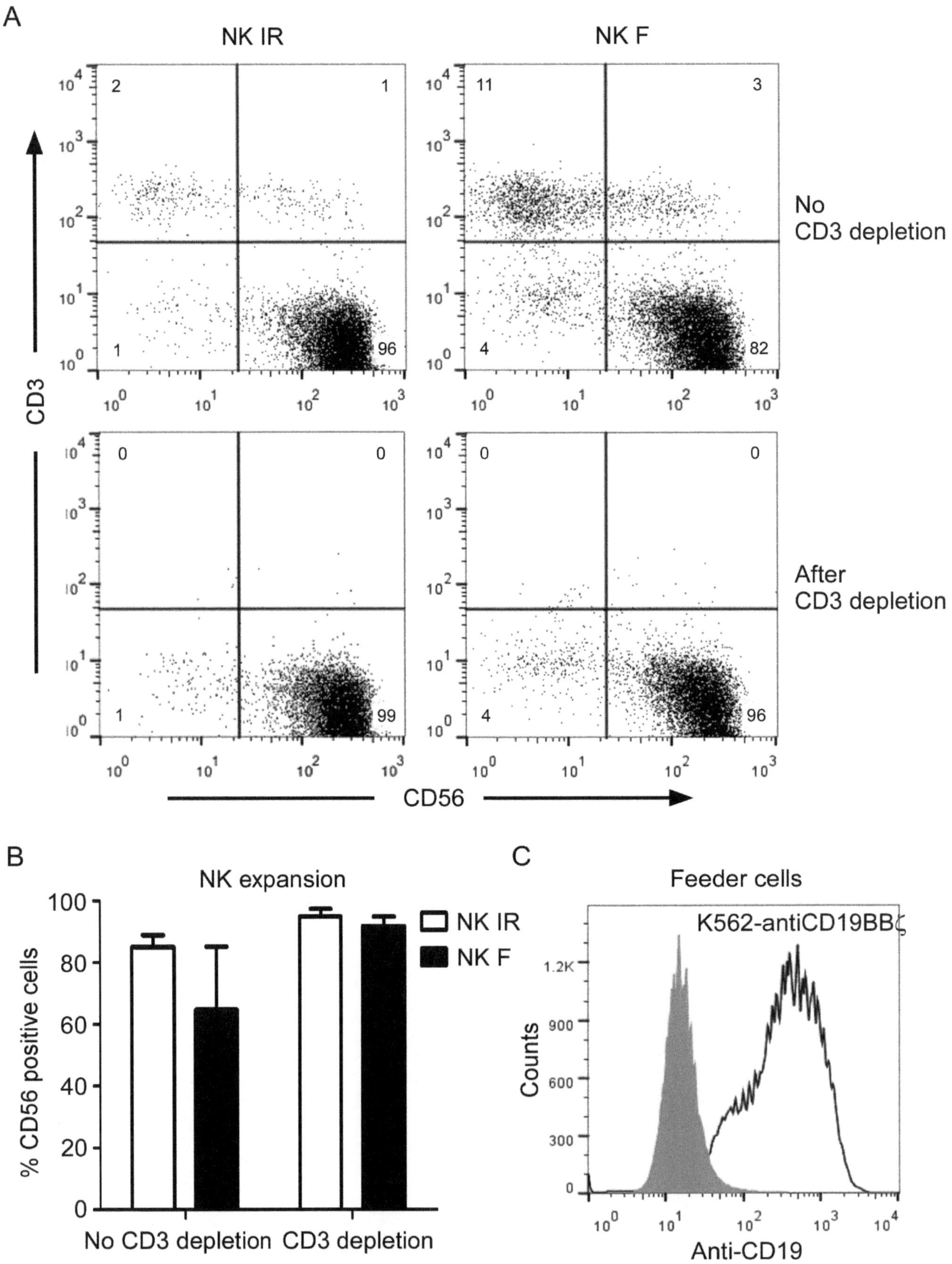
A
NK IR
NK F
No
CD3 depletion
After
CD3 depletion
CD3
CD56
B
NK expansion
% CD56 positive cells
NK IR
NK F
No CD3 depletion
CD3 depletion
C
Feeder cells
K562-antiCD19BBζ
Counts
Anti-CD19

Figure 1. Immunophenotypic features of expanded NK cells (acceptor cells) and K562-antiCD19BBζ cells (feeder cells). A. Expression of CD56 and CD3 on peripheral blood mononuclear cells from a healthy donor was examined after 1 week (top row) of co-culture with irradiated (IR, left column) or freeze/thaw-treated (F, right column) K562-mb15-41BBL cells at a low dose (10 U/mL) of IL-2. The T cells were removed using CD3 Dynabeads, generating cell populations comprising >95% CD56+CD3- NK cells (bottom row). B. Percentage of CD56-positive cells within NK cells expanded by co-culturing with irradiated (IR) or freeze/thaw-treated (F) K562-mb15-41BBL cells prior to and after CD3 depletion on day 7. The data are presented as the mean of values obtained using 3 unrelated NK donors. Error bars represent the SD. C. Histogram illustrating the anti-CD19 expression on K562 cells (control, shaded histogram) and K562-antiCD19BBζ cells (feeder cells, open histogram).

in PBS for 15 min at room temperature. After washing, the cells were stained with biotin-SP-conjugated AffiniPure goat anti-mouse IgG,F(ab')2 fragment-specific antibody (Jackson ImmunoResearch 115-065-072), followed by PE-conjugated streptavidin (Jackson ImmunoResearch 016-110-084) and CD56-FITC. The percentage of the NK cells that acquired anti-CD19-BB-ζ CARs through trogocytosis was determined by flow cytometry.

Immunofluorescence analysis

The NK cells co-cultured with donor cells were fixed using 4% paraformaldehyde in PBS, permeabilized with 0.4% Triton X-100, and blocked with 2% goat serum for 15 min. The cells were then incubated with biotin-SP-conjugated AffiniPure goat anti-mouse IgG,F(ab')2 fragment-specific antibody (Jackson ImmunoResearch 115-065-072) at 4°C overnight. After washing, the cells were incubated with Alexa Fluor 568-conjugated streptavidin (Invitrogen, Carlsbad, CA) and CD56-FITC antibodies (BD Biosciences, San Jose, CA). DAPI was used for staining the nucleus. The localization of fluorescently labeled protein was visualized using a fluorescence microscope (Carl Ziess, Jena, Germany).

Cytotoxicity assay

The target cells were suspended in RPMI-1640 containing 10% FBS, labeled with calcein AM (BD Biosciences, San Jose, CA), and plated onto 96-well flat-bottom plates (Costar, Corning, NY). The NK cells, suspended in RPMI-1640 containing 10% FBS, were then added at various E:T ratios and co-cultured with the target cells for 4 h. Following this, the cells were stained with propidium iodide (Sigma-Aldrich, St. Louis, MO), and the cytotoxicity was assessed by flow cytometry on a FACSCalibur (Becton Dickinson) instrument enumerating the number of viable target cells (calcein AM-positive, propidium-iodide negative, and light scattering properties of viable cells) [3,26].

Degranulation assays

NK cells (1×10^5) were plated in each well of a 96-well flat-bottom plate and incubated with RS4;11 cells at an E:T ratio of 1:4. A phycoerythrin-conjugated anti-human CD107a antibody (BD Biosciences, San Jose, CA) was added at the beginning of the cultures. After 1 h of incubation, GolgiStop (0.15 µL; BD Biosciences, San Jose, CA) was added. The cells were then stained with fluorescein isothiocyanate-conjugated anti-human CD56 antibody (eBiosciences, San Diego, CA) and were analyzed by flow cytometry.

Results

Freeze/thaw treated K562-mbIL15-41BBL cells can be used in NK cell expansion

The NK cells were expanded from PBMCs by co-culturing with irradiated (Figure 1A, left panels) or freeze/thaw-treated (Figure 1A, right panels) K562-mb15-41BBL cells. The freeze/thaw cycle compromised the membrane integrity of K562-mb15-41BBL cells, which allowed trypan blue staining, but intact cell morphology was maintained. After 7 days of expansion, the PBMCs co-cultured with irradiated K562-mbIL15-41BBL cells produced 95.8% CD56+CD3- NK cells (Figure 1A, left upper panel), whereas K562-mbIL15-41BBL cells subjected to one freeze/thaw cycle yielded 81.8% CD56+CD3- cells (Figure 1A, right upper panel). After CD3 depletion, the percentages of CD56+CD3- NK cells were 98.7% from PBMCs co-cultured with irradiated K562-mbIL15-41BBL cells and 95.5% using a freeze/thawed procedure (Figure 1A, lower panels). The relatively poor expansion of NK cells in co-cultures of PMBCs and freeze/thaw-treated K562-mbIL15-41BBL cells was likely due to the reduced stimulation from mbIL15 and 41BB ligands as a result of freeze/thaw-induced cell damage and lysis.

By day 7 in cultures, approximately 85% (n = 3; Figure 1B, NK IR) of the NK cells expanded from PBMCs co-cultured with irradiated K562-mb15-41BBL cells were CD56-positive, and this number increased to 95% following CD3 depletion. These results were comparable to those of NK cells expanded by co-culturing with freeze/thaw-treated K562-mb15-41BBL cells (Figure 1B, NK F; 92% CD56-positive cells following CD3 depletion). These results indicated that both freeze/thaw-treated and irradiated K562-mbIL15-41BBL cells can be used for NK cell expansion.

To generate donor cells (K562-antiCD19BBζ cells) for trogocytosis, K562 cells were transduced with anti-CD19-BB-ζ chimeric antigen receptors (CARs). After single cell sorting, we chose the clone stably expressing high levels of anti-CD19 CARs. The mean fluorescence intensity (MFI) of the K562-antiCD19BBζ cells was 301, which was substantially higher than that (12) of control K562 cells (Figure 1C).

Expanded NK cells acquired anti-CD19 CARs from K562-based donor cells via trogocytosis

To examine whether the NK cells were able to acquire anti-CD19 CARs from donor cells via trogocytosis, we cultured NK cells with live or freeze/thaw-treated donor K562-antiCD19BBζ cells. FACS analysis of NK cells co-cultured with live donor cells revealed that 47.0±16.4% (n = 3, ± s.d.) of the NK cells acquired anti-CD19 CARs (data not shown), and 24.1% of NK cells acquired anti-CD19 CARs from freeze/thawed donor K562-antiCD19BBζ cells (Figure 2A). Although a higher efficiency of trogocytosis was observed in co-cultures with live donor cells, separation of NK cells from live donor cells is challenging. In contrast, donor cells subjected to a freeze/thaw cycle could facilitate the separation of NK cells from nonviable donor cells using Ficoll-Paque centrifugation. Therefore, the NK cells co-cultured with freeze/thaw-treated K562-antiCD19BBζ cells were used for subsequent experiments.

Using these two types of expanded NK cells (Figure 1A, NK IR and NK F), we evaluated the uptake of anti-CD19 CARs by NK cells from donor K562-antiCD19BBζ cells. We found that approximately 19% (n = 3) of the NK F cells expressed anti-CD19 CARs, whereas only 11% (n = 3) of the NK IR cells expressed anti-CD19 CARs (Figure 2B). Therefore, we used NK cells expanded with freeze/thaw-treated K562-mb15-41BBL cells (NK F cells) in subsequent trogocytosis experiments.

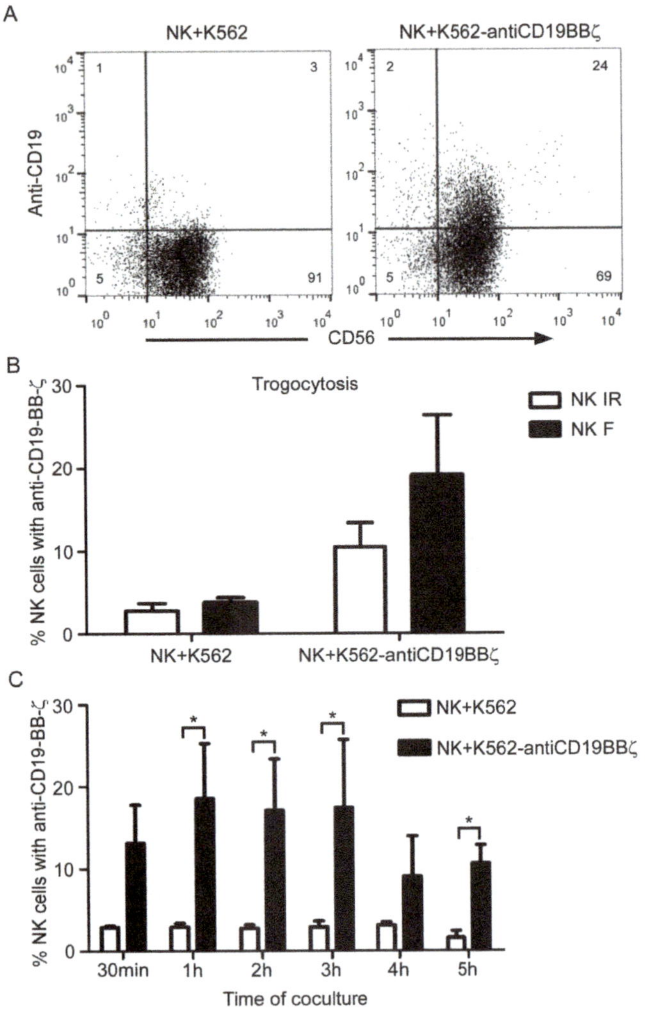

Figure 2. Acquisition of anti-CD19 chimeric antigen receptors (CARs) by NK cells from donor cells via trogocytosis. A. Flow cytometry dot plots illustrating the uptake of anti-CD19 CARs by NK cells via trogocytosis. NK cells co-cultured with K562 cells (control) or K562-antiCD19BBζ cells were stained with an anti-CD56-FITC antibody and a biotin-SP-conjugated AffiniPure goat anti-mouse IgG,F(ab′)2 fragment specific antibody, followed by PE-conjugated streptavidin. B. Uptake of anti-CD19 CARs by NK cells expanded by co-culturing with irradiated (IR) or freeze/thaw-treated (F) K562-mb15-41BBL cells. The data are presented as the mean ± SD of values obtained using three unrelated NK donors. C. Kinetics of anti-CD19 CAR uptake by NK cells from K562-antiCD19BBζ cells (black bars) and control K562 cells (white bars). The uptake of anti-CD19 CARs by NK cells was analyzed after co-culturing with feeder cells for the indicated time and was compared with that of NK cells co-cultured with control K562 cells. The data are presented as the mean ± SD of values obtained using 3 unrelated NK cell donors. *: significant increase compared with control cells ($p<0.05$; two-tailed paired Student's *t*-tests).

We assessed the optimal duration for the co-culture of NK cells with donor cells. After 1 h of co-culture, approximately 18.6% of the NK cells were anti-CD19-positive (Figure 2C). The uptake of anti-CD19 CARs peaked at 1 h and decreased after 4–5 h of co-culture with K562-antiCD19BBζ cells. These results suggested that the efficiency of NK cell trogocytosis peaked at 1 h of co-culture. Therefore, we used 1 h as the standard duration of co-culture in subsequent experiments.

Immunofluorescence analysis of trogocytosis

To verify that trogocytosis was the mechanism of uptake of anti-CD19 CARs by NK cells, we examined the interaction between NK cells and donor cells during trogocytosis using immunofluorescence microscopy. The NK cells stained positively for CD56 (Figure 3A, green), and the donor cells expressed anti-CD19 CARs (Figure 3B, red). Because the NK cells interacted with donor cells and formed immune synapses over a period of 15 min, these cells were imaged 15 min after co-culture initiation. During trogocytosis, small patches of acquired anti-CD19 CARs were observed on the surface of NK cells (Figure 3C). The NK cells continued to express anti-CD19 CARs (Figure 3D, yellow) on their surfaces after detaching from the donor cells following pipetting and Ficoll-Paque separation.

Acquiring anti-CD19 CARs via trogocytosis enhanced NK cell degranulation and cytotoxicity against B-ALL cell lines

To assess the degranulation of NK cells after stimulating with a B-ALL cell line, NK cells co-cultured with freeze/thaw-treated donor K562-antiCD19BBζ or K562 (control) cells were incubated with RS4;11 cells to induce degranulation. CD107a staining was performed to detect degranulation. The percentage of CD107a-positive NK cells co-cultured with donor K562-antiCD19BBζ cells (9.9±1.2%, n = 3) was significantly higher ($p<0.05$) than that of CD107a-positive NK cells co-cultured with control cells (1.9±0.6%, n = 3) (Figure 4A).

To examine whether the gain of anti-CD19 CARs via trogocytosis improved the cytotoxicity of NK cells against B-ALL cells, three B-ALL cell lines (RS4;11, OP-1, and SUP-B15)

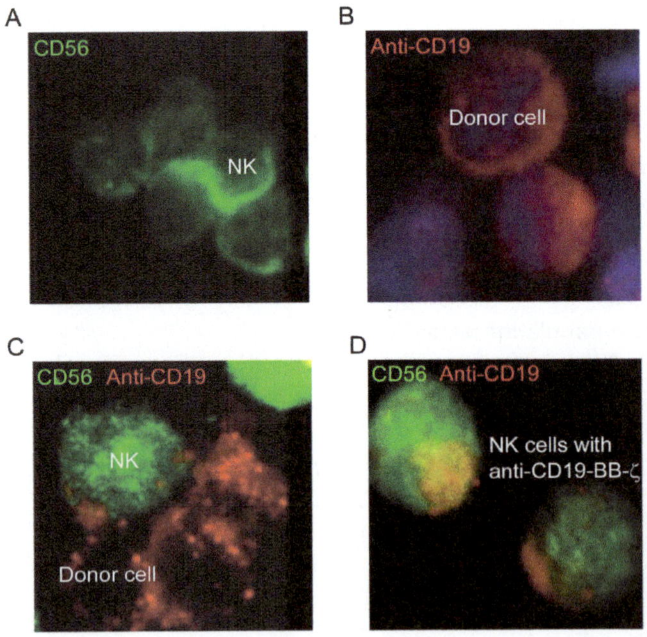

Figure 3. Immunofluorescence analysis for trogocytosis. A. NK cells stained with anti-CD56-FITC antibody. B. K562-antiCD19BBζ cells were stained with a biotin-SP-conjugated AffiniPure goat anti-mouse IgG,F(ab′)2 fragment specific antibody, followed by Alexa Fluor 568-conjugated streptavidin. The nucleus was stained with DAPI (blue). C. NK cells co-cultured with K562-antiCD19BBζ cells were stained for CD56 and anti-CD19 CARs, as previously described. D. Acquisition of anti-CD19-BB-ζ by NK cells via trogocytosis was observed.

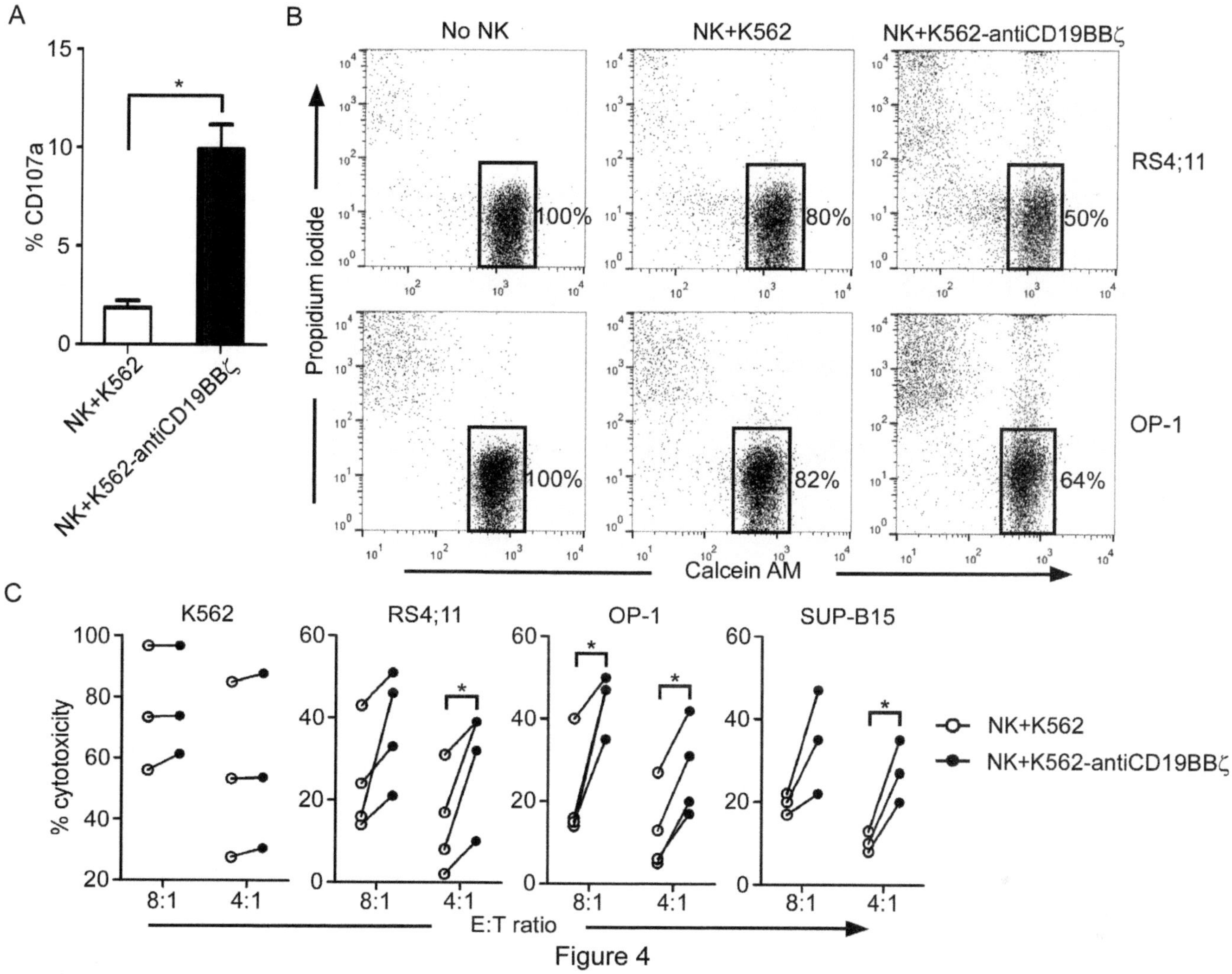

Figure 4. Degranulation and cytotoxicity assays. A. Incubation of NK cells with RS4;11 cells induced degranulation, which was significantly higher in NK cells co-cultured with K562-antiCD19BBζ cells than in NK cells cultured with control K562 cells. Percentages of CD56-positive cells from 3 donors expressing CD107a after a 4-h RS4;11 stimulation are shown. The data are presented as the mean ± SD of values obtained using 3 unrelated NK cell donors. B. Flow cytometric dot plots illustrating the assay used to measure cell killing. Results for RS4;11 and OP-1 cell lines are shown. Tumor cells were either cultured alone (left panel), with NK cells previously co-cultured with K562 cells (middle panel), or with NK cells previously co-cultured with K562-antiCD19BBζ cells (right panel). Residual viable target cells, which were defined as calcein AM-positive and propidium iodide (PI)-negative, are shown in the bottom right corner of each panel, and the percentages of viable cells are shown. C. Cytotoxicity of the non-B leukemia cell line (K562) and B-ALL cell lines (RS4;11, OP1, and SUP-B15) after 4-h co-culture with NK cells previously co-cultured with K562 cells (white circles) and K562-antiCD19BBζ cells (black circles) at the indicated E:T ratios. The equation [(tumor co-culture with NK cells)/(tumor alone)] × 100%, represents the quantitative percentage of viable tumor cells. The cytotoxicity was calculated according the following equation: 100% − quantitative percentage of viable tumor cells. Each symbol corresponds to the mean of three values. *: significant increase compared with control cells ($p<0.05$; two-tailed paired Student's *t*-tests).

were targeted. We conducted 4-h cytotoxicity assays with NK cells expanded from three donors at effector:target (E:T) ratios of 4:1 and 8:1. As shown in Figure 4B and 4C, the gain of anti-CD19 CARs via trogocytosis significantly increased the cytotoxicity of NK cells against RS4;11, OP-1, and SUP-B15 cells at a 4:1 E:T ratio. Similarly, gains of cytotoxicity were also observed against the three B-ALL cell lines at an 8:1 E:T ratio. In contrast, there was no increase in cytotoxicity against the non-B cell lines, K562 cells (Figure 4C) and U937 cells (data not shown).

Gain of anti-CD19 CARs via trogocytosis increased NK cytotoxicity against primary B-ALL cells from patients

To determine whether the acquisition of anti-CD19 CARs via trogocytosis improved the cytotoxicity of NK cells against the patient-derived primary B-ALL cells, three samples of bone marrow from B-ALL patients were tested. Compared to the mock NK cells, the NK cells that acquired anti-CD19 CARs consistently showed enhanced cytotoxicity against primary B-ALL cells (Figure 5). The p-value was 0.069 ($n = 3$) at an 8:1 E:T ratio.

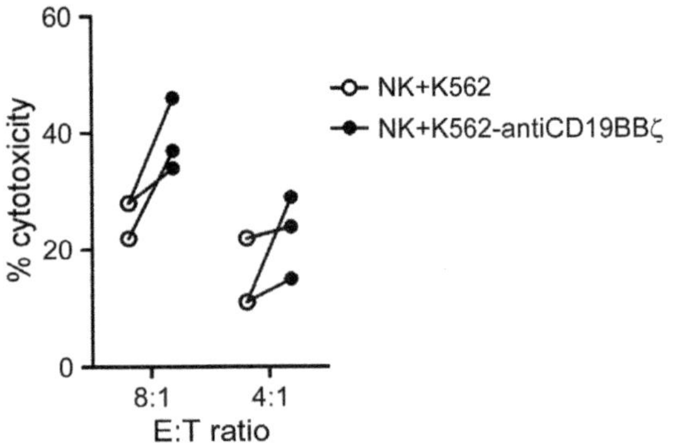

Figure 5. NK cells that acquired anti-CD19 CARs were more cytotoxic against patient-derived primary B-ALL cells. Cytotoxicity against B-ALL cells after 4-h co-culture with NK cells previously co-cultured with K562 cells (white circles) and K562-antiCD19BBζ cells (black circles) at the indicated E:T ratios. Each symbol corresponds to the mean of three values.

Discussion

Because of the safety concerns regarding the use of viral transduction methods, we endeavored to improve the cytotoxicity of adoptively transferred NK cells against B-cell leukemia cells using a simple co-culture method. For NK cell therapy, the clinical protocol required up to 5×10^7 NK cells/kg and CD3 depletion before transfusing into patients, indicating that approximately 10×10^9 NK cells are needed for the treatment of hematological malignancies [27]. If the MSCV retroviral methods were to be used to modify NK cells, a large quantity of viral supernatant would be required to transduce 10×10^9 NK cells [4]. Additionally, there is a safety concern associated with viral integration into the NK cell genome. Further, the application of viral transduction is also very limited because of the complexity of the procedures and high costs. In this study, we evaluated the usefulness of trogocytosis as a relatively simple non-viral method that can be readily scaled up to modify large numbers of NK cells with a single K562-based donor cell line.

The K562 cell line is a typical NK cell target because it lacks major histocompatibility complex class I expression, which triggers inhibitory signals to abolish NK cell activation. Additionally, K562 cells have been genetically modified to express 41BBL and the membrane-bound IL15 or IL21, which allowed the establishment of co-culture methods for the *in vitro* expansion of NK cells [25,28]. Here, we showed that K562 cells were able to donate anti-CD19 CARs to expanded NK cells in co-cultures via trogocytosis. The rapid expression of anti-CD19 CAR on the NK cell surface when co-cultured with donor K562-antiCD19BBζ cells and the lack of expression in NK cells co-cultured with control K562 cells was suggestive of an acquisition from donor cells. This was confirmed by the positive anti-CD19 staining of the NK cell surface (Figure 2A, Figure 4). Previous studies have shown that trogocytosis occurs during cell-cell interactions between target cells and stimulated NK cells [9–11,29,30].

Although the extent of trogocytosis was greater when NK cells were co-cultured with live donor cells (47%, data not shown) than when cultured with nonviable donor cells (19%), the ease of separating NK cells from nonviable donor cells using Ficoll-Paque centrifugation prompted us to use the latter method. The uptake of CCR7 by NK cells from nonviable donor cells subjected to one freeze/thaw cycle was also lower (50%) than that from live donor cells (80%) [10]. Further, the anti-CD19 CARs acquired by NK cells were rapidly lost, and only less than 20% of the acquired anti-CD19 CARs remained on the NK cell surface after 2 h (data not shown). CCR7 uptake has been reported to decline to baseline levels by 72 h [10]. The rapid loss of the acquired anti-CD19 CARs observed in our study might be due to the degradation of the low amount of anti-CD19 CARs acquired. Fas signaling was reported to promote trogocytosis in T cells [31]. Further studies are needed to improve anti-CD19 CAR uptake and enhance its stability on the NK cell surface.

The CAR, anti-CD19-BB-ζ, comprising an anti-CD19 single-chain variable fragment (scFv), a 41BB signaling domain, and a CD3ζ signaling domain, was transduced into the NK cell genome, inducing powerful anti-leukemic effects [4]. Additionally, expanded NK cells electroporated with anti-CD19-BB-ζ mRNA also exerted significantly higher cytotoxicity against B-cell malignancies than mock NK cells [23]. In this study, we showed enhanced cytotoxicity of NK cells following the acquisition of anti-CD19-BB-ζ protein molecules via trogocytosis.

A reduction in NK cell cytotoxicity was observed after the intercellular transfer of NK Group 2 member D (NKG2D) and MHC class I chain-related molecule (MIC) B proteins at the NK cell immune synapse [32]. The amount of NKG2D, a key activating receptor on NK cell surface, is reduced after trogocytosis because the transfer of NK cell-derived NKG2D to target cells and internalization of NKG2D contribute to receptor down-modulation after interaction, leading to impaired NK cell cytotoxicity [7,32]. In addition, activated NK cells acquired HLA-G, an immunosuppressive molecule, from tumor cells via trogocytosis, leading to impaired cytotoxicity [9]. On the contrary, our study showed enhanced NK cytotoxicity because of expanded NK cells acquiring anti-CD19-BB-ζ from feeder cells via trogocytosis.

To our knowledge, this is the first report that describes the increased cytotoxicity of NK cells following the acquisition of anti-CD19 CARs from donor cells via trogocytosis. Our findings could potentially be extended to develop safer and effective therapeutic strategies for treating B-cell tumors. Our model, which used anti-CD19 CAR for targeting a B-cell tumor, could also be relevant to other tumor types that are targeted by the tumor-directed chimeric antigen receptors [33]. Therefore, further studies are warranted to examine the utility of our method for treating diverse tumor types *in vivo*.

Author Contributions

Conceived and designed the experiments: FNC THC MCK YHC. Performed the experiments: MCK CHC YHC. Analyzed the data: FNC THC CWS MCK SKL CHH YHC. Contributed reagents/materials/analysis tools: KHW MSY SJL YCH. Wrote the paper: FNC THC CWS YHC.

References

1. Ljunggren HG, Malmberg KJ (2007) Prospects for the use of NK cells in immunotherapy of human cancer. Nat Rev Immunol 7: 329–339.
2. Caligiuri MA (2008) Human natural killer cells. Blood 112: 461–469.
3. Chang YH, Connolly J, Shimasaki N, Mimura K, Kono K, et al. (2013) A chimeric receptor with NKG2D specificity enhances natural killer cell activation and killing of tumor cells. Cancer Res 73: 1777–1786.

4. Imai C, Iwamoto S, Campana D (2005) Genetic modification of primary natural killer cells overcomes inhibitory signals and induces specific killing of leukemic cells. Blood 106: 376–383.
5. Rechavi O, Goldstein I, Kloog Y (2009) Intercellular exchange of proteins: the immune cell habit of sharing. FEBS Lett 583: 1792–1799.
6. Caumartin J, Lemaoult J, Carosella ED (2006) Intercellular exchanges of membrane patches (trogocytosis) highlight the next level of immune plasticity. Transpl Immunol 17: 20–22.
7. Roda-Navarro P, Reyburn HT (2007) Intercellular protein transfer at the NK cell immune synapse: mechanisms and physiological significance. FASEB J 21: 1636–1646.
8. Williams GS, Collinson LM, Brzostek J, Eissmann P, Almeida CR, et al. (2007) Membranous structures transfer cell surface proteins across NK cell immune synapses. Traffic 8: 1190–1204.
9. Caumartin J, Favier B, Daouya M, Guillard C, Moreau P, et al. (2007) Trogocytosis-based generation of suppressive NK cells. EMBO J 26: 1423–1433.
10. Somanchi SS, Somanchi A, Cooper LJ, Lee DA (2012) Engineering lymph node homing of ex vivo-expanded human natural killer cells via trogocytosis of the chemokine receptor CCR7. Blood 119: 5164–5172.
11. Marcenaro E, Cantoni C, Pesce S, Prato C, Pende D, et al. (2009) Uptake of CCR7 and acquisition of migratory properties by human KIR+ NK cells interacting with monocyte-derived DC or EBV cell lines: regulation by KIR/HLA-class I interaction. Blood 114: 4108–4116.
12. Domaica CI, Fuertes MB, Rossi LE, Girart MV, Avila DE, et al. (2009) Tumour-experienced T cells promote NK cell activity through trogocytosis of NKG2D and NKp46 ligands. EMBO Rep 10: 908–915.
13. Bene MC, Kaeda JS (2009) How and why minimal residual disease studies are necessary in leukemia: a review from WP10 and WP12 of the European LeukaemiaNet. Haematologica 94: 1135–1150.
14. Rawstron AC, Villamor N, Ritgen M, Bottcher S, Ghia P, et al. (2007) International standardized approach for flow cytometric residual disease monitoring in chronic lymphocytic leukaemia. Leukemia 21: 956–964.
15. Imai C, Mihara K, Andreansky M, Nicholson IC, Pui CH, et al. (2004) Chimeric receptors with 4-1BB signaling capacity provoke potent cytotoxicity against acute lymphoblastic leukemia. Leukemia 18: 676–684.
16. Milone MC, Fish JD, Carpenito C, Carroll RG, Binder GK, et al. (2009) Chimeric receptors containing CD137 signal transduction domains mediate enhanced survival of T cells and increased antileukemic efficacy in vivo. Mol Ther 17: 1453–1464.
17. Porter DL, Levine BL, Kalos M, Bagg A, June CH (2011) Chimeric antigen receptor-modified T cells in chronic lymphoid leukemia. N Engl J Med 365: 725–733.
18. Brentjens RJ, Riviere I, Park JH, Davila ML, Wang X, et al. (2011) Safety and persistence of adoptively transferred autologous CD19-targeted T cells in patients with relapsed or chemotherapy refractory B-cell leukemias. Blood 118: 4817–4828.
19. Grupp SA, Kalos M, Barrett D, Aplenc R, Porter DL, et al. (2013) Chimeric antigen receptor-modified T cells for acute lymphoid leukemia. N Engl J Med 368: 1509–1518.
20. Brentjens RJ, Davila ML, Riviere I, Park J, Wang X, et al. (2013) CD19-targeted T cells rapidly induce molecular remissions in adults with chemotherapy-refractory acute lymphoblastic leukemia. Sci Transl Med 5: 177ra138.
21. Altvater B, Landmeier S, Pscherer S, Temme J, Schweer K, et al. (2009) 2B4 (CD244) signaling by recombinant antigen-specific chimeric receptors costimulates natural killer cell activation to leukemia and neuroblastoma cells. Clin Cancer Res 15: 4857–4866.
22. Li L, Liu LN, Feller S, Allen C, Shivakumar R, et al. (2009) Expression of chimeric antigen receptors in natural killer cells with a regulatory-compliant non-viral method. Cancer Gene Ther.
23. Shimasaki N, Fujisaki H, Cho D, Masselli M, Lockey T, et al. (2012) A clinically adaptable method to enhance the cytotoxicity of natural killer cells against B-cell malignancies. Cytotherapy 14: 830–840.
24. Manabe A, Coustan-Smith E, Kumagai M, Behm FG, Raimondi SC, et al. (1994) Interleukin-4 induces programmed cell death (apoptosis) in cases of high-risk acute lymphoblastic leukemia. Blood 83: 1731–1737.
25. Fujisaki H, Kakuda H, Shimasaki N, Imai C, Ma J, et al. (2009) Expansion of Highly Cytotoxic Human Natural Killer Cells for Cancer Cell Therapy. Cancer Res.
26. Wu KH, Sheu JN, Wu HP, Tsai C, Sieber M, et al. (2013) Cotransplantation of umbilical cord-derived mesenchymal stem cells promote hematopoietic engraftment in cord blood transplantation: a pilot study. Transplantation 95: 773–777.
27. Lapteva N, Durett AG, Sun J, Rollins LA, Huye LL, et al. (2012) Large-scale ex vivo expansion and characterization of natural killer cells for clinical applications. Cytotherapy 14: 1131–1143.
28. Denman CJ, Senyukov VV, Somanchi SS, Phatarpekar PV, Kopp LM, et al. (2012) Membrane-bound IL-21 promotes sustained ex vivo proliferation of human natural killer cells. PLoS One 7: e30264.
29. Carlin LM, Eleme K, McCann FE, Davis DM (2001) Intercellular transfer and supramolecular organization of human leukocyte antigen C at inhibitory natural killer cell immune synapses. J Exp Med 194: 1507–1517.
30. Poupot M, Fournie JJ, Poupot R (2008) Trogocytosis and killing of IL-4-polarized monocytes by autologous NK cells. J Leukoc Biol 84: 1298–1305.
31. Luchetti F, Canonico B, Arcangeletti M, Guescini M, Cesarini E, et al. (2012) Fas signalling promotes intercellular communication in T cells. PLoS One 7: e35766.
32. Roda-Navarro P, Vales-Gomez M, Chisholm SE, Reyburn HT (2006) Transfer of NKG2D and MICB at the cytotoxic NK cell immune synapse correlates with a reduction in NK cell cytotoxic function. Proc Natl Acad Sci U S A 103: 11258–11263.
33. Sadelain M, Brentjens R, Riviere I (2009) The promise and potential pitfalls of chimeric antigen receptors. Curr Opin Immunol 21: 215–223.

16

Potential Therapeutic Benefit of C1-Esterase Inhibitor in Neuromyelitis Optica Evaluated *In Vitro* and in an Experimental Rat Model

Lukmanee Tradtrantip, Nithi Asavapanumas, Puay-Wah Phuan, A. S. Verkman*

Departments of Medicine and Physiology, University of California San Francisco, San Francisco, California, United States of America

Abstract

Neuromyelitis optica (NMO) is an autoimmune demyelinating disease of the central nervous system in which binding of anti-aquaporin-4 (AQP4) autoantibodies (NMO-IgG) to astrocytes causes complement-dependent cytotoxicity (CDC) and inflammation resulting in oligodendrocyte and neuronal injury. There is compelling evidence for a central role of complement in NMO pathogenesis. Here, we evaluated the potential of C1-esterase inhibitor (C1-inh) for complement-targeted therapy of NMO. C1-inh is an anti-inflammatory plasma protein with serine protease inhibition activity that has a broad range of biological activities on the contact (kallikrein), coagulation, fibrinolytic and complement systems. C1-inh is approved for therapy of hereditary angioedema (HAE) and has been studied in a small safety trial in acute NMO relapses (NCT 01759602). In vitro assays of NMO-IgG-dependent CDC showed C1-inh inhibition of human and rat complement, but with predicted minimal complement inhibition activity at a dose of 2000 units in humans. Inhibition of complement by C1-inh was potentiated by ~10-fold by polysulfated macromolecules including heparin and dextran sulfate. In rats, intravenous C1-inh at a dose 30-fold greater than that approved to treat HAE inhibited serum complement activity by <5%, even when supplemented with heparin. Also, high-dose C1-inh did not reduce pathology in a rat model of NMO produced by intracerebral injection of NMO-IgG. Therefore, although C1r and C1s are targets of C1-inh, our in vitro data with human serum and in vivo data in rats suggest that the complement inhibition activity of C1-inh in serum is too low to confer clinical benefit in NMO.

Editor: Jun-ichi Kira, Kyushu University, Japan

Funding: This work was supported by grants EY13574, DK35124, EB00415 and DK72517 from the National Institutes of Health, a grant from the Guthy-Jackson Charitable Foundation, and a sponsored research agreement from CSL Behring (Kankakee, IL). The funders had no role in study design, data collection and analysis, decision to publish, or preparation of the manuscript.

Competing Interests: This work described in this manuscript was funded, in part, by a sponsored research agreement from a commercial source (CSL Behring).

* Email: Alan.Verkman@ucsf.edu

Introduction

Neuromyelitis optica (NMO) is autoimmune disease of the central nervous system in which inflammatory demyelinating lesions cause optic neuritis and transverse myelitis [1], [2]. Most NMO patients are seropositive for immunoglobulin G autoantibodies (NMO-IgG) against aquaporin-4 [3], [4], a water channel expressed on the plasma membrane of astrocytes [5]. NMO pathogenesis in patients seropositive for NMO-IgG involves NMO-IgG binding to astrocyte AQP4, causing cytotoxicity with secondary inflammation, oligodendrocyte injury and demyelination [6], [7]. Currently used NMO therapies include immunosuppressives, B-cell depletion and plasma exchange [8]–[10].

There is strong evidence for a central role of complement in NMO pathogenesis and hence for the utility of complement-targeted therapeutics. Inflammatory lesions in human NMO show prominent vasculocentric deposition of activated complement [11]–[13]. In vitro, addition of NMO-IgG and complement to AQP4-expressing cells, including astrocytes, produces complement-dependent cytotoxicity (CDC) [14]–[17]. Characteristic NMO pathology with demyelination is produced in ex vivo spinal cord slice cultures exposed to NMO-IgG and human complement [18], and in mice in vivo following intracerebral injection of NMO-IgG and human complement [19]. In rats, which have an active complement system similar to that in humans, administration of NMO-IgG alone causes complement-dependent NMO pathology, as pathology is not seen when complement is inactivated by cobra venom toxin or when NMO-IgG is mutated to block its complement effector function [20]. An open-label clinical trial of eculizumab, a monoclonal antibody inhibitor targeting complement protein C5, showed reduced recurrence rate in NMO patients with severe disease [21].

Though further clinical evaluation of eculizumab in NMO is awaited, there is interest in the development of alternative complement-targeted therapeutics in NMO as eculizumab is very costly and associated with significant infectious complications including meningococcal meningitis [22]. Our lab recently introduced complement-targeted monoclonal therapeutics that target C1q and C1s in the classical complement pathway [23]. Selective inhibition of early steps in the classical complement pathway has potential benefit over inhibition of later steps because the lectin activation pathway, which is involved in bacterial killing, remains intact. There has been interest in the therapeutic potential of C1-esterase inhibitor (C1-inh), an anti-inflammatory plasma protein with serine protease inhibition activity and a wide range of biological activities on the contact (kallikrein), coagulation and fibrinolytic systems, and on the complement pathway (Fig. 1) [24], [25]. Purified C1-inh from human serum is approved for use in hereditary angioedema (HAE) based on its kallikrein inhibition activity, and recently, based on its known complement inhibition activity, a safety trial (NCT 01759602) has been completed for acute NMO relapses [26]. In that trial safety was demonstrated in ten patients administered 2000 units of C1-inh daily for three days. Here, utilizing in vitro and rat model systems, we evaluated the potential utility of C1-inh therapy for NMO.

Materials and Methods

Cell culture and antibodies

Chinese hamster ovary (CHO) cells stably expressing M23-AQP4 were used, as described [27]. CHO cells were cultured in F-12 Ham's Nutrient mix medium supplemented with 10% fetal bovine serum, 100 units/ml penicillin, 100 µg/ml streptomycin, and 200 µg/ml geneticin (as selection marker). Cells were grown at 37°C in 5% CO_2/95% air. Recombinant monoclonal NMO antibody rAb-53 (referred to as NMO-IgG) was generated from clonally expanded plasma blasts from cerebrospinal fluid of NMO patients and purified, as described [17].

Rats

Lewis rats were purchased from Charles River Lab (Wilmington, MA). Experiments were done using weight-matched rats (250–350 g). Protocols were approved by Institutional Animal Care and Use Committee (IACUC) at the University of California San Francisco (Project Number: AN89748). All surgery was performed under ketamine/xylazine anesthesia, and analgesics were used post-operatively to minimize pain.

Complement-dependent cytotoxicity (CDC)

M23-AQP4-expressing CHO cells were plated onto 96-well microplates (Costar, Corning Inc., Corning, NY) at 25,000 cells/well and grown for 24 h. Cells were washed with phosphate-buffered saline (PBS) and incubated at room temperature for 60 min with NMO-IgG or purified IgG from NMO patient serum in live cell buffer (PBS containing 6 mM glucose and 1 mM sodium pyruvate) containing pooled normal human complement serum (Innovative Research, Novi, MI) with or without C1-inh (Berinert, CSL Behring, King of Prussia, PA) in a final volume of 50 µl. Some experiments were done with rat serum, prepared from freshly obtained rat blood and then frozen, in place of the human complement serum. In some experiments, a solution containing pooled normal human complement serum was incubated with C1-inh for specified times before addition to the CHO cells. To measure cytotoxicity, cells were washed twice with PBS, and incubated with 50 µl of a 20% Alamar Blue solution (Invitrogen) for 45 min at 27°C. Cytotoxicity was measured from resorufin fluorescence using a TECAN Infinite M1000 plate reader (TECAN Groups Ltd, Mannedorf, Switzerland) (excitation/emission 560/590 nm).

C1-inh administration in rats and serum collection

Adult rats (250–350 g) were injected by tail vein with C1-inh (600 units/kg) in saline in a total volume of 500 µl. Blood was collected through the tail vein at 0.25, 0.5, 1, 2, 4, 6, and 8 h, left for 30 min at room temperature to allow clotting, and centrifuged for 10 min at 3000 g, 4°C. Serum was collected and bioassayed for CDC by addition of 2% rat serum and 1 µg/ml NMO-IgG to M23-AQP4-expressing CHO cells for 60 min. Cytotoxicity was measured by the Alamar Blue assay. In some experiments heparin (1.5 mg/kg) was injected by tail vein with or without C1-inh (600 units/kg) in saline in a total volume of 500 µl.

Rat model of NMO

A rat intracerebral injection was used, as described [20]. Briefly, adult rats were anesthetized with intraperitoneal ketamine (75–100 mg/kg) and xylazine (5–10 mg/kg) and mounted in a stereotaxic frame. Following a midline scalp incision, a burr hole of diameter 1 mm was made in the skull 3.5 mm to the right of the bregma. A 30-gauge needle attached to 50-µl gas-tight glass syringe was inserted 5 mm deep to infuse 10 µg NMO-IgG in a total volume of 10 µl (at 2 µl/min). Rats were injected intravenously with 300 units/kg C1-inh in saline (or saline alone) in a total volume of 300 µl just before and 1 day after intracerebral injection of NMO-IgG. After 2 days, rats were anesthetized and

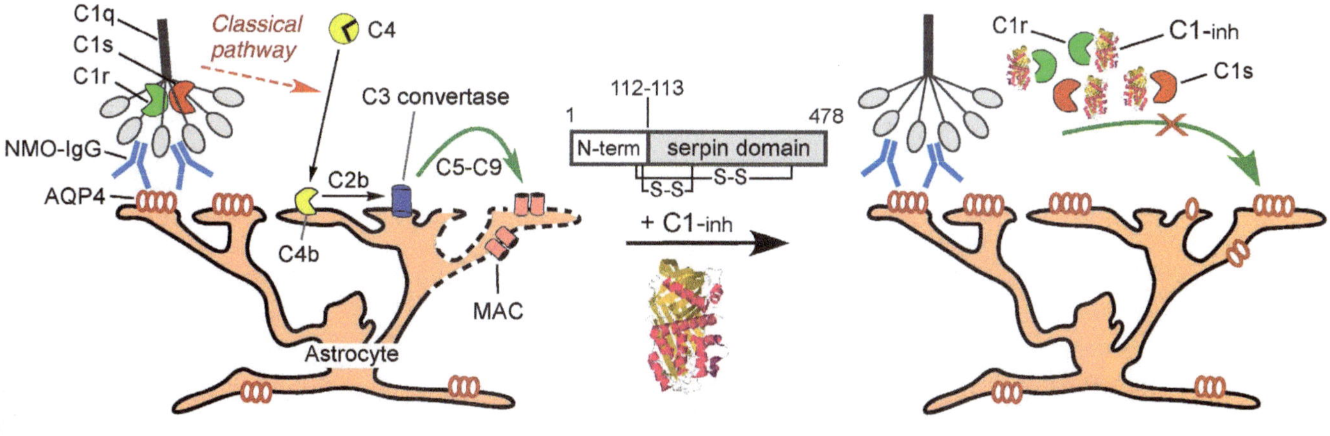

Figure 1. Action of C1-inh on NMO-IgG-dependent cytotoxicity involve the classical complement pathway. NMO-IgG binds to cell surface AQP4 on astrocytes, which activates the classical complement pathway by C1q binding to the Fc region of NMO-IgG leading to formation of the membrane attack complex (MAC) (left). C1-inh binds to and inactivates C1r and C1s, preventing C1q action on C4 and downstream complement activation.

perfused through the left cardiac ventricle with 100 ml PBS and then 25 ml of PBS containing 4% paraformaldehyde. Five micrometer-thick paraffin sections were immunostained for 1 h with antibodies against rat AQP4 (1:200, Santa Cruz Biotechnology), GFAP (1:100, Millipore) and myelin basic protein (MBP) (1:200, Santa Cruz Biotechnology), followed by the appropriate fluorescent secondary antibody (1:200, Invitrogen). Tissue sections were examined with a Leica (Wetzlar, Germany) DM 4000 B microscope at 2.5× magnification. AQP4, GFAP and MBP immunonegative areas were defined by hand and quantified using ImageJ. Data are presented as area (mm^2) of immunonegative area.

Statistical analysis

Data are presented as mean ± S.E. Comparisons between two groups were performed using Student's t-test.

Results

C1-inh inhibits NMO-IgG-dependent CDC

To study C1-inh inhibition of NMO-IgG-dependent CDC, cytotoxicity was assayed in CHO cells expressing AQP4 after incubation with NMO-IgG and human complement, as diagrammed in Fig. 2A. Fig. 2B shows concentration-dependent C1-inh inhibition of cytotoxicity in cells incubated with 2% human complement and a recombinant monoclonal NMO-IgG (rAb-53) (left) and sera from two NMO patients and pooled NMO patient sera (right). C1-inh inhibited cytotoxicity in a concentration-dependent manner, with 50% protection at ~ 2.5–5 μM C1-inh at 2% human complement. Fig. 2C shows C1-inh protection from cytotoxicity in cells incubated with rAb-53 and 2% rat serum.

Because C1-inh action involves blocking the classical complement pathway by irreversible inactivation of enzymes C1r and C1s in the C1 complex [28], [29], we investigated the efficacy of C1-inh inhibition on NMO-IgG-dependent CDC as a function of complement concentration. Fig. 3A, B shows concentration-dependent C1-inh inhibition with increasing complement, shown for two different NMO antibody concentrations. At an antibody concentration of 1 μg/ml, 7 μM C1-inh produced ~50% inhibition of CDC at 10% human complement. Though it is not possible to test undiluted human serum or higher concentrations of human complement in this assay, a much higher concentration of C1-inh, perhaps as high as 70 μM (estimated by multiplying 7 μM IC_{50} at 10% human complement by ten to extrapolate to undiluted serum), may be required to produce 50% inhibition of CDC with undiluted (100%) human serum. Given the equivalence of 1 unit/mL C1-inh to 2.5 μM, a 2000 unit dose of C1-inh administrated to a human with 5000 mL blood volume could produce a maximum increase of serum C1-inh of 0.4 units/mL (or 1 μM), much less than 70 μM. The inset to Fig. 3A shows that the inhibition of CDC by C1-inh is rapid.

Polysulfated macromolecules increase C1-inh inhibition potency

Based on the in vitro data showing weak C1-inh complement inhibition, we attempted to augment and potentiate the complement inhibition action of C1-inh using macromolecules, such as heparin, that inhibit complement directly [30], possibly by multivalent interaction with C1q that impair its binding to the antibody Fc region [31], [32]. Polysulfated macromolecules are also reported to potentiate the action of C1-inh [33], possibly by direct binding to the contact area of the protease – C1-inh complex [34]. We studied the complement inhibition action and potentiation of C1-inh activity in vitro by three polysulfated macromolecules – heparin, dextran sulfate, and the nutraceutical fucoidan. Fig. 4A shows that each of the three polysulfated macromolecules potentiated C1-inh inhibition of NMO-IgG-dependent CDC, reducing the C1-inh concentration needed to

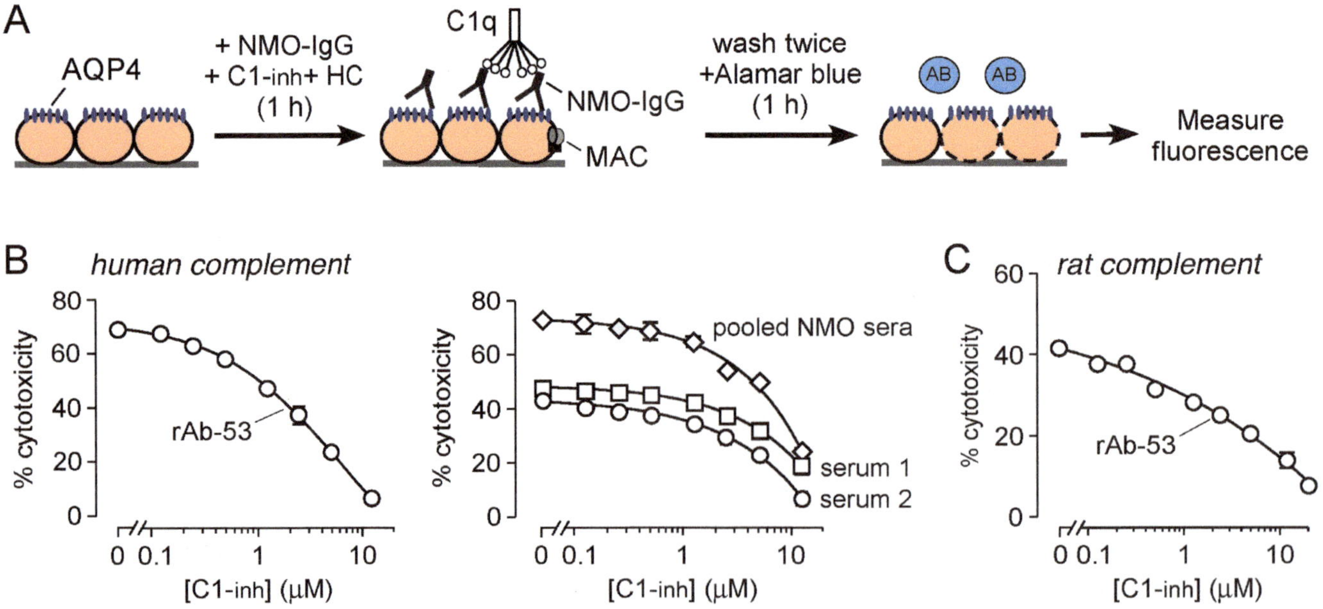

Figure 2. C1-inh inhibits complement-dependent cytotoxicity produced by NMO-IgG in AQP4-expressing cell cultures. A. CDC assay showing 1 h incubation of NMO-IgG, human complement (HC) and C1-INH to CHO cells expressing human M23-AQP4, followed by assay of cytotoxicity with AlamarBlue. B. Percentage cytotoxicity as a function of C1-inh concentration in cells incubated with human complement (2%) and recombinant NMO-IgG rAb-53 (1 μg/ml) (left) or individual or pooled human NMO sera (100 μg/ml) (right). C. Study as in B in cells incubated with rAb-53 (0.5 μg/ml) and rat serum (2%). Data are mean ± S.E. for 4 measurements per condition.

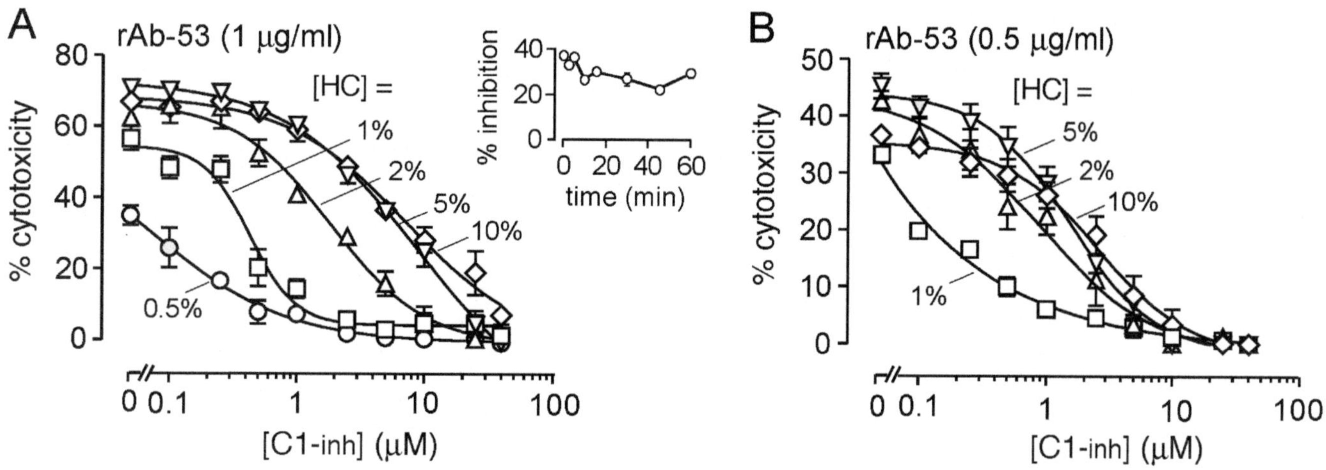

Figure 3. C1-inh inhibition of NMO-IgG-dependent cytotoxicity depends on complement and NMO-IgG concentration. A. Percentage cytotoxicity as a function of C1-inh concentration in cells incubated with 1 μg/ml rAb-53 and indicated concentration of human complement. Data are mean ± S.E. for 4 measurements per condition. Inset shows percentage inhibition of cytotoxicity as a function of time after pre-incubation of 2 μM C1-inh and 2% human complement prior to addition AQP4-expressing CHO-cell that were pre-incubated with 1 μg/ml rAb-53. B. Study as in A but with 0.5 μg/ml rAb-53.

produce inhibition, by nearly 10-fold for heparin and dextran sulfate. Fig. 4B shows that each of the polysulfated macromolecules alone inhibited CDC, with substantial potentiation seen at a low concentration of C1-inh. These results suggest the potential synergistic use of C1-inh and heparin to enhance complement inhibition activity in vivo, as investigated further below.

C1-inh administration to rats produces little inhibition of serum complement activity

To investigate the complement inhibition activity of C1-inh in rats, C1-inh was administered intravenously, and serum obtained at different times was assayed for complement activity. Initial studies done using doses used to treat HAE in humans (20 units/kg) did not show inhibition of complement activity in rat serum. We therefore increased the C1-inh dose by 30-fold (to 600 units/kg). However, no significant inhibition of rat serum complement activity was found at any time after C1-inh administration at this high dose (Fig. 5). The presence of multiple C1-inh substrates in serum could contribute the poor inhibition potency of C1-inh in vivo (see Discussion).

Given the results above showing that heparin and related polysulfated macromolecules can potentiate the inhibitory effect of C1-inh on CDC in vitro, we administered rats high-dose C1-inh and heparin together. However, as shown in Fig. 5, there was only minimal and transient reduction in complement activity in rat serum in both heparin-treated groups, with no potentiation produced by C1-inh.

C1-inh administration to rats does not reduce NMO pathology

We used an established rat model of NMO to test the in vivo efficacy of C1-inh. Though the lack of complement inhibition in rat serum would predict no significant clinical benefit in a rat model of NMO, the complement data do not rule out other actions of C1-inh that might be beneficial in NMO. In our model NMO pathology in brain is produced by a single intracerebral injection of NMO-IgG, which results in characteristic pathological features of NMO including astrocyte damage (loss of AQP4 and GFAP immunoreactivity), inflammation and demyelination by 5 days [20]. Pathology in this model was fully prevented by complement inhibition with cobra venom toxin [20], indicating that rat complement activity is required to produce pathology. Also, NMO pathology in this rat model was partially (~50%) reduced by intravenous immunoglobulin G [35], indicating its sensitivity to disease-modifying interventions. Here, C1-inh (or saline as control) was administered intravenously at high dose (300 units/kg daily for 2 days after intracerebral injection with NMO-IgG). Fig. 6 (left) shows that C1-inh did not protect against NMO pathology produced by NMO-IgG injection at day 2, as seen by large areas with loss of AQP4 and GFAP (white lines) or AQP4 alone ('penumbra region', dashed white lines). Fig. 6 (right) summarizes areas of loss of AQP4 and GFAP immunofluorescence.

Discussion

The complement system is an attractive target for therapy of NMO, as complement-dependent astrocyte cytotoxicity is an early initiating event in the neuroinflammation and demyelination that occurs in NMO. Selective inhibition of early steps in the classical complement pathway is of particular interest because NMO pathogenesis involves NMO-IgG complement effector function in which the antibody Fc region binds C1q to initiate complement activation. We found previously that CDC requires physical clustering of NMO-IgG on astrocytes [27], which is facilitated by the supramolecular assembly of AQP4 molecules in orthogonal arrays of particles [36]. Currently used therapies in NMO, immunosuppressives, B-cell depletion and plasma exchange, do not target the underlying pathogenic events in NMO. Alternative NMO therapeutics under evaluation or development include monocloncal antibodies targeting IL-6 receptors on plasma cells (toculizumab, [37]) and NMO-IgG binding to AQP4 (aquaporumab, [38]), and small molecule inhibitors of neutrophils [39] and eosinophils [40]. We previously found that a novel, C1q-targeted monoclonal antibody blocked CDC in vitro and prevented the development of pathology in an in vivo mouse model of NMO [23]. Evaluation of C1-inh in NMO is motivated by its complement inhibition action on the serine protease activities of C1r and C1s in the classical complement pathway [25], [28], [29];

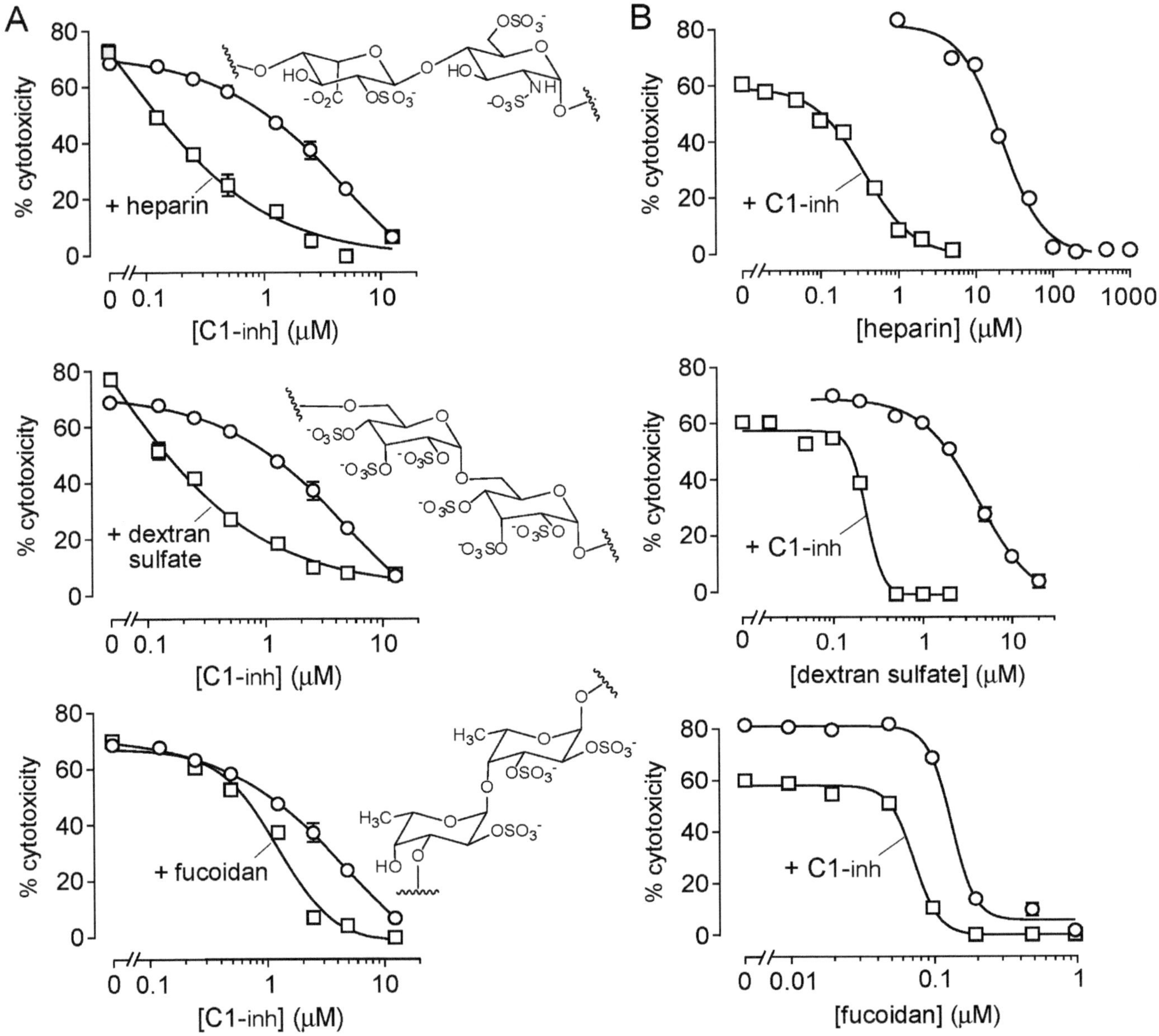

Figure 4. Polysulfated macromolecules potentiate C1-inh inhibition of NMO-IgG-dependent CDC. A. Percentage cytotoxicity as a function of C1-inh concentration in cells incubated with human complement (2%) and rAb-53 (1 μg/ml) in the absence or presence of 2 μM heparin, 0.5 μM dextran sulfate or 0.02 μM fucoidan (mean ± S.E., n = 4). Insets show chemical structures. B. Percentage cytotoxicity as a function of concentration of polysulfated macromolecules in cells incubated with human complement (2%) and rAb-53 (1 μg/ml) in the absence or presence of 1 μM C1-inh.

however, C1-inh inhibition of the complement lectin pathway has also been reported [41] and the serine protease inhibition action of C1-inh affects multiple other biological activities, including the coagulation system, the fibrinolytic system, T-lymphocyte activation and the kallikrein contact activation system [25], [26]. The therapeutic efficacy of C1-inh in HAE involves its action on the kallikrein system [42], [43].

We found here that C1-inh produced modest inhibition of NMO-IgG-dependent CDC in vitro, and that intravenous administration of large amounts of C1-inh in rats did not inhibit rat complement activity nor did it reduce brain pathology in a rat model of NMO. The studies here were done using rats rather than mice because of the weak activity of the mouse classical complement pathway [44]. C1-inh concentration in human serum is ~2–3 μM, and can increase by 2–2.5 fold with inflammation [4]. HAE patients have a broad range of C1-inh concentrations of 5–38% of normal levels [45]. After treatment with 25 units/kg C1-inh, C1-inh concentration increases to 60–70% of normal levels shortly and is maintained at 40% of normal levels over 24 h [46]. We found here that intravenous administration to rats of substantially greater amounts of C1-inh, 600 units/kg, did not inhibit serum complement activity or reduce NMO pathology produced by intracerebral NMO-IgG administration. These data suggest that C1-inh is unlikely to be of therapeutic benefit in NMO based on its complement inhibition activity.

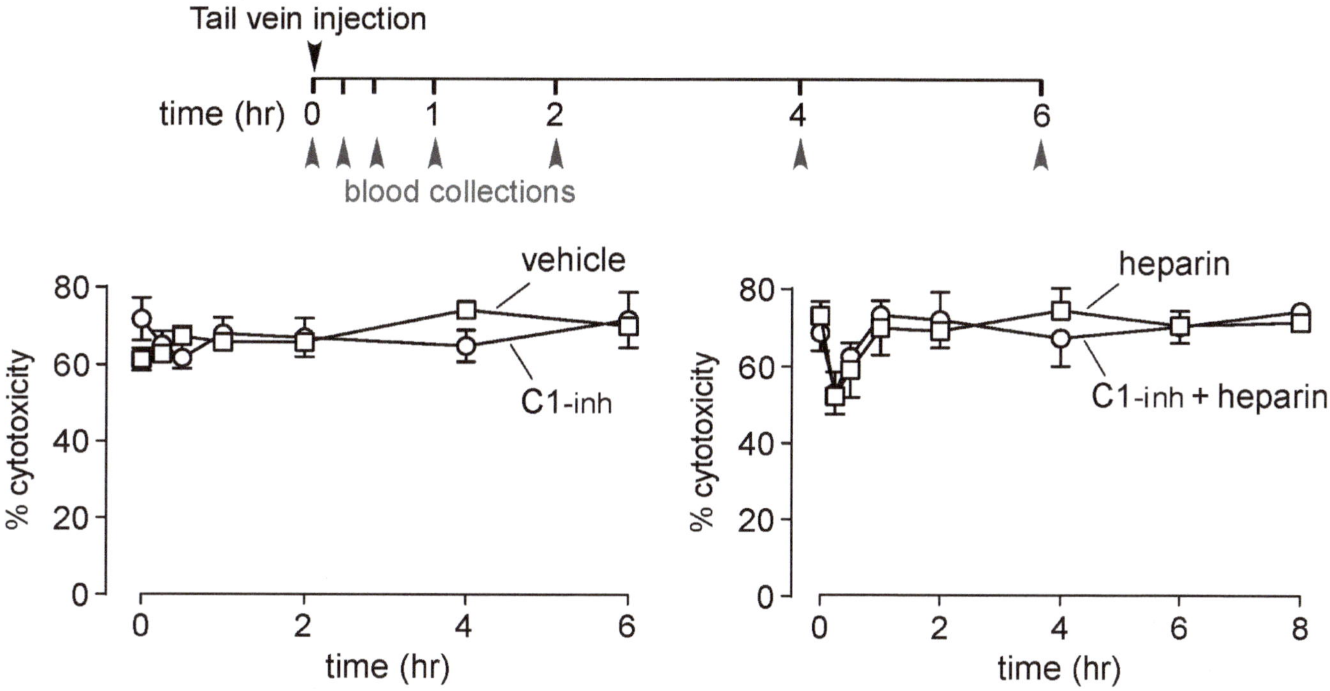

Figure 5. Intravenous administration of high-dose C1-inh in rats does not inhibit complement activity. A. C1-inh was administered to rats by tail vein injection, and serum samples obtained at specified times were bioassayed for NMO-IgG-dependent CDC by incubation of AQP4-expressing CHO-cells with 2% rat serum with 1 µg/ml rAb-53 on percentage cytotoxicity produced by rat serum at different times after administration of 600 units/kg C1-inh (or vehicle control) (left), and 1.5 mg/kg heparin, without or with 600 units/kg C1-inh (right) (mean ± S.E., 4 rats per condition).

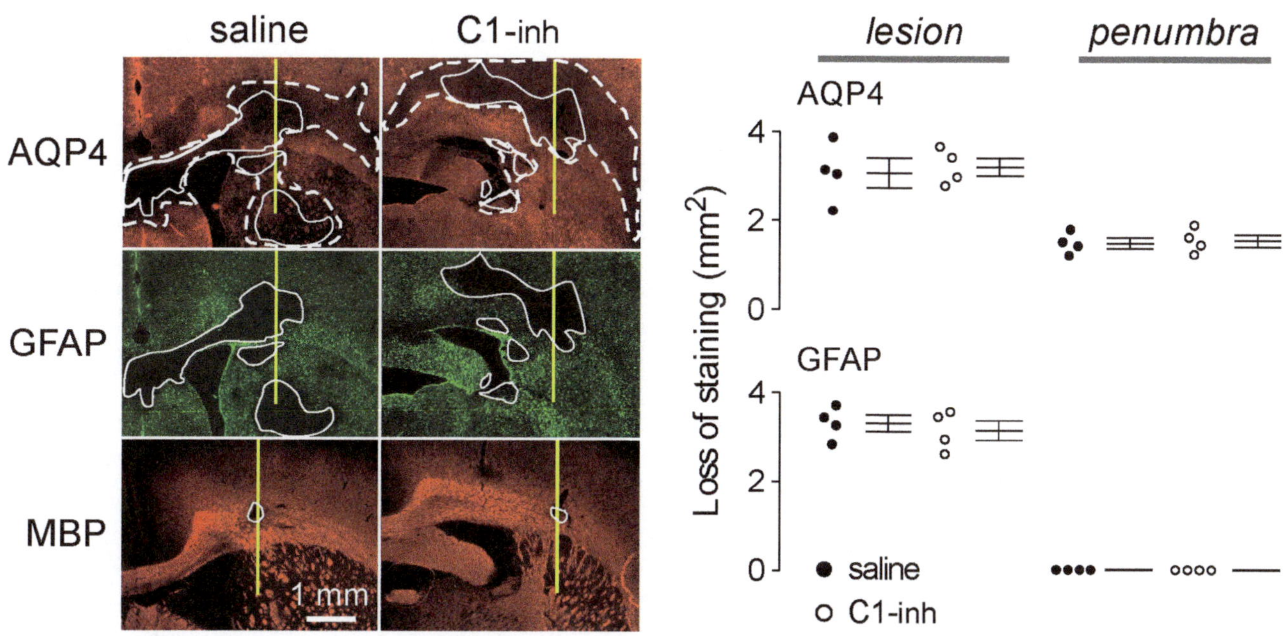

Figure 6. High-dose C1-inh does not reduce brain pathology in a rat model of NMO produced by intracerebral injection NMO-IgG. AQP4, GFAP and MBP immunofluorescence at 2 days after injection of 10 µg rAb-53. C1-inh (300 units/kg) (or saline control) was administered intravenously just before and 1 day after intracerebral injection of NMO-IgG. The needle tract is shown in yellow line, the white line demarcates lesion areas with loss of AQP4 and GFAP immunofluorescence, and the white dashed line demarcates the penumbra areas with reduced AQP4 but normal GFAP immunofluorescence (left). Summary of lesion areas showing data for individual rats (mean ± S.E., n = 4). Differences not significant.

The recently completed C1-inh safety trial (NCT 01759620) involved intravenous administration of CINRYZ, 2000 units per day for 3 days, in NMO patients undergoing an acute exacerbation [26]. For a 70 kg person this corresponds to 28 units/kg per dose, similar to that used to treat HAE. As discussed above, this dose of C1-inh is predicted to produce no significant inhibition of complement activity. In a single case report, a substantially higher dose of C1-inh (6000 units, followed by a total of 7000 units over 2 days) was administered to a 64-year old female with autoimmune hemolytic anemia associated with non-Hodgkin lymphoma [47]. An apparent clinical response was reported, with improved transfusion efficiency and reduced erythrocyte C3 deposition. Potentially such very larges doses of C1-inh could have therapeutic benefit in NMO, though the in vitro and rat data here provide evidence against significant complement inhibition by C1-inh even with such large doses. Human studies are needed to quantify the inhibition of NMO-IgG-dependent cytotoxicity of serum from humans administered high doses of C1-inh.

Given the weak complement inhibition activity of C1-inh, as a possible strategy to potentiate its activity, we investigated the efficacy of polysulfated macromolecules, including heparin, dextran sulfate, and fucoidan. Assays of NMO-IgG-dependent CDC in vitro showed that each of the three polysulfated macromolecules enhanced the inhibitory effect of C1-inh, with ~10-fold potentiation produced by heparin and dextran sulfate. However, administration of C1-inh and heparin to rats showed no C1-inh -dependent serum complement inhibition.

Though modest C1-inh inhibition of human and rat complement were found in in vitro assays, no significant inhibition of serum complement activity was found in rats in vivo, even after high-dose intravenous C1-inh administration, which translated to lack of efficacy in a rat model of NMO. There are several possible reasons for the lack of C1-inh efficacy on vivo. Whole blood contains many substrates of C1-inh, in addition to C1r and C1s, that are present relatively large amounts, including mannan-binding lectin-associated proteins (MASP-1 and MASP-2), kallikrein, FXIIa and plasmin, which compete for inactivation by C1-inh. C1-inh is a member of a suicidal family of inhibitors, in which a covalent complex is irreversibly formed between C1-inh and its protease target [28], [29]. Though C1-inh can inhibit C1-auto-inactivation in the fluid-phase [48], it is a poor inhibitor of C1-activated immune complexes on membranes [49], [50], as is relevant in NMO. One group reported that the concentration of C1-inh for inhibition of C1 binding to the antibody-sensitized sheep erythrocyte surface was 1000-fold higher than that required for C1 inhibition in the fluid phase [49]. To achieve C1-inh inhibition in NMO, one approach might be direct targeting of C1-inh to AQP4-expressing astrocytes in the brain, perhaps by C1-inh / AQP4 antibody fusions.

In conclusion, our in vitro and rat data in NMO models suggest that the complement inhibition activity of C1-inh in serum is far too low to produce significant complement inhibition and hence therapeutic benefit in NMO. Notwithstanding the similar activities of rat and human complement, and the similar C1-inh inhibition efficacy, studies done in experimental rat models should be extrapolated to humans with caution. Definitive evaluation of the potential efficacy of C1-inh in human NMO will require controlled clinical trials.

Acknowledgments

We thank Dr. Jeffrey Bennett (Univ. Colorado Denver, Aurora, CO) for providing recombinant monoclonal NMO antibody and Dr. Mel Berger (CSL Behring) for advice.

Author Contributions

Conceived and designed the experiments: LT ASV. Performed the experiments: LT NA PP. Analyzed the data: LT NA PP. Contributed to the writing of the manuscript: LT NA PP ASV.

References

1. Wingerchuk DM, Lennon VA, Lucchinetti CF, Pittock SJ, Weinshenker BG (2007) The spectrum of neuromyelitis optica. Lancet Neurol 6: 805–815.
2. Jacob A, McKeon A, Nakashima I, Sato DK, Elsone L, et al. (2013) Current concept of neuromyelitis optica (NMO) and NMO spectrum disorders. J Neurol Neurosurg Psychiatry 84: 922–930.
3. Jarius S, Paul F, Franciotta D, Waters P, Zipp F, et al. (2008) Mechanism of disease: aquaporin-4 antibodies in neuromyelitis optica. Nat Clin Pract Neurol 4: 202–214.
4. Lennon VA, Kryzer TJ, Pittock SJ, Verkman AS, Hinson SR (2005) IgG marker of optic-spinal multiple sclerosis binds to the aquaporin-4 water channel. J Exp Med 202: 473–477.
5. Papadopoulos MC, Verkman AS (2012) Aquaporin water channels in the nervous system. Nat Rev Neurosci 14: 265–277.
6. Papadopoulos MC, Verkman AS (2013) Aquaporin 4 and neuromyelitis optica. Lancet Neurol 11: 535–544.
7. Kira J (2011) Autoimmunity in neuromyelitis optica and opticospinal multiple sclerosis: astrocytopathy as a common denominator in demyelinating disorders. J Neurol Sci 311: 69–77.
8. Collongues N, de Seze J (2011) Current and future treatment approaches for neuromyelitis optica. Ther Adv Neurol Disord 4: 111–121.
9. Cree B (2008) Neuromyelitis optica: diagnosis, pathogenesis, and treatment. Curr Neurol Neurosci Rep 8: 427–433.
10. Kim SH, Kim W, Huh SY, Lee KY, Jung IJ, et al. (2013) Clinical efficacy of plasmapheresis in patients with neuromyelitis optica spectrum disorder and effects on circulating anti-aquaporin-4 antibody level. J Clin Neurol 9: 36–42.
11. Lucchinetti CF, Mandler RN, McGavern D, Bruck W, Gleich G, et al. (2002) A role for humoral mechanisms in the pathogenesis of Devic's neuromyelitis optica. Brain 125: 1450–1461.
12. Misu T, Fujihara K, Kakita A, Konno H, Nakamura M, et al. (2007) Loss of aquaporin 4 in lesions of neuromyelitis optica: distinction from multiple sclerosis. Brain 130: 1224–1234.
13. Roemer SF, Parisi JE, Lennon VA, Benarroch EE, Lassmann H, et al. (2007) Pattern-specific loss of aquaporin-4 immunoreactivity distinguishes neuromyelitis optica from multiple sclerosis. Brain 130: 1194–1205.
14. Sabater L, Giralt A, Boronat A, Hankiewicz K, Blanco Y, et al. (2009) Cytotoxic effect of neuromyelitis optica antibody (NMO-IgG) to astrocytes: an in vitro study. J Neuroimmunol 215: 31–35.
15. Hinson SR, Pittock SJ, Lucchinetti CF, Roemer SF, Fryer JP, et al. (2007) Pathogenic potential of IgG binding to water channel extracellular domain in neuromyelitis optica. Neurology 69: 2221–2231.
16. Hinson SR, McKeon A, Fryer JP, Apiwattanakul M, Lennon VA, et al. (2009) Prediction of neuromyelitis optica attack severity by quantitation of complement-mediated injury to aquaporin-4-expressing cells. Arch Neurol 66: 1164–1167.
17. Bennett JL, Lam C, Kalluri SR, Saikali P, Bautista K, et al. (2009) Intrathecal pathogenic anti-aquaporin-4 antibodies in early neuromyelitis optica. Ann Neurol 66: 617–629.
18. Zhang H, Bennett JL, Verkman AS (2011) Ex vivo spinal cord slice model of neuromyelitis optica reveals novel immunopathogenic mechanisms. Ann Neurol 70: 943–954.
19. Saadoun S, Waters P, Bell BA, Vincent A, Verkman AS, et al. (2010) Intra-cerebral injection of neuromyelitis optica immunoglobulin G and human complement produces neuromyelitis optica lesions in mice. Brain 133: 349–361.
20. Asavapanumas N, Ratelade J, Verkman AS (2014) Unique neuromyelitis optica pathology produced in naöve rats by intracerebral administration of NMO-IgG. Acta Neuropathol 127: 539–551.
21. Pittock SJ, Lennon VA, Mckeon A, Mandrekar J, Weinshenker BG, et al. (2013) Eculizumab in AQP4-IgG-positive relapsing neuromyelitis optica spectrum disorders: an open-label pilot study. Lancet Neurol 12: 554–562.
22. Parker C (2009) Eculizumab for paroxysmal nocturnal haemoglobinuria. Lancet 373: 759–767.
23. Phuan PW, Zhang H, Asavapanumas N, Leviten M, Rosenthal A, et al. (2013) C1q-targeted monoclonal antibody prevents complement-dependent cytotoxicity and neuropathology in in vitro and mouse models of neuromyelitis optica. Acta Neuropathol 125: 829–840.
24. Caliezi C, Wuillemin WA, Zeerleder S, Redondo M, Eisele B, et al. (2000) C1-esterase inhibitor: an anti-inflammatory agent and its potential use in the treatment of disease other than hereditary angioedema. Pharmacol Rev 52: 91–112.

25. Karnaukhova E (2013) C1-esterase inhibitor: biological activities and therapeutic applications. J Hematol Thromb Dis 1: 3. Available: http://dx.doi.org/10.4172/2329-8790.1000113.
26. Levy M, Mealy MA (2014) Purified human C1-esterase inhibitor is safe in acute relapses of neuromyelitis optica. Neurol Neuroimmunol Neutoinflamm 1: doi 10.1212/NXI.0000000000000005
27. Phuan PW, Ratelade J, Rossi A, Tradtrantip L, Verkman AS (2012) Complement-dependent cytotoxicity in neuromyelitis optica requires aquaporin-4 protein assembly in orthogonal arrays. J Biol Chem 287: 13829–13839.
28. Gettins PG (2002) Serpin structure, mechanism, and function. Chem Rev 102: 4751–4804.
29. Ziccardi RJ, Cooper NR (1979) Active disassembly of the first complement component, C1, by C1 inactivator. J Immunol 123: 788–792.
30. Weiler JM, Edens RE, Linhardt RJ, Kapelanski DP (1992) Heparin and modified heparin inhibit complement activation in vivo. J Immunol 148: 3210–3215.
31. Garlatti V, Chouquet A, Lunardi T, Vives R, Paidassi H, et al. (2010) Cutting edge: C1q binds deoxyribose and heparin sulfate through neighboring sites of its recognition domain. J Immunol 185: 808–812.
32. Hughes-Jones NC, Gardner B. (1978) The reaction between the complement subcomponent C1q, IgG complexes and polyionic molecules. Immunol 34: 459–463.
33. Wuillemin WA, Velthuis HT, Lubbers YTP, de Ruig CP, Eldering E, et al. (1997) Potentiation of C1 inhibitor by glycosaminoglycans. J Immunol 159: 1953–1960.
34. Beinrohr L, Harmat V, Dobo J, Lorincz Z, Gal P, et al. (2007) C1 inhibitor serpin domain structure reveals the likely mechanism of heparin potentiation and conformational disease. J Biol Chem 282: 21100–21109.
35. Ratelade J, Smith AJ, Verkman AS (2014) Human immunoglobulin G reduces the pathogenicity of aquaporin-4 autoantibodies in neuromyelitis optica. Exper Neurol 255: 145–153.
36. Crane JM, Lam C, Rossi A, Gupta T, Bennett JL, et al (2011) Binding affinity and specificity of neuromyelitis optica autoantibodies to aquaporin-4 M1/M23 isoforms and orthogonal arrays. J Biol Chem 286: 16516–16524.
37. Ayzenberg I, Kleiter I, Schruder A, Hellwig K, Chan A, et al. (2013) Interleukin 6 receptor blockade in patients with neuromyelitis optica nonresponsive to anti-CD20 therapy. JAMA Neurol 70: 394–397.
38. Tradtrantip L, Zhang H, Saadoun S, Phuan PW, Lam C, et al. (2012) Anti-aquaporin-4 monoclonal antibody blocker therapy for neuromyelitis optica. Ann Neurol 71: 314–322.
39. Saadoun S, Waters P, MacDonald C, Bell BA, Vincent A, et al. (2012) Neutrophil protease inhibition reduces neuromyelitis optica-immunoglobulin G-induced damage in mouse brain. Ann Neurol 71: 323–33.
40. Zhang H, Verkman AS (2013) Eosinophil pathogenicity mechanisms and therapeutics in neuromyelitis optica. J Clin Invest 123: 2306–2316.
41. Kerr FK, Thomas AR, Wijeyewickrema LC, Whisstock JC, Boyd SE, et al. (2008) Elucidation of the substrate specificity of the MASP-2 protease of the lectin complement pathway and identification of the enzyme as a major physiological target of the serpin, C1-inhibitor. Mol Immunol 45: 670–677.
42. Prematta MJ, Prematta T, Craig TJ (2008) Treatment of hereditary angioedema with plasma-derived C1 inhibitor. Ther Clin Risk Manag 4: 975–982.
43. Cugno M, Zanichelli A, Foieni F, Caccia S, Cicardi M (2009) C1-inhibitor deficiency and angioedema: molecular mechanisms and clinical progress. Trend Mol Med 15: 69–78.
44. Ratelade J, Verkman AS (2014) Inhibitors of the classical complement pathway in mouse serum limit the utility of mice as experimental models of neuromyelitis optica. Mol Immunol 62: 104–113.
45. Cicardi M, Bergamaschini L, Cugno M, Beretta A, Zingale LC, et al. (1998) Pathogenetic and clinical aspects of C1 inhibitor deficiency. Immunobiol 199: 366–376.
46. Kunschak M, Engl W, Maritsch F, Rosen FS, Eder G, et al. (1998) A randomized, controlled trial to study the efficacy and safety of C1 inhibitor concentrate in treating hereditary angioedema. Transfusion 38: 540–549.
47. Wouters D, Stephan F, Strengers P, Hass MD, Brouwer C, et al. (2013) C1-esterase inhibitor concentrate rescues erythrocyte from complement-mediated destruction in autoimmune hemolytic anemia. Blood 121: 1242–1244.
48. Bianchino AC, Poon PH, Schumaker VN (1988) A mechanism for the spontaneous activation of the first component of complement, C1, and its regulation by C1-inhibitor. J Immunol 141: 3930–3936.
49. Nielsen EW, Waage C, Fure H, Brekke OL, Sfyroera G. (2007) Effect of supraphysiologic levels of C1-inhibitor on the classical, lectin and alternative pathways of complement. Mol Imunol 44: 1819–1826.
50. Tenner AJ, Frank MM. (1986) Activator-bound C1 is less susceptible to inactivation by C1 inhibition than is fluid-phase C1. J Immunol 137: 625–630.

17

Selective DNA Delivery to Tumor Cells Using an Oligoarginine-LTVSPWY Peptide

Cheng Gong[1,2], Deng Pan[1,2], Fengwu Qiu[1,2], Pei Sun[1,2], Yu-Hui Zhang[1,2]*

1 Britton Chance Center for Biomedical Photonics, Wuhan National Laboratory for Optoelectronics-Huazhong University of Science and Technology (HUST), Wuhan, Hubei, China, **2** Key Laboratory of Biomedical Photonics of Ministry of Education, Department of Biomedical Engineering, Huazhong University of Science and Technology, Wuhan, Hubei, China

Abstract

DNA therapy for cancer requires efficient, selective and safe DNA delivery systems. Compared with other non-viral methods such as lipid or polymer-based DNA delivery vectors, peptide-based DNA delivery systems are biocompatible and biodegradable, which leads to lower immunogenicity and lower toxicity. Moreover, peptide vectors are easier to produce and their compositions easier to control because solid-phase peptide synthesis has been extensively developed. However, peptide-based systems for DNA delivery toward special tumor cells or tissues are still lacking. In this study, we constructed a non-viral 9rR-LTVSPWY peptide-based DNA delivery system and showed that it is able to efficiently and selectively transfect DNA into targeted tumor cells. This work presents a novel strategy for tumor cell-specific DNA delivery and a reference for designing more efficient DNA delivery systems targeted towards various types of cancer.

Editor: Joseph C. Glorioso, University of Pittsburgh School of Medicine, United States of America

Funding: This work was supported by the Science Fund for Creative Research Group of China (Grant No. 61121004), the National Natural Science Foundation of China (Grant No. 31270906), and the seed project of Wuhan National Laboratory for Optoelectronics. The funders had no role in study design, data collection and analysis, decision to publish, or preparation of the manuscript.

Competing Interests: The authors have declared that no competing interests exist.

* Email: zhangyh@mail.hust.edu.cn

Introduction

DNA therapy for cancer requires efficient and safe systems that can deliver the therapeutic DNA selectively into targeted tumor cells. Viral vectors have been used to transfer genes into cancer cells successfully [1–4]. However, these vectors have serious disadvantages, such as limited loading capacity, complexity of production, innate immunogenicity, and the risk of inflammatory responses and toxicity, that limit their clinical applications [5,6]. To avoid these problems, various non-viral carriers have been developed; these carriers display low immunogenicity, relative safety, ease of production, and no cargo size limitation. Of the existing non-viral vectors, cationic lipids and cationic polymers are the most intensively studied and frequently employed. However, they have dose-dependent toxicities in *in vivo* applications [7,8].

Cationic peptides have also been explored as gene delivery systems due to several advantages: biodegradability, biocompatibility, less toxicity, and ease of synthesis compared with polymeric carriers [9,10]. Moreover, the composition of peptides is easy to control. By altering the composition of a given peptide, various functions can be achieved. Self-Assembling Peptides (SAPs) have been investigated as a tool for targeted tumor drug delivery based on the enhanced permeability and retention (EPR) method [11,12]. However, EPR is not very effective, and the size dependency, slow time frame, variability from tumor to tumor and relative inability to operate in non-tumor vascular beds limit the clinical applications of the EPR method [13]. Furthermore, SAPs themselves usually cannot bind DNA; therefore, this method cannot be used as a DNA delivery system.

Cell-penetrating peptides (CPPs) are short peptides that can deliver cell-impermeable compounds into living cells and have been successfully employed to translocate various bulky cargos (including peptides, proteins, siRNA, DNA, and nanoparticles) across cellular plasma membranes [14,15]. Among the CPPs, oligopeptides based on arginine are frequently used because of their ease of synthesis and cell-penetrating ability compared with other peptides based on other amino acids such as lysine [16]. With some modifications, cationic oligoarginines have successfully been employed to transport DNA into cells [17–19]. However, these systems lack selectivity toward the tumor cells, which is regarded as a prerequisite for safe and successful gene therapy [20]. Thus, developing an arginine-rich peptide that can mediate tumor cell-specific DNA delivery is very relevant for cancer gene therapy.

Targeted delivery of drugs into tumor cells using specific extracellular receptors has the following advantages in cancer therapy: (1) limiting adverse side effects caused by the drug absorption of normal cells; (2) enhancing drug internalization by tumor cells; (3) solving the resistance problem based on the active drug efflux from tumor cells [21]. LTVSPWY, a 7-residue peptide, has been shown to specifically bind to and be absorbed by certain types of cancer cells, possibly via receptor-mediated endocytosis [22]. Moreover, certain LTVSPWY-attached nanoparticles have successfully been absorbed by these tumor cells [23].

In this study, we explored the possibility of using an oligoarginine-LTVSPWY peptide as a non-viral vehicle to deliver DNA selectively into tumor cells. The peptide has a tri-block design composed of nona-arginine (rRrRrRrRr, r: D-Arg, R: L-Arg) for binding DNA through electrostatic interactions, four histidine residues as a spacer and for enhancing endosomal escape [10], and the LTVSPWY sequence, which is used for tumor cell targeting and cell adsorption.

Materials and Methods

Materials

The plasmid pEGFP-N1 was obtained from Clontech (CA, USA), and the pGL3 control vector was from Promega (WI, USA). The peptides 9rR-LTVSPWY (rRrRrRrRrHHHHLTVSPWY) and 9rR (rRrRrRrRr) [24] were prepared using solid-phase peptide synthesis and purified to homogeneity by preparative high performance liquid chromatography (HPLC) to achieve >95% purity. Their appropriate masses were confirmed by electrospray ionization (ESI) mass spectrometry (Figures S1 and S2).

Cell culture

5–8F cells (a human nasopharyngeal cancer cell line) [25] were a kind gift from Prof. Yi-Xin Zeng and Prof. Mu-Sheng Zeng (Sun Yat-sen University Cancer Center, Guangzhou, China). SKOV-3 cells (a human ovarian cancer cell line) were from the American Type Culture Collection (ATCC). SKOV-3, MCF-7 and 5–8F cells were cultured in RPMI1640 supplemented with 10% fetal bovine serum (FBS). HeLa cells were cultured in Dulbecco's modified Eagle's medium (DMEM) containing 10% FBS. The cells were grown in a CO_2 (5%, v/v) atmosphere at 37°C.

Determination of zeta potential and particle size of complexes

The peptide/pDNA complexes were formed via self-assembly. The N/P ratios represent the molar ratio of the arginines in 9rR (or the arginines in the nona-arginine fragment of the 9rR-LTVSPWY) to the phosphorus content in the DNA [19]. The full procedure was performed as previously described [24].

DNA migration assay

Peptide/pDNA complexes at the indicated N/P ratios were prepared in PBS (phosphate buffered saline), and the pDNA amount was kept constant (0.1 μg). Agarose gel electrophoresis (0.8%) was performed in Tris-acetate (TAE) running buffer at 150 V for 10 min.

Plasmid labeling. The procedure was performed as previously described [24]. 200 μg of pEGFP-N1 was incubated with 5 μg ethidium monoazide (EMA) in 2 mL of ddH_2O for 10 min at room temperature. The solution was exposed to UV light for 5 min and then visible light for 10 min. The plasmid was precipitated by the addition of ethanol, collected by centrifugation at above 10,000 rpm for 15 min, and redissolved in ddH_2O.

Confocal microscopy

Cells were seeded into eight-well Lab-Tek chamber slides (Nunc) at a density of 1.0×10^4 cells/well in 200 μL RPMI1640 (for SKOV-3, MCF-7 and 5–8F cells) or DMEM (for HeLa cells) supplemented with 10% FBS. The rest of the procedure was performed as previously described [24].

To track the internalization of the plasmids, 5–8F cells were incubated with the indicated peptide/EMA-labeled pEGFP-N1 complex (N/P 6:1, 240 μL, 1 μg pDNA) and LysoTracker Green (100 nM, Invitrogen, USA) at 37°C or 4°C for 30 min. (For the 4°C experiments, the cells were pre-incubated at 4°C for 15 min). Then, the media was replaced with RPMI1640 containing 10% FBS. The rest of the procedure was performed as previously described [24]. The co-localization analysis was performed using the Manders Coefficients plugin of Image J.

Flow cytometry

Cells were seeded into a 24-well plate at a density of 5.0×10^4 cells/well in 400 μL RPMI1640 (for SKOV-3, MCF-7 and 5–8F cells) or DMEM (for HeLa cells) containing 10% FBS. Cells in the logarithmic phase were incubated with the indicated peptide/pEGFP-N1 complexes (480 μL, 2 μg pDNA) at the indicated N/P ratios in full serum medium without antibiotics for 5 h. Subsequently, the media was replaced by medium containing 10% FBS. The rest of the procedure was performed as previously described [24].

To track the intracellular uptake of the plasmids, 5–8F cells were incubated with the peptide/EMA-labeled pEGFP-N1 complexes (N/P 6:1, 480 μL, 2 μg pDNA) at 37°C for 30 min. For the 4°C experiments, the cells were pre-incubated at 4°C for 15 min and then incubated with the complexes for 30 min. Subsequently, the media was replaced by RPMI1640 containing 10% FBS. The rest of the procedure was performed as previously described [24].

Luciferase expression assay

Cells were seeded into a 96-well plate at a density of 5×10^3 cells/well in 100 μL RPMI1640 (for SKOV-3, MCF-7 and 5–8F cells) or DMEM (for HeLa cells) containing 10% FBS. Cells in the logarithmic phase were incubated with the indicated peptide/pGL3 complexes (120 μL, 0.5 μg pDNA) at an N/P ratio of 6:1 in full serum medium without antibiotics for 5 h. Subsequently, the media was replaced by media containing 10% FBS. The rest of the procedure was performed as previously described [24].

Cell toxicity

The cytotoxicity of 9rR-LTVSPWY was evaluated using a 3-(4, 5-dimethylthiazol-2-yl)-2, 5-diphenyltetrazolium bromide (MTT) assay. 5–8F cells were incubated in a 96-well plate at a density of 5×10^3 cells/well with the 9rR-LTVSPWY/pEGFP-N1 complex (N/P 6:1, 120 μL, 0.5 μg pDNA). The rest of the procedure was performed as previously described [24].

Results

Characterization of 9rR-LTVSPWY/pDNA complexes

Our previous study has indicated that the replacement of L-Arg residues with D-Arg residues imparts greater proteolytic resistance of oligoarginine peptides [26]. We used a heterogeneous nona-arginine (rRrRrRrRr, r: D-Arg, R: L-Arg) peptide because the heterogeneous peptide is more protease resistant than the L-oligoarginine, which is more bioavailable. Additionally, the heterogeneous peptide is more biodegradable than the D-oligoarginine, leading to reduced toxicity [26]. An LTVSPWY segment was added into the peptide sequence as a tumor cell-targeting moiety.

The formation of the 9rR-LTVSPWY/pDNA complexes was evaluated by agarose gel electrophoresis. Figure 1 shows that 9rR-LTVSPWY was able to bind DNA in a manner that depended on the N/P ratio and yielded complete retardation of the DNA at an N/P ratio ≥ 3, indicating that 9rR-LTVSPWY encapsulated the DNA completely at those N/P ratios. The N/P ratios represent the molar ratio of the arginines in 9rR (or the arginines in the

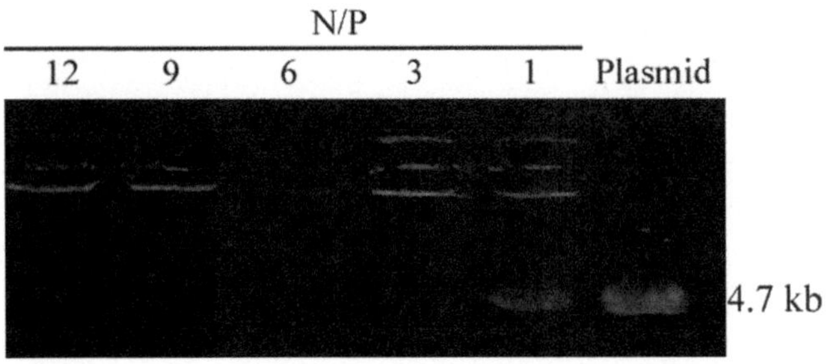

Figure 1. Agarose gel electrophoresis of the 9rR-LTVSPWY/pEGFP-N1 complexes at the indicated N/P ratios.

nona-arginine fragment of the 9rR-LTVSPWY) to the phosphorus content in the DNA [19].

Figure 2A shows the particle size values of the 9rR-LTVSPWY/pDNA complexes at various N/P ratios measured using dynamic light scattering (DLS). At an N/P ratio of 2, the particle sizes of the 9rR-LTVSPWY/pDNA mixture were ~126 nm, which was smaller than that of the naked pDNA (195±24 nm) and larger than that of the mixture at N/P ratios ≥ 3; this demonstrated that the pDNA was partly condensed by the peptide. When the N/P ratio rose above 3, the particle sizes of the 9rR-LTVSPWY/pDNA complexes were decreased to approximately 75 nm, suggesting that the pDNA was efficiently encapsulated and condensed by 9rR-LTVSPWY. Figure 2B indicates that the zeta potential of the 9rR-LTVSPWY/pDNA mixture was negative at N/P ratio ≤ 2; at higher N/P ratios, the zeta potential of the 9rR-LTVSPWY/pDNA complexes turned positive. When the N/P ratio rose above 3, the zeta potential of the complexes was increased to approximately 21 mV. The particle size and zeta potential measurement results were consistent with the agarose gel electrophoresis assay and suggested that, at N/P ratios ≥ 3, 9rR-LTVSPWY encapsulates the DNA completely.

Gene transfection into cells using 9rR-LTVSPWY/pDNA complexes

Because the LTVSPWY peptide has been used as a specific ligand for the cell lines 5–8F and SKOV-3 [23,27], these two cell lines were used to evaluate the ability of the 9rR-LTVSPWY/pDNA complexes to mediate DNA transfection, while MCF-7 and HeLa cells were used as negative controls. To optimize the peptide/pDNA stoichiometry (N/P ratio) for the best transfection efficiency, the transfection efficiencies of the 9rR-LTVSPWY/pEGFP-N1 complexes at various N/P ratios were measured using GFP as a reporter protein. Confocal microscopy (Figure 3A) showed significant pDNA expression (green fluorescence) in 5–8F and SKOV-3 cells transfected with 9rR-LTVSPWY/pEGFP-N1 at N/P ratios of 6, 9, 12 and 15. N/P = 6 yielded the highest number of cells expressing GFP (Figures 3A, 3B), suggesting that N/P = 6 is the optimal condition for transfection. In contrast, no pDNA expression was observed in MCF-7 and HeLa cells under the same conditions (Figure 3C), demonstrating the cell selectivity of the peptide/pDNA complexes. In contrast, although 9rR encapsulated DNA completely at N/P ratios ≥ 3 [24], there was no obvious GFP expression in the 5–8F cells transfected with the 9rR/pEGFP-N1 complex (Figure 3A and Figure S3), suggesting that the LTVSPWY sequence is essential for the cellular uptake of the 9rR-LTVSPWY/pDNA complexes.

Because the confocal imaging and FACS results suggested that N/P = 6 results in the highest transfection efficiency, we conducted further experiments with the peptide/pDNA complexes prepared at an N/P ratio of 6. To quantify the specificity and efficiency of transfection in various types of cells, we performed quantitative analysis using flow cytometry. Figure 4A showed that the percentages of 5–8F, SKOV-3, MCF-7 and HeLa cells expressing GFP were approximately 30.33, 8.23, 1.04 and 1.05, respectively, implying that the plasmids were expressed in 5–8F and SKOV-3 cells, with significantly different expression levels in the experimental groups and the blank group, but not to MCF-7 or HeLa, which exhibited no significant difference between the experimental groups and the blank group. Next, we performed luciferase expression assays using the four cell lines. Figure 4B indicates that 9rR-LTVSPWY delivered the plasmid into 5–8F and SKOV-3 cells but not MCF-7 or HeLa cells, verifying the excellent selectivity of this method.

The cellular uptake of the 9rR-LTVSPWY/pDNA complex

The cellular uptake mechanism of the 9rR-LTVSPWY/pDNA complex was then investigated. 5–8F cells were incubated with EMA-labeled pEGFP-N1 (red) encapsulated with 9rR-LTVSPWY or 9rR. Figure 5A reveals that the EMA-labeled 9rR-LTVSPWY/pDNA complexes (red) were obviously absorbed by the cells in 30 min after incubation and mainly distributed in dots. These puncta co-localized well with the LysoTracker Green probe (green), with a Pearson's coefficient of 0.794, suggesting that the complexes were mainly located in endosomes and/or lysosomes. In contrast, the 9rR/pDNA complex could not enter the cells at all (Figure S3). When the incubation temperature was decreased from 37°C to 4°C, which typically inhibits endocytosis, the cellular uptake of the 9rR-LTVSPWY/pDNA complex was significantly decreased (Figure 6). These results suggest that the 9rR-LTVSPWY/pDNA complex entered the cells via receptor-mediated endocytosis, consistent with the previously reported cellular uptake mechanism of LTVSPWY [27]. One hour post-incubation, the pDNA (red fluorescence) had escaped from the

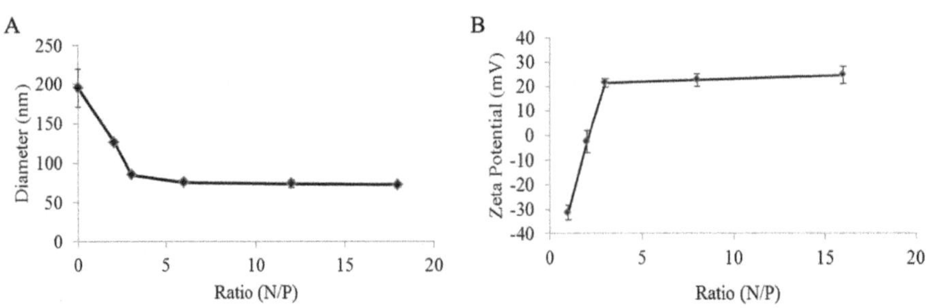

Figure 2. Characterization of the 9rR-LTVSPWY/pEGFP-N1 complexes at the indicated N/P ratios. (A) Particle size, (B) zeta potential (n = 3, mean ± SD). The polydispersity index of each sample was below 0.3.

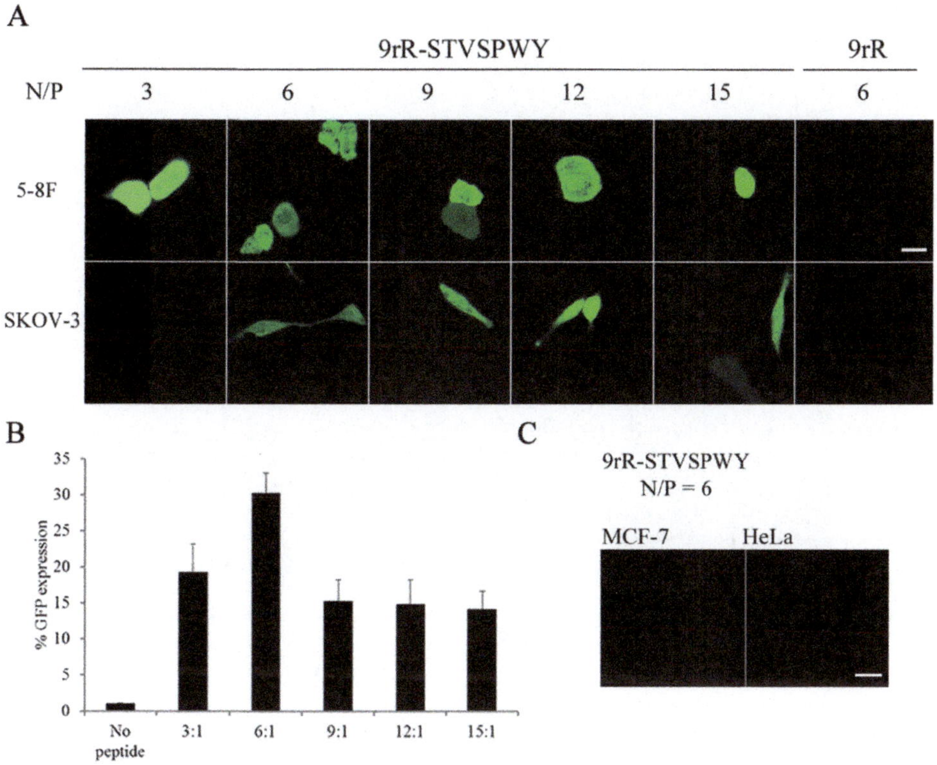

Figure 3. Cellular uptakes of 9rR-LTVSPWY at different N/P ratios in different cells. (A) Confocal microscopy images of 5-8F and SKOV-3 cells at 48 h after transfection with the 9rR-LTVSPWY/pEGFP-N1 or 9rR/pEGFP-N1 complexes at the indicated N/P ratios. (B) Percentage of 5-8F cells expressing GFP at 48 h after transfection with the 9rR-LTVSPWY/pEGFP-N1 complexes at the indicated N/P ratios or with a vehicle control (no peptide). Cellular fluorescence was examined by FACS. The data are shown as the means ± SD. All measurements were performed in triplicate. (C) Confocal microscopy images of MCF-7 and HeLa cells at 48 h after transfection with the 9rR-LTVSPWY/pEGFP-N1 complexes at an N/P ratio of 6. Green fluorescence represents pEGFP-N1 expression. (Scale bar = 20 μm).

endosomes (green fluorescence) and was diffusely distributed in the cytosol (Figure 5B), with a resulting decrease in the Pearson's coefficient (to 0.303). The results showed that pDNA transfected into cells using 9rR-LTVSPWY can escape from endocytic vesicles quickly.

Cytotoxicity of the 9rR-LTVSPWY/pDNA complex in 5–8F cells

A MTT assay was conducted to measure the cytotoxicity of the 9rR-LTVSPWY/pDNA complex in 5–8F cells. Our result (Figure S4) illustrated that the viability of the cells incubated with 9rR-LTVSPWY/pDNA complexes was approximately 93%, suggesting that the cytotoxicity of the 9rR-LTVSPWY/pDNA complex is low.

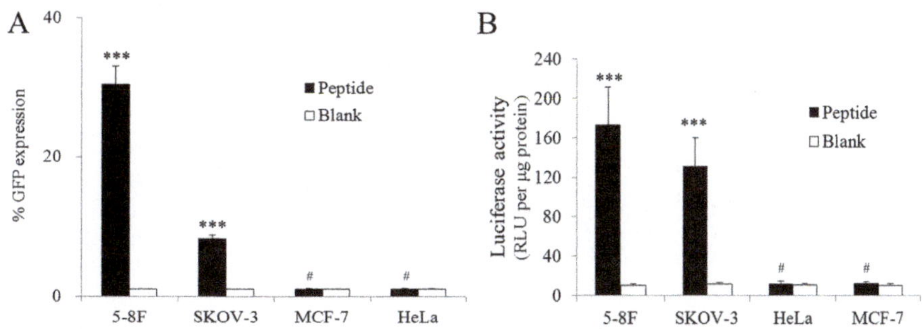

Figure 4. Comparison of the cellular uptakes of 9rR-LTVSPWY in different cells. (A) Percentage of 5-8F, SKOV-3, MCF-7 and HeLa cells expressing GFP at 48 h after transfection with the 9rR-LTVSPWY/pEGFP-N1 complexes (N/P 6:1, black bars) or a vehicle control (blank, white bars). (B) Quantitative evaluation of luciferase expression in 5-8F, SKOV-3, MCF-7 and HeLa cells at 48 h after transfection with the 9rR-LTVSPWY/pGL3 complex (N/P 6:1, black bars) or a vehicle control (blank, white bars). The data are shown as the means ± SD. All measurements were performed in triplicate. The "***" symbol indicates a significant difference between the two groups based on three independent experiments ($p<0.001$). The "#" symbol indicates no significant difference between the two groups based on three independent experiments ($p>0.05$).

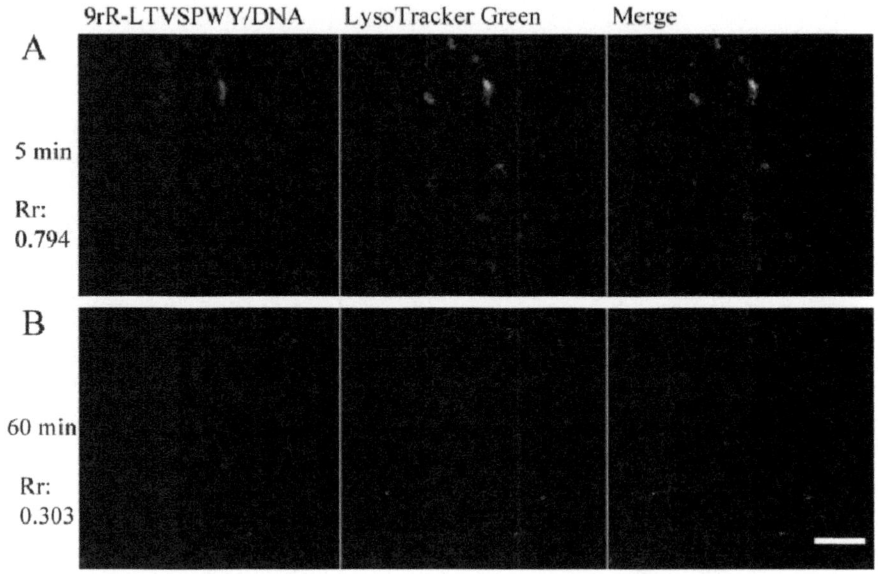

Figure 5. Confocal microscopy analysis of the intercellular distributions of 9rR-LTVSPWY/plasmid complexes. Living 5-8F cells were incubated with the 9rR-LTVSPWY/EMA-labeled pEGFP-N1 complex (N/P 6:1) and LysoTracker Green (100 nM) for 30 min at 37°C and post-incubation in RPMI1640 containing 10% FBS for (A) 5 min and (B) 60 min. Rr: Pearson's coefficient, Red: EMA-labeled pDNA, green: LysoTracker Green, yellow: merge. (Scale bar = 10 μm).

Discussion

The development of tumor-targeted DNA therapeutics is still hindered by the lack of efficient, selective and safe DNA delivery systems. Although viral vectors have high transfection efficiencies, they present serious safety issues in therapeutic applications. Because of this concern, non-viral vectors have been developed as safer alternatives for clinical applications. Of the non-viral vectors, cationic lipids or polymers are frequently used due to several advantages, such as their low immunogenicity and ease of use. However, their toxicity and transfection efficiency problems still limit the tumor-targeting DNA delivery applications.

Compared with lipid- or polymer-based DNA delivery vectors, peptide-based DNA delivery systems are more biocompatible and biodegradable and therefore have lower immunogenicity and lower toxicity. Moreover, peptides are easier to design and produce because solid-phase peptide synthesis has been extensively developed. Thus, the development of peptides that can deliver DNA specifically to cancer cells is very relevant for cancer gene therapy.

In this study, we tested a peptide-based selective DNA delivery system in specific tumor cells. We demonstrated that at N/P ratios ≥ 3, the 9rR-LTVSPWY peptide efficiently condensed extended pDNA (~195 nm in diameter), leading to the formation of compacted nanoparticles of ~75 nm in diameter with a positive charge (~ 21 mV) (Figure 3). Compared to the particle size of polymer/pDNA (> 100 nm) or liposome/pDNA complexes (> 100 nm) [28], the ~75 nm diameter of the 9rR-LTVSPWY/pDNA is much smaller, which may lend the particle greater efficiency and penetration into tumor tissue [29]. This size is also large enough to avoid rapid renal clearance [30]. Certain types of nanoparticles work less effectively due to aggregation in the blood. Aggregation of 9rR-LTVSPWY/pDNA complexes is avoided by their positive charge, which exerts an electrostatic repulsion force [31,32].

LTVSPWY can bind specifically to certain tumor cells and be absorbed via receptor-mediated endocytosis. Our results showed that 9rR-LTVSPWY successfully delivered pDNA into these tumor cells. The peptide/pDNA stoichiometry (N/P ratio) was optimized to obtain the highest transfection efficiency, which was then achieved with an N/P ratio of 6 (Figure 3). Quantitative analysis based on FACS and luciferase expression revealed that 9rR-LTVSPWY can deliver DNA into 5–8F cells and SKOV-3 cells but not into receptor-negative MCF-7 cells or HeLa cells (Figures 4), suggesting that 9rR-LTVSPWY successfully mediates cell-specific DNA delivery. By altering the receptor sequence, more peptide-based DNA delivery systems can be developed for various types of cancers.

The results of our experiments studying the cellular uptake and intracellular distribution of the peptide/pDNA complexes (Figure 5 and 6) suggested that the LTVSPWY-9rR/pDNA complex enters the cells via endocytosis, like LTVSPWY alone. The findings that 5–8F cells could not take up the 9rR/pDNA complex and that 9rR-LTVSPWY could not deliver DNA into receptor-negative MCF-7 cells or HeLa cells (Figure 3, 4, and S3) imply that the cellular uptake of the 9rR-LTVSPWY/pDNA complex is mediated by an LTVSPWY receptor. The significant intracellular uptake of the 9rR-LTVSPWY/pDNA complex observed within 30 min of treatment shows that this endocytosis process was fairly rapid, most likely because the electrostatic interactions between the positive charge (~21 mV) of the complex and the negatively charged cell membrane enhanced the effective local concentration of the complex at the cell surface, increasing the efficiency of cellular uptake [33]. Release from endocytic vesicles is one of the prerequisites for efficient DNA therapy [26]. The pDNA/9rR-LTVSPWY complex escaped from the endocytic vesicles quickly (within 1.5 h). This effect was most likely due to proton scavenging effects of the histidine residues in the 9rR-LTVSPWY sequence, which can disrupt endocytic vesicles [10], and to fusion effects of the arginine residues (9rR), which can enhance endosomal escape through liposomal fusion to the endosomal membrane at both acidic and neutral pH [26,34]. These results may aid in the designing of a more efficient DNA delivery system.

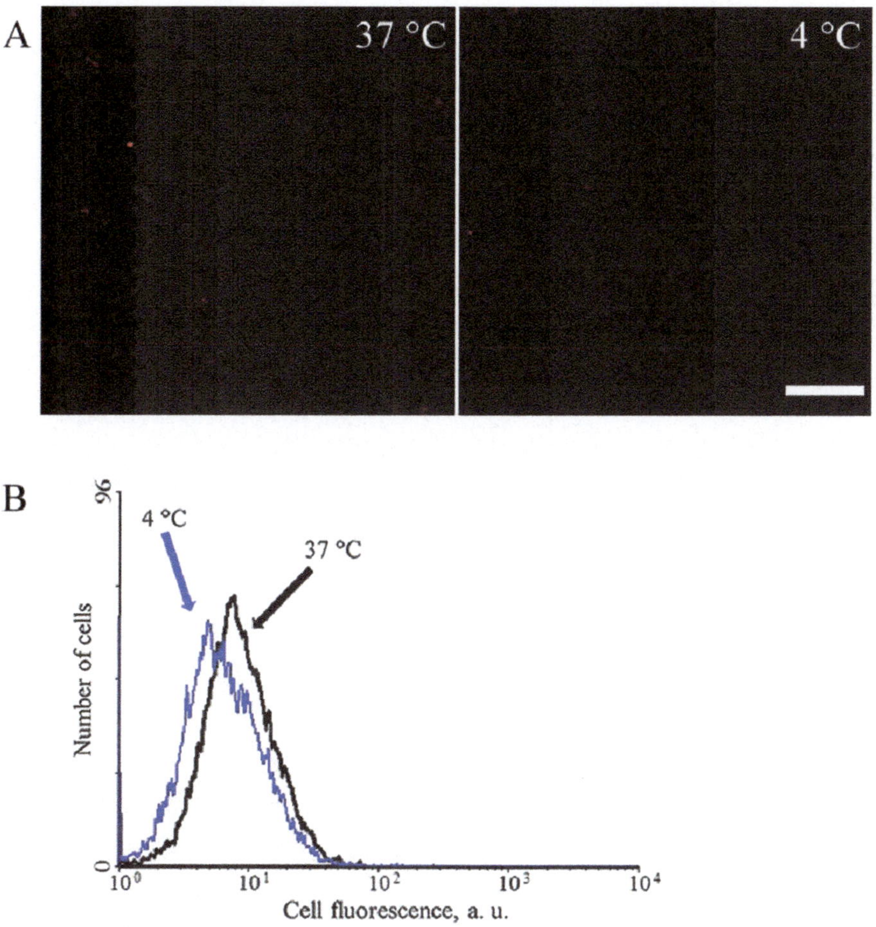

Figure 6. Comparison of the cellular uptakes of 9rR-LTVSPWY/plasmid complexes at different temperatures. Living 5-8F cells were incubated with the 9rR-LTVSPWY/EMA-labeled pEGFP-N1 complex (N/P 6:1) for 30 min at 37°C or 4°C and then examined by (A) Confocal microscopy or (B) flow cytometry analyses. (Scale bar = 10 μm).

In summary, a non-viral 9rR-LTVSPWY peptide-based DNA delivery system was constructed that can mediate efficient DNA expression in only selected tumor cells. The advantages of this system include its ease of synthesis, low immunogenicity, biocompatibility, low cytotoxicity and selectivity. The results suggest a strategy for tumor cell-specific DNA delivery and provide a reference for designing more efficient DNA delivery systems targeted towards various types of cancers. However, for *in vivo* applications, the systemic toxicity, the *in vivo* dose, and the *in vivo* transfection efficiency of 9rR-LTVSPWY DNA delivery system need further careful evaluated. *In vivo* studies on our system are currently underway.

Supporting Information

Figure S1 HPLC analysis of 9rR-LTVSPWY.

Figure S2 ESI-MS analysis of 9rR-LTVSPWY.

Figure S3 Confocal microscopy images of 5–8F cells after incubation with the 9rR/EMA-labeled pDNA complex (N/P 6:1) for 30 min at 37°C. (Scale bar = 10 μm).

Figure S4 Cell viability of 5-8F cells at 48 h after transfection with the 9rR-LTVSPWY/pEGFP-N1 complex (N/P 6:1). The data are shown as the means ± SD.

Acknowledgments

We thank Prof. Yi-Xin Zeng and Prof. Mu-Sheng Zeng (Sun Yat-sen University Cancer Center, Guangzhou, China) for providing the 5–8F cell line. We also thank the Analytical and Testing Center (Huazhong University of Science and Technology) for spectral measurements.

Author Contributions

Conceived and designed the experiments: YZ . Performed the experiments: CG DP FQ PS. Analyzed the data: CG DP FQ PS YZ. Contributed to the writing of the manuscript: YZ CG.

References

1. Münch RC, Janicki H, Völker I, Rasbach A, Hallek M, et al. (2013) Displaying High-affinity Ligands on Adeno-associated Viral Vectors Enables Tumor Cell-specific and Safe Gene Transfer. Mol Ther 21: 109–118.
2. Naldini L (2001) Viral vectors for gene therapy: the art of turning infectious agents into vehicles of therapeutics. Nat Med 7: 33–39.

3. Huang S, Kamihira M (2012) Development of hybrid viral vectors for gene therapy. Biotechnol Adv 31: 208–223.
4. He H, Kong SK, Chan KT (2011) Transfection and cell fusion by femtosecond lasers. J Innov Opt Health Sci 04: 113–125.
5. Lehrman S (1999) Virus treatment questioned after gene therapy death. Nature 401: 517–518.
6. Kim PH, Kim T, Yockman JW, Kim SW, Yun CO (2010) The effect of surface modification of adenovirus with an arginine-grafted bioreducible polymer on transduction efficiency and immunogenicity in cancer gene therapy. Biomaterials 31: 1865–1874.
7. Lv H, Zhang S, Wang B, Cui S, Yan J (2006) Toxicity of cationic lipids and cationic polymers in gene delivery. J Control Release 114: 100–109.
8. Zhang Y, Satterlee A, Huang L (2012) In Vivo Gene Delivery by Nonviral Vectors: Overcoming Hurdles&quest. Mol Ther 20: 1298–1304.
9. Martin ME, Rice KG (2007) Peptide-guided gene delivery. AAPS J 9: 18–29.
10. Seow WY, Yang YY, George AJT (2009) Oligopeptide-mediated gene transfer into mouse corneal endothelial cells: expression, design optimization, uptake mechanism and nuclear localization. Nucleic Acids Res 37: 6276–6289.
11. Sadatmousavi P, Soltani M, Nazarian R, Jafari M, Chen P (2011) Self-assembling peptides: potential role in tumor targeting. Curr Pharm Biotechnol 12: 1089–1100.
12. Standley SM, Toft DJ, Cheng H, Soukasene S, Chen J, et al. (2010) Induction of cancer cell death by self-assembling nanostructures incorporating a cytotoxic peptide. Cancer Res 70: 3020–3026.
13. Ruoslahti E (2012) Peptides as targeting elements and tissue penetration devices for nanoparticles. Adv Mater 24: 3747–3756.
14. Fonseca SB, Pereira MP, Kelley SO (2009) Recent advances in the use of cell-penetrating peptides for medical and biological applications. Adv Drug Deliver Rev 61: 953–964.
15. Nakase I, Akita H, Kogure K, Graslund A, Langel Ul, et al. (2012) Efficient intracellular delivery of nucleic acid pharmaceuticals using cell-penetrating peptides. Account Chem Res 45: 1132–1139.
16. Wender PA, Galliher WC, Goun EA, Jones LR, Pillow TH (2008) The design of guanidinium-rich transporters and their internalization mechanisms. Adv Drug Deliver Rev 60: 452–472.
17. Wang HY, Chen JX, Sun YX, Deng JZ, Li C, et al. (2010) Construction of cell penetrating peptide vectors with N-terminal stearylated nuclear localization signal for targeted delivery of DNA into the cell nuclei. J Control Release 155: 26–33.
18. Khalil I, Futaki S, Niwa M, Baba Y, Kaji N, et al. (2004) Mechanism of improved gene transfer by the N-terminal stearylation of octaarginine: enhanced cellular association by hydrophobic core formation. Gene Ther 11: 636–644.
19. Seow WY, Yang YY (2009) A Class of Cationic Triblock Amphiphilic Oligopeptides as Efficient Gene-Delivery Vectors. Adv Mater 21: 86–90.
20. Liu L, Zheng M, Renette T, Kissel T (2012) Modular Synthesis of Folate Conjugated Ternary Copolymers: Polyethylenimine-graft-Polycaprolactone-block-Poly (ethylene glycol)-Folate for Targeted Gene Delivery. Bioconjugate Chem 23: 1211–1220.
21. Minko T (2012) Receptor Mediated Delivery Systems for Cancer Therapeutics. Fundamentals and Applications of Controlled Release Drug Delivery. US: Springer 329–355 p.
22. Shadidi M, Sioud M (2003) Identification of novel carrier peptides for the specific delivery of therapeutics into cancer cells. FASEB J 17: 256–258.
23. Luo H, Yang J, Jin H, Huang C, Fu J, et al. (2011) Tetrameric far-red fluorescent protein as a scaffold to assemble an octavalent peptide nanoprobe for enhanced tumor targeting and intracellular uptake in vivo. FASEB J 25: 1865–1873.
24. Gong C, Li X, Xu L, Zhang YH (2012) Target delivery of a gene into the brain using the RVG29-oligoarginine peptide. Biomaterials 33: 3456–3463.
25. Song L, Yan J, Jian S, Zhang L, Li M, et al. (2002) Molecular mechanisms of tumorgenesis and metastasis in nasopharyngeal carcinoma cell sublines. Ai Zheng 21: 158–162.
26. Ma Y, Gong C, Ma Y, Fan F, Luo M, et al. (2012) Direct Cytosolic Delivery of Cargoes in vivo by a Chimera Consisting of D-and L-Arginine Residues. J Control Release 162: 286–294.
27. Wang XF, Birringer M, Dong LF, Veprek P, Low P, et al. (2007) A peptide conjugate of vitamin E succinate targets breast cancer cells with high ErbB2 expression. Cancer Res 67: 3337–3344.
28. Won YW, Kim HA, Lee M, Kim YH (2009) Reducible poly (oligo-D-arginine) for enhanced gene expression in mouse lung by intratracheal injection. Mol Ther 18: 734–742.
29. Jiang W, Kim BYS, Rutka JT, Chan WCW (2008) Nanoparticle-mediated cellular response is size-dependent. Nat Nanotechnol 3: 145–150.
30. Choi HS, Liu W, Misra P, Tanaka E, Zimmer JP, et al. (2007) Renal clearance of quantum dots. Nat Biotechnol 25: 1165–1170.
31. Wang S, Song H, Ong WY, Han MY, Huang D (2009) Positively charged and pH self-buffering quantum dots for efficient cellular uptake by charge mediation and monitoring cell membrane permeability. Nanotechnology 20: 425102–425109.
32. Chi EY, Krishnan S, Randolph TW, Carpenter JF (2003) Physical stability of proteins in aqueous solution: mechanism and driving forces in nonnative protein aggregation. Pharm Res 20: 1325–1336.
33. Henriques ST, Melo MN, Castanho MARB (2006) Cell-penetrating peptides and antimicrobial peptides: how different are they? BioChem J 399: 1–7.
34. El-Sayed A, Khalil IA, Kogure K, Futaki S, Harashima H (2008) Octaarginine- and octalysine-modified nanoparticles have different modes of endosomal escape. J Biol Chem 283: 23450–23461.

18

Human Adipose-Derived Mesenchymal Stem Cells as a New Model of Spinal and Bulbar Muscular Atrophy

Marta Dossena[1☯], Gloria Bedini[1☯], Paola Rusmini[2], Elisa Giorgetti[2,3], Alessandra Canazza[1], Valentina Tosetti[1], Ettore Salsano[4], Anna Sagnelli[4], Caterina Mariotti[5], Cinzia Gellera[5], Stefania Elena Navone[1], Giovanni Marfia[1], Giulio Alessandri[1], Fabio Corsi[6], Eugenio Agostino Parati[1], Davide Pareyson[¶4*], Angelo Poletti[¶2*]

1 Cellular Neurobiology Laboratory, Unit of Cerebrovascular Disease, Fondazione IRCCS Istituto Neurologico Carlo Besta, Milan, Italy, **2** Dipartimento di Scienze Farmacologiche e Biomolecolari, Centro Interdipartimentale sulle Malattie Neurodegenerative, Università degli Studi di Milano, Milan, Italy, **3** Department of Pathology, University of Michigan, Ann Arbor, Michigan, 48109, United States of America, **4** Clinic of Central and Peripheral Degenerative Neuropathies Unit, Department of Clinical Neurosciences, Fondazione IRCCS Istituto Neurologico Carlo Besta, Milan, Italy, **5** Unit of Genetics of Neurodegenerative and Metabolic Diseases, Department of Diagnostic and Applied Technology, Fondazione IRCCS Istituto Neurologico Carlo Besta, Milan, Italy, **6** Surgery Division, Department of Clinical Sciences, University of Milan, "Luigi Sacco" Hospital, Milan, Italy

Abstract

Spinal and bulbar muscular atrophy (SBMA) or Kennedy's disease is an X-linked CAG/polyglutamine expansion motoneuron disease, in which an elongated polyglutamine tract (polyQ) in the N-terminal androgen receptor (ARpolyQ) confers toxicity to this protein. Typical markers of SBMA disease are ARpolyQ intranuclear inclusions. These are generated after the ARpolyQ binds to its endogenous ligands, which promotes AR release from chaperones, activation and nuclear translocation, but also cell toxicity. The SBMA mouse models developed so far, and used in preclinical studies, all contain an expanded CAG repeat significantly longer than that of SBMA patients. Here, we propose the use of SBMA patients adipose-derived mesenchymal stem cells (MSCs) as a new human *in vitro* model to study ARpolyQ toxicity. These cells have the advantage to express only ARpolyQ, and not the wild type AR allele. Therefore, we isolated and characterized adipose-derived MSCs from three SBMA patients (ADSC from Kennedy's patients, ADSCK) and three control volunteers (ADSCs). We found that both ADSCs and ADSCKs express mesenchymal antigens, even if only ADSCs can differentiate into the three typical cell lineages (adipocytes, chondrocytes and osteocytes), whereas ADSCKs, from SBMA patients, showed a lower growth potential and differentiated only into adipocyte. Moreover, analysing AR expression on our mesenchymal cultures we found lower levels in all ADSCKs than ADSCs, possibly related to negative pressures exerted by toxic ARpolyQ in ADSCKs. In addition, with proteasome inhibition the ARpolyQ levels increased specifically in ADSCKs, inducing the formation of HSP70 and ubiquitin positive nuclear ARpolyQ inclusions. Considering all of this evidence, SBMA patients adipose-derived MSCs cultures should be considered an innovative *in vitro* human model to understand the molecular mechanisms of ARpolyQ toxicity and to test novel therapeutic approaches in SBMA.

Editor: Udai Pandey, Children's Hospital of Pittsburgh, University of Pittsburgh Medical Center, United States of America

Funding: This work was funded by Regione Lombardia (to D.P., A.P., E.A.P.); AriSLA Foundation Italy (ALS_HSPB8 to A.P), Telethon - Italy (GGP14039 to A.P.); Italian Ministry of Labour, Health and Social Affairs (Convenzione Fondazione Mondino/UNIMI to A.P.); Università degli Studi di Milano (to A.P.); Fondazione CARIPLO (2008-2307 to A.P.); Fondation Thierry Latran, France (to A.P.), Association Française contre les Myopathies (to A.P.) The funders had no role in study design, data collection and analysis, decision to publish, or preparation of the manuscript.

Competing Interests: The authors have read the journal's policy and the authors of this manuscript have the following competing interests: DP received research grants to his Institution from Telethon Italy, Regione Lombardia, ACMTRete (patients' association), CMTA (patients' association); DP received travel grants from Pfizer Italy and Kedrion. AP received research grants to his Institution by: Regione Lombardia, AriSLA Foundation Italy, Telethon - Italy, Fondazione CARIPLO, Fondation Thierry Latran France, Association Française contre les Myopathies.

* Email: davide.pareyson@istituto-besta.it (DP); angelo.poletti@unimi.it (AP)

☯ These authors contributed equally to this work.

¶ These authors also contributed equally to this work.

Introduction

Spinal and bulbar muscular atrophy (SBMA) or Kennedy's disease, an X-linked disorder affecting adult males, is characterized by wasting and weakness of facial, bulbar and limb muscles associated with motoneuron degeneration in brainstem and spinal cord. Mild sensory signs occur related to abnormalities of dorsal root ganglia neurons [1]. Muscle atrophy results from both denervation and direct involvement of muscle cells [2]. Signs of androgen insensitivity (gynecomastia, hypogonadism, and reduced fertility) can be also observed. No treatment or cure for SBMA is available.

SBMA is linked to a CAG repeat expansion in the androgen receptor (AR) gene, which is translated into an elongated polyglutamine tract (polyQ) in the AR protein (ARpolyQ) [3]. The ARpolyQ alters AR behaviour, conferring neurotoxicity responsible for motoneuron death [3–5]. In fact, the polyQ induces AR misfolding and its aggregation into cytoplasmic and nuclear inclusions. This is triggered by testosterone and dihydrotestosterone, which activate AR [6–8] inducing the AR nuclear neurotoxicity [9,10].

Different SBMA mouse models have been developed and used in preclinical studies until now, which demonstrated the prominent role of androgens in symptoms appearance, disease progression and death. These mice have been generated using a CAG repeat of a size markedly higher than that found in the human disease [10–16]. In addition, in most mouse models the AR transgene expression is driven by constitutive promoters (such as actin or prion promoters), with the only exception of a knock-in SBMA mouse model, in which ARpolyQ expression is driven by endogenous promoter to maintain normal AR synthesis and localization. Alternative SBMA mice models have been developed using a human AR promoter by using either YAC or BAC constructs to insert the entire human AR gene. Despite of being under the control of an "exogenous" promoter, and the possible differences in transcriptional regulation between species, these mice should also mimic the tissue distribution of the AR protein found in human [12,13,17]. However, the use of longer AR CAG repeats dramatically accelerates the disease phenotype in these SBMA animal models, which instead is normally characterized by a very slow progression rate in patients. This aspect has not been taken into account in all murine models [18]. Therefore, it is important to develop a new model closer to human pathological condition to test innovative drug treatments designed to reduce cytotoxic aggregates.

Induced pluripotent stem cells (iPSCs) have been recently developed from SBMA patients. Their relevant value is to be cells of human origin that can be successfully differentiated toward a motoneuronal phenotype, to produce reliable cell models that mimic disease in this particular cell type affected in SBMA [19]. However, muscle tissue is another target of ARpolyQ toxicity, and, to the best of our knowledge, all attempts to generate muscle cells from iPSCs failed so far. In addition, iPSCs are produced by genetic transformation of fibroblasts, using four oncogenic or differentiating agents that may impact on cell behavior. Thus, other cell models of human origin may be of value to complement the data obtained in iPSCs.

Mesenchymal stem cells (MSCs), originally identified in bone marrow stroma, can be isolated from different tissues (e.g.: umbilical cord blood, adipose tissue), expanded and differentiated *ex vivo* into multiple cell types [20]. Moreover, compared to iPSCs, MSCs are not retro-induced with genes involved in oncogenic cell transformation.

Adipose tissue is an abundant, accessible source of adipose-derived MSCs (ADSCs) [20], which contains a population of mesenchymal stem cells with no tumorigenic or telomerase activities [21,22], with marked neuro-immunomodulatory properties and the capability to migrate to sites of injury, thereby conferring them as a possible contributor in tissue repair. AR is highly expressed in adipose tissue [21,23], where androgens modulate ADSCs commitment to pre-adipocytes [21,24]. Moreover, ADSCs from SBMA patients present the advantage that they express only the ARpolyQ, and not the wild type allele. Hence, we evaluated the potential of ADSCs to be used as novel human SBMA *in vitro* models to better understand ARpolyQ-toxicity with the overall goal of finding novel future therapeutic approaches.

We found that AR is highly expressed in normal ADSCs while it was reduced in SBMA ADSCs. Testosterone induced AR nuclear translocation and, in a limited SBMA cellular population, AR nuclear inclusions after proteasome inhibition. Therefore, ADSCs could be considered as an innovative SBMA human model useful for clarifying molecular mechanisms underlying SBMA pathophysiology.

Materials and Methods

Cell isolation and culture

The study was approved by the local institutional review board of the Fondazione IRCCS Istituto Neurologico "C. Besta" (Milan, Italy). Informed written consent was obtained from all volunteers and SBMA patients. The study conformed with the 2013 WMA Declaration of Helsinki.

Specimens of fat from periumbilical regions of three male controls undergoing surgery for ventriculoperitoneal shunt (ADSC samples) (aged 55, 69, 73 years; specimen CAG repeat number = 22, 23, 24) and three SBMA patients (ADSC Kennedy, ADSCK, samples) (aged 49, 57, 76 years; leukocyte CAG repeat number = 46, 44, 44), were mechanically dissociated, washed in PBS 1X (EuroClone, Milan, Italy) and centrifuged at 1300×g for 10 min; the upper phase was plated into T75-cm^2 flasks, allowed to dry and then Stem Cells Medium was added [SCM: DMEM-F12 with 10% Fetal Bovine Serum (Gibco, Grand Island, NY, USA), 1% penicillin/streptomycin solution (Sigma-Aldrich, Basel, Switzerland)] [25]. Cells were seeded in T75-cm^2 flasks at 1×10^4 cells/cm^2, and passed weekly, for expansion or freezing procedures. Freezing was performed in FBS with 10% of dimethyl sulfoxide (DMSO) (–80°C freezer for 24 hours, then stored in liquid nitrogen). After de-freezing by quickly thawing at 37°C, cells were plated in T75-cm^2 flasks with SCM for 24 hours, and grown in fresh SCM. For the experiments, cells were used before passage nine. Cell viability was assessed by Trypan Blue dye exclusion assay (EuroClone).

Growth curve

Cell growth was analysed by direct cell counts and by calculation of cumulative population doublings at each passage (three to nine) with the formula:

$\log_{10}$ (harvested cells/seeded cells)/$\log_{10}$ (2) [26].

Cells (2×10^5) were seeded in a T25-cm^2 flask with 3.5 ml of SCM. After 4 days of culture, cells were harvested, counted and re-seeded for next passage growth.

Flow cytometry analysis

Flow cytometry (FC) was performed to evaluate the mesenchymal phenotype: CD105 (AbDSerotec, Raleigh, NC, USA), CD90 (Millipore Temecula, CA, USA), CD73, CD14, CD19, CD31, CD34, CD45, HLA-DR (BD Pharmingen, San Jose, CA, USA). Briefly, 10^5 cells/tube were stained with fluorochrome conjugated monoclonal antibodies for 30 min at 4°C in the dark. After centrifugation at 1300×g for 10 min and a PBS wash, cells were fixed with 4% paraformaldehyde (Sigma Aldrich). Fluorescence-activated cell sorting was performed with Cell Quest software (BD Pharmingen). Non-viable cells were excluded according to the side scatter vs. forward scatter parameters, and 5,000 events were acquired for each sample.

Multipotency characterization of ADSCs

ADSCs and ADSCKs were tested for their ability to differentiate into adipocytes, chondrocytes and osteocytes using Human Mesenchymal Stem Cell Functional Identification Kit (R&D

Systems), according to the manufacturer instructions. Adipogenic differentiation was performed starting from 2.1×10^4 cells/cm^2 in SCM up to confluence, when SCM was replaced with Adipogenic Differentiation Medium (and changed every 3–4 days). After 21 days, cells were fixed with 4% paraformaldehyde for 1 hour and visualized by Oil Red O Staining (Sigma Aldrich).

Chondrogenic differentiation was performed with 2.5×10^5 cells seeded in 15 ml conical tubes in Chondrogenic Differentiation Medium (replaced every 2–3 days); after 21 days, chondrocyte pellet was fixed with 4% paraformaldehyde for 20 min and immunostained for aggrecan (R&D System).

Osteogenic differentiation was done starting form 4.2×10^3 cells/cm^2 in SCM up to 70–80% confluence, when SCM was replaced with Osteogenic Differentiation Medium (changed every 3–4 days). After 21 days, cells were fixed 10 min with 70% ethanol and processed for Alizarian Red Staining (Sigma Aldrich). Images were obtained at 20X magnification, using Nikon Eclipse TE300 equipped with the Axiovision device camera (Zeiss Instr., Oberkochen, Germany). Images were processed using Axiovision release 4.6.3 (Zeiss Instr., Oberkochen, Germany).

Immunofluorescence analysis

Cells (4000 cell/mL) plated on glass coverslips in 24-well multiwell plates in SCM with 10% charcoal stripped-FBS (CS-FBS; basal condition to eliminate endogenous steroids) were grown to confluence in absence (–T) or presence (+T) of 10 nM of testosterone (T) for 48 hours, with or without a proteasome inhibitor (MG132, 10 μM; Sigma-Aldrich, St Louis, MO, USA) for 24 hours. Cells were fixed using 4% paraformaldehyde and AR detected using an anti-AR rabbit antibody (D6F11, Cell Signaling Technology, Inc., Danvers, MA, USA) 1:200 in milk followed by Alexa 488 anti-rabbit (Molecular Probes), 1:1,000 in milk. Double immunofluorescence (IF) analyses were done with H280 anti-AR (H280) (Santa Cruz Biotech, SantaCruz, CA, USA) 1:100 in milk with a) mouse monoclonal Hsp70 (sc-24, Santa Cruz Biotech) or b) mouse monoclonal anti-Ubiquitin (sc-8017, Santa Cruz Biotech) 1:100 in milk, followed by Alexa 488 anti-rabbit (to visualize AR) or Alexa 594 anti-mouse (Molecular Probes, to visualize Hsp70 or Ubiquitin) (1:1,000 in milk).

Nuclei were visualized with DAPI. Images were obtained at 63X magnification, using an Axiovert 200 microscope (Zeiss Instr., Oberkochen, Germany) with Photometric Cool-Snap CCD camera (Ropper Scientific, Trenton, NJ, USA). Images were processed using Metamorph software version 7.7.7.0 (Universal Imaging, Downingtown, PA, USA).

Western blot analysis

Western blot analysis (WB) was performed as previously described [27]. Cells were grown in SCM with 10% CS-FBS in the absence (–T) or in the presence (+T) of 10 nM of testosterone (T) for 48 hours, then harvested, centrifuged 10 min at 1300×g. Pellets were resuspended in 150 μl RIPA buffer containing protease inhibitors cocktail (Sigma-Aldrich) homogenized using slight sonication and total protein concentration determined with bicinchoninic acid kit (BCA assay, Thermo Scientific Pierce, IL, USA). WB was performed using 10% SDS-PAGE with 15 μg total proteins. Electrotransfer on nitrocellulose membrane was done with Transblot Turbo Transfer System (Bio-Rad). Membranes were treated with 5% nonfat dried milk powder (Euroclone, Italy) in Tween-TBS (TBS-T, 20 mM TrisHCl, pH 7.5, 0.5 M NaCl, 0.05% Tween-20) for 1 hour and incubated with the following primary antibodies: (a) rabbit polyclonal AR-H280 (Santa Cruz, 1:1,000) to detect wtAR and ARpolyQ; (b) mouse monoclonal anti-α-tubulin (Sigma Aldrich, 1:3,000). The following secondary peroxidase-conjugated antibodies were used: goat anti-rabbit to identify the anti-AR (sc-2004, Santa Cruz, dilution 1:5,000) and goat anti-mouse to identify the anti-α-tubulin (sc-2005, Santa Cruz, dilution 1:5,000). The immunoreactivity was visualized with enhanced chemiluminescence detection kit (Amersham ECL Prime Western Blotting Detection Reagent). WB images were obtained with ChemiDoc XRS System (Bio-Rad).

mRNA expression analysis

For real-time PCR (RT-qPCR), cells were plated into T75-cm^2 flasks in SMC+10% FBS. Total RNA was isolated with TRIreagent protocol (Sigma-Aldrich) as previously described [27]. Total RNA (1 μg), treated with DNAse, was reverse-transcribed into cDNA using High-Capacity cDNA Archive Kit (Applied Biosystems).

RT-qPCR for human AR and GAPDH mRNAs were designed using the program Primer Express 3 and synthesized by MWGBiotech (Ebersberg, Germany) with the following sequences: hARforward: 5′-ATCCCAGTCCCACTTGTGTC-3′; hARreverse: 5′-GGTCTTCTGGGGTGGAAAGT-3′; hGAPDHforward: 5′-GAAGGTGAAGGTCGGAGTC-3′, hGAPDHreverse: 5′-TTGATGGCAACAATATCCACTT-3′. Primer efficiencies was close to 100% for both target and reference gene. RT-qPCR was performed using CFX 96 Real Time System (Bio-Rad) in a 10 μl total volume with iTaq SYBR Green Supermix (Bio-Rad), and 500 nmol primers, in the following conditions: 94°C for 10 min, 35 cycles at 94°C for 15 s and 60°C for 1 min. Melting curve analysis was always performed at the end of each PCR assay to control specificity. Data were expressed as Ct values and used for relative quantification of targets with ΔΔCt calculation. Potential bias, due to averaging data transformed through the equation $2^{-\Delta\Delta Ct}$ to give N-fold changes in gene expression, were excluded performing all statistics with ΔCt values, and hAR values normalized with hGAPDH values.

Statistical analysis

Statistical analysis was performed using one-way analysis of variance (ANOVA) for group comparisons followed by Bonferroni *post hoc* test, using PRISM software (GraphPad, San Diego, CA, USA). Data were expressed as mean±SD or mean±SEM of three independent samples. $P<0.05$ was considered statistically significant.

Results

ADSCs and ADSCKs characterization

ADSCs derived from fat specimens of the periumbilical regions of controls (n = 3, ADSCs) and SBMA patients (n = 3, ADSCKs) were characterized by evaluating their growth in adhesion and by direct cell count resulting in the cumulative population doubling [26]. The analyses (Figure 1A) show that the two cell populations have different growth rates since, at passage 3, the mean cumulative population doubling value was 1.74±0.41 for ADSCs and 0.5±0.6 for ADSCKs, confirmed also at advanced passages. Moreover, after thawing the viability measured with Trypan Blue exclusion dye assay (Figure 1B) was 79.43±6.12% for ADSCs and only 58.85±4.30% for ADSCKs.

Immunophenotypic analysis, performed on cells collected at passage 3 (Figure 1C), showed that both cell cultures (n = 3, ADSCs; n = 3, ADSCKs) express mesenchymal markers with some differences. ADSCs were positive for mesenchymal antigens CD73 (99.6±0.2%), CD90 (98.1±1.4%) and CD105 (99.0±1.2%), negative for lymphocytic markers CD14 (0.5±0.6%) and CD45 (21.0±7.6%); ADSCs were slightly positive

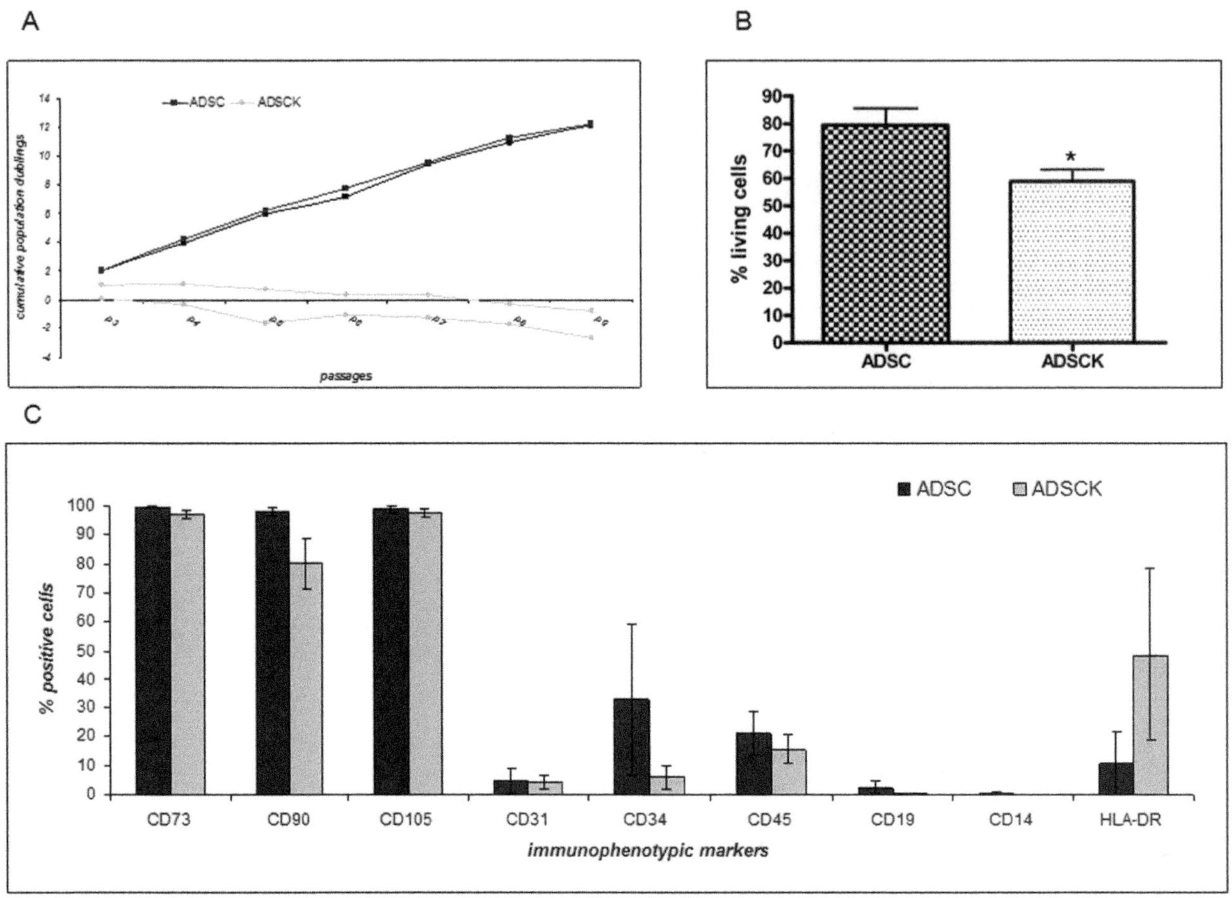

Figure 1. Characterization of ADSC and ADSCK cell cultures. (A) Growth curve of two representative ADSC and ADSCK primary cultures. Values were expressed as cumulative population doublings calculated with the formula reported by Avanzini et al. [26,44]. (B) Cell viability after freezing and thawing evaluated by trypan blue dye exclusion assay comparing the number of viable cells after thawing to the number of cells previously frozen. Data are expressed as mean±SEM (n = 9 per group; three independent experiments for each cell culture; *$P<0.05$ vs ADSCs). (C) FACs analysis performed at passage 3 on ADSC (black bar) and on ADSCK (grey bar) to evaluate immunophenotypically mesenchymal markers. Data are expressed as mean±SD (n = 3 per group).

for early hematopoietic marker CD34 (32.8±26.3%) and negative for CD19 (2.5±2.2%), for endothelial typical marker CD31 (4.8±4.1%), and for immunological marker HLA-DR (10.6±10.9%). ADSCKs were positive for CD73 (97.2±1.4%), CD90 (80.1±8.6%) and CD105 (97.6±1.3%) and negative for CD14 (0.1±0.0%), CD45 (15.6±5.0%), CD34 (6.0±4.1%), CD19 (0.3±0.1%), CD31 (4.2±2.6%); only the immunological marker HLA-DR was slightly higher than expected (48.4±29.9%).

Immunophenotypic characterization performed after repeated passages (passage 8), showed that all the three different cultures of ADSCs preserved mesenchymal markers expression CD73 (99.2±0.5%), CD90 (98.9±1.2%) and CD105 (98.6±2.1%), remaining negative for lymphocytic (CD14 0.4±0.2%; CD45 5.2±3.3%), early hematopoietic (CD34 19.0±13.7%; CD19 2.2±1.91%), endothelial (CD31 3.9±4.0%) and immunological (HLA-DR 9.0±8.7%) markers. The three different cultures of ADSCKs were not analysed because of their reduced viability at high passages (see Figure 1A).

Multipotent ability of ADSCs and ADSCKs

In undifferentiated conditions, ADSCs and ADSCKs were plastic adherent with typical small, spindle-shape morphology (Figure 2A). When induced to differentiate into adipocytes, osteocytes and chondrocytes (passage 4), they lost their typical aspect. In particular, ADSCs differentiated in all three cell lineages, forming lipid-droplets stained with Oil Red O (Figure 2B) typical of adipocytes, calcium deposition stained with Alzarin Red (Figure 2C), usually present in osteoblasts, and expressing aggrecan (Figure 2D) a typical marker specific for chondroblast cell lineage. Conversely, ADSCKs differentiated only into adipocytes (Figure 2F), since cells stained were positive for lipid-droplets (Oil Red O), but negative for calcium deposits (Figure 2G) and aggrecan expression (Figure 2H).

Characterization of AR expression and its effect on ADSCs and ADSCKs growth and survival

ADSCs and ADSCKs both maintain AR expression. CAG repeat number in cells at passage 7 did not change in controls (22, 23, 24) and in one SBMA patient (44), increased by one repeat in another patient (45), whereas it increased by 5 repeats in the third patient (51). In RT-qPCR, performed using primers amplifying human AR (Figure 3A) we found that both cell types contained considerable amounts of AR mRNA. However, ADSCKs expressing mutant ARpolyQ were characterized by lower AR mRNA levels when compared to ADSCs expressing wtAR. The reduction of AR expression was similar in all different ADSCKs

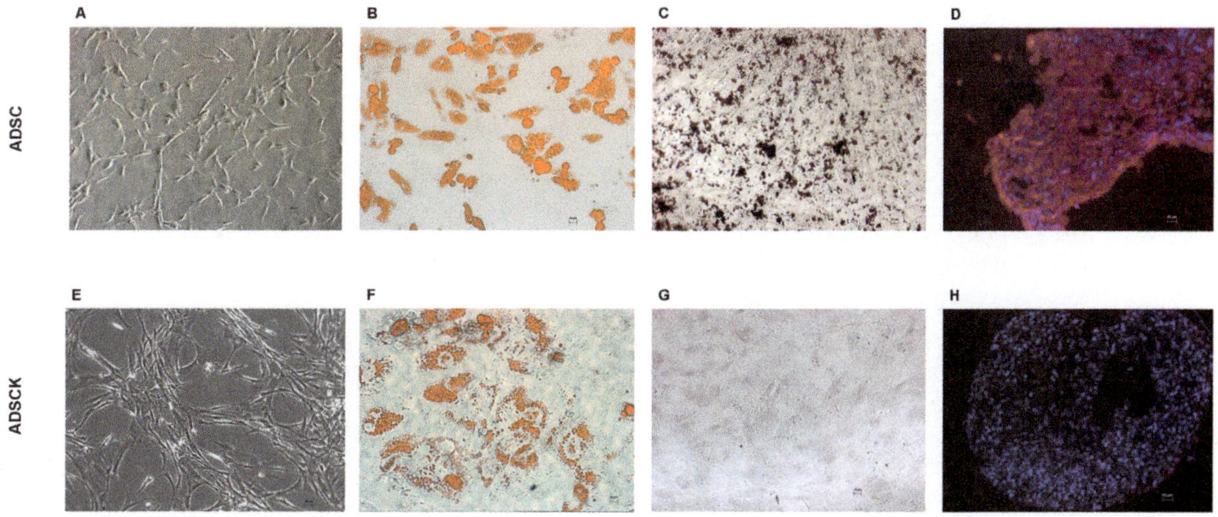

Figure 2. Multipotency characterization of ADSCs and ADSCKs. (A–E) Representative images of normal morphology in absence of differentiation stimuli. (B–F) Adipogenic differentiation was confirmed by Oil Red O Staining. (C–G) Osteogenic differentiation was displayed by Alizarin Red staining. (D–H) Chondrogenic differentiation was visualized by Aggrecan immunofluorescence (red); nuclei were stained with DAPI (blue). (20X magnification).

clones from SBMA patients. This might be due to a negative pressure generated by toxicity of mutant ARpolyQ [28] on patient-derived cells, resulting in clonal selection of cells expressing lower AR levels, thus less sensitive to its toxicity, as also suggested by the decreased ADSCKs growth/survival (Figure 1). WB analysis, performed with anti-AR antibody (Figure 3B), showed that both wtAR and ARpolyQ proteins were correctly translated, being detectable at molecular weights (M.W.) expected for them. As expected, the band of the mutant ARpolyQ had an apparent higher M.W. (slower gel mobility migration) than that of wtAR, because of the presence of the elongated polyQ tract which slightly increases its total mass [3]. In line with our previous observations [28,29,30], testosterone stabilized both wt and SBMA AR, thus increasing their intracellular levels, but no relevant amounts of high M.W. SDS-resistant species were observed in the stacking gel. Thus, the ADSC protein quality control system correctly handles potentially misfolded ARpolyQ species induced by testosterone. Even if WB is a qualitative analysis, the levels of mutant ARpolyQ appeared lower than that of wtAR (as seen with quantitative RT-qPCR analysis).

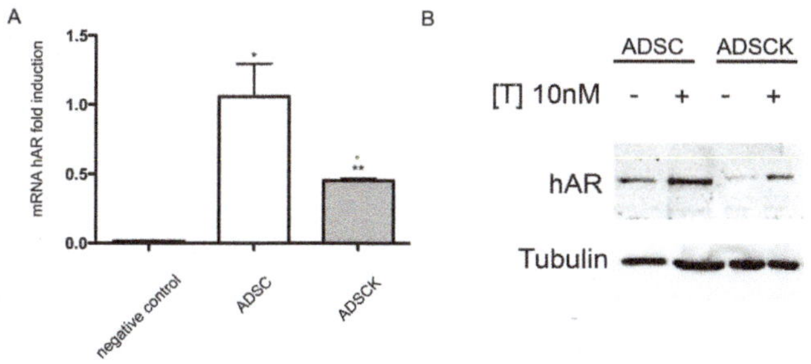

Figure 3. Characterization of AR expression in ADSCs and ADSCKs. (A) AR mRNA levels in ADSCs and ADSCKs were determined by real time quantitative PCR. SHSY-5Y cells were used as negative control. (n = 3 per group; *P<0.05, **P<0.01 *vs* negative control; °P< 0.05 *vs* ADSCs). (B) Western Blot on ADSCs and ADSCKs in absence (–T) or in presence (+T) of 10 nM of testosterone for 48 hours. Alpha-tubulin was used to normalize for protein loading.

Biochemical behaviour of wtAR and SBMA ARpolyQ in ADSCs and ADSCKs

To evaluate whether testosterone normally induces wtAR or ARpolyQ translocation into the nuclei of ADSCs, we performed immunofluorescence analysis using an anti-AR antibody. Figure 4A shows that without testosterone both ARs were cytoplasmic; testosterone treatment (10 nM) for 48 hours induced complete wtAR translocation to the nucleus, while ARpolyQ did not completely translocate to the nucleus, as a small amount remained cytoplasmic. When we evaluated the effect of proteasome blockage on AR clearance, we found that treatment with proteasome inhibitor MG132 induced an overall increase of both wtAR and ARpolyQ levels in cytoplasm and nucleus. Interestingly, we noted that in ADSCKs the co-treatment with testosterone and MG132 induced formation of ARpolyQ nuclear inclusions in a few cells (approximately 5% of total analysed cells); the autophagy inhibitor 3-MA increased ARpolyQ levels, but did not induce intranuclear inclusions (not shown). Thus, the proteasome appears to be one of the preferential pathways to clear the misfolded fraction of ARpolyQ, capable to aggregate, present in ADSCKs. Therefore, we analysed whether MG132-induced ARpolyQ nuclear inclusions directly resulted from inhibition of degradative systems, by analysing possible sequestration of two proteins essential for proteasomal degradation. We found that testosterone- and MG132-induced nuclear inclusions sequestered the chaperone HSP70 (responsible for recognition of misfolded protein species to be degraded by ubiquitin-proteasome pathway) (Figure 4B) and were also positive for ubiquitin (Figure 4C). This suggests that some misfolded species of ARpolyQ present in ADSCKs require a functional proteasome for their clearance, even if an indirect effect can also be postulated. In fact, several stressor stimuli resulting from proteasome blockage (which limits the clearance also of other ubiquitinated proteins) are added to a system, which is possibly almost overwhelmed by the presence of mutant ARpolyQ.

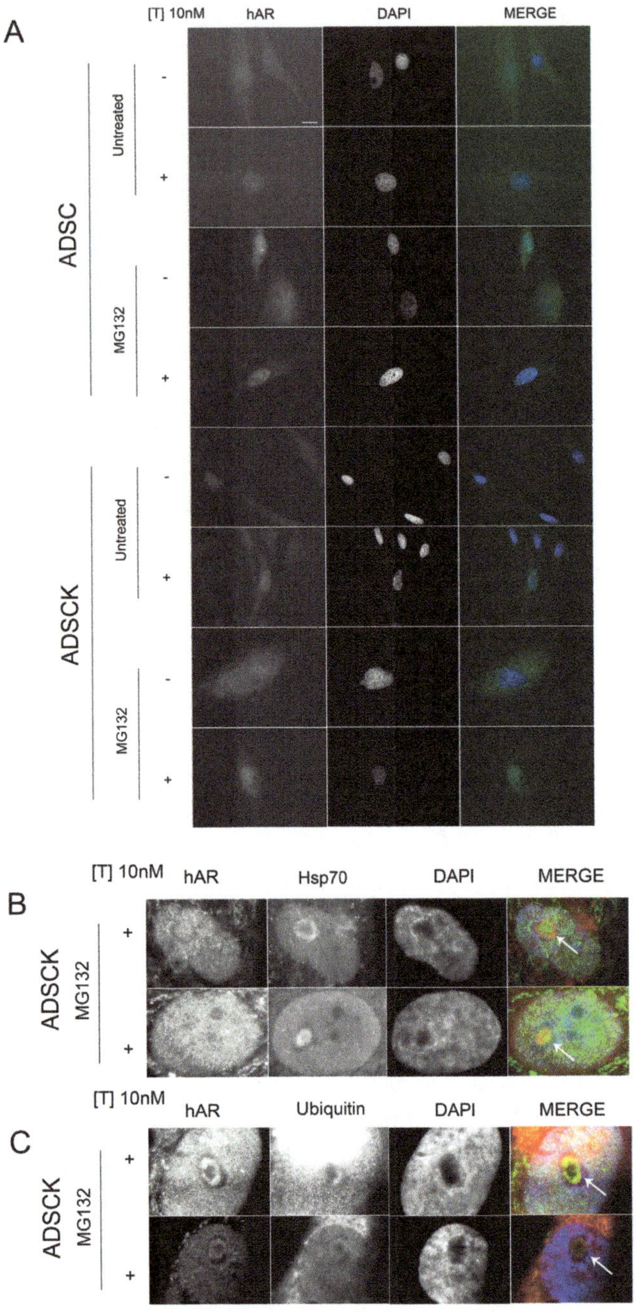

Figure 4. Biochemical behaviour of AR in ADSCs and ADSCKs. (A) High resolution fluorescence microscopy analysis (63X magnification) performed on ADSCs and ADSCKs in absence (–T) or in presence (+T) of 10 nM of testosterone for 48 hours in basal condition or after treatment with 10 μM of MG132 for 24 hours. Nuclei were stained with DAPI. Scale bar 10 μm. (B) High resolution fluorescence microscopy analysis (63X magnification) performed on ADSCKs in the presence (+T) of 10 nM of testosterone for 48 hours after treatment with 10 μM of MG132 for 24 hours. Fluorescence microscopy localization of AR (green) and Hsp70 (red). Nuclei were stained with DAPI. (C) High resolution fluorescence microscopy analysis (63X magnification) performed on ADSCKs in presence (+T) of 10 nM of testosterone for 48 hours after treatment with 10 μM of MG132 for 24 hours. Fluorescence microscopy localization of AR (green) and ubiquitin (red). Nuclei were stained with DAPI. Panel B and C represent only the nuclear area of the cells where the aggregates were observed. The arrows indicate the ARpolyQ inclusions co-stained with Hsp70 and ubiquitin, respectively.

Collectively, these data demonstrate that ADSCKs cells are valuable tool to model pathogenic events taking place in cells of SBMA patients.

Discussion

In this study we explored the use of adipose-derived stem cells isolated from periumbilical fat as a new human *in vitro* model to investigate AR behaviour and toxicity in SBMA.

Our results demonstrate that the isolation and expansion of ADSCs from healthy volunteers and from SBMA patients is feasible, and the different cell populations obtained from three controls and three SBMA patients were comparable morphologically. However, the growth capacity of ADSCKs was significantly lower than ADSCs, evidenced by a reduced survival rate of ADSCKs as compared to normal ADSCs. This apparently contrasts a recent study by Huang and colleagues showing that if AR functions are suppressed, murine ADSCs have higher growth rate [31]. In fact, they found that AR plays a suppressive role in self-renewal of the bone marrow MSCs isolated from AR knock-out mice (ARKO), by inactivating Akt and Erk signalling [31]. It is possible that the reduced growth rate and lower AR expression of mutant cells is due to selective pressure caused by polyQ toxicity, although this idea remains to be explored. In fact, both cell types express high amounts of AR mRNA and protein, but at different levels, and AR expression is significantly lower in ADSCKs than in ADSCs, in agreement with other studies performed by comparing wt and SBMA cell models [28,32,33].

An immunophenotypic marker profile in ADSCs and ADSCKs confirmed their mesenchymal identity as required by guidelines [34] and was maintained by ADSCs in culture at passage 3 and 8. Despite this, ADSCs showed only a mild positivity for CD34, and this might depend on the tissue collection procedure, degree of bleeding, vascular isolation technique or chopping [35]. Conversely, in ADSCKs we observed expression of HLA-DR. In MSC from bone marrow, this marker is induced by interferon-gamma treatment [36], thus an abnormal inflammatory response associated with mutant ARpolyQ could induce HLA-DR expression in ADSCKs, as ADSCs usually do not express this marker.

It is unclear whether the toxic effect of ARpolyQ is also responsible for the different potentiality of ADSCs and ADSCKs, since ADSCs differentiated into all the three cell lineages possibly deriving from MSC (adipocytes, osteocytes and chondrocytes), while ADSCKs only generated adipocytes. We believe that this differentiation inability might depend on mutant ARpolyQ, since ARKO mice have a significant decrease in bone mass, and bone marrow MSCs derived from these mice have a reduced expression of osteogenic markers during the lineage differentiation without affecting the MSC identity [37].

ADSCKs may be used to study pathophysiology of SBMA. In fact, both wt and mutated AR maintained the expected response to their ligand, i.e., when exposed to testosterone AR translocated to the nucleus. Moreover, under particular conditions we could reproduce the formation of inclusions, a specific (though possibly protective) cellular reaction in SBMA, as well as in other misfolded protein diseases. Indeed, we observed that when proteasome function is pharmacologically blocked, a fraction of ARpolyQ (normally cleared via this pathway) accumulates into ubiquitinated HSP70-positive nuclear inclusions, similar to those usually present in spinal cord motoneurons of SBMA patients. This confirms that proteasome may be responsible for ARpolyQ clearance, and demonstrates that our novel cellular model recapitulates a typical feature present in the tissue of affected SBMA individuals. However, it is also possible that our stressor stimuli linked to proteasome blockage, which also impairs the clearance of other ubiquitinated proteins, when added to a system which is possibly almost overwhelmed by the presence of mutant ARpolyQ, may limit the clearance of all the misfolded species including ARpolyQ.

Interestingly, SBMA patient-derived iPSCs have been recently used to study some molecular alteration occurring as a consequence of ARpolyQ expression. The great advantage of these SBMA iPSCs is to be of human origin, and with a great potential to be differentiated to "bona fide" motoneurons [19]. Using the iPSCs models, Grunseich and coll. showed that testosterone activated ARpolyQ induced an increase of acetylated alpha-tubulin and reduced HDAC6 [19], with consequently a reduction of the perinuclear accumulation of lysosomes.

A limitation of the iPSCs could be that, as far as we know, there are no established procedures to differentiate them into muscle cell types. In our view, this is relevant, since muscle cells have been recently recognized as targets of ARpolyQ toxicity [17,38,39]. However, to the best of our knowledge, all attempts to generate muscle cells from iPSCs failed so far. In addition, iPSCs are produced by genetic transformation of fibroblasts, using four oncogenic or differentiating agents that may impact on cell behavior. Thus, other cell models of human origin may be of value to complement the data obtained in iPSCs. Conversely, despite the fact that ADSCKs are difficult to maintain in culture for many passages, there are studies demonstrating the possibility to differentiate ADSC into muscle cell lines [40–43]. It must be taken into account also that, compared to iPSCs, ADSCs have the advantages of no retro transduction and manipulation needing, and thus do not express exogenous genes.

In conclusion, ADSCs represent a potential novel model of patient-derived cell populations useful to study the SBMA disease mechanism. Although these cells are not differentiable in motoneurons, at present, ADSCKs express AR and mimic some pathogenic SBMA mechanisms. Moreover, ADSCKs can be easily obtained with minimally invasive approach. Therefore, they have an interesting and still unexplored potential in studying disease mechanisms, and in designing and testing therapeutic approaches in SBMA and other disorders.

Acknowledgments

We are grateful to Dr. Kenneth Fischbeck for helpful reading of the manuscript. **Funding**: Regione Lombardia (to D.P., A.P., E.A.P.); AriSLA Foundation Italy (ALS_HSPB8 to A.P), Telethon - Italy (GGP14039 to A.P.); Italian Ministry of Labour, Health and Social Affairs (Convenzione Fondazione Mondino/UNIMI to A.P.); Università degli Studi di Milano (to A.P.); Fondazione CARIPLO (2008-2307 to A.P.); Fondation Thierry Latran, France (to A.P.), Association Française contre les Myopathies (to A.P.).

Author Contributions

Conceived and designed the experiments: MD GB PR ES SEN GM GA EAP DP AP. Performed the experiments: MD GB PR EG AC VT SEN GM FC. Analyzed the data: MD GB PR EG VT ES AS CM CG SEN GM GA EAP DP AP. Contributed reagents/materials/analysis tools: CG FC EAP AP. Contributed to the writing of the manuscript: MD GB PR AS DP AP.

References

1. Fischbeck KH (1997) Kennedy disease. J Inher Metab Dis 20: 152–158.
2. Jordan CL, Lieberman AP (2008) Spinal and bulbar muscular atrophy: a motoneuron or muscle disease? Curr Opin Pharmacol 8: 752–758.
3. La Spada AR, Wilson EM, Lubahn DB, Harding AE, Fischbeck KH (1991) Androgen receptor gene mutations in X-linked spinal and bulbar muscular atrophy. Nature 352: 77–79.
4. Poletti A (2004) The polyglutamine tract of androgen receptor: from functions to dysfunctions in motor neurons. Front Neuroendocrinol 25: 1–26.
5. Grunseich C, Kats IR, Bott LC, Rinaldi C, Kokkinis A, et al. (2014) Early onset and novel features in a spinal and bulbar muscular atrophy patient with a 68 CAG repeat. Neuromuscul Disord. 70: 12–20.
6. Poletti A, Rampoldi A, Piccioni F, Volpi S, Simeoni S, et al. (2001) 5Alpha-reductase type 2 and androgen receptor expression in gonadotropin releasing hormone GT1-1 cells. J Neuroendocrinol 13: 353–357.
7. Pozzi P, Bendotti C, Simeoni S, Piccioni F, Guerini V, et al. (2003) Androgen 5-alpha-reductase type 2 is highly expressed and active in rat spinal cord motor neurones. J Neuroendocrinol 15: 882–887.
8. Grunseich C, Rinaldi C, Fischbeck KH (2014) Spinal and bulbar muscular atrophy: pathogenesis and clinical management. Oral Dis 20: 6–9.
9. Katsuno M, Adachi H, Doyu M, Minamiyama M, Sang C, et al. (2003) Leuprorelin rescues polyglutamine-dependent phenotypes in a transgenic mouse model of spinal and bulbar muscular atrophy. Nat Med 9: 768–773.
10. Katsuno M, Adachi H, Kume A, Li M, Nakagomi Y, et al. (2002) Testosterone reduction prevents phenotypic expression in a transgenic mouse model of spinal and bulbar muscular atrophy. Neuron 35: 843–854.
11. Adachi H, Kume A, Li M, Nakagomi Y, Niwa H, et al. (2001) Transgenic mice with an expanded CAG repeat controlled by the human AR promoter show polyglutamine nuclear inclusions and neuronal dysfunction without neuronal cell death. Hum Mol Genet 10: 1039–1048.
12. La Spada AR, Peterson KR, Meadows SA, McClain ME, Jeng G, et al. (1998) Androgen receptor YAC transgenic mice carrying CAG 45 alleles show trinucleotide repeat instability. Human Mol Genet 7: 959–967.
13. Sopher BL, Thomas PS Jr, LaFevre-Bernt MA, Holm IE, Wilke SA, et al. (2004) Androgen receptor YAC transgenic mice recapitulate SBMA motor neuronopathy and implicate VEGF164 in the motor neuron degeneration. Neuron 41: 687–699.
14. Abel A, Walcott J, Woods J, Duda J, Merry DE (2001) Expression of expanded repeat androgen receptor produces neurologic disease in transgenic mice. Hum Mol Genet 10: 107–116.
15. Yu Z, Dadgar N, Albertelli M, Scheller A, Albin RL, et al. (2006) Abnormalities of germ cell maturation and sertoli cell cytoskeleton in androgen receptor 113 CAG knock-in mice reveal toxic effects of the mutant protein. The American journal of pathology 168: 195–204.
16. Chevalier-Larsen ES, O'Brien CJ, Wang H, Jenkins SC, Holder L, et al. (2004) Castration restores function and neurofilament alterations of aged symptomatic males in a transgenic mouse model of spinal and bulbar muscular atrophy. J Neurosci 24: 4778–4786.
17. Cortes CJ, Ling SC, Guo LT, Hung G, Tsunemi T, et al. (2014) Muscle expression of mutant androgen receptor accounts for systemic and motor neuron disease phenotypes in spinal and bulbar muscular atrophy. Neuron 82: 295–307.
18. Rocchi A, Pennuto M (2013) New routes to therapy for spinal and bulbar muscular atrophy. J Mol Neurosci 50: 514–523.
19. Grunseich C, Zukosky K, Kats IR, Ghosh L, Harmison GG, et al. (2014) Stem cell-derived motor neurons from spinal and bulbar muscular atrophy patients. Neurobiol Dis 70: 12–20.
20. Pittenger MF, Mackay AM, Beck SC, Jaiswal RK, Douglas R, et al. (1999) Multilineage potential of adult human mesenchymal stem cells. Science 284: 143–147.
21. Casteilla L, Dani C (2006) Adipose tissue-derived cells: from physiology to regenerative medicine. Diabetes Metab 32: 393–401.
22. Ogura F, Wakao S, Kuroda Y, Tsuchiyama K, Bagheri M, et al. (2014) Human adipose tissue possesses a unique population of pluripotent stem cells with non-tumorigenic and low telomerase activities: potential implications in regenerative medicine. Stem Cells Dev. 23: 717–728.
23. Ibrahim MM (2010) Subcutaneous and visceral adipose tissue: structural and functional differences. Obes Rev 11: 11–18.
24. Chazenbalk G, Singh P, Irge D, Shah A, Abbott DH, et al. (2013) Androgens inhibit adipogenesis during human adipose stem cell commitment to preadipocyte formation. Steroids 78: 920–926.
25. Alessandri G, Pagano S, Bez A, Benetti A, Pozzi S, et al. (2004) Isolation and culture of human muscle-derived stem cells able to differentiate into myogenic and neurogenic cell lineages. Lancet 364: 1872–1883.
26. Avanzini MA, Bernardo ME, Cometa AM, Perotti C, Zaffaroni N, et al. (2009) Generation of mesenchymal stromal cells in the presence of platelet lysate: a phenotypic and functional comparison of umbilical cord blood- and bone marrow-derived progenitors. Haematologica 94: 1649–1660.
27. Rusmini P, Crippa V, Giorgetti E, Boncoraglio A, Cristofani R, et al. (2013) Clearance of the mutant androgen receptor in motoneuronal models of spinal and bulbar muscular atrophy. Neurobiol Aging 34: 2585–2603.
28. Simeoni S, Mancini MA, Stenoien DL, Marcelli M, Weigel NL, et al. (2000) Motoneuronal cell death is not correlated with aggregate formation of androgen receptors containing an elongated polyglutamine tract. Hum Mol Genet 9: 133–144.

29. Piccioni F, Pinton P, Simeoni S, Pozzi P, Fascio U, et al. (2002) Androgen receptor with elongated polyglutamine tract forms aggregates that alter axonal trafficking and mitochondrial distribution in motor neuronal processes. Faseb J 16: 1418–1420.
30. Rusmini P, Sau D, Crippa V, Palazzolo I, Simonini F, et al. (2007) Aggregation and proteasome: the case of elongated polyglutamine aggregation in spinal and bulbar muscular atrophy. Neurobiol Aging 28: 1099–1111.
31. Huang CK, Tsai MY, Luo J, Kang HY, Lee SO, et al. (2013) Suppression of androgen receptor enhances the self-renewal of mesenchymal stem cells through elevated expression of EGFR. Biochim Biophys Acta 1833: 1222–1234.
32. Brooks BP, Merry DE, Paulson HL, Lieberman AP, Kolson DL, et al. (1998) A cell culture model for androgen effects in motor neurons. J Neurochem 70: 1054–1060.
33. Brooks BP, Paulson HL, Merry DE, Salazar-Grueso EF, Brinkmann AO, et al. (1997) Characterization of an expanded glutamine repeat androgen receptor in a neuronal cell culture system. Neurobiol Dis 3: 313–323.
34. Dominici M, Le Blanc K, Mueller I, Slaper-Cortenbach I, Marini F, et al. (2006) Minimal criteria for defining multipotent mesenchymal stromal cells. The International Society for Cellular Therapy position statement. Cytotherapy 8: 315–317.
35. Bourin P, Bunnell BA, Casteilla L, Dominici M, Katz AJ, et al. (2013) Stromal cells from the adipose tissue-derived stromal vascular fraction and culture expanded adipose tissue-derived stromal/stem cells: a joint statement of the International Federation for Adipose Therapeutics and Science (IFATS) and the International Society for Cellular Therapy (ISCT). Cytotherapy 15: 641–648.
36. Le Blanc K, Tammik C, Rosendahl K, Zetterberg E, Ringden O (2003) HLA expression and immunologic properties of differentiated and undifferentiated mesenchymal stem cells. Exp Hematol 31: 890–896.
37. Tsai MY, Shyr CR, Kang HY, Chang YC, Weng PL, et al. (2011) The reduced trabecular bone mass of adult ARKO male mice results from the decreased osteogenic differentiation of bone marrow stroma cells. Biochem Biophys Res Commun 411: 477–482.
38. Chua JP, Reddy SL, Merry DE, Adachi H, Katsuno M, et al. (2014) Transcriptional activation of TFEB/ZKSCAN3 target genes underlies enhanced autophagy in spinobulbar muscular atrophy. Hum Mol Genet 23: 1376–1386.
39. Lieberman AP, Yu Z, Murray S, Peralta R, Low A, et al. (2014) Peripheral androgen receptor gene suppression rescues disease in mouse models of spinal and bulbar muscular atrophy. Cell Rep 7: 774–784.
40. Desiderio V, De Francesco F, Schiraldi C, De Rosa A, La Gatta A, et al. (2013) Human Ng2+ adipose stem cells loaded in vivo on a new crosslinked hyaluronic acid-Lys scaffold fabricate a skeletal muscle tissue. J Cell Physiol 228: 1762–1773.
41. de la Garza-Rodea AS, van der Velde-van Dijke I, Boersma H, Goncalves MA, van Bekkum DW, et al. (2012) Myogenic properties of human mesenchymal stem cells derived from three different sources. Cell Transplant 21: 153–173.
42. Marra KG, Brayfield CA, Rubin JP (2011) Adipose stem cell differentiation into smooth muscle cells. Methods Mol Biol 702: 261–268.
43. Wang C, Yin S, Cen L, Liu Q, Liu W, et al. (2010) Differentiation of adipose-derived stem cells into contractile smooth muscle cells induced by transforming growth factor-beta1 and bone morphogenetic protein-4. Tissue Eng Part A 16: 1201–1213.
44. Kotaja N, Karvonen U, Janne OA, Palvimo JJ (2002) The nuclear receptor interaction domain of GRIP1 is modulated by covalent attachment of SUMO-1. J Biol Chem 277: 30283–30288.

Prognoses and Long-Term Outcomes of Nasopharyngeal Carcinoma in Han and Uyghur Patients Treated with Intensity-Modulated Radiotherapy in the Xinjiang Autonomous Region of China

Ruozheng Wang[1,2], Yao Tan[2], Xiyan Wang[2], Lingling Ma[2], Duoming Wang[2], Yunhui Hu[2], Yonghui Qin[2], Kai Liu[2], Cheng Chang[2], Jinming Yu[3]*

1 Tianjin Medical University, Tianjing, People's Republic of China, **2** Department of Radiation Oncology, The Affiliated Tumor Hospital of Xinjiang Medical University, Urumqi, Xinjiang Uygur Autonomous Region, People's Republic of China, **3** Department of Radiation Oncology, Tumor Hospital of Shandong, Jinnan, Shandong province, People's Republic of China

Abstract

Objectives: The objective of this study was to investigate the long-term outcomes and prognostic factors for nasopharyngeal carcinoma (NPC) in Han and Uyghur patients treated with intensity-modulated radiotherapy (IMRT) in the Xinjiang region of China.

Materials and Methods: One hundred twenty-one Han and 60 Uyghur patients with newly diagnosed NPC without distant metastasis received IMRT at the Affiliated Tumor Hospital of Xinjiang Medical University between 2005 and 2008. The Kaplan-Meier method was used to estimate survival rates, and the log-rank test was used to evaluate differences in survival.

Results: Comparing Han and Uyghur patients, the 5-year overall survival (OS), disease-free survival (DFS), local control (LC), regional control (RC), and distant metastasis-free survival (DMFS) rates were 81.9% vs 77.6% ($P=0.297$), 72.1% vs 65.6% ($P=0.493$), 88.3% vs 86.5% ($P=0.759$), 95.0% vs 94.6% ($P=0.929$), and 79.1% vs 75.2% ($P=0.613$), respectively. Multivariate Cox proportional hazards regression identified the following independent prognostic factors in Han patients: N stage ($P=0.007$) and age ($P=0.028$) for OS, and age ($P=0.028$) for DFS. OS differed significantly between Han and Uyghur patients >60 years old group ($P=0.036$). Among Uyghur patients, the independent prognostic factors were age for OS ($P=0.033$), as well as N stage ($P=0.037$) and age ($P=0.021$) for DFS. Additionally, Uyghur patients were less likely to experience mucositis and dermatitis than Han patients.

Conclusion: Han and Uyghur patients with NPC had statistically significant differences in age, smoking history, and N staging. There was no significant difference in overall treatment outcomes with IMRT between these 2 ethnic populations in Xinjiang, China.

Editor: Ke Lan, Institut Pasteur of Shanghai, Chinese Academy of Sciences, China

Funding: This study received funding from the Department of International Cooperation Project (2012DFA31560), the Chinese Nature Science Foundation (81160327, 30960432), the Xinjiang International Collaboration Fund (101141038, 20126024), the Special Fund on Urumqi Talent Project (P08131002), the Youth Science Foundation of Xinjiang Uygur Autonomous Region (2011211B22), and the Xinjiang Science and Technology Infrastructure Project (PT1401). The funders had roles in the study design, data collection and analysis, decision to publish, or preparation of the manuscript.

Competing Interests: The authors have declared that no competing interests exist.

* Email: wrz8526@163.com

These authors contributed equally to this work.

Introduction

Nasopharyngeal carcinoma (NPC) is among the most common malignancies in China, especially in southeast coastal areas. Studies have demonstrated that NPC has unique pathogenic factors, such as genetic susceptibility, Epstein-Barr virus (EBV) infection, chemical carcinogens, and environmental factors. Patients with NPC often share common features, such as ethnic susceptibility, family history, and regional aggregation [1].

NPC commonly demonstrates extensive invasion of adjacent tissues with poorly defined and large tumors in close proximity to critical structures, such as the brain stem, spinal cord, and optic chiasm. These features of NPC can complicate adequate surgical resection. Chemotherapy alone has not been very effective. Radiotherapy is generally accepted as the first-line therapy for NPC. Studies have demonstrated that intensity-modulated radiation therapy (IMRT) can substantially concentrate the radiation dose to target volumes while avoiding or reducing unnecessary radiation to normal tissues and organs, thereby leading to gains in the therapeutic ratio. As obtained using IMRT, the differential dose painting to multiple target volumes may also confer a better radiobiological effect. Previous studies have demonstrated that the implementation of IMRT improves quality of life for patients with NPC, in addition to tumor control and survival [2,3].

Xinjiang Uyghur Autonomous Region is a multinational area with 13 ethnic groups. Uyghur and Han residents constitute the major ethnic populations in Xinjiang [4]. Uyghur and Han populations vary in terms of genetics, lifestyles, and dietary habits. Uyghur and Han residents also tend to use different languages and/or dialects. Accordingly, cases of NPC may differ between Uyghur and Han patients, in terms of both disease and patient characteristics. Therefore, we sought to investigate and compare clinicopathologic factors and outcomes in Uyghur and Han patients who were diagnosed with NPC and received IMRT in Xinjiang, China. These aims were accomplished in the present study.

Patients and Methods

Subjects

We collected and analyzed the medical records of 181 patients who had NPC without distant metastasis. All patients were treated with IMRT at Xinjiang Tumor Hospital between January 2005 and November 2008. This study was approved by the institutional ethics committee of The Affiliated Tumor Hospital of Xinjiang Medical University. All patients provided written consent for inclusion in this study.

All patients had pathologically confirmed NPCs. The routine pretreatment evaluations included a complete medical history, physical examination, dental examination, hematologic and biochemical profile, chest radiography, abdominal ultrasonography, and magnetic resonance imaging (MRI) or computed tomography (CT) of the head and neck. All patients had Karnofsky Performance Status scores of at least 80 and adequate renal and liver functions. Data on the patients' general histories were also collected. A patient was identified as Han provided the patient's self-reported family history did not include intermarriage with any other ethnicities in the past three generations and that the patient had lived in Xinjiang or immigrated to Xinjiang. A patient was identified as Uyghur provided that the patient's self-reported family history did not include intermarriage with any other ethnicities in the past three generations, and that the patient had always lived in Xinjiang. Of the 181 patients, 121 were Han (88 men and 33 women, constituting a gender ratio of 2.67:1), 60 were Uyghur (41 men and 19 women, constituting a radio of 2.16:1). Patients' ages ranged from 9 to 81 years old, with a mean age of 45 years and a median age of 44 years. None of the patients had World Health Organization (WHO) type I differentiation, 81 had type II differentiation, and 100 had type III differentiation. According to the Union for International Cancer Control–American Joint Committee on Cancer (UICC/AJCC) 2002 staging criteria, 8, 16, 83, and 74 patients had clinical stage I, II, III, and IVa disease, respectively. Comparisons of the clinical characteristics of the Uyghur and Han patients are presented in Table 1.

Radiation Therapy

All patients were treated with IMRT. The head and shoulders were immobilized using a thermal plastic mask. A contrast-enhanced planning CT scan (Philips MX8000, Philips Healthcare, Cleveland, USA) was obtained with a 3-mm slice thickness, with coverage beginning from the skull vertex and ending 1 cm below the clavicles. Following the guidelines of the International Commission on Radiation Units & Measurements (ICRU) Reports 50 and 62, the CT images were retrieved on the IMRT

Table 1. Patient characteristics.

Characteristic		Han n (%)	Uyghur n (%)	χ^2	*P*
Age (year)	Range	11–81	9–67	2.71[a]	0.052
	Median	45	43		
Gender	Male	88 (72.7)	41 (68.3)	0.38	0.602
	Female	33 (27.3)	19 (31.7)		
Smoking history	Yes	54 (44.7)	15 (25.0)	6.55	0.010
	No	67 (55.3)	45 (75.0)		
WHO type	II	59 (48.9)	28 (46.7)	0.07	0.791
	III	62 (51.1)	32 (53.3)		
T stage	T1–2	29 (24.0)	14 (23.3)	0.01	0.925
	T3–4	92 (76.0)	46 (76.7)		
N stage	N0–1	55 (45.5)	14 (23.3)	8.32	0.004
	N2–3	66 (54.5)	46 (76.7)		
Clinical stage	I–II	19 (15.7)	5 (8.3)	1.89	0.388
	III	54 (44.6)	29 (48.3)		
	IV	48 (39.7)	26 (43.3)		
Treatment	Radiotherapy	12 (9.9)	11 (18.3)	2.56	0.110
	Radiotherapy + Chemotherapy	109 (90.1)	49 (81.7)		
Total patients		121	60		

[a]t-test.
WHO, World Health Organization.

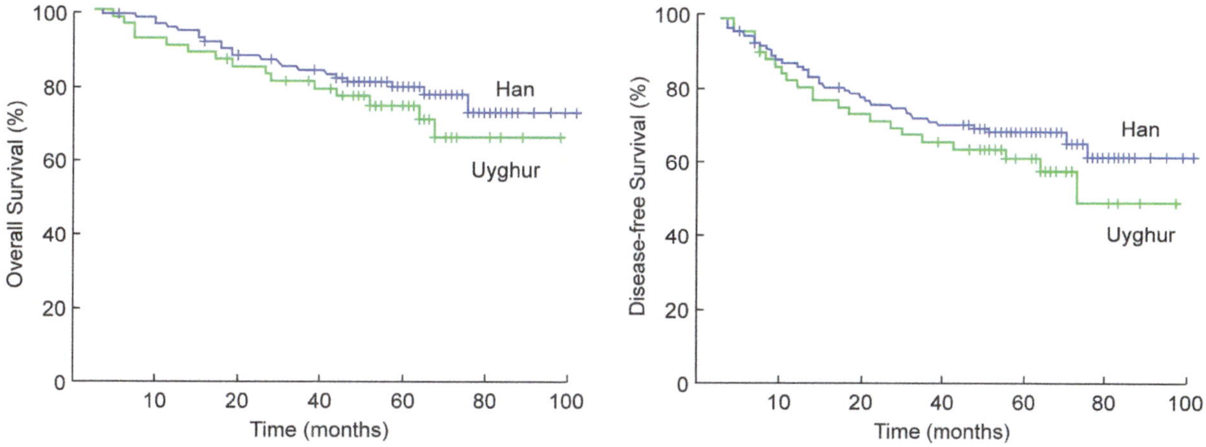

Figure 1. Survival curves for Han and Uyghur patients who received intensity-modulated radiation therapy for nasopharyngeal carcinoma.

workstation to delineate target volumes and normal tissues. The target volumes were defined as follows. The primary tumor volume (GTVnx) and the volume for nodal disease (GTVnd) were delineated by clinically and/or radiographically evident disease. There were 2 clinical target volumes: CTV1 and CTV2. CTV1 included the GTV plus a 5-mm margin, the bilateral nasopharyngeal meatuses, the clivus, the base of the skull, the pterygopalatine fossae, the medial and lateral pterygoids, the nasal cavity, the area behind 1/3–1/2 of the maxillary sinus, part of the ethmoid sinus, the sphenoid sinus, the pharynx clearance, the oropharynx, the upper lymphatic drainage area, and the hyoid. CTV2 included the lower lymphatic drainage area. The planning target volume (PTV) 1 was defined as the CTV1 plus a 5-mm margin, and the PTV2 was defined as the CTV2 plus a 3-mm margin.

Simultaneous modulated accelerated radiation therapy (SMART) was applied. The initial 28 radiation treatments were administered over 5.5–6 weeks at doses of 2.12–2.36 Gy/f (59–66 Gy in total) to the GTVnx, 2.12 Gy/f (59 Gy in total) to the GTVnd, 1.82–2.00 Gy/f (51–56 Gy in total) to the PTV1, and 1.82–2.00Gy/f (51–56 Gy in total) to PTV2. Five additional radiation doses were delivered after the initial treatment: 2.12–2.36 Gy/f (74–78 Gy in total) to the GTVnx, 2.12 Gy/f (70 Gy in total) to the GTVnd, and 1.82–2.00Gy/f (60–66 Gy in total) to the PTV1. The maximum point doses to the adjacent critical organs were restricted as follows: 54 Gy to the brainstem, 45 Gy to the spinal cord, 54 Gy to the optic nerves/optic chiasm, 9 Gy to the lenses, 50 Gy to the hypophysis, 60 Gy to the temporal lobe and temporomandibular joints, and 70 Gy to the mandible. The 50% volume dose to the parotid glands was limited to less than 30 Gy.

A 9-field IMRT comprehensive treatment planning system was used (Eclipse, Varian Medical Systems, Seattle, WA, USA). Dose-volume histograms (DVH) were used to evaluate the treatment plans. The evaluation criteria for the treatment plans were as follows. The volume of PTV that received >110% of the prescription dose was required to be ≤20%. Less than 3% of the PTV could receive <93% of the prescription dose. Outside the PTV, the dose was required to be <110% of the prescription dose. Each case had to be verified for quality control and quality assurance before treatment. A MatriXX evolution system (IBA Dosimetry, Schwarzenbruck, Germany) was used for validation. Patients could receive treatment if the delivery error was less than 5% when compared with the same plan's dose distribution, and as analyzed using OmniPro - IMRT software (IBA Dosimetry).

All patients were re-examined using a flexible nasopharyngoscope and MRI at 3–4 weeks after completion of radiotherapy. Stereotactic Body Radiation Therapy (SBRT) with an X-knife machine (Brainlab AG, Feldkirchen, Germany) was used to deliver an additional boost (dose: 9–12 Gy) for patients with residual disease in the nasopharynx, skull base, or oropharynx. Patients who had >2-cm residual cervical lymph nodes after 3 to 6 months of follow-up underwent interval neck dissections.

Chemotherapy

Of the 181 patients, 23 were treated with radiotherapy only, and the remaining 158 patients were treated with concurrent chemoradiotherapy, of whom 60 received induction chemotherapy, and 20 received induction plus adjuvant chemotherapy. Concurrent chemoradiotherapy consisted of weekly cisplatin (40 mg/m^2) for 4 to 6 weeks. However, 2 patients suspended concurrent chemotherapy due to poor tolerance. Induction chemotherapy mainly consisted of 1–4 cycles of cisplatin at 180–100 mg/m^2 on day 1 and 5-fluorouracil at 800–1000 mg/m^2 on days 1–5, or cisplatin at 180–100 mg/m^2 on day 1 and paclitaxel at 135–175 mg/m^2 on day 1 and the same regimen for adjuvant chemotherapy.

Follow-up

The duration of follow-up began at the completion of treatment. Acute adverse effects were evaluated weekly during the treatment period. All patients were followed every 3 months through the first 2 years. Each follow-up included basic serum detection, biochemistry profiles, chest X-ray, abdominal ultrasound, and MRI of the nasopharynx, head, and neck areas. Patients who were considered at high risk for distant metastasis received additional CT scans of the chest and abdomen, as well as bone scans. Toxicities were scored according to the toxicity criteria of the Radiation Therapy Oncology Group (RTOG) at each follow-up visit. Follow-ups were carried out by re-examinations, mailings, and/or telephone calls.

Statistical Methods

SPSS 17.0 software (IBM SPSS, Inc., Chicago, IL, USA) was used for all data analyses. Statistical significance was set at $P < 0.05$. Between-group differences were evaluated using the χ^2 test. Survival time was interpreted as beginning with the initiation of

IMRT treatment. Overall survival (OS), disease-free survival (DFS), local control (LC), regional control (RC), and distant metastasis-free survival (DMFS) were estimated using the Kaplan-Meier method. The log-rank test was employed to determine statistical significance. The Kruskal-Wallis analysis for continuous variables was used to compare the distributions of toxicities in different groups.

Results

Patient Characteristics

Patient characteristics are presented in Table 1. We observed significant differences between Han and Uyghur patients in terms of smoking history ($\chi^2 = 6.55$, $P = 0.010$) and N staging ($\chi^2 = 8.32$, $P = 0.004$). There were no statistically significant differences in gender, pathological classification, T stage, clinical stage, or chemotherapy use between the 2 groups. We found that 84.3% of the Han patients and 91.6% of Uyghur patients had stage III or IV disease.

Follow-up

As of November 1st, 2012, 10 patients had been lost to follow-up. The median follow-up time was 61 months (range, 1 to 90 months). Overall, 150 patients were followed for more than 3 years, and 92 patients were followed for more than 5 years. The 5-year OS, DFS, LC, RC, and DMFS rates for the entire group were 80.2%, 67.9%, 87.9%, 94.9%, and 77.6%, respectively (Table 2, Figure 1).

Outcomes: Death, Recurrence, and Metastasis

Thirty-eight patients died (23 Han and 15 Uyghur patients). Among them, distant metastasis was the cause of death for 5 Han and 8 Uyghur patients. One Han and 3 Uyghur patients died of local recurrence. One Han and 1 Uyghur patient died of second malignancies. Two mortalities among the Uyghur patients resulted from nasal cavity bleeding. The 2 remaining deaths resulted from cardiac disease in a Han patient and an unknown cause in a Uyghur patient.

Overall, 22 patients developed recurrences. The median time of recurrence was 20 months (range, 6–58 months). Among Han patients, there were 9 local recurrence (5 with stage T3 and 4 with stage T4 disease) and 3 regional lymph node recurrences. Two patients developed local and lymph node recurrences. Among Uyghur patients, there were 5 local recurrences (3 with stage T3 and 2 with stage T4 disease) and 3 regional lymph nodes recurrences.

Overall, 34 patients developed distant metastases. The median time to distant metastasis was 13 months (range, 3–72 months). Among Han patients, there were 10 bone, 3 lung, and 5 liver metastases. Four patients developed metastases to multiple organs. Among the Uyghur patients, there were 5 bone, 2 lung, and 2 liver metastases, as well as multi-organ metastases. One patient suffered a parotid lymph node metastasis.

Analyses of Prognosis

Univariate analyses of the Han and Uyghur patients indicated that age was a prognostic factor for OS in both groups ($P = 0.031$ and $P = 0.001$, respectively) and that N stage was a prognostic factors for OS among Han patients ($P = 0.043$). Patient age ($P = 0.036$) and N stage ($P = 0.047$) were the main factors associated with DFS among Uyghur patients. Among Han patients, T stage ($P = 0.043$) and clinical stage ($P = 0.020$) were the main factors associated with DFS. Furthermore, age and N stage were significant prognostic factors for DFS (Table 3).

Table 2. Comparisons of 3-year and 5-year survival rates for Han and Uyghur patients.

		OS (%)	P	DFS (%)	P	LC (%)	P	RC (%)	P	DMFS (%)	P
3-year	Han	88.3	0.43	78.2	0.29	88.9	0.66	95.1	0.91	83.5	0.38
	Uyghur	83.2		71.3		92.4		94.8		74.2	
5-year	Han	81.9	0.30	72.1	0.49	88.3	0.76	95.0	0.93	79.1	0.61
	Uyghur	77.6		65.6		86.5		94.6		75.2	

OS, overall survival; DFS, disease-free survival; LC, local control; RC, regional control; DMFS, distant metastasis-free survival.

Table 3. Univariate analysis of OS and DFS for Han and Uyghur patient.

Factors	Han n (%)	Uyghur n (%)	OS		DFS	
			χ^2	*P*	χ^2	*P*
Age (years)						
<45	57 (47.1)	34 (56.7)	0.211	0.649	0.009	0.926
45–59	13 (10.7)	19 (31.7)	0.597	0.440	2.031	0.154
≥60	51 (42.2)	7 (11.6)	4.387	0.036	1.177	0.278
N stage						
N_{0-1}	55 (45.5)	14 (23.3)	0.560	0.454	1.490	0.222
N_{2-3}	66 (54.5)	46 (76.7)	0.321	0.571	1.441	0.230
T stage						
T_{0-1}	27 (22.3)	11 (18.3)	1.441	0.230	0.000	0.996
T_{2-3}	94 (77.7)	49 (81.7)	1.415	0.234	0.352	0.553
Clinical stage						
I–III	71 (58.7)	32 (53.3)	0.514	0.473	0.138	0.710
Iva	50 (41.3)	28 (46.7)	2.455	0.117	0.254	0.614

OS, overall survival; DFS, disease-free survival.

In log-rank analyses of outcomes with age, N stage, T stage, clinical stage as the strata factors, utcomes differed significantly between the Han and Uyghur groups for patients older than 60 years ($P = 0.036$). However, stratifying by the various levels N stage, T stage, and clinical stage groups did not reveal any statistically significant differences in outcome between the Han and Uyghur groups (Table 3).

Adverse Events

All patients completed treatment. Two patients had treatment interruptions due to side effects, but eventually were able to complete the treatments after symptomatic management. The primary toxicities were xerostomia, mucositis, dermatitis, and neutropenia. The acute toxicities are summarized in Table 4. There were significant differences between the Uyghur and Han groups in terms of dermatitis and mucositis ($\chi^2 = 4.87$, $P = 0.03$ for dermatitis; $\chi^2 = 9.82$, $P = 0.00$ for mucositis). There was no significant difference in terms of xerostomia ($\chi^2 = 0.68$, $P = 0.878$) or neutropenia ($\chi^2 = 0.00$, $P = 0.99$) (Table 4).

One hundred and fifty patients survived more than 3 years (105 Han patients and 45 Uyghur patients). The main late toxicity was xerostomia, which had similar incidences in both groups (Table 5). Thirty-eight Han patients (36.2%) and 10 Uyghur patients (22.2%) had grade 1 neck fibrosis. Six Han patients (5.7%) and 2 Uyghur patients (4.4%) had grade 2 or 3 neck fibrosis, but the difference was not statistically significant ($\chi^2 = 2.89$, $P = 0.09$). Four patients had radiation damage to the temporal lobes, each of whom had skull base tumor involvement prior to the radiation therapy. Three patients had cranial nerve neuropathy. No grade 4 toxicity was observed in this cohort of patients.

Discussion

In China, NPC is among the most common malignancies of the head and neck. According to the WHO International Agency for Research on Cancer, the annual absolute incidence of NPC is 64,796 worldwide, of which 28,022 cases (43.1%) are from China. The regions of China that have high prevalences of NPC are mainly the southern provinces, including Guangdong, Guangxi, Hunan, Jiangxi, Fujian, Yunnan, Guizhou, and Sichuan. It has been reported that the prevalence of NPC is lower in Xinjiang than in Southern China, probably due to differences in lifestyle, environmental factor, climate, and ethnicities [5,6]. The Affiliated Tumor Hospital of Xinjiang Medical University is the main hospital in this region and has been designated for investigations of the clinical characteristics and prognostic factors of NPC in patients of different ethnicities. Therefore, studies of patients from this hospital may play a critical role in the efforts to curtail NPC in Xinjiang minorities.

The true prevalence of NPC may be obscured by the lack of significant symptoms in early stage disease. It has been reported that 60–70% of newly diagnosed patients have locally advanced disease [7]. As demonstrated in our study, however, 84.3% of Han and 91.6% of Uyghur patients were diagnosed with locally advanced disease. These higher percentages may relate to the differences in the level of health consciousness and economic development in Xinjiang [8]. IMRT has been implemented in our hospital since 2003, which has greatly benefited patients with NPC. IMRT's improved ability to spare normal tissues has allowed us to administer higher doses of radiation for locally advanced disease. When compared with conventional radiotherapy, IMRT has improved quality of life and treatment efficacy.

In a study of 865 patients with NPC, Su et al. [8] observed 5-year OS, DFS, LC, RC, and DMFS rates of 80.2%, 67.9%, 87.9%, 94.9%, and 77.6% respectively. As compared with this report from the southeast coastal region, which has a higher incidence of NPC, we also found similar rates of OS, LC, and RC in the present study. The high proportion of locally advanced disease may contribute to the lower rates of DFS and DMFS in this population. In our study, the 5-year OS, DFS, LC, RC, and DMFS rates were 81.9%, 72.1%, 88.3%, 95.0%, and 79.1% in Han patients and 77.6%, 65.6%, 86.5%, 94.6%, and 75.2% in Uyghur patients, respectively. Although there were no statistically significant differences between the survival rates of these ethnic groups, the percentages provided a non-statistically significant indication that Han patients might have had better survival than Uyghur patients. With longer follow-up periods and a larger sample size of the Uyghur patients, these differences may have

Table 4. Acute toxicities during the treatment of 181 patients with nasopharyngeal carcinoma.

Acute toxicities (RTOG grade) n (%)		0	1	2	3	χ^2	*P*
Mucositis	Han	-	22 (18.2)	59 (48.8)	40 (33.1)	9.82	0.00
	Uyghur	-	20 (33.3)	32 (53.3)	8 (13.3)		
Dermatitis	Han	-	77 (63.6)	37 (30.6)	7 (5.8)	4.87	0.03
	Uyghur	-	48 (80.0)	10 (16.7)	2 (3.3)		
Xerostomia	Han	10 (8.3)	41 (33.9)	66 (54.5)	4 (3.3)	0.48	0.49
	Uyghur	6 (10.0)	22 (36.7)	31 (51.7)	1 (2.1)		
Neutropenia	Han	45 (37.2)	40 (33.1)	30 (24.8)	6 (5.0)	0.00	0.99
	Uyghur	23 (38.3)	19 (31.7)	14 (23.3)	4 (6.7)		

RTOG, Radiation Therapy Oncology Group.
Kruskal-Wallis analysis.

been statistically significant. However, additional studies of larger patient cohorts would be necessary to resolve this issue.

In the present study, the age of onset was younger among Uyghur patients than it was among Han patients. We have previously reported a similar result in a study of conventional radiotherapy in 143 patients with NPC [9]. The present study also showed that smoking history differed significantly between Han and Uyghur patients with NPC. As compared with Han patients, a smaller proportion of Uyghur patients were smokers, which may be related to differences in religion, culture, and lifestyle. Previous studies have demonstrated that the frequency of genetic mutations in Han people resembles that in other Asian populations, but significantly differs from that in Caucasians [10]. The frequency distributions of genetic mutations in Uyghur and Kazak people fell between the Asian and Caucasian distributions, but were much closer to the Caucasian distribution. Thus, the characteristics and outcomes of some diseases may differ according to ethnic background.

It has been reported that the pathological type of NPC in high-incidence areas in the inland was mainly WHO III. Particularly, Han et al. found that 98.7% of cases were WHO III type [11]. The percentage of WHO II cases was 71.7% according to a report from Northwest China [12], indicating that there were differences between the pathological distributions of NPC in high- and non-high-incidence areas in China. According to the statistics from Northwest China, the percentages of WHO II and III disease were nearly the same in Han and Uyghur patients (Han, 48.9% and 51.1%; Uyghur, 46.7% and 53.3%). However, the percentage of WHO III NPC was distinctly lower in Northwest China than it was in inland high-incidence areas. The patients in our study were lifelong residents of Xinjiang. As the pathological distribution of NPC patients in Xinjiang differed from the distributions in high-incidence areas, such as Guangdong, Fujian, and Hunan, further studies are warranted to determine whether there are differences in outcomes among the different ethnic populations in China.

We conducted multivariate analyses to identify factors that were independently associated with survival in Han and Uyghur patients with NPC. For Han patients, our analysis revealed that age and N stage were independent prognostic factors for OS, and that T stage and clinical stage were independent predictors of DFS. For Uyghur patients, age and N stage were independent prognostic factors for OS, and age and N stage were independent predictors for DFS. Gender, smoking history, T stage, and clinical stage did not have significant associations with OS or DFS in either Han or Uyghur patients. Further analysis showed that OS was better in Han patients older than 60 years; this difference was statistically significant ($P = 0.036$). However, N stage, T stage, and clinical stage did not differ significantly between the two groups. Previously, Yeh et al. [13] reported that age was independently associated with survival. In a study of 990 cases of NPC in elderly patients, Sze et al. [14] reported that age was an important and independent prognostic risk factor. In the present study, we found that age was independently associated with OS and DFS for both Han and Uyghur patients. This may indicate that elderly patients would have a poor prognosis regardless of ethnicity, perhaps as a result of differences in endocrine metabolism and tissue repair capacity [15]. In Han patients, N stage was associated with OS, but did not have an obvious influence on DFS. In Uyghur patients, age was associated with OS, and age and N stage were associated with DFS. The elderly Uyghur patients showed poor OS, which may be associated with their different attitudes towards ongoing treatment in the face of tumor recurrence or metastasis.

Our conclusions regarding gender and smoking history were not entirely consistent with some previous reports. Indeed, it has been

Table 5. Late toxicities in patients with >3 years follow-up.

Type n (%)		RTOG grade)					
		0	1	2	3	χ^2	*P*
Xerostomia	Han	35 (33.3)	59 (56.2)	11 (10.5)	-	0.10	0.75
	Uyghur	13 (28.9)	28 (62.2)	4 (8.9)	-		
Hearing	Han	54 (51.4)	46 (43.8)	4 (3.8)	1 (1.0)	0.05	0.83
loss	Uyghur	24 (53.3)	19 (42.2)	2 (4.4)	-		
Neck	Han	61 (58.1)	38 (36.2)	6 (5.7)	-	2.89	0.09
fibrosis	Uyghur	33 (73.3)	10 (22.2)	2 (4.4)	-		
Mandible	Han	100 (95.2)	5 (4.8)	-	-	0.01	0.93
necrosis	Uyghur	43 (95.6)	2 (4.4)	-	-		
Temporal lobe	Han	102 (97.1)	3 (2.9)	-	-	0.05	0.83
necrosis	Uyghur	44 (97.8)	1 (2.2)	-	-		
Cranial nerve	Han	103 (98.1)	1 (1.0)	1 (1.0)	-	0.01	0.91
damage	Uyghur	44 (97.8)	1 (2.2)	-	-		

RTOG, Radiation Therapy Oncology Group.
Kruskal-Wallis analysis.

reported that women with NPC had a better prognosis than men with NPC [16,17,18]. Further, Xue et al. [19] reported that the risk of developing NPC was 1.6 times higher in smokers than that in non-smokers. In the present study, the OS and DFS were higher for non-smoking and female Han patients. Among the Uyghur patients, women had better OS and DFS than men, as was seen among the Han patients. However, these differences were not statistically significant, perhaps because of the limited sample size.

In a prospective study of 323 cases of NPC, Lin et al. [20] reported that T stage was not a prognostic factor in patients who received IMRT. However, N stage was a significant predictor of DMFS and OS. Our multivariate analysis did not suggest that T stage or clinical stage were independently associated with the survival rate. While the sample size of our study was small, IMRT has overcome some of the deficits of conventional radiation therapy by establishing a more accurate dose distribution with improved conformity to the treatment targets. These benefits of IMRT may have led to better treatment outcomes for our patients with locally advanced disease. IMRT could have mitigated the adverse effects that T stage and clinical stage would otherwise have had on survival. In the present study, however, N stage was a significant independent prognostic factor for OS among Han patients, and was almost significant among Uyghur patients ($P = 0.064$). As previously mentioned, this lack of statistical significance may relate to the relatively small number of Uyghur patients in this study. There were also differences in the adverse events experienced by Han and Uyghur patients. Both oral mucosal and skin reactions were less common in Uyghur patients than they were in Han patients, but the causes of these differences require further study.

Overall, the treatment of NPC at our hospital has greatly improved since the launch of IMRT. Both Han and Uyghur patients obtained excellent outcomes with long-term follow-up. The survival rate has improved significantly, as compared with previous findings [11]. In this study, the distribution of pathological types differed from that in high-incidence areas in China. We also observed some differences between Han and Uyghur patients in terms of acute skin and mucosal reactions and the OS of the elderly Uyghur patients. Further studies should be performed to investigate the possibility of genetic differences in NPC between the patients in Xinjiang and other high-incidence inland regions of China. Additionally, further research may elucidate differences in NPC among other ethnic groups.

Author Contributions

Conceived and designed the experiments: RW JY YT XW. Performed the experiments: RW YT LM YH DW YQ KL CC. Analyzed the data: RW YT XW LM. Contributed reagents/materials/analysis tools: YT LM. Wrote the paper: RW YT LM. Manuscript review: RW JY.

References

1. Cao SM, Simons MJ, Qian CN (2011) The prevalence and prevention of nasopharyngeal carcinoma in China. Chin J Cancer 30: 114–119.
2. Kam MK, Teo PM, Chau RM, Cheung KY, Choi PH, et al. (2004) Treatment of nasopharyngeal carcinoma with intensity-modulated radiotherapy: the Hong Kong experience. Int J Radiat Oncol Biol Phys 60: 1440–1450.
3. Kwong DL, Pow EH, Sham JS, McMillan AS, Leung LH, et al. (2004) Intensity-modulated radiotherapy for early-stage nasopharyngeal carcinoma: a prospective study on disease control and preservation of salivary function. Cancer 101: 1584–1593.
4. Jin J, editor (2010) Xinjiang Uyghur Autonomous Region Bureau of Statistics, Xinjiang Statistical Yearbook. Beijing: China Statistics Press. 640 p
5. Su MK, Wang RZ (2004) Distribution variant of nasopharyngeal carcinoma in different ethnics and times in Xinjiang area [in Chinese]. J Mod Oncol 12: 589–591.
6. Su MK, Zhang JR (2001) Retrospective analysis in 218 ethnic nasopharyngeal carcinoma patients in Xinjiang area [in Chinese]. J Xinjiang Med University 24: 371–372.
7. Ji X, Xie C, Hu D, Fan X, Zhou Y, et al. (2013) Survival benefit of adding chemotherapy to intensity modulated radiation in patients with locoregionally advanced nasopharyngeal carcinoma. PLoS One 8: e56208.
8. Su SF, Han F, Zhao C, Huang Y, Chen CY, et al. (2011) Treatment outcomes for different subgroups of nasopharyngeal carcinoma patients treated with intensity-modulated radiation therapy. Chin J Cancer 30: 565–573.
9. Wang DM, Wang RZ, Zhong XL (2008) Prognostic analysis of nasopharyngeal carcinoma in Han and Uygur ethnic people in Xinjiang [in Chinese]. J Xinjiang Med University 31: 1504–1507.
10. Chen R, Ye G, Geng Z, Wang Z, Kong F, et al. (1993) Revelations of the origin of Chinese nation from clustering analysis and frequency distribution of HLA polymorphism in major minority nationalities in mainland China [in Chinese]. Yi Chuan Xue Bao 20: 389–398.
11. Han L, Lin SJ, Pan JJ, Chen CB, Zhang Y, et al. (2010) Prognostic factors of 305 nasopharyngeal carcinoma patients treated with intensity-modulated radiotherapy. Chin J Cancer 29: 145–150.
12. Wang J, Shi M, Hsia Y, Luo S, Zhao L, et al. (2012) Failure patterns and survival in patients with nasopharyngeal carcinoma treated with intensity modulated radiation in Northwest China: a pilot study. Radiat Oncol 7: 2.
13. Yeh SA, Tang Y, Lui CC, Huang YJ, Huang EY (2005) Treatment outcomes and late complications of 849 patients with nasopharyngeal carcinoma treated with radiotherapy alone. Int J Radiat Oncol Biol Phys 62: 672–679.
14. Sze HC, Ng WT, Chan OS, Shum TC, Chan LL, et al. (2012) Radical radiotherapy for nasopharyngeal carcinoma in elderly patients: the importance of co-morbidity assessment. Oral Oncol 48: 162–167.
15. Itzhaki O, Kaptzan T, Skutelsky E, Sinai J, Michowitz M, et al. (2004) Age-adjusted antitumoral therapy based on the demonstration of increased apoptosis as a mechanism underlying the reduced malignancy of tumors in the aged. Biochim Biophys Acta 1688: 145–159.
16. Xiao G, Cao Y, Qiu X, Wang W, Wang Y (2013) Influence of gender and age on the survival of patients with nasopharyngeal carcinoma. BMC Cancer 13: 226.
17. Lai SZ, Li WF, Chen L, Luo W, Chen YY, et al. (2011) How does intensity-modulated radiotherapy versus conventional two-dimensional radiotherapy influence the treatment results in nasopharyngeal carcinoma patients? Int J Radiat Oncol Biol Phys 80: 661–668.
18. Ng WT, Lee MC, Hung WM, Choi CW, Lee KC, et al. (2011) Clinical outcomes and patterns of failure after intensity-modulated radiotherapy for nasopharyngeal carcinoma. Int J Radiat Oncol Biol Phys 79: 420–428.
19. Xue WQ, Qin HD, Ruan HL, Shugart YY, Jia WH (2013) Quantitative association of tobacco smoking with the risk of nasopharyngeal carcinoma: a comprehensiv meta-analysis of studies conducted between 1979 and 2011. Am J Epidemiol 178: 325–338.
20. Lin S, Pan J, Han L, Zhang X, Liao X, et al. (2009) Nasopharyngeal carcinoma treated with reduced-volume intensity-modulated radiation therapy: report on the 3-year outcome of a prospective series. Int J Radiat Oncol Biol Phys 75: 1071–1078.

Whole Brain Radiotherapy Plus Concurrent Chemotherapy in Non-Small Cell Lung Cancer Patients with Brain Metastases

Hong Qin, Feng Pan, Jianjun Li, Xiaoli Zhang, Houjie Liang*, Zhihua Ruan*

Department of Oncology, Southwest Hospital, the Third Military Medical University, Chongqing, PR China

Abstract

Objective: The aim of the present meta-analysis is to evaluate the response rate, median survival time (MST) and toxicity in patients with brain metastases (BM) originating from non-small cell lung cancer (NSCLC) and who were treated using either whole brain radiotherapy (WBRT) plus concurrent chemotherapy or WBRT alone.

Methods: PubMed, EMBASE, Web of Science, The Cochrane Library, clinical trials and current controlled trials were searched to identify any relevant publications. After screening the literature and undertaking quality assessment and data extraction, the meta-analysis was performed using Stata11.0 software.

Results: In total, six randomized controlled trials (RCT) involving 910 participants were included in the meta-analysis. The results of the analysis indicate that WBRT plus concurrent chemotherapy was more effective at improving response rate (RR = 2.06, 95% CI [1.13, 3.77]; P = 0.019) than WBRT alone. However, WBRT plus concurrent chemotherapy did not improve median survival time (MST) (HR = 1.09, 95%CI [0.94, 1.26]; P = 0.233) or time of neurological progression (CNS-TTP) (HR = 0.93, 95%CI [0.75, 1.16]; P = 0.543), and increased adverse events (Grade≥3) (RR = 2.59, 95% CI [1.88, 3.58]; P = 0.000). There were no significant differences in Grade 3–5 neurological or hematological toxicity between two patient groups (RR = 1.08, 95%CI [0.23, 5.1]; P = 0.92).

Conclusion: The combination of chemotherapy plus WBRT in patients with BM originating from NSCLC may increase treatment response rates of brain metastases with limited toxicity. Although the therapy schedule did not prolong MST or CNS-TTP, further assessment is warranted.

Editor: Matthew B. Schabath, H. Lee Moffitt Cancer Center, United States of America

Funding: These authors have no support or funding to report.

Competing Interests: The authors have declared that no competing interests exist.

* Email: zhihuaruan@gmail.com (ZR); lianghoujie@sina.com (HL)

Introduction

Approximately 20% to 40% of patients with cancer develop brain metastases (BM) during their disease course. Patients with solid tumors, such as lung and, breast cancer or melanoma, are at high risk for BM. In particular, it has been estimated that approximately 50% of primary lung cancers develop into BM [1]. Furthermore, non-small cell lung cancer (NSCLC) accounts a large percentage of lung cancer cases. It has also been estimated that 25% to 30% of newly diagnosed NSCLC patients also suffer from brain metastases [2]. NSCLC patients who develop BM often have poor prognoses, severe neurological symptoms, poor quality of life and dismal survival rates. The overall survival time (OS) for NSCLC patients with BM is less than 3–6 months when left untreated [3]; effective treatment options for NSCLC patients with BM are needed urgently.

Whole brain radiotherapy (WBRT) has been the standard therapy for most patients with multiple BM.WBRT can palliate neurological symptoms and control the local disease. However, it has been difficult to eradicate the tumors due to the limitations of radiation therapy. One study reported that one-third of included patients had uncontrollable localized tumors following WBRT treatment and that 50% of patients died of intracranial tumor progression [4]. Systemic chemotherapy has also been used to reduce tumor burden in patients with BM originating from NSCLC. However, the treatment's effectiveness is limited due to the brain-blood barrier (BBB). Clinical doctors, therefore, faced a dilemma when treating NSLCL patients with BM. Some researchers have suggested that chemical drugs can infiltrate the brain tissue when radiation destroys the BBB, and several clinical trials have indicated that WBRT combined with chemotherapy is not only more effective than WBRT alone, but also improves the response rate and prolongs survival [5–7]. Other studies have failed to confirm the efficacy of chemotherapy and suggest that chemotherapy concurrent with WBRT increases the incidence of adverse events and does not benefit NSCLC patients with BM [8–10]. The role of chemotherapy concurrent with WBRT for the treatment of patients with BM originating from NSCLC is controversial. We have therefore conducted a meta-analysis assessing the efficacy and safety of chemotherapy combined with WBRT versus treatment with WBRT alone.

Materials and Methods

Search strategy

PubMed, EMBASE, the Cochrane Library, Web of Science, clinical trials and current controlled trials were searched to identify relevant studies in the published literature. The search was performed on September 25, 2013, using both Mesh and free text words. The following basic search terms were used: lung neoplasms, lung tumor, lung cancer, brain metastasis, brain neoplasms, radiotherapy and chemotherapy. The search was performed without any language limitations.

Inclusion criteria

All articles which met the following criteria were eligible: (1) randomized controlled trials (RCT) with voluntarily enrolled patients; (2) patients had histologically or cytologically confirmed NSCLC and had been diagnosed with multiple brain metastases using CT or MRI; (3) the trials compared WBRT plus chemotherapy with WBRT alone; (4) trials did not include patients with chemotherapy contraindications or serious vital organ dysfunction and Karnofsky performance status (KPS) scores ≥70; (5) the analyses included response rate, median survival time (MST), the time to neurological progression (CNS-TTP), adverse events (Grade≥3) or hematological toxicity (Grade≥3); (6) response rate was determined using the Response Evaluation Criteria in Solid Tumors (RECIST) or WHO evaluation criteria on solid tumors. complete remission (CR) was defined as tumor completely disappearing for at least four weeks without any new lesions, partial response (PR) was defined as more than 50% tumor regression for at least for four weeks without new lesions, Progressive disease (PD) was defined as an increase in the sum of the longest diameters (LD) of the target lesions by 25% or higher, using as reference the smallest sum LD recorded since treatment started or the appearance of one or more new lesions. Stabilized disease (SD) was defined as a ≤50% tumor regression or an increase ≤25%. (7) Toxicity was evaluated according to the National Cancer Institute Common Terminology Criteria for Adverse Events.

Study selection

The eligibility assessment was first performed by screening titles and abstracts and subsequently reviewing the full text of articles. The selection of all studies was performed independently, according to the inclusion criteria, by two reviewers. Disagreement on whether an article should be included was resolved using a third reviewer.

Data extraction

Two authors independently extracted data from all the eligible studies. When the extracted data were not uniform, consultation was needed to make a final determination. All of the studies included in the analysis contain the following data: first author's name, published year, type of study, trial phase, country of origin study, percentage of men, performance status, number of patients, average ages, interventions and outcomes.

Quality assessment

All of the selected studies were evaluated by two reviewers according to the Cochrane Handbook for RCT, based on the following criteria: (1) randomized method; (2) allocation concealment; (3) blinding of participants, personnel and outcome assessment; and (4) intention-to-treat analysis if the trials lost participants to follow-up or if participants quit. Each trial for bias based on the criteria listed above was marked as 'low risk', 'high risk' or 'unclear risk'. Trials judged as low risk of bias (i.e. A rating) when all criteria are assessed as low risk; Trials judged as moderate risk of bias (i.e. B rating) when one or more criteria are assessed as unclear risk; Trials judged as high risk of bias (i.e. C rating) when one or more criteria are assessed as high risk.

Statistical analysis

Statistical analyses were performed using Stata software11.0. Chi-square and I-square tests were used to test the heterogeneity of different studies [11]; no heterogeneity was considered to exist when $P>0.1$ and $I^2<50\%$. A fixed-effect model was applied to pool the study results. Significant heterogeneity was found if $P<0.1$ and $I^2>50\%$, and a random-effects statistical model was used [12]. Response rate, severe hematological toxicity and advent events were analyzed using dichotomous variables. MST and CNS-TTP were calculated using effect variables.

Results

Selection of studies

In total, we identified 2104 studies that met our selection criteria after searching the relevant databases; 236 of these studies were excluded due to duplication. By verifying related terms in the titles and abstracts, we excluded 1847 irrelevant articles, and another 15 articles were excluded after the full text was read. Finally, six RCTs [10,13–17] were selected for the present meta-analysis. A flowchart depicting the study selection is shown in figure 1.

General characteristics of included studies

There were 910 patients with BM originating from NSCLC in the six selected RCT trials, with 478 patients having received WBRT concurrent with chemotherapy and 432 patients having received only WBRT; these results are summarized in table 1. Of the six RCTs, three were phase III clinical trials [10,14,16], two were phase II studies [15,17], and one was a study [13] that did not mention a trial phase. The analyzed interventions were WBRT plus chemotherapy and WBRT alone, except in the case

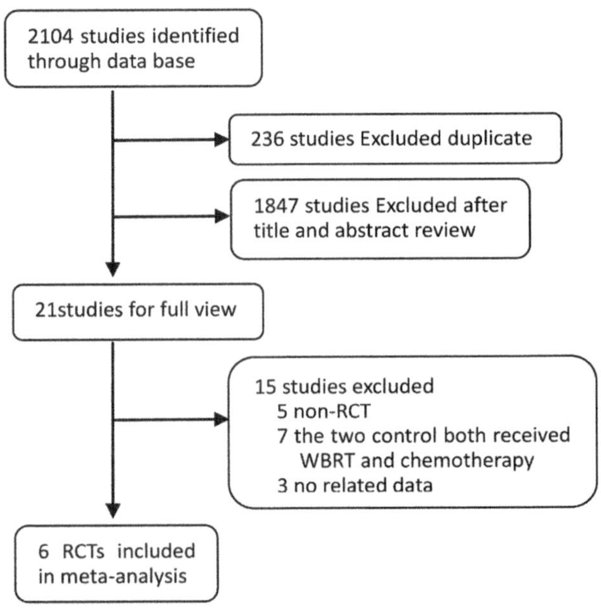

Figure 1. A flow chart on selection included of trials in the Meta-analysis.

of Sperduto, P. W.2013, which compared the combination treatment WBRT, stereotactic radiotherapy (SRS) and chemotherapy with WBRT+SRS treatment. Among all of the included studies, chemotherapy drugs included temozolomide (TMZ), carboplatin, motexafin gadolinium (MGD), chloroethylnitrosoureas and tegafur. TMZ was used in three of the trials. Outcomes included response rate, adverse events, hematological toxicity, median survival time (MST) and time to central nervous system progression (CNS-TTP).

Methodological quality

In accordance with the recommendations of the Cochrane Handbook for Systematic Reviews, two authors evaluated the eligible studies using the four aspects mentioned above. Four studies [10,15–17] mentioned the use of random allocation, but only two articles discussed the methods [13,14]. None of the studies performed or reported their allocation concealment and blinding methods. The Hassler, M.R.2013 [15] trial reported follow-up information, but the other studies did not. All of the articles applied the intent-to-treat analysis. The six eligible studies all received B quality scores, as shown in table 1.

Response rate

Three of the included studies [13,15,16] reported the efficacy of treatment using WBRT plus concurrent chemotherapy and WBRT alone. Ushio, Y.1991 [13] reported tumor response rates in the WBRT and WBRT plus chemotherapy groups were 36% and 71%, respectively. Hassler, M.R.2013 [15] reported two cases of PR in the WBRT arm, and two CR and three PR cases in the WBRT plus chemotherapy arm. Guerrieri, M.2004 [16] reported response rates were 10% and 29% in the WBRT and WBRT plus carboplatin arms, respectively. There was no heterogeneity ($P = 0.801$, $I^2 = 0.0\%$) among the three studies, and as a result, the fixed effect model was used for the meta-analysis. The results indicate that WBRT plus concurrent chemotherapy resulted in superior response rates when compared with WBRT alone (RR = 2.06, 95%CI [1.13, 3.77]; $P = 0.019$) (figure 2).

Adverse events

Three studies [10,15,17] reported the occurrence of drug-related hematological toxicity (Grade≥3). A random effects model was used for the meta-analysis of these studies based on the heterogeneity values ($P = 0.041$, $I^2 = 68.8\%$). The results indicate no significant difference in hematological toxicity between WBRT plus chemotherapy and WBRT alone (RR = 1.08, 95% CI [0.23, 5.1]; $P = 0.92$) (figure 3). However, another four studies [10,14,15,17] described adverse events (Grade≥3) and included both hematological and non-hematological toxicity. A fixed effect model was used for the meta-analysis of these studies because heterogeneity did not exist ($P = 0.500$, $I^2 = 0.0\%$). The results indicate that the incidence of severe adverse events was higher in the group treated using WBRT concurrent with chemotherapy (RR = 2.59, 95% CI [1.88, 3.58]; $P = 0.000$) (figure 4).

Survival

Five of the studies [10,14–17] reported MST for both patient groups; the studies were not heterogeneous ($P = 0.425$, $I^2 = 0.0\%$). Analysis using a fixed effect model suggests that in NSCLC patients diagnosed with BM, there was no significant MST difference between those who were treated with chemotherapy and those who were not (HR = 1.09, 95% CI [0.94, 1.26]; $P = 0.233$) (figure 5). The most meaningful outcome was the time to neurological progression (CNS-TTP). Three studies [10,14,17]

Table 1. Characteristics of trials included in the Meta-analysis.

Studies	Country	Trial phase	N (T/C)	Male (T/C,%)	Ages (T/C, years)	ECOG	Interventions T	Interventions C	Outcomes	Study quality
Ushio, Y.1991	Japan	NM	69/31	80/87	57/63	NM	WBRT+methyl-CCNU/ACNU/tegafur	WBRT	Response rate	B
Sperduto, P.W.2013	American	3	40/44	NM	63/64	0–1	WBRT+SRS+TMZ	WBRT+SRS	MST, OS, CNS-TTP, toxicity	B
Mehta, M.P.2009	France	3	279/275	59/55	59/55	0–2	WBRT+MGD	WBRT	MST, CNS-TTP, Adverse event	B
Hassler, M.R.2013	Austria	2	22/13	59/62	69/64	0–2	WBRT+TMZ	WBRT	Response rate, MST, Toxicity, Adverse events	B
Guerrieri, M.2004	Australia	3	21/21	71/71	60/63	0–2	WBRT+ carboplatin	WBRT	MST, Response rate	B
Chua, D.2010	China	2	47/48	64/67	59/62	0–2	WBRT+TMZ	WBRT	MST, CNS-TTP, Toxicity, Adverse event	B

T: treatment group (chemotherapy plus WBRT group); C: control group (WBRT group); NM: not mentioned; ECOG: ECOG/WHO performance status WBRT: Whole brain radiotherapy; TMZ: temozolomide; MST: median survival time; OS: overall survival; CNS: central nervous system; MGD: motexafin gadolinium; TTP: time to progression.

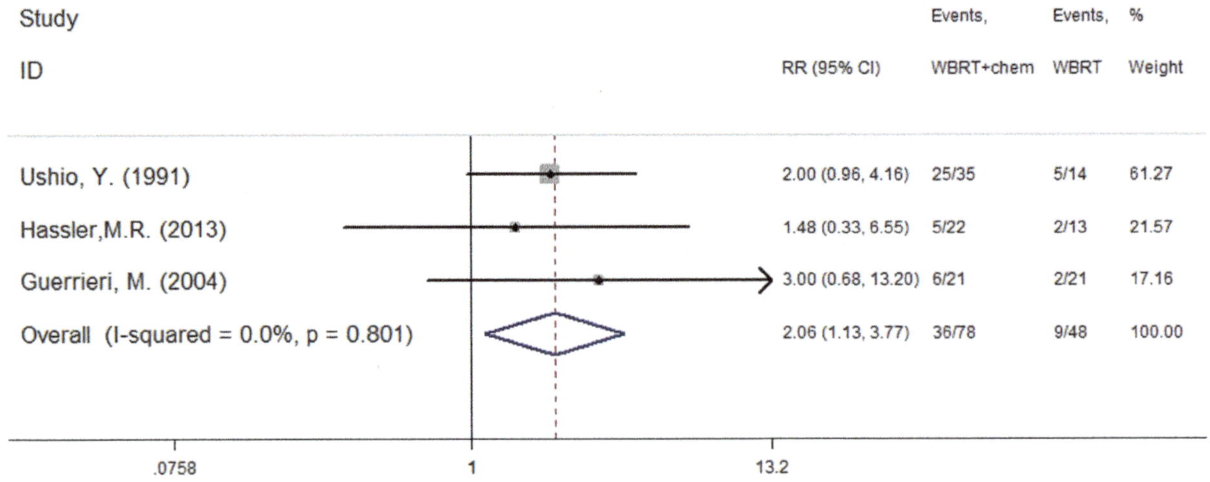

Figure 2. Response rate (P = 0.019).

reported CNS-TTP, and there was no significant heterogeneity between them (P = 0.186, I^2 = 40.5%); accordingly, a fixed effect model was used for the meta-analysis of CNS-TTP. The results suggest that combining chemotherapy with WBRT could prolong the time of neurological progression (HR = 0.93, 95% CI [0.75, 1.16]; P = 0.543) (figure 6). In conclusion, this meta-analysis suggests that WBRT concurrent with chemoradiotherapy significantly increased response rate and potentially prolonged the time of neurologic progression for patients with BM originating from NSCLC. However, more hypotoxic chemotherapy drugs still need to be explored in future clinical research.

Discussion

Currently, WBRT is the standard therapy for NSCLC patients whose disease has metastasized to the brain. Several studies have verified that WBRT palliates the neurological symptom associated with BM. However, because radiotherapy doses are limited, the treatment has been unsuccessful at curing malignant lesions. Furthermore, the brain-blood barrier (BBB) prevents the transport of most anticancer agents to the central nervous system and restricts the delivery of drugs to infiltrating BM. These additional barriers restrict the use of chemotherapy for patients with BM. Results of several trails have indicated that chemotherapy combined with WBRT benefits NSCLC patients with BM. Some clinicians found that WBRT could allow chemotherapy drugs to pass through the BBB. Additionally, chemotherapy had the potential to make brain tumor cells more sensitive to radiotherapy. Several studies have indicated that WBRT plus concurrent chemotherapy is playing an increasing role in the treatment of BM. Mehta, M. P. 2009 [14] reported that treatment using motexafin gadolinium (MGd) improved the neurologic progression interval when compared with WBRT alone (15 months vs.10 months). Verger, E. 2005 [18] reported results of patients who received the combination treatment of WBRT with TMZ, noting that they exhibited good tolerance and significantly better progression-free survival of BM at 90 days (54% vs. 72%; P = 0.03). Nonetheless, some studies have suggested that WBRT had a minor effect on promoting chemotherapy drugs across the BBB. Adding chemotherapy to WBRT treatment did not confer any benefits to the patients, but did increase the incidence of adverse events. For instance, Neuhaus, T.2009 [19] reported that concurrent radiochemotherapy (WBRT+topotecan) did not achieve significant curative effects in patients with lung cancer.

A total of six RCTs were included in present meta-analysis. The response rate was significantly improved in patients treated with

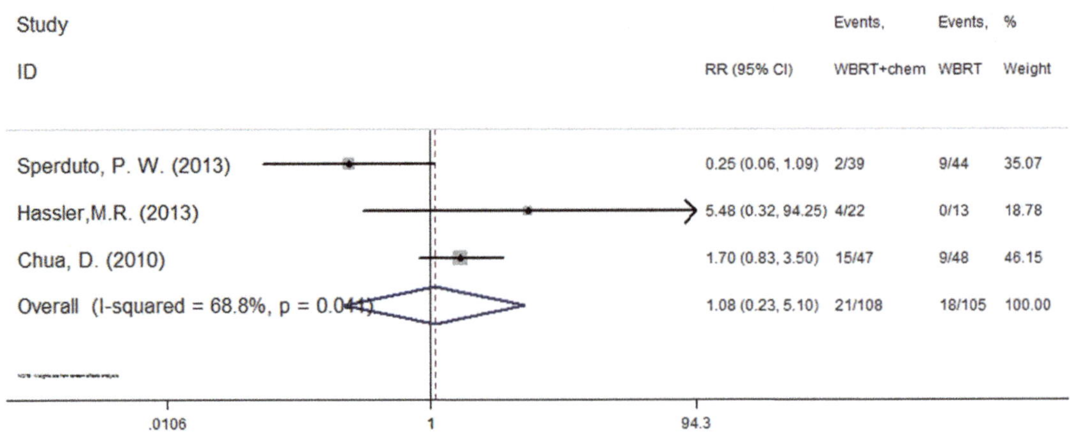

Figure 3. Severe haematological toxicity (P = 0.92).

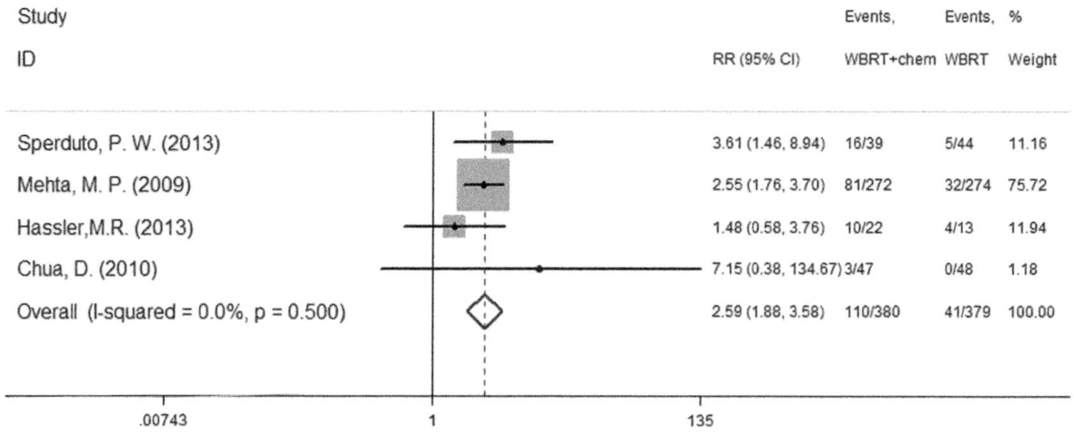

Figure 4. Severe adverse event (P = 0.000).

WBRT plus chemotherapy. However, WBRT plus chemotherapy did not improve MST and CNS-TTP for patients with malignant lesions. WBRT plus chemotherapy increased the incidence of adverse reactions, such as asthenia, fatigue, nausea, vomiting, infection, thrombocytopenia, anemia and neutropenia, but there were no significant differences in the rates of severe hematological toxicity. Each group of patients in the Sperduto, P. W.2013 study received SRS, which might have influenced the final outcomes of the meta-analysis; therefore, we extracted these data and re-analyzed MST and CNS-TTP. Although no significant statistical differences were found, treatment with both WBRT with concurrent chemotherapy tended to prolong MST and CNS-TTP (MST: HR = 1.06, 95% CI [0.91, 1.23], P = 0.462; CNS-TTP: HR = 0.84, 95% CI [0.65, 1.08], P = 0.183).

Researchers hold the opinion that the favorable therapeutic effects of combining chemotherapy with WBRT depend on the drugs crossing the BBB [20]. Temozolomide (TMZ) protocols have been recommended for the favorable distribution of the chemotherapy drug through the BBB and to achieve an effective concentration in the brain tissue [21]. A number of trials have shown that TMZ was well tolerated by patients with BM and achieved high release rates [22,23]. By contrast, some chemotherapy drugs, such as etoposide and cisplatin, had difficulty reaching the intracranial environment because of the BBB. Those agents did not improve response rates and only increased the incidence of neurological toxicity [8]. Due to a lack of response rates and CNS-TTP data, we did not analyze the differences between TMZ and non-TMZ treatments.

Our present systematic review suggests that the combination of chemotherapy and WBRT does not obviously improve MST and CNS-TTP. Moreover, we found a tendency toward prolonged CNS-TTP intervals in the concurrent chemoradiotherapy group.

Our results suggest that the combination of chemotherapy and WBRT significantly increased adverse events of Grade 3 or higher, although there was insufficient evidence to indicate that the treatment resulted in severe nervous-system toxicity. Sperduto, P. W.2013 [10] reported that rates of Grade 3–5 toxicity for WBRT/SRS and WBRT/SRS/TMZ were 11% and 41%, respectively. However, most of the adverse events were fatigue, dehydration and other aspecific symptoms. Chua, D.2010 [17] reported that three patients suffered adverse events (≥Grade 3), and none of these events were related to neurologic toxicity.

Meta-analysis is based on the results of published articles and several steps of integration; thus, certain biases are inevitable. Also, 6 studies included in the meta-analysis were published in the period 1991–2013. Although the dose of WBRT for the treatment of multiple brain metastases did not change during this 22-year period, more precisely delineated target regional and advanced equipment would affect the effectiveness of treatment as technology advances. In addition, as more and more researchers focused their attention on the investigation of targeted drug combine with WBRT in the treatment of multiple brain metastases, and many chemoradiotherapy-related studies were retrospective study and single-arm study, limited RCT were included in this study. Furthermore, the methods of randomization, allocation concealment, and blinding in most of the included studies are not clear. As

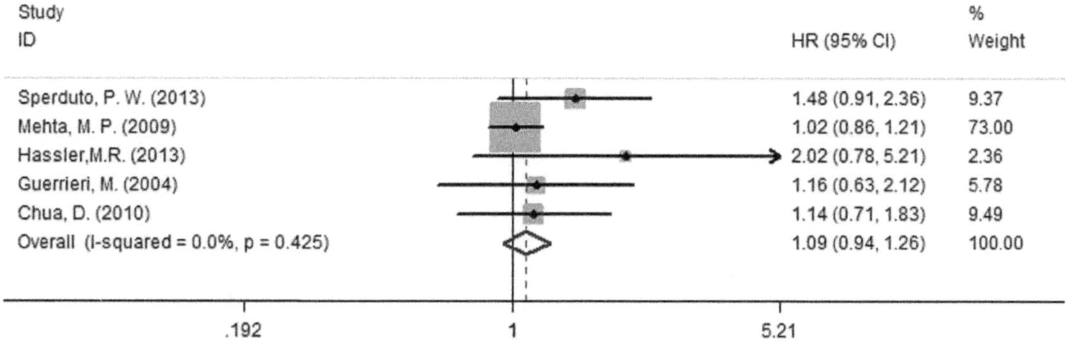

Figure 5. Median survival time (MST) (P = 0.233).

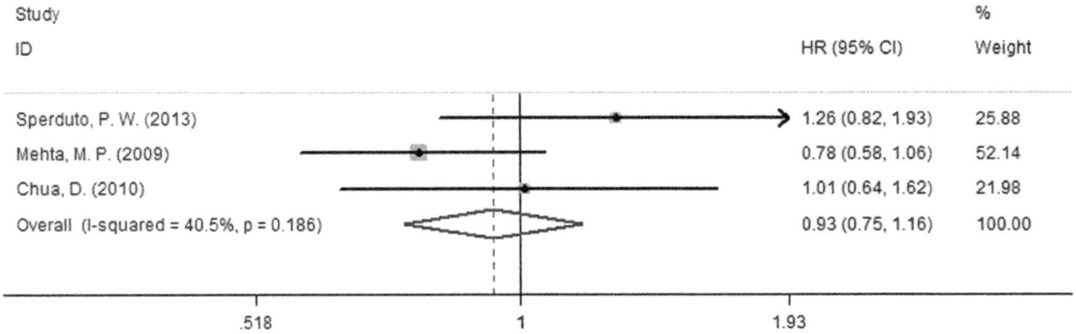

Figure 6. The time to CNS-progression (P = 0.543).

a result, the quality of the 6 RCTs was not high. The different strategies used to divide groups and the dose of WBRT and chemotherapy drugs were not clear; these procedures may have influenced the final outcomes. More high quality and large scale trials are necessary to confirm the efficacy and safety of WBRT plus concurrent chemotherapy for the treatment of patients with BM originating from NSCLC.

In conclusion, this meta-analysis reports that concurrent chemoradiotherapy significantly increased response rates and had the potential to prolong neurologic progression time NSCLC patients with BM. However, additional future clinical research is needed to explore the use of more hypotoxic chemotherapy drugs.

Supporting Information

Checklist S1 PRISMA Checklist.

Figure S1 PRISMA Flow Diagram.

Acknowledgments

The authors thank all the participants in this study.

Author Contributions

Conceived and designed the experiments: HQ HL ZR. Performed the experiments: HQ FP JL XZ. Analyzed the data: HQ ZR. Wrote the paper: HQ FP.

References

1. Seoane J and De Mattos-Arruda L (2014) Brain metastasis: New opportunities to tackle therapeutic resistance. Mol Oncol 8: 1120–1131.
2. Berger LA, Riesenberg H, Bokemeyer C and Atanackovic D (2013) CNS metastases in non-small-cell lung cancer: Current role of EGFR-TKI therapy and future perspectives. Lung Cancer 80: 242–248.
3. Sperduto PW, Kased N, Roberge D, Xu Z, Shanley R, et al. (2012) Summary report on the graded prognostic assessment: an accurate and facile diagnosis-specific tool to estimate survival for patients with brain metastases. J Clin Oncol 30: 419–425.
4. Gijtenbeek JM, Ho VK, Heesters MA, Lagerwaard FJ, de Graeff A, et al. (2011) [Practice guideline Brain metastases' (revision)]. Ned Tijdschr Geneeskd 155: A4141.
5. Siu TL, Jeffree RL and Fuller JW (2011) Current strategies in the surgical management of cerebral metastases: an evidence-based review. J Clin Neurosci 18: 1429–1434.
6. Bailon O, Chouahnia K, Augier A, Bouillet T, Billot S, et al. (2012) Upfront association of carboplatin plus pemetrexed in patients with brain metastases of lung adenocarcinoma. Neuro-Oncology 14: 491–495.
7. Barlesi F, Gervais R, Lena H, Hureaux J, Berard H, et al. (2011) Pemetrexed and cisplatin as first-line chemotherapy for advanced non-small-cell lung cancer (NSCLC) with asymptomatic inoperable brain metastases: a multicenter phase II trial (GFPC 07-01). Ann Oncol 22: 2466–2470.
8. Kaba SE, Kyritsis AP, Hess K, Yung WK, Mercier R, et al. (1997) TPDC-FuHu chemotherapy for the treatment of recurrent metastatic brain tumors. J Clin Oncol 15: 1063–1070.
9. Nieder C (1999) Front-line chemotherapy with cisplatin and etoposide for patients with brain metastases from breast carcinoma, nonsmall cell lung carcinoma, or malignant melanoma. A prospective study. Cancer 86: 900–903.
10. Sperduto PW, Wang M, Robins HI, Schell MC, Werner-Wasik M, et al. (2013) A phase 3 trial of whole brain radiation therapy and stereotactic radiosurgery alone versus WBRT and SRS with temozolomide or erlotinib for non-small cell lung cancer and 1 to 3 brain metastases: Radiation Therapy Oncology Group 0320. Int J Radiat Oncol Biol Phys 85: 1312–1318.
11. Higgins JPT, Thompson SG, Deeks JJ and Altman DG (2003) Measuring inconsistency in meta-analyses. BMJ (Clinical research ed) 327: 557–560.
12. Ford AC, Forman D, Hunt RH, Yuan Y and Moayyedi P (2014) Helicobacter pylori eradication therapy to prevent gastric cancer in healthy asymptomatic infected individuals: systematic review and meta-analysis of randomised controlled trials. BMJ (Clinical research ed) 348: g3174.
13. Ushio Y, Arita N, Hayakawa T, Mogami H, Hasegawa H, et al. (1991) Chemotherapy of brain metastases from lung carcinoma: a controlled randomized study. Neurosurgery 28: 201–205.
14. Mehta MP, Shapiro WR, Phan SC, Gervais R, Carrie C, et al. (2009) Motexafin gadolinium combined with prompt whole brain radiotherapy prolongs time to neurologic progression in non-small-cell lung cancer patients with brain metastases: results of a phase III trial. Int J Radiat Oncol Biol Phys 73: 1069–1076.
15. Hassler MR, Pfeifer W, Knocke-Abulesz TH, Geissler K, Altorjai G, et al. (2013) Temozolomide added to whole brain radiotherapy in patients with multiple brain metastases of non-small-cell lung cancer: a multicentric Austrian phase II study. Wien Klin Wochenschr 125: 481–6.
16. Guerrieri M, Wong K, Ryan G, Millward M, Quong G, et al. (2004) A randomised phase III study of palliative radiation with concomitant carboplatin for brain metastases from non-small cell carcinoma of the lung. Lung Cancer 46: 107–111.
17. Chua D, Krzakowski M, Chouaid C, Pallotta MG, Martinez JI, et al. (2010) Whole-brain radiation therapy plus concomitant temozolomide for the treatment of brain metastases from non-small-cell lung cancer: a randomized, open-label phase II study. Clin Lung Cancer 11: 176–181.
18. Verger E, Gil M, Yaya R, Vinolas N, Villa S, et al. (2005) Temozolomide and concomitant whole brain radiotherapy in patients with brain metastases: a phase II randomized trial. Int J Radiat Oncol Biol Phys 61: 185–191.
19. Neuhaus T, Ko Y, Muller RP, Grabenbauer GG, Hedde JP, et al. (2009) A phase III trial of topotecan and whole brain radiation therapy for patients with CNS-metastases due to lung cancer. Br J Cancer 100: 291–297.
20. Biswas G, Bhagwat R, Khurana R, Menon H, Prasad N, et al. (2006) Brain metastasis-evidence based management. J Cancer Res Ther 2: 5–13.
21. Addeo R and Caraglia M (2011) Combining temozolomide with other antitumor drugs and target-based agents in the treatment of brain metastases: An unending quest or chasing a chimera? Expert Opin Investig Drugs 20: 881–895.
22. Adonizio CS, Babb JS, Maiale C, Huang C, Donahue J, et al. (2002) Temozolomide in non-small-cell lung cancer: preliminary results of a phase II trial in previously treated patients. Clin Lung Cancer 3: 254–258.
23. Addeo R, Caraglia M, Faiola V, Capasso E, Vincenzi B, et al. (2007) Concomitant treatment of brain metastasis with Whole Brain Radiotherapy WBRT and Temozolomide TMZ is active and improves Quality of Life. BMC cancer 7: doi: 10.1186/1471-2407-7-18.

21

Further Insights into Metal-DOM Interaction: Consideration of Both Fluorescent and Non-Fluorescent Substances

Huacheng Xu[1,2]*, Jicheng Zhong[2], Guanghui Yu[2,3], Jun Wu[4], Helong Jiang[2], Liuyan Yang[1]*

1 State Key Laboratory of Pollution Control and Resources Reuse, School of the Environment, Xianlin Campus, Nanjing University, Nanjing, China, **2** State Key Laboratory of Lake Science and Environment, Nanjing Institute of Geography and Limnology, Chinese Academy of Sciences, Nanjing, China, **3** College of Resources and Environmental Sciences, Nanjing Agricultural University, Nanjing, China, **4** Key Laboratory of Soil Environment and Pollution Remediation, Institute of Soil Science, Chinese Academy of Sciences, Nanjing, China

Abstract

Information on metal binding with fluorescent substances has been widely studied. By contrast, information on metal binding with non-fluorescent substances remains lacking despite the dominance of these substances in aquatic systems. In this study, the metal binding properties of both fluorescent and non-fluorescent substances were investigated by using metal titration combined with two-dimensional correlation spectroscopy (2D–COS) analysis. The organic matters in the eutrophic algae-rich lake, including natural organic matters (NOM) and algae-induced extracellular polymeric substances (EPS), both contained fluorescent and non-fluorescent substances. The peaks in the one-dimensional spectra strongly overlapped, while 2D–COS can decompose the overlapped peaks and thus enhanced the spectral resolution. Moreover, 2D FTIR COS demonstrated that the binding susceptibility of organic ligands in both NOM and algal EPS matrices followed the order: 3400>1380>1650 cm^{-1}, indicative the significant contribution of non-fluorescent ligands in metal binding. The modified Stern-Volmer equation also revealed a substantial metal binding potential for the non-fluorescent substances ($\log K_M$: 3.57~4.92). As for the effects of organic ligands on metal binding, EPS was characterized with higher binding ability than NOM for both fluorescent and non-fluorescent ligands. Algae-induced EPS and the non-fluorescent substances in eutrophic algae-rich lakes should not be overlooked because of their high metal binding potential.

Editor: Dariush Hinderberger, Martin-Luther-Universitat Halle-Wittenberg, Germany

Funding: This work was funded by the National Natural Science Foundation of China (51209192, 51379199 and 41301241), the Natural Science Foundation of Jiangsu province, China (BK2012501), the China Postdoctoral Science Foundation (2013M540438), the Jiangsu Planned Projects for Postdoctoral Research Funds (1301007A), the State Key Laboratory of Pollution Control and Resource Reuse Foundation (PCRRF13011), and the State Key Laboratory of Lake Science and Environment, Chinese Academy of Sciences (2012SKL007). The funders had no role in study design, data collection and analysis, decision to publish, or preparation of the manuscript.

Competing Interests: The authors have declared that no competing interests exist.

* Email: hcxu@niglas.ac.cn (HCX); yangly@nju.edu.cn (LYY)

Introduction

Dissolved organic matter (DOM) in aquatic environments can interact with metals and consequently affect their toxicity, mobility and bioavailability [1,2]. Investigation into the binding properties of DOM is necessary to clearly understand the biogeochemical behavior of metals. DOM in waters is generally characterized with heterogeneous organic compositions including fluorescent (i.e., protein-, humic-, and fulvic-like) and non-fluorescent (i.e., carbohydrates, lipids, and lignins) substances [3–5]. Thus, a comprehensive study on metal-DOM interaction should include related information on both fluorescent and non-fluorescent substances.

Fluorescence quenching titration is a popular method in the study of metal-ligand interaction [6–8]. Saar and Weber [9] compared synchronous fluorescence (SF) quenching titration and ion-selective electrode potentiometry and proposed the former. Ghatak et al. [10] later studied the binding heterogeneity of sediment humic substances and reported more energy binding sites for Fe(III). More recently, through SF quenching titration technique, the heterogeneous metal binding behaviors between leaf litter and soil DOM can be explored [7,8]. Although widely used, SF technique is not applicable for non-florescent substances, even with these making up a significant proportion of aquatic DOMs [11].

By contrast, Fourier transform infrared spectroscopy (FTIR) can be used to explore non-fluorescent substances [12,13]. Binding information on non-fluorescnt substances can be obtained by investigating spectral variations in response to metal addition. Thus, the metal binding properties of both fluorescent and non-fluorescent substances can be obtained through the combination of the SF and FTIR techniques.

However, one-dimensional SF and FTIR spectra usually exhibited many overlapped peaks because of the heterogeneous nature of the investigated DOM [6,10,14]. Hence, further analysis based solely on the traditional visual inspection approach is laborious and difficult. Recent studies showed that two-dimensional correlation spectroscopy (2D–COS) can be used to solve the peak overlapping problems by extending peaks along the second dimension [12,14]. This technique can also be used to explore the sequence of metal-sensitive spectral changes through asynchronous map, which benefits the analysis of binding mechanisms [7,8,15,16]. To date, only one study on the application of 2D–COS for exploring the metal binding properties of soil organic matters has been published [16].

Anthropogenically induced eutrophication in recent years has caused the increasing occurrence of cyanobacterial blooms in lakes, especially in eutrophic shallow lakes [17]. Our previous studies have demonstrated that DOM in such algal-rich lakes should consist of natural organic matter (NOM) and algal extracellular polymeric substances (EPS). EPS can be further fractionated into loosely bound EPS (LB–EPS) and tightly bound EPS (TB–EPS) in terms of its binding strength with algal cells [18,19]. Due to the high reactively, both NOM and algae-induced EPS can bind with metals and consequently play an important role in the toxicity, mobility, and bioavailability of metals.

The present study were therefore aims (1) to investigate the metal-induced spectral variations of both fluorescent and non-fluorescent substances in NOM and algal EPS fractions, and (2) to explore the detailed metal binding behavior and mechanisms by using 2D–COS combined with SF and FTIR techniques. For these purposes, the interactions of a commonly found metal Cu(II) in eutrophic shallow lakes [20] with NOM, LB–EPS and TB–EPS were respectively studied. The results of this study would provide a comprehensive picture of metal-DOM interaction in aquatic environments.

Materials and Methods

Ethics statement

No specific permits were required for the described field studies. The location studied is not privately-owned or protected in any way and our studies did not involve any endangered or protected species.

Sample preparation and EPS extraction

A total of 2.4 L of surface water samples (30 cm from the surface) containing 1.0 g L^{-1} algal biomass were collected from Meiliang Bay, which is located in the northern region of Lake Taihu (Figure S1 in File S1). The samples were collected in the summer-fall period because typical algal blooms occurred during this period [21]. All samples were transported to a laboratory within several hours and firstly filtered through the 0.45 μm polytetrafluoroethylene membranes (Xingya Purification Materials Co., Shanghai, China) for NOM measurement. The residues were carefully scraped and dissolved with a 0.05% NaCl solution to the original volume and centrifuged at 5000 *g* for 15 min, with the liquid collected carefully for measurement of LB–EPS. Afterward, the harvested samples were re-suspended in NaCl solution, heated at 60°C for 30 min and centrifuged at 15,000 *g* for 20 min, with the liquid collected as TB–EPS fraction. More information on sample collection and fractionation procedure can be found elsewhere [18,19].

Cu(II) titration and complexation modeling

Titration experiments were carried out by adding 0.1 mol L^{-1} $Cu(NO_3)_2$ solutions to a series of sealed brown vials containing 50 ml of diluted solution [dissolved organic carbon (DOC): 10 mg L^{-1}] using an automatic syringe [1,2,19,22]. Metal concentrations ranging from 0 to 100 μmol L^{-1} were obtained in the final solutions by adding no more than 50 μl metal titrant. The pH was maintained at 6.0 through adding 0.1 M HCl or NaOH solution under which no precipitate was formed [1]. All solution samples were shaken for 24 h at room temperature to ensure complexation equilibrium [2]. Afterward, the titrated solutions and freeze-dried samples underwent SF and FTIR analyses, respectively. Each titration experiment was performed in duplicate.

The modified Stern-Volmer equation [23] was applied to estimate the conditional stability constants between metals and SF/FTIR-derived peaks:

$$\frac{F_o}{F_o - F} = \frac{1}{f \cdot K_M \cdot C_M} + \frac{1}{f} \quad (1)$$

Here F and F_0 are the measured SF or FTIR intensities at the metal concentration C_M and the beginning of the titration (i.e., no metal addition), respectively. The parameters K_M and f represents the conditional stability constant and the fraction of the initial spectral intensities which correspond to the metal binding.

Two–dimensional correlation spectroscopy

SF was measured using the Hitachi F–7000 fluorescence spectrometer (Hitachi High Technologies, Tokyo, Japan) in synchronous mode at room temperature by ranging the excitation wavelengths from 200 to 450 nm with a constant offset (60 nm) [7,24]. The freeze-dried samples were mixed with KBr (sample to KBr ratio of 1:100, w/w) and then compressed to form a disc. FTIR spectra were obtained by collecting 64 scans at a resolution of 2 cm^{-1} with a Nicolet Nexus 870 FTIR spectrometer [25].

The 2D–COS was produced according to the method of Noda and Ozaki [14]. The metal addition was used as the external perturbation, and thus a set of concentration–dependent SF/FTIR spectra were obtained.

For the perturbation–induced spectral variation y(x, t) as a function of a spectral variable (x) and a perturbation variable (t), the dynamic spectrum $\tilde{y}(x,t)$ is formally defined as follows:

$$\tilde{y}(x,t) = \begin{cases} y(x,t) - \bar{y}(x) \; for \; T_{\min} \le t \le T_{\max} \\ 0 \quad otherwise \end{cases} \quad (2)$$

where $\bar{y}(x)$ denotes the reference spectra, which is typically the average spectrum and is expressed as $\bar{y}(x) = \frac{1}{T_{\max} - T_{\min}} \int_{T}^{T_{\max}} y(x,t)dt$. The synchronous correlation spectroscopy can be written as follows:

$$\phi(x_1,x_2) = \frac{1}{T_{\max} - T_{\min}} \int_{T}^{T_{\max}} y(x_1,t) \cdot z(x_2,t)dt \quad (3)$$

The asynchronous correlation spectroscopy can be obtained directly from the dynamic spectrum $\tilde{y}(x_2,t)$and its orthogonal spectrum$\tilde{z}(x_2,t)$.

$$\varphi(x_1,x_2) = \frac{1}{T_{\max} - T_{\min}} \int_{T}^{T_{\max}} \tilde{y}(x_1,t) \cdot \tilde{z}(x_2,t)dt \quad (4)$$

Prior to 2D–COS analysis, the SF and FTIR spectra were normalized by the summed absorbance from 200 to 450 nm and

400 to 4000 cm^{-1}. The noise components were removed by principal component analysis [16,22]. Then the reconstructed data matrix was progressed using the "2D shige" software (Kwansei Gakuin University, Japan).

Results and Discussion

Determination of the organic matters in the eutrophic algae-rich lake

Figure 1 shows the typical SF and FTIR spectra for the NOM and algal EPS matrices. NOM was characterized with two prominent peaks (230, 280 nm) and one shoulder peak (300~390 nm), whereas the EPS matrix showed only the prominent peaks (Fig. 1a). Peaks located at 230 and 280 nm were ascribed to tyrosine- and tryptophan-like substances, respectively, whereas peaks ranging from 300 to 390 nm were attributed to humic-like substances [7]. The SF results indicated that the algal EPS mainly contained fluorescent protein-like substances, while not only protein-like but also humic-like substances were observed in NOM. In the previous studies aiming at lab-cultured cynobacterium *M. aeruginosa*, Xu et al. [18] detected the protein-like substances in all EPS fractions but the humic-like substances in only the soluble EPS fractions. And Qu et al. [25] demonstrated that the mainly components in bound and dissolved extracellular organic matters were protein-like substances and protein-like + humic-like substances, respectively. These similar results indicated that the distribution patterns of organic compositions were not form-specific.

As for the FTIR spectroscopy, the NOM and algal EPS matrices were both characterized with four obvious absorption peaks at 3400, 1650, 1380, and 1100 cm^{-1} (Fig.1b). We assigned these bands as follows: the band at 1650 cm^{-1} was ascribed to the CO stretching of amide I in protein compounds, the band at 1380 cm^{-1} to the CH deformations in aliphatic groups, and the bands at 3400 and 1100 cm^{-1} to the stretching vibration of OH and CO in carbohydrate-like substances [16,26]. The combined results of SF and FTIR showed that DOM in the eutrophic algae-rich lakes, including NOM and algal EPS, definitely contained both fluorescent (230, 280, and 300~390 nm, and 1650 cm^{-1}) and non-fluorescent (3400, 1380, and 1100 cm^{-1}) substances.

It should be noted that the organic compositions of EPS matrix were highly dependent on the extraction methods. For example, Liu et al. [27] compared five methods for *M. aeruginosa* EPS extraction and found different fluorescent properties among the five methods. A further study also demonstrated that EPS using cation exchange resin extraction contained more aromatic condensed structures with less acidic functional groups when compared with formaldehyde/NaOH extraction [28]. The heterogeneity in organic compositions exhibits different metal binding ability, which will be discussed in the following section.

Metal binding properties of the fluorescent substances

Changes in one dimensional SF spectra with metal addition. The effects of Cu(II) addition on the variations of SF spectra are shown in Fig.2. Fluorescence quenching generally occurred with increasing metal addition. This result revealed strong metal binding and the marked modification of the electronic structures of DOM [7,29]. Further analysis showed that the behavior of spectral changes between NOM and EPS exhibited obvious differences. Monotonic fluorescence quenching was observed for NOM along the entire wavelength region (200~450 nm), whereas both fluorescence quenching (280 nm) and enhancement (230 nm) were identified in the EPS matrix. The complex formation and the fluorophore-quencher collision were responsible for fluorescence quenching [30,31], while the fluorescence enhancement was attributed to the changes in quantum yields of tyrosine-like substances in algal EPS matrix caused by high levels of Cu(II) and/or interactions with inorganic and other organic components [2]. Metal-induced fluorescence enhancement was previously reported in other studies but via fluorescence emission and excitation matrix analysis [1,19,32].

In addition, the quenching rates and degrees were highly related to the organic ligands. Metal addition caused a sharp and gradual reduction at 250~300 nm and 300~370 nm, respectively for NOM. However, the algal EPS matrix exhibited a more pronounced quenching degree than NOM at 250~300 nm. This result indicated that the protein-like substances in NOM might not be as involved in metal binding as those in the algal EPS matrix [7]. The different quenching rates and degrees observed here indicated that the functional sites for metal binding in the NOM and algal EPS matrices were unevenly distributed.

Metal binding sequencing investigated by 2D fluorescence COS. Although SF can capture heterogeneous

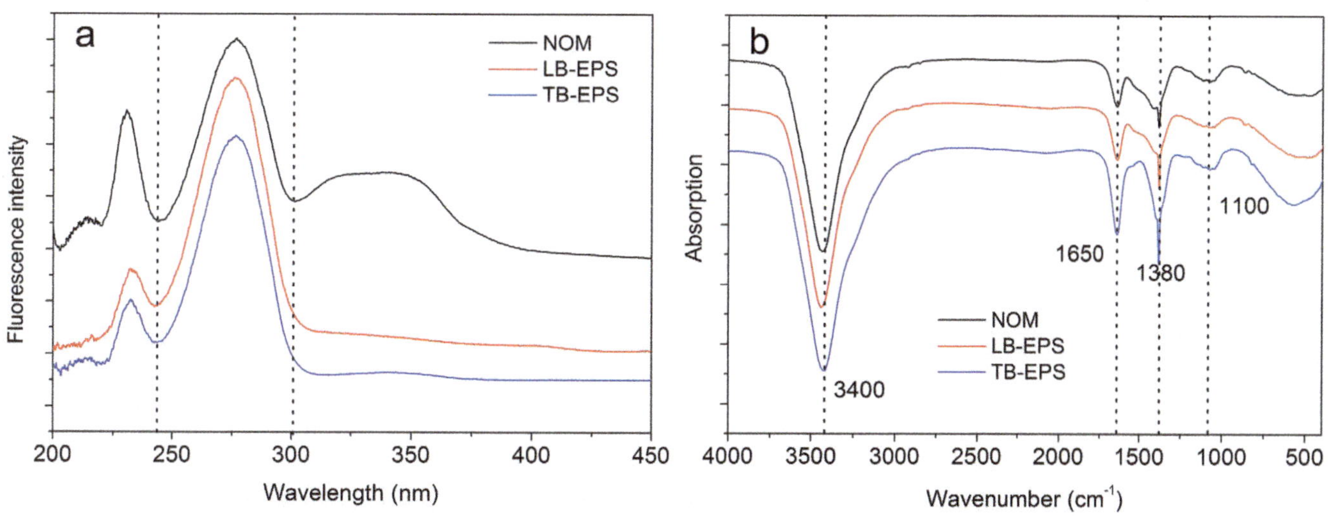

Figure 1. Typical spectral shapes of SF (a) and FTIR (b) for NOM and algal EPS matrix in the eutrophic algae-rich lake.

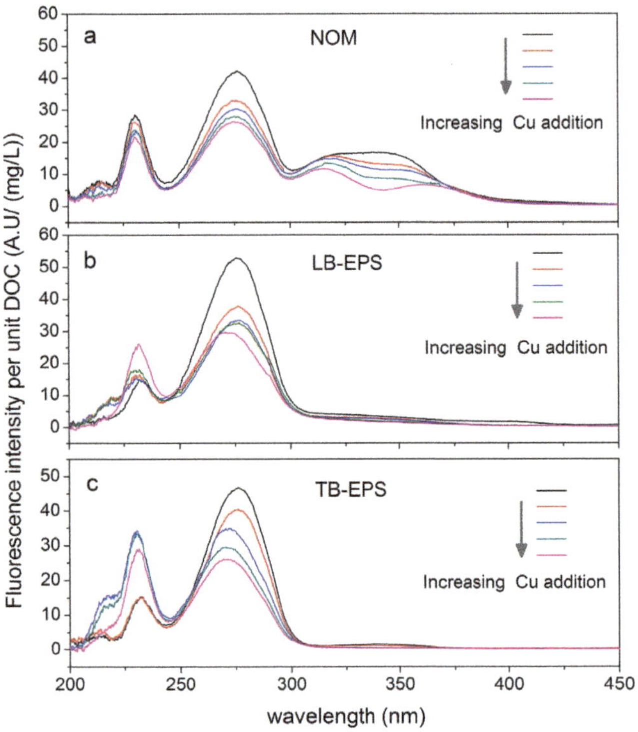

Figure 2. Changes in the one-dimensional SF intensities with Cu(II) addition. The arrows refer to the direction of the increasing metal concentrations.

quenching degrees and rates, extensive information on binding sites and binding sequence cannot be obtained via the one-dimensional spectral technique. 2D–COS was applied in the present study to explore binding behavior in detail (Fig. 3). The synchronous map displayed two autopeaks at 230/230 and 280/280 nm along the diagonal line for both NOM and algal EPS. The synchronous map is a symmetric spectrum with respect to the diagonal line. The correlation peaks include autopeaks and crosspeaks on the diagonal and off-diagonal lines, respectively. An autopeak represents the overall susceptibility of the corresponding spectral region to the change in spectral intensity as an external perturbation is applied to the system [14,16,22]. The intensities of the autopeaks followed the order: 230>280 nm for NOM and 280>230 nm for the EPS matrix. This suggested that tyrosine-like substances in NOM were more susceptible to metal addition compared with the tryptophan-like substances, whereas the tryptophan-like substances in algal EPS matrix exhibited higher metal sensitivity than the tyrosine-like substances [8,14].

A crosspeak in synchronous map represents the simultaneous or coincidental changes of two different spectral variables. NOM and algal EPS were respectively characterized with positive and negative crosspeak at 280/230 nm. This characterization implies that tyrosine- and tryptophan-like substances in NOM covary in their interaction with metal addition, whereas those in algal EPS might vary with opposite directions. Consistent with our previous observations (Fig. 2), NOM demonstrated monotonic quenching effects, whereas the algal EPS matrix demonstrated both fluorescence quenching and enhancement.

An asynchronous map reveals the sequential or successive changes in spectral intensities with increasing metal addition. As shown, one and three negative crosspeaks were observed above the diagonal line for the NOM and algal EPS matrices, respectively. Based on Noda's rule [14], the sequences of binding affinities followed the order: 230>275 nm for NOM, 265>280>225>230 nm for LB–EPS and 270>280>230>232 nm for TB–EPS. The asynchronous maps revealed that metal binding in NOM took place at the shorter wavelengths (~230 nm) prior to that at the longer wavelength range (>265 nm); while for the algal EPS matrix, meals binding first occurred at the longer wavelengths. As indicated by these results, the preferential metal ligands in NOM and algal EPS were tyrosine- and tryptophan-like substances, respectively. In addition, the asynchronous maps also showed that the wavelength 230 nm in the one-dimensional SF spectra overlapped with 230 and 232 nm, and the wavelength 280 nm overlapped with 270, 275 and 280 nm. Unlike the one-dimensional spectral technique, 2D–COS can enhance spectral resolution and yield extensive detailed binding information.

Information on metal binding with both fluorescent and non-fluorescent substances

FTIR can be used to characterize non-fluorescent substances [12,13]. The absorption variations of the normalized one-dimensional FTIR spectra of NOM and algal EPS in response to Cu(II) addition are given in Figure S2 in File S1. Although four absorption peaks were obviously observed at 3400, 1650, 1380 and 1100 cm^{-1}, the one-dimensional FTIR was strongly overlapped and information on the interrelationship among metal-sensitive functional groups cannot be obtained.

Therefore, 2D–COS was applied to explore in detail the binding information on fluorescent and non-fluorescent substances (Fig.4). We mainly focused on the region between 1700~1300 cm^{-1} because this spectral region usually contains the major bands of the amide, carboxylic, ester, and carbohydrate functional groups [12,16]. Both NOM and algal EPS matrices were characterized with two autopeaks located at 1380/1380 and 1650/1650 cm^{-1}, with the former exhibiting higher intensity than the latter. Combined with the synchronous maps of the 4000–400 cm^{-1} region (Figure S3 in File S1), the peak at 3400 cm^{-1} was more susceptible to Cu(II) addition, followed by the peak at 1380 cm^{-1} and 1650 cm^{-1}. This result revealed the substantial contributions of non-fluorescent substances (3400, 1380 cm^{-1}) in metal binding. Zhang et al. [31] also emphasized the metal binding potential of non-fluorescent carboxyl and hydroxyl functional groups in biofilm EPS, although no further evidence was given in their studies.

By contrast, the asynchronous maps revealed that the sequencing of binding affinities of the peaks followed the order: amide I in proteins (1650 cm^{-1})> lignin and aliphatic CH (1380 cm^{-1}) for NOM and LB–EPS, and amide I in proteins (1650 cm^{-1})> lignin and aliphatic CH (1380 cm^{-1})> cellulose CH_2 (1420 cm^{-1}) for TB–EPS [14]. Therefore, metal Cu(II) preferentially bound with the fluorescent protein-like substances, followed by the non-fluorescent liquid- and carbohydrate-like substances, regardless of the difference in DOM fractions (i.e., NOM, LB–EPS and TB–EPS).

Metal binding abilities of both fluorescent and non-fluorescent substances

The conditional stability constants ($\log K_M$) based on the modified Stern-Volmer equation are tabulated in Table 1. The $\log K_M$ values for EPS matrix were calculated within the range from 4.66 to 5.03, which were slightly higher than those reported in Lee et al. [28] but lower than those in McIntyre and Gueguen [33]. The heterogeneous binding capacities were partly attributed to the different organic compositions and functional groups, which

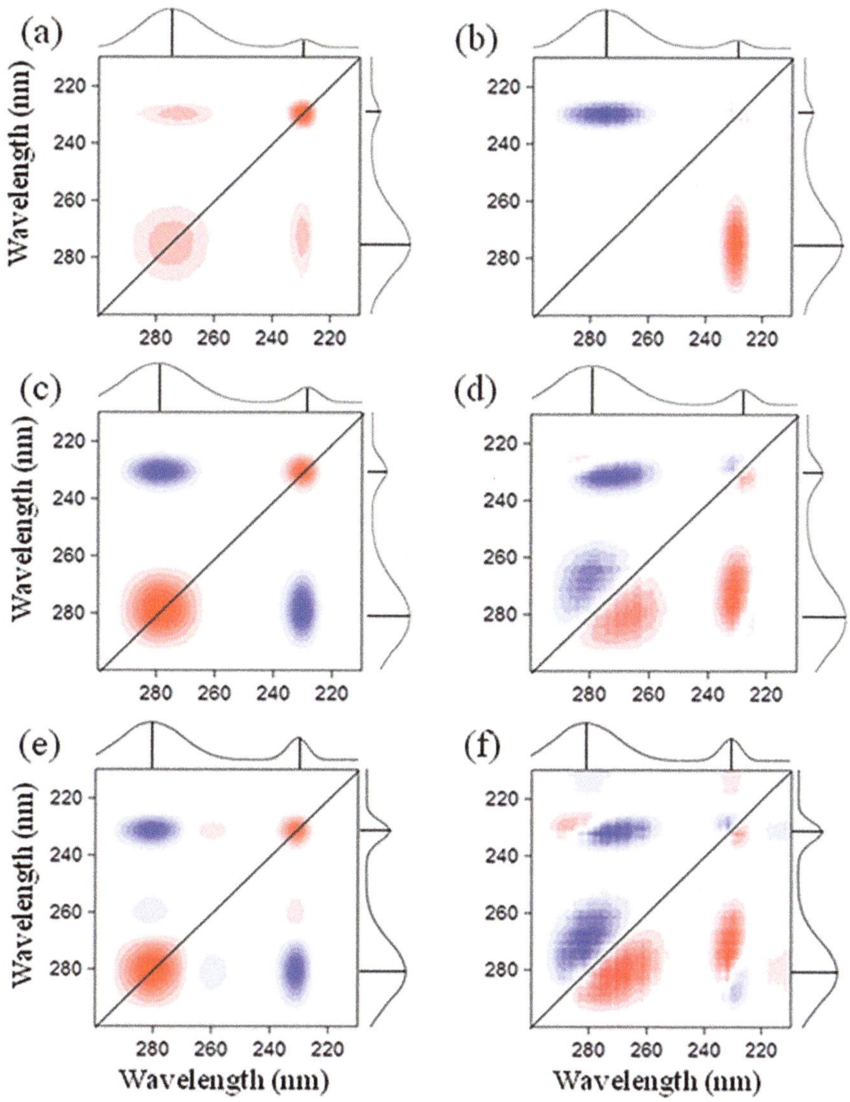

Figure 3. The 2D fluorescence correlation maps generated from 210 to 300 nm region for NOM and EPS matrix with increasing Cu(II) addition. (a) synchronous map for NOM; (b) asynchronous for NOM; (c) synchronous map for LB–EPS; (d) asynchronous for LB–EPS; (e) synchronous map for TB–EPS; (f) asynchronous for TB–EPS. Red represents positive correlations and blue represents negative correlations; higher color intensity indicates a stronger positive or negative correlation.

were highly dependent on the extraction methods [28]. Further analysis showed that the stability constants between NOM and EPS exhibited obvious differences. Specifically, the logK_M of tryptophan-like substances in NOM (4.51±0.06) was lower than that in the LB–EPS and TB–EPS fractions (>4.66±0.05). The results indicated that the tryptophan-like substances in the algal EPS exhibited higher binding abilities than those in NOM [19].

For the FTIR-derived peaks, the peak at 1650 cm^{-1} showed a higher logK_M value than the peaks at 1100, 1380 and 1420 cm^{-1} for NOM (4.62±0.08 vs. <4.27±0.05), LB–EPS (4.79±0.04 vs. < 4.59±0.08) and TB–EPS (4.96±0.07 vs. <4.92±0.03). This result implies that the binding ability of fluorescent proteins is stronger than that of the non-fluorescent liquids and carbohydrates. The calculated logK_M values were consistent with the results of asynchronous correlation spectroscopy, thus providing evidence for the feasibility of 2D–COS in the investigation of metal-DOM interaction. Further analysis showed that the algal EPS matrix exhibited a higher logK_M value than NOM for both the fluorescence proteins (>4.79±0.04 vs. 4.62±0.08) and the non-fluorescence liquids and carbohydrates (>4.26±0.10 vs. < 3.57~4.27). Therefore, the algal EPS matrix is expected to play a more important role than NOM in the migration and detoxification of heavy metals in eutrophic algae-rich lakes.

Significance of this study

Both fluorescent and non-fluorescent substances were detected in the DOM found in lakes (Fig. 1). Previous studies on metal-DOM interaction mainly focused on fluorescent substances [1,34,35], lacking the related information on non-fluorescent substances. Non-fluorescent substances may sometimes even occupy a significant proportion in aquatic environments [11]. The contribution of non-fluorescent substances in metal binding was investigated for the first time in this study, which is an

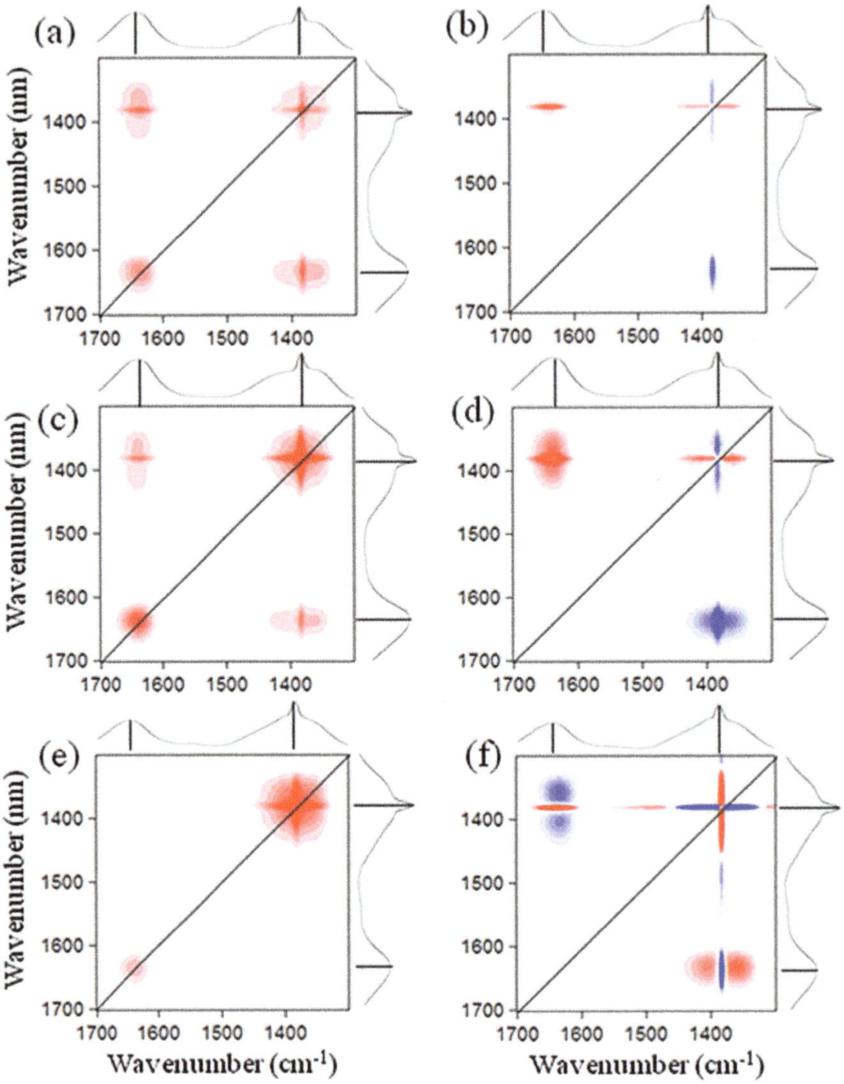

Figure 4. The 2D infrared correlation maps generated from 1300 to 1700 cm^{-1} region for NOM and EPS matrix with increasing Cu(II) addition. (a) synchronous map for NOM; (b) asynchronous for NOM; (c) synchronous map for LB–EPS; (d) asynchronous for LB–EPS; (e) synchronous map for TB–EPS; (f) asynchronous for TB–EPS. Red represents positive correlations and blue represents negative correlations; higher color intensity indicates a stronger positive or negative correlation.

important complement to the research scope of metal-DOM interaction.

Algal blooms have become common in recent years because of climatic changes and nutrient enrichment [17]. This phenomenon has made the algae-induced EPS an important contributor to the DOM pool [33]. Although studies on metal binding with NOM have been carried out [2,8], information on the metal-EPS interactions in eutrophic algae-rich lakes remains insufficient. This information would help provide a comprehensive picture of metal-DOM interactions in aquatic environments. The higher metal-EPS binding capacities observed in this work highlighted the ecological importance of algal EPS in the toxicity, mobility, and bioavailability of metals because the toxicity of metals is related to free iron activity [36].

Although the metal binding degrees and rates can be explored through a one-dimensional spectral analysis, the interrelationships among different organic ligands cannot be obtained. The 2D–COS applied in this study clearly demonstrated the heterogeneous metal binding sites between NOM and algal EPS. More importantly, 2D–COS can capture the binding sequencing of metal-sensitive spectral variations, which could improve our understanding about metal binding mechanisms. This also allows the scope of 2D–COS to be unlimited in the fields of metal binding within lake ecosystems [7]. The interaction of DOM with other objects (i.e., nanoparticles and persistent organic pollutants) in selected aquatic environments (i.e., groundwater and wastewater) can also be explored by 2D–COS in future studies. External perturbation need not be limited to metal concentration but may include other parameters, such as temperature, pressure, and pH.

Conclusions

DOM in eutrophic algae-rich lakes was fractionated into NOM, LB–EPS, and TB–EPS fractions. Both fluorescent and non-fluorescent substances were determined in these fractions. Although fluorescence titration was an effective method for

Table 1. Conditional stability constants ($\log K_M$) for SF/FTIR-derived peaks binding to Cu(II) as calculated using the modified Stern-Volmer equation.

Samples	Wavelength (nm)	$\log K_M$	R^2	Samples	Wavenumber (cm^{-1})	$\log K_M$	R^2
	230	4.69±0.04	>0.87		1100	3.57±0.04	>0.93
NOM	275	4.51±0.06	>0.72	NOM	1380	4.27±0.05	>0.88
					1650	4.62±0.08	>0.71
	230	Not modeled			1100	4.35±0.03	>0.92
LB-EPS	265	4.94±0.07	>0.51	LB-EPS	1380	4.59±0.08	>0.71
	280	4.66±0.05	>0.84		1650	4.79±0.04	>0.86
	230	Not modeled			1100	4.74±0.05	>0.80
	233	Not modeled			1380	4.92±0.03	>0.86
TB-EPS	270	5.03±0.04	>0.88	TB-EPS	1420	4.26±0.10	>0.62
	280	4.74±0.03	>0.95		1650	4.96±0.07	>0.76

investigating metal-ligand interaction, FTIR was able to determine both fluorescent and non-fluorescent substances. 2D–COS can enhance spectral resolution and thus can be used to explore detailed binding properties such as binding sites and binding sequencing. The metal binding properties of non-fluorescent substances were quantified, and the significance of the algal EPS matrix in metal detoxification was emphasized. In sum, 2D–COS combined with SF and FTIR analysis is an effective approach to explore the metal binding properties of both fluorescent and non-fluorescent substances.

Supporting Information

File S1 Supporting figures. Figure S1. The map of Lake Taihu with the location of the sampling station; Figure S2 The variations of the normalized one-dimensional FTIR with metal addition; Figure S3 The 2D synchronous correlation maps generated from 4000–400 cm^{-1} region of the FTIR spectra for NOM and algal EPS.

Author Contributions

Conceived and designed the experiments: HCX LYY. Performed the experiments: JCZ GHY. Analyzed the data: JW. Contributed reagents/materials/analysis tools: HLJ. Wrote the paper: HCX.

References

1. Ohno T, Amirbahman A, Bro R (2008) Parallel factor analysis of excitation–emission matrix fluorescence spectra of water soluble soil organic matter as basis for the determination of conditional metal binding parameters. Environ Sci Technol 42: 186–192.
2. Yamashita Y, Jaffe R (2008) Characterizing the interactions between trace metals and dissolved organic matter using excitation-emission matrix and parallel factor analysis. Environ Sci Technol 42: 7374–7379.
3. Wang Y, Zhang D, Shen ZY, Feng CH, et al. (2013) Revealing sources and distribution changes of dissolved organic matter (DOM) in pore water of sediment from the Yangtze estuary. PLoS ONE 8: e76633.
4. Yao X, Zhang YL, Zhu GW, Qin BQ, Feng LQ et al. (2011) Resolving the variability of CDOM fluorescence to differentiate the sources and fate of DOM in Lake Taihu and its tributaries. Chemosphere 82: 145–155.
5. Zhang YL, Yin Y, Feng LQ, Zhu GW, Shi ZQ, et al. (2011) Characterizing chromophoric dissolved organic matter in Lake Tianmuhu and its catchment basin using excitation–emission matrix fluorescence and parallel factor analysis. Water Res 45: 5110–5122.
6. Cabaniss SE (1992) Synchronous fluorescence spectra of metalfulvic acid complexes. Environ Sci Technol 26: 1133–1139.
7. Hur J, Lee BM (2011). Characterization of binding site heterogeneity for copper within dissolved organic matter fractions using two-dimensional correlation fluorescence spectroscopy. Chemosphere 83: 1603–1611.
8. Hur J, Lee BM (2011) Comparing the heterogeneity of copper-binding characteristics for two different-sized soil humic acid fractions using fluorescence quenching combined with 2D-COS. Scientific World Journal 11: 1865–1876.
9. Saar RA, Weber JH (1980) Comparison of spectrofluorometry and ion-selective electrode potentiometry for determination of complexes between fulvic acid and heavy-metal ions. Anal Chem 52: 2095–2100.
10. Ghatak H, Mukhopadhyay SK, Jana TK, Sen BK, Sen S (2004) Interactions of Cu(II) and Fe(III) with mangal humic substances studied by synchronous fluorescence spectroscopy and potentiometric titration. Wetlands Ecol Manage 12: 145–155.
11. Leenheer JA, Croue JP (2003) Characterizing aquatic dissolved organic matter. Environ Sci Technol 37: 18–26.
12. Hussain ANA, Elizabeth CM, Patrick GH (2010) Using two-dimensional correlations of ^{13}C NMR and FTIR to investigate changes in the chemical composition of dissolved organic matter along an estuarine transect. Environ Sci Technol 44: 8044–8049.
13. Landry C, Tremblay L (2012) Compositional differences between size classes of dissolved organic matter from freshwater and seawater revealed by an HPLC-FTIR system. Environ Sci Technol 46: 1700–1707.
14. Noda I, Ozaki Y (2004) Two-dimensional correlation spectroscopy: Applications in vibrational and optical spectroscopy. John Wiley and Sons Inc., London.
15. Li XM, Shen QR, Zhang DQ, Mei XL, Ran W, et al. (2013) Functional groups determine biochar properties (pH and EC) as studied by two-dimensional ^{13}C NMR correlation spectroscopy. PLoS ONE 8: e65949.

16. Yu GH, Wu MJ, Wei GR, Luo YH, Ran W, et al. (2012) Binding of organic ligands with Al(III) in dissolved organic matter from soil: Implications for soil organic carbon storage. Environ Sci Technol 46: 6102–6109.
17. Paerl HW, Paul VJ (2011) Climate change: Links to global expansion of harmful cyanobacteria. Water Res 46: 1349–1363.
18. Xu HC, Cai HY, Yu GH, Jiang HL (2013) Insights into extracellular polymeric substances of cyanobacterium *Microcystis aeruginosa* using fractionation procedure and parallel factor analysis. Water Res 47: 2005–2014.
19. Xu HC, Yan ZS, Cai HY, Yu GH, Yang LY, et al. (2013) Heterogeneity in metal binding by individual fluorescent components in a eutrophic algae-rich lake. Ecotox Environ Safe 98: 266–272.
20. Shen J, Liu EF, Zhu YX, Hu SY, Qu WC (2007) Distribution and chemical fractionation of heavy metals in recent sediments from Lake Taihu, China. Hydrobiologia 581: 141–150.
21. Qin BQ, Zhu GW, Gao G, Zhang YL, Li W, et al. (2010) A drinking water crisis in lake Taihu, China: linkage to climatic variability and lake management. Environ Manage 45: 105–112.
22. Xu HC, Yu GH, Yang LY, Jiang HL (2013) Combination of two-dimensional correlation spectroscopy and parallel factor analysis to characterize the binding of heavy metals with DOM in lake sediments. J Hazard Mater 263: 412–421.
23. Hays MD, Ryan DK, Pennell S (2004) A modified multisite Stern-Volmer equation for the determination of conditional stability constants and ligand concentrations of soil fulvic acid with metal ions. Anal Chem 76: 848–854.
24. Xu HC, Jiang HL (2013) UV-induced photochemical heterogeneity of dissolved and attached organic matter associated with cyanobacterial blooms in a eutrophic freshwater lake. Water Res 47: 6506–6515.
25. Qu FS, Liang H, He JG, Ma J, Wang ZZ, et al. (2012) Characterization of dissolved extracellular organic matter (dEOM) and bound extracellular organic matter (bEOM) of *Microcystis aeruginosa* and their impacts on UF membrane fouling. Water Res 46: 2881–2890.
26. Xu HC, Cai HY, Yu GH, Jiang HL (2013) Investigation on extracellular polymeric substances from mucilaginous cyanobacterial blooms in eutrophic freshwater lakes. Chemosphere 93: 75–81.
27. Liu LZ, Qin BQ, Zhang YL, Zhu GW, Gao G (2014) Extraction and characterization of bound extracellular polymeric substances from cultured pure cyanobacterium (*Microcystis wesenbergii*). J Environ Sci 26: 1725–1732.
28. Lee BM, Shin HS, Hur J (2013) Comparison of the characteristics of extracellular polymeric substances for two different extraction methods and sludge formation conditions. Chemosphere 90: 237–244.
29. Plaza C, Brunetti G, Senesi N, Polo A (2006) Molecular and quantitative analysis of metal ion binding to humic acids from sewage sludge and sludge-amended soils by fluorescence spectroscopy. Environ Sci Technol 40: 917–923.
30. Pan XL, Liu J, Zhang DY (2010) Binding of phenanthrene to extracellular polymeric substances (EPS) from aerobic activated sludge: A fluorescence study. Colloids and Surfaces B: Biointerfaces 80: 103–106.
31. Zhang DY, Pang XL, Mostofaa KMG, Chen X, Mu GJ, et al. (2010) Complexation between Hg(II) and biofilm extracellular polymeric substances: An application of fluorescence spectroscopy. J Hazard Mater 175: 359–365.
32. Hernández D, Plaza C, Senesi N, Polo A (2006) Detection of copper(II) and zinc(II) binding to humic acids from pig slurry and amended soils by fluorescence spectroscopy. Environ Pollut 143: 212–220.
33. McIntyre AM, Gueguen C (2013) Binding interactions of algal-derived dissolved organic matter with metal ions. Chemosphere 90: 620–626.
34. Wu J, Zhang H, Shao LM, He PJ (2012) Fluorescent characteristics and metal binding properties of individual molecular weight fractions in municipal solid waste leachate. Environ Pollut 162: 63–71.
35. Chai XL, Liu GX, Zhao X, Hao YX, Zhao YC (2012) Complexion between mercury and humic substances from different landfill stabilization processes and its implication for the environment. J Hazard Mater 209–210: 59–66.
36. Sheng GP, Xu J, Luo HW, Li WW, Li WH, et al. (2013) Thermodynamic analysis on the binding of heavy metals onto extracellular polymeric substances (EPS) of activated sludge. Water Res 47: 607–614.

22

Usnic Acid Potassium Salt: An Alternative for the Control of *Biomphalaria glabrata* (Say, 1818)

Mônica C. B. Martins[1*], Monique C. Silva[1], Luanna R. S. Silva[2], Vera L. M. Lima[1], Eugênia C. Pereira[3*], Emerson P. S. Falcão[4], Ana M. M. A. Melo[2], Nicácio Henrique da Silva[1]

1 Departamento de Bioquímica, Universidade Federal de Pernambuco, Pernambuco, Brazil, **2** Departamento de Radiobiologia, Universidade Federal de Pernambuco, Pernambuco, Brazil, **3** Departamento de Ciências Geográficas, Universidade de Pernambuco, Pernambuco, Brazil, **4** Centro Acadêmico de Vitória de Santo Antão, Universidade Federal de Pernambuco, Pernambuco, Brazil

Abstract

In Brazil, the snail *Biomphalaria glabrata* is the most important vector of schistosomiasis due to its wide geographical distribution, high infection rate and efficient disease transmission. Among the methods of schistosomiasis control, the World Health Organization recommends the use of synthetic molluscicides, such as niclosamide. However, different substances of natural origin have been tested as alternatives for the control or eradication of mollusks. The literature describes the antitumor, antimicrobial and antiviral properties of usnic acid as well as other important activities of common interest between medicine and the environment. However, usnic acid has a low degree of water solubility, which can be a limiting factor for its use, especially in aquatic environments, since the organic solvents commonly used to solubilize this substance can have toxic effects on aquatic biota. Thus, the aim of the present study was to test the potassium salt of usnic acid (potassium usnate) with regard to molluscicidal activity and toxicity to brine shrimp (*Artemia salina*). To obtain potassium usnate, usnic acid was extracted with diethyl ether isolated and purified from the lichen *Cladonia substellata*. Biological assays were performed with embryos and adult snails of *B. glabrata* exposed for 24 h to the usnate solution solubilized in dechlorinated water at 2.5; 5 and 10 µg/ml for embryos, 0.5; 0.9; 1;5 and 10 µg/ml for mollusks and 0.5; 1; 5; 10 µg/ml for *A. salina*. The lowest lethal concentration for the embryos and adult snails was 10 and 1 µg/ml, respectively. No toxicity to *A. salina* was found. The results show that modified usnic acid has increased solubility (100%) without losing its biological activity and may be a viable alternative for the control of *B. glabrata*.

Editor: Peter P. Fong, Gettysburg College, United States of America

Funding: MCBM received a doctoral grant by Coordenação de Aperfeiçoamento de Pessoal de Nível Superior (CAPES) (www.capes.gov.br). ECP has a grant in research productivity by Conselho Nacional de Desenvolvimento Cientifico e Tecnológico (CNPq) (www.cnpq.br). The funders had no role in study design, data collection and analysis, decision to publish or preparation of the manuscript.

Competing Interests: The authors have declared that no competing interests exist.

* Email: monicabarmartins@hotmail.com (MCBM); verticillaris@gmail.com (ECP)

These authors contributed equally to this work.

Introduction

Schistosomiasis is a parasitic disease distributed in 74 countries on the African, Asian and American continents. In Latin America, this disease is found in Venezuela, the Caribbean islands and Brazil. In the latter country, public health agencies reported 0.28 deaths per 100 million inhabitants in 2009, which corresponds to 1.1% of deaths related to infectious diseases [1]. While these data from 2009 seem to indicate a reduction in the transmission of this disease in Brazil, there were 820 cases of the most severe forms between 2000 and 2010, with approximately 505 deaths in the period [2]. It is therefore clear that access of infected individuals to diagnosis and treatment remains limited [3].

Disease transmission is complex due to the many factors involved, e.g., the presence definitive host in the environment, low socio economic status of the population, deficient or absent environmental or household basic sanitation systems, low level of health education among the population at risk and the presence of intermediate hosts (snails) infected with *Schistosoma mansoni* [1]. According to the World Health Organization, control methods for this parasitic disease should involve chemotherapy, health education, basic sanitary services, the interruption of the lifecycle of the parasite and the elimination or reduction in the mollusk intermediate host population in endemic areas using molluscicidal agents [4,5].

It is estimated that more than 7000 chemicals have been tested for mollusk control, such as copper sulfate, gramoxone, calcium hydroxide, N-tritylmorpholin (Frescon) and niclosamide [6,7]. Niclosamide is highly efficient in all stages of the mollusk lifecycle, killing 100% of adults and embryos at a concentration of less than 1.5 µg/ml after 2 h of exposure [8]. However, its high cost, sensitivity to sunlight and high degree of toxicity to fish, amphibians and aquatic plants [9] has limited its use, especially in developing countries, which often have serious basic sanitation problems [10]. The use of niclosamide in *B. glabrata* eradications programs is common in Brazil. However, the re-colonization of mollusks is observed after months of application, possibly due to resistance mechanisms developed by these organisms. This makes biological control programs expensive and operationally difficult, especially in poor municipalities [11]. Thus, there is a need for studies aimed at developing specific, low cost alternatives to combating snails. Accordingly, the search for molluscicides derived from plants or natural extracts has gained prominence worldwide, the aim of which is to obtain cheaper, biodegradable, safe, efficient products for controlling snail populations [12].

A lichen is the association of a heterotrophic organism (mycobiont) and an autotrophic organism (photobiont). Lichens produce a variety of substances with phenolic traits, known as lichen substances, which are considered natural protectors of the thallus. The antibacterial [13], antifungal [14], antiviral [15], antitumor [16], insecticidal [17] and sunscreen [18] properties of these substances are known. Phenols derived from lichen substances (some unique to the class) comprise four distinct, well characterized, core chemical structures: depsides, depsidones, depsonesand dibenzofurans (usnic acid) [19]. Usnic acid (Figure 1) is widely distributed among the genera *Cladonia* and *Usnea*, accounting for as much as 6% of the dry weight of the lichen thallus [20]. However, usnic acid has a low degree of water solubility, which can be a limiting factor for its use, especially in aquatic environments, since the organic solvents commonly used to solubilize this substance can have toxic effects on aquatic biota. Potassium and sodium salts derived from usnic acid are completely soluble in water and can be obtained from the purified molecule extracted from different lichens, such as *Cladonia substellata*, or chemically synthesized in the laboratory [21]. The literature describes the antibacterial and fungal properties of these salts [22].

Although usnic acid has been investigated for its potential in different fields of research, including its insecticidal or antiherbivory action on snails, the molluscicidal action of its potassium salt against vectors of parasitic diseases has not yet been described. Thus, the aim of the present study was to investigate the effect of potassium usnate chemically derived from usnic acid purified from *Cladonia substellata* (Vainio) on adults and embryos (blastocyst stage) of *B. glabrata* as well as its toxicity to brine shrimp (*Artemia salina* Leach). The simple bench top bioassay involving the brine shrimp lethality test was used as an environmental bioindicator to probe the pharmacological activity of potassium usnate.

Material and Methods

Lichen

Samples of *Cladonia substellata* (Vainio) (300 g) were collected in Mamanguape, Paraíba (northeastern Brazil), at the margins of a Federal Highway BR-101, Coordinates 6° 8′ 19″ S/35° 8′ 11″ W, in August 2012. This way, permission for lichen collection was not required, as well as field studies did not involve endangered or protected species. Thallus identification was performed by MSc Maria de Lourdes Buril and a voucher specimen was deposited at the UFP Herbarium, Dept. of Botany of the Federal University of Pernambuco, Brazil (voucher n° 34402).

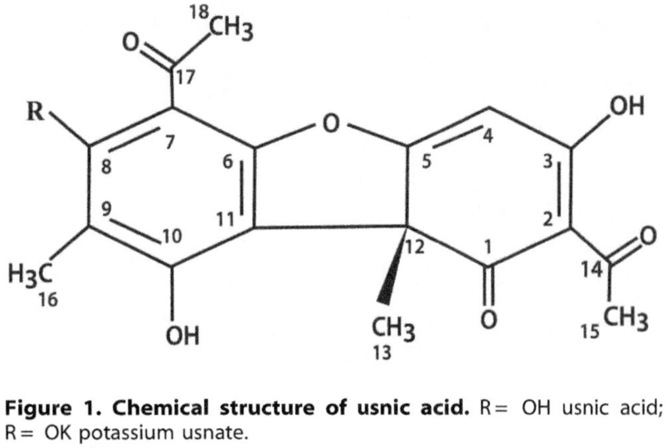

Figure 1. Chemical structure of usnic acid. R= OH usnic acid; R= OK potassium usnate.

Ether extract to obtain purified usnic acid

The thallus of *C. substellata in natura* (200 g) was maceratedand submitted to four successive extractions with diethyl ether in a Soxhlet apparatus at 40°C for 16 h, as described elsewhere [23].

Isolation and purification of usnic acid and acquisition of potassium usnate

For the separation, isolation and purification of usnic acid, the ether extract was fractionated in a column of silica gel (70–230 mesh) eluted in the chloroform:hexane (80:20 v/v) solvent system [24]. After the purification process, 500 mg of the acid were partially dissolved in 40 ml of water at 40°C and subsequently sprayed with previously prepared KOH 10% until complete solubilization of the compound. The solution was frozen at −80°C and then lyophilized.

Identification of usnic acid and potassium usnate

Usnic acid purity (>95%) was determined using high performance liquid chromatography (HPLC). The HPLC assays involved the use of a Hitachi chromatograph coupled to a UV detector at 254 nm and a reverse phase column (RP 18). The samples (usnic acids purified and usnic standard Merck) were injected at a concentration of 0.1 mg ml. The mobile phase was methanol/water/acetic acid (80:19.5:0.5 v/v/v). The flow was 1.0 ml min^{-1}. The substances were identified based on their retention times and comparisons with standard usnic acid [25]. The confirmation of the molecular structure was performed by analyzing the spectra of proton nuclear magnetic resonance (^1H NMR) obtained at 300 MHz in $CDCl_3$ (Varian UNITY spectrometer) and infrared spectroscopy. Potassium usnate was analyzed by infrared spectroscopy (Bruker- IF566).

Biomphalaria glabrata (Say, 1818)

A pigmented wild type strain of *B. glabrata* obtained from São Lourenço da Mata (state of Pernambuco, Brazil) reared in the Laboratory of Radiobiology at the Biophysics and Radiobiology Department at the Federal University of Pernambuco, was used. Adult animals, *S. mansoni* negative, were maintained in plastic aquaria with filtered, dechlorinated water (changed once a week), pH 7.0, at room temperature 25±2°C and fed with fresh lettuce (*Lactuca sativa*).

Toxicity of potassium usnate to *B. glabrata* embryos

Embryotoxicity assay were conducted with five intact egg masses containing 100 embryos (n = 100) in the blastula stage (15 h after spawning) were selected in a stereomicroscope (Leica MZ6; Leica Microsystems, Wetzlar, Germany) and placed in Petri dishes (10 mm). The potassium usnate (2.5; 5 and 10 µg/ml) or filtered water as control were added to the Petri dishes (10 ml) and, after incubation (24 h). The egg masses were washed under filtrated water and maintained in Petri dishes containing filtered and dechlorinated water at 25±2°C in BOD (Eletrolab). Embryos were observed for 8 days from the beginning of the development until nearly hatching. The mortality and hatched were determined by observation under a stereomicroscope (Leica MZ6; Leica

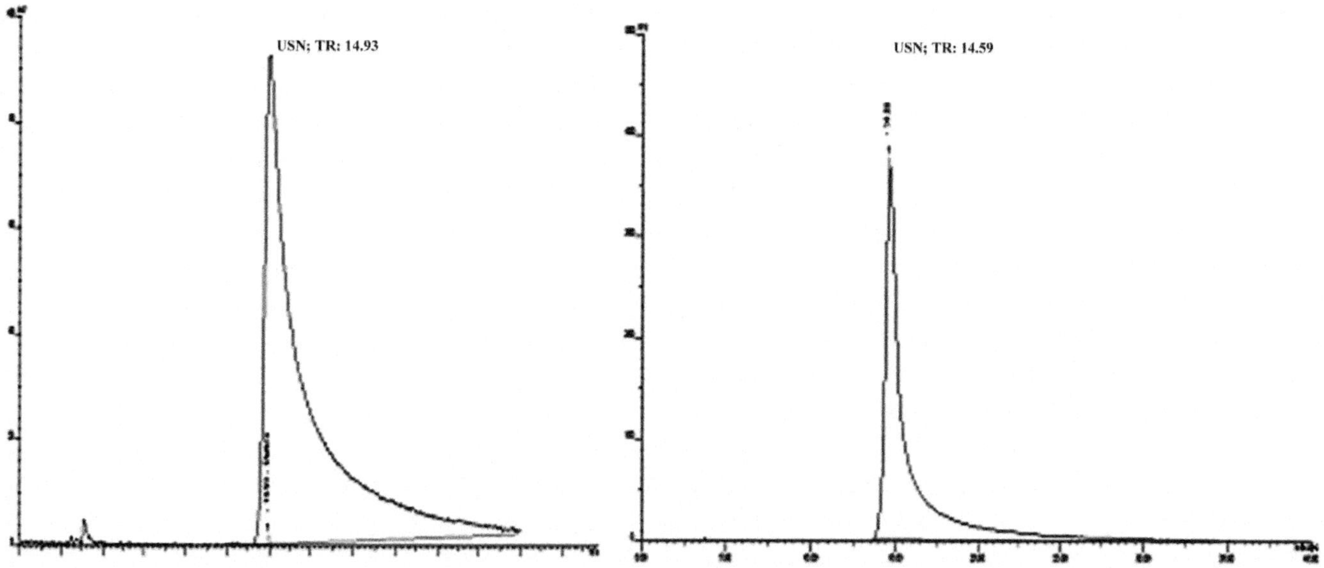

Figure 2. High performance liquid chromatograms. A- Merk usnic acid standard (RT-14.93), B- usnic acid purified from *Cladonia substellata* (RT-14.59).

Microsystems, Wetzlar, Germany). The experiments were performed in triplicate for each potassium usnate concentrations.

Toxicity of potassium usnate to adults of *Biomphalaria glabrata*

Bioassays were performed following the method described by the WHO Expert Committee on the Bilharzia [26]. A population of 200 snails was pre-selected and kept in reproductive isolation for five days to confirm sexual maturity. For the experiments, healthy adult snails (n = 5) of uniform size (shell diameter = 10±1 mm) were exposed for 24 h to potassium usnate solubilized in filtered water at concentrations of 0.25; 0.5; 0.9; 1; 2.5; 5; 10; 50 and 100 µg/ml with a final volume of 500 ml of solution. The assay was performed in 500 ml glass beakers, and copper (II) carbonate (50 µg/ml) and filtered tap water were used as positive and negative controls, respectively. The snails were considered dead when they showed inactivity, the body and shell were discoloured and absence of heart beats. The bioassays were conducted in triplicate for each of the potassium usnate concentrations.

Environmental toxicity

Toxicity assay using brine shrimp (encysted eggs of *A. salina* - 25 mg) were hatched in a beaker filled with natural seawater (pH 8), artificial light at 30°C, under constant aeration. After 48 h, the newly hatched larvae were collected and the lethality assay according to procedures described by elsewhere [5]. Groups of 10 (n = 10) larvae were exposed the potassium usnate (0.5; 1, 5; 10 µg/ml) diluted in natural seawater (5 ml) and, after 24 h, the survival rates (%) were recorded. In the control group, larvae were incubated in seawater. The experiments were conducted in quadruplicate for each of the potassium usnate concentrations.

Statistical analysis

The statistical analysis and standard deviations (SD) were performed using GraphPad Prism 5.0 for Windows (GraphPad Software San Diego, CA, USA) and the data were expressed as replicate means ± SD. Significant differences were established using one-way analysis of variance (ANOVA) and Tukey's test, and $p<0.05$ was considered significant for both analysis. The lethal concentrations required for 50% killing (LC_{50}) of *A. salina* larvae, embryos or *B. glabrata* snails were calculated using probit

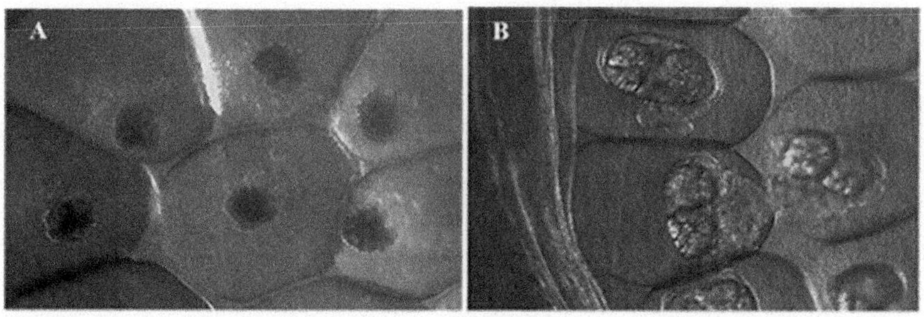

Figure 3. Survival and mortality of *B. glabrata* embryos. A- eggs masses treated at 10 µg ml of potassium usnate showing 100% mortality, B – eggs masses not treated with 100% survival in control group 1 (dechlorinated water).

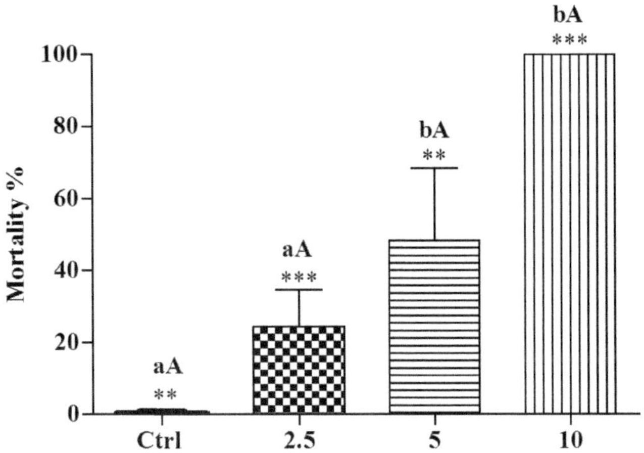

Figure 4. Toxicity of potassium usnate to *B. glabrata* embryos. Mortality of blastula stage embryos subjected to 24 h of treatment with potassium usnate at 2.5; 5 and 10 μg/ml and control group (Crtl 1 – dechlorinated water). Letter A to indicate bars are the same (ANOVA ***p<0.0001); and letter B to indicate that the bars are significantly different from each other (Tukey's test **p<0.05).

analysis with a reliability interval of 95% using the StatPlus1 168 2006 software (Analyst Soft, Canada).

Results

Chemical analysis

Purified usnic acid was obtained from the ether extract. The data on the purified substance was analyzed by HPLC (Figure 2), which revealed a retention time (RT) of 14.59 min and 99% purity. The molecular structure confirmed by ^{1}H NMR and infrared spectroscopy may be compared with data reported by Huneck and Yoshimura [27], with the following description: ^{1}H NMR and IR analysis of purified usnic acid: this substances is a yellow crystalline solid, MP, 201–203°C, $[\alpha]_D$ at 25°C + 493° ($CHCl_3$ c, 1.00), ^{1}H NMR (300 MHz, DMSO-d_6) δH (H; *mult.*; int.): 1.73 (3H; *s*, CH_3-13), 2.08 (3H; *s*, CH_3-16), 2.64 (3H; *s*, CH_3-15), 2.65 (3H; *s*, CH_3-18), 5.92 (1H; *s*, C-4-H), 11.04 (1H; *s*, C-10-OH), 13.29 (1H; *s*, C-8-OH), 18.89 (1H; *s*, C-3-OH). IR (KBr): 3090, 3007, 2930, 1692, 1632, 1560, 1454, 1370, 1357, 1340, 1330, 1289, 1230, 1190, 1143, 1118, 1070, 1039, 959, 940, 840, 818, 700 cm^{-1}.

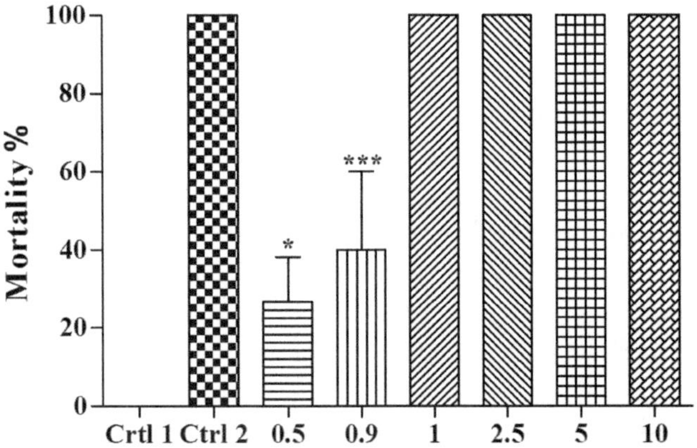

Figure 5. Toxicity of potassium usnate to *B. glabrata* adults. Mortality *B. glabrata* subjected to 24 h of treatment with usnate potassium at 0.5; 0.9; 1; 2.5; 5 and 10 μg/ml and control groups (Crtl 1 dechlorinated water; Crtl 2–cupric carbonate). ***p<0.0001 compared to Ctrl 1 (negative control) group.

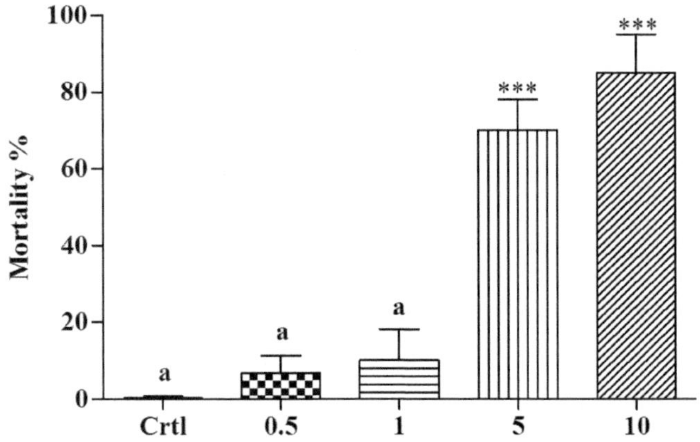

Figure 6. Toxicity of potassium usnate to *A. salina*. Mortality of *A. salina* subjected to 24 h of treatment with potassium usnate at 0.5; 1; 5; 10 μg/ml. ***p<0.0001 for concentrations higher than 5 μg/ml when compared to Ctrl (sea water) group. Letter "a" to indicate that the bars are not significantly different from each other (Tukey's test **p<0.05).

After modification of the usnic acid to form the potassium salt, the sample was analyzed by infrared spectroscopy (IR) and the following data were obtained: (KBr): 3454, 2982, 2929, 1687, 1631, 1540, 1480, 1460, 1382, 1330, 1291, 1240, 1140, 1188, 1136, 1073, 1038, 980, 940, 880, 840, 820, 770, 730, 700 cm^{-1}.

Toxicity of potassium usnate to *B. glabrata* embryos, adult snails and *A. salina*

The embryos in the blastula stage of development were treated with potassium usnate at 2.5, 5 and 10 μg/ml and the solutions were lethal at all concentrations tested when compared to control groups ($p<0.0001$, $F=42.46$). The mortality rate was 100% at 10 μg/ml (Figure 3). The treatments were significant between the groups: Ctrl vs 5 μg/ml (95% CI of diff −77,61 to −18.39 **p<0.005), Ctrl vs 10 μg/ml (95% CI of diff −129.3 to −70.05 ***p<0.0001), 2.5 vs 10 μg/ml (95% CI of diff −105.3 to −46.05 ***p<0.0001) and 5 vs 10 μg/ml (95% CI of diff −81.28 to −22.05 **p<0.05) (Figure 4).

Adult snails were more sensitive to the treatments than embryos (***p<0.0001, $F=78.25$). At 0.5 μg/ml (95% CI of diff −49.75 to −3.582 *p<0.05) and 0.9 μg/ml (95% CI of diff −63.08 to −16.92 ***p<0.0001), mean mortality was less than 100%, but significantly different from the negative control (Ctrl 1) (**p<0.05). The mortality rate was 100% at 1to 10 μg/ml (Figure 5).

The LC_{50} was 5.77 μg/ml and 0.92 μg/ml for adults and embryos, respectively. In the negative control, no mortality occurred for either embryos or adults, whereas 100% mortality occurred for adult snails treated with cupric carbonate (positive control - Ctrl 2). These results demonstrate the effectiveness of potassium usnate for both embryos and adults of *B. glabrata*. The results of the toxicity on environmental indicator show that potassium usnate was not toxic to *A. salina* at a concentration of 1 μg/ml (95% CI of diff −25.33 to −5.828), but significant results were found at concentrations above 5 μg/ml(95% CI of diff −85.33 to −54.17) and 10 μg/ml (95% CI of diff −100.3 to −69.17) (***p<0.0001) when compared to control group (Figure 6). The LC_{50} for *A. salina* was 2.6 μg/ml.

Discussion

The mortality rate of *B. glabrata* embryos and adults treated with potassium usnate was higher than that reported for plant extracts. Saponins derived from *Phytolacca icosandra* and *Cartunare gam milotica* are reported to be active at concentrations of 200 to 25 μg/ml and 26 to 3 μg/ml, respectively [28]. The ethanol extract of the latex from *Euphorbia conspicua* (Euphorbiaceae) at a concentration of 4.87 μg/ml is reported to kill 90% of adults within 96 h; the ethanol extract from the leaves of *Solanum jabrense* (Solonaceae) at a concentration of 39.8 μg/ml is reported to kill 90% within at 24 h [29]; and lectins obtained from *Dioclea guianensis* achieve 100% mortality at 10 μg ml^{-1} [5]. However, the whole plant extract of *Agave filifera*, which is rich in flavonoids and steroids [30], is inactive to embryos, likely due to the high molecular weight of their components, which prevents penetration into the gelatinous embryo membrane [31].

A number of studies in the literature report the biological activity of usnic acid isolated from lichens, but few studies have investigated usnate. A line of research has been focused on the antimicrobial activity of sodium usnate against methicillin-resistant *Staphylococcus aureus* [32], whereas no previous reports are found on the molluscicidal activity of potassium usnate. Some studies describe insecticidal and anti-herbivore activities, with records of a deterrent role against herbivores as the most likely ecological function of lichen substances [33,18]. While lichens do not produce these substances with exclusive deterrent activity against herbivore attacks, some animals, such as the slug *Pallifera varia* (Hubricht.), avoid feeding on lichen species, such as *Xanthoparmelia cumberlandia* (Gyelnik) Hale, which contains usnic, norstictic and stictic acids, and *Huilia albocaerulescens* (Wulfen) Hertel, which produces constictic and stictic acids [34].

Different authors have described the preferences of terrestrial snails that feed on certain thallus parts of lichen and report that invertebrates prefer parts that are free of secondary metabolites [35,36,37,38]. The action mechanisms of lichen substances on snails are not well understood, but the poor palatability and direct toxicity to the intestinal microbiota of invertebrates are likely involved [35,39]. The studies cited are in agreement with Bustinza [40], who offered pioneering descriptions of the antimicrobial activity of usnic acid. With the observed preference of terrestrial mollusks, it has been hypothesized that the action of the modified molecule in salt form, which is completely water soluble, could have the same toxic effect on aquatic mollusks, as confirmed by the present results.

It is estimated that more than 7000 chemicals have been tested for mollusk control, such as copper sulfate, gramoxone, calcium hydroxide, N-tritylmorpholin (Frescon) and niclosamide [6]. Niclosamide is the agent recommended by the World Health Organization [7] and is highly efficient in all stages of the mollusk lifecycle, killing 100% of adults and embryos at a concentration of less than 1.5 μg/ml after 2 h of exposure [8]. However, its high cost, sensitivity to sunlight and high degree of toxicity to fish, amphibians and aquatic plants [9] has limited its use, especially in developing countries, which often have serious basic sanitation problems [10]. This demonstrates the importance of investigating novel substances of a natural origin that may have the same potential as synthetic products.

The potassium usnate seems to be as effective against mollusk as niclosamide, since 100% of mortality was observed at 1 μg/ml. Therefore, due the high cost and toxic effects of niclosamide on the environment as well as the possible buildup of resistance among snails to this compound have become a constant concern to public health authorities [8]. This underscores the importance of the present findings with regard to combating schistosomiasis.

The *A. salina* test was performed to determine the environmental toxicity of potassium usnate. This microcrustacean is highly sensitive to a number of substances, which favors its use as an experimental model for the evaluation of the toxicity of organic residues that might pose risks to the environment and mammals in general [6]. The assays demonstrated that potassium usnate was non-toxic at the moluscicidal concentration, indicating that usnate offers an extremely low risk to aquatic organisms. These results are in agreement to those reported by Lima et al. [9] on potassium salt (lapachol), which was also found to be non-toxic to brine shrimp.

An extract or active component of a plant molluscicide is only considered effective when achieving 90% mortality at 20 μg/ml for the active ingredient and 100 μg/ml for the extract [4]. The present findings demonstrate that potassium salt from usnic acid has increased water solubility (100%), does not lose its biological activity and may be an alternative drug and prototype for a new class of compounds with moluscicidal activity for the control of *B. glabrata* while exhibiting low toxicity to *A. salina*.

These findings are of considerable importance to countries in which the prevalence of schistosomiasis is high and synthetic chemicals are used to eradicate snails, which are costly and cause serious environmental problems, including the development of other types of diseases in mammals due to their high degree of toxicity. Brazil has a rich lichen flora, particularly in endemic regions for *Schistosoma mansoni*. Thus, further investigation into lichen substances is merited.

Acknowledgments

The authors are grateful to the Brazilian fostering agencies Conselho Nacional de Pesquisa (CNPq), Fundação de Amparo à Ciência do Estado de Pernambuco (FACEPE) and Coordenação de Aperfeiçoamento de Pessoal de Nível Superior (CAPES).

Author Contributions

Conceived and designed the experiments: MCBM VLML ECP NHS. Performed the experiments: MCBM MCS LRSS. Analyzed the data: MCBM ECP EPSF NHS. Contributed reagents/materials/analysis tools: AMMAM ECP NHS. Wrote the paper: MCBM ECP EPSF NHS.

References

1. Brasil. Ministério da Saúde (2010) Guia de Vigilância Epidemiológica. 7ª ed. Brasília. Cad 10: 19–29.
2. Brasil. Programa de Vigilância e Controle da Esquistossomose (2011) Coordenação Geral de Doenças Negligenciadas. Situação Epidemiológica de *Esquistossomose mansoni* no Brasil. SUS. Governo Federal.
3. Reis DC, Kloos H, King C, Quites HFO, Matoso LF, et al. (2010) Accessibility to and utilization of schistosomiasis related health services in a rural area of state of Minas Gerais Brazil. Mem Inst Oswaldo Cruz 105: 587–597.
4. World Health Organization (1993) The control of schistosomiasis. WHO Technical Report Series 830 Geneva.
5. Santos AF, Cavada BS, Rocha BAM, Nascimento KS, Sant'Ana AEG (2010) Toxicity of some glucose/manose-binding lectins to *Biomphalaria glabrata* and *Artemia salina*. BioresourTechnol 101: 794–798.
6. Cantanhede SPD, Marques AM, Silva-Souza N, Valverde AL (2010) Atividade moluscicida de plantas: uma alternativa profilática. Braz J Pharmacogn 20: 282–288.
7. Oliveira-Filho EC, Geraldino BR, Coelho DR, De-Carvalho RR, Paumgertten FJR (2010) Comparative toxicity of *Euphorbia milii* látex and synthetic molluscicides to *Biomphalaria glabrata* embryos. Chemosphere 81: 218–227.
8. Ribeiro KAL, Carvalho CM, Molina MT, Lima EP, Lópes-Monteiro E, et al. (2009) Activities of naphthoquinonas against *Aedes aegypti* (Linnaeus, 1762) (Diptera: Culicidae), vector of dengue and *Biomphalaria glabrata* (Say, 1818), intermediate host *Schistosoma mansoni*. Acta Trop 111: 44–50.
9. Lima NMF, Santos AF, Porfírio Z, Goulart MOF, Sant'Ana AEG (2002) Toxicity of lapachol and isolapachol and their potassium salts against

Biomphalaria glabrata, Schistosoma mansoni carcarie, *Artemia salina* and *Tilapia nilotica*. Acta Trop 83: 43–47.
10. Giovanelli A, Vieira MV, Coelho da Silva CLPA (2002) Interaction between the intermediate host of schistosomiasis in Brazil *Biomphalaria glabrata* (Planorbidae) and a possible competitor *Melanoides tuberculata* (Thiaridae): I. Laboratory experiments. Mem Inst Oswaldo Cruz 90: 261–269.
11. Schall VT, Vasconcellos MC, Rocha SR, Souza CP, Mendes NM (2001) The control of the schistosome-transmitting snail *Biomphalaria glabrata* by the plant molluscicide *Euphorbia splendens* var. hislopii (synmilli Des. Moul): a longitudinal field study in an endemic area in Brazil. Acta Trop 79: 165–170.
12. Luna JS, Santos AF, Lima MRF, Omena MC, Mendonça FAC, et al. (2005) A study of the larvicidal and molluscicidal activities of some medicinal plants from northeast Brazil. J Ethnopharmacol 97: 199–206.
13. Martins MCB, Lima MJG, Silva FP, Azevedo-Ximenes E, Silva NH, et al. (2010) *Cladia aggregata* (lichen) from Brazilian Northeast: chemical characterization and antimicrobial Activity. Braz Arch Biol Technol 53: 115–122.
14. Wei X, Jeon HS, Han KS, Koh YJ, H JS (2008) Antifungal Activity of Lichen-forming Fungi against *Colletotrichum acutatum* on hot pepper. Plant Pathol J 24: 202–206.
15. Campanella L, Delfini M, Ercole A, Iacoangeli E, Risuleo G (2002) Molecular chracterization and action of usnic acid a drug that inhibits proliferation of mouse polyomavirus *in vitro* and whose main target is RNA transcription. Biochim 84: 329–334.
16. Einarsdóttir E, Groeneweg J, Björnsdóttir G, Hardardóttir G, Omarsdóttir S, et al. (2010) Cellular mechanisms of the anticancer effects of the lichen compound usnic acid. Planta Med 76: 969–974.
17. Nimis PL, Skert N (2006) Lichen chemistry and selective grazing by the coleopteran *Lasioderma serricorne*. Environ Exp Bot 55: 175–182.
18. Solhaug KA, Lind M, Nybakken L, Gauslaa Y (2009) Possible functional roles of cortical depsides and medullary depsidones in the foliose lichen *Hypogymnia physodes*. Flora 204: 40–48.
19. Howell GME, Newton EM, Williams-Wynn DD (2003) Molecular structural studies of lichen substances II: atranorin, gyrophoric acid, fumarprotocetraric acid, rhizocarpic acid, calycin, pulvinic dilactone and usnic acid. J Mol Struct 651: 27–57.
20. Ingólfsdóttir K (2002) Molecules of interest usnic acid. Phytochemistry 61: 729–736.
21. Stanković M, Stanković S, Dordević Z, Dordević S (1989) Process for obtaining natrium salt of (+) -usnic acid. Pat. Yu. 43978.
22. Savić I, Nikolić G, Savić I, Zlatković S, Djokić D (2010) New, simple and validated uv-spectrophotometric methods for the estimation of sodium usnate in preparations. Maced J Chem En 29: 157–164.
23. Asahina Y, Shibata S (1954) Chemistry of lichen substances, first ed. Japan Society for the Promotion of Science, Tokyo.
24. Odabasoglu F, Cakir A, Suleyman H, Aslan A, Bayir Y, et al. (2006) Gastroprotective and antioxidant effects of usnic acid on indomethacin-indiced gastric ulcer in rats. J Ethnopharmacol 103: 59–65.
25. Legaz ME, Vicente C (1983) Endogenous inactivators of arginase, arginine descarboxilase and agmatine amidinohydrolase in *Evernia prunastri* thallus. Plant Physiol Lancaster 71: 300–301
26. World Health Organization (1965) WHO Expert Committee on Bilharziasis. Geneva, Tech Rep Ser 299: 1–56.
27. Huneck S, Yoshimura I (1996) Identification of lichen substances, first ed. Springer, Verlag, Berlin.
28. Treyvaud V, Marston A, Dyatmiko W, Hostettmann K (2000) Molluscicidal saponins from *Phytolacca icosandra*. Phytochemistry 55, 603–609.
29. Santos MS, Azevedo DPL, Mata RCS, Mendonça DIMD, Sant'Ana AEG (2007) The lethality of *Euphorbia conspicua* to adults of *Biomphalaria glabrata*, cercaria of *Schistosoma mansoni* and larvae of *Artemia salina*. Bioresour Technol 98: 135–139.
30. Rawi SM, Al-Hazmi M, Seif Al Nassr F (2011) Comparative study of the molluscicidal activity of some plants extracts on the snail vector *Schistosoma mansoni*, *Biomphalaria alexandrina*. Int J Zool Res 7: 169–189.
31. Lemma A, Yau P (1974) Studies on the molluscicidal properties of endod (*Phylolacca dodecandra*), III. Ethiopian Med J 12: 109–114.
32. Elo H, Matikainen J, Pelttari E (2007) Potent activity of the lichen antibiotic (+)-usnic acid against clinical isolates of vacomycin-resisten enterococci and methicillin-resistant *Staphylococcus aureus*. Naturwissenschaften 94: 465–468.
33. Asplund J, Solhaug KA, Gauslaa Y (2009) Fungal depsidones – an inducible or constitutive defense against herbivores in the lichen *Lobaria pulmonaria*? Basic Appl Ecol 10: 273–278.
34. Lawrey JD (1980). Correlations between lichen secondary chemistry and grazing activity by *Pallifera varia*. Bryologist 23: 128–134.
35. Benesperi R, Tretiach M (2004) Differential land snail damage to selected species of the lichen genus *Peltigera*. Biochem Sys Ecol 32: 127–138.
36. Gauslaa Y (2005) Lichen palatability depends on investments in herbivore defense. Oecologia 143: 94–105.
37. Pöykkö H, Hyvärinen M, Backor M (2005) Removal of lichen secondary metabolites affects food choice and survival of lichenivorous moth larvae. Ecology 86: 2623–2632.
38. Pöykkö H, Backor M, Bencúrová E, Molcanová V, Backorová M, et al. (2010) Host use of a specialist lichen-feeder: dealing with lichen secondary metabolites Oecologia 164: 423–430.
39. Lawrey JD (1989) Lichen secondary compounds: evidence for a correspondence between antiherbivore and antimicrobial function. Bryologist 92: 326–328.
40. Bustinza F (1951) Antibacterial substances from lichens. Endeavour 10: 95–99. 1951.

Iron Overload and Apoptosis of HL-1 Cardiomyocytes: Effects of Calcium Channel Blockade

Mei-pian Chen[1], Z. Ioav Cabantchik[2], Shing Chan[1], Godfrey Chi-fung Chan[1]*, Yiu-fai Cheung[1]*

1 Department of Pediatrics and Adolescent Medicine, The University of Hong Kong, Hong Kong, China, **2** Department of Biological Chemistry, Alexander Silberman Institute of Life Sciences, Hebrew University of Jerusalem, Safra Campus at Givat Ram, Jerusalem, Israel

Abstract

Background: Iron overload cardiomyopathy that prevails in some forms of hemosiderosis is caused by excessive deposition of iron into the heart tissue and ensuing damage caused by a raise in labile cell iron. The underlying mechanisms of iron uptake into cardiomyocytes in iron overload condition are still under investigation. Both L-type calcium channels (LTCC) and T-type calcium channels (TTCC) have been proposed to be the main portals of non-transferrinic iron into heart cells, but controversies remain. Here, we investigated the roles of LTCC and TTCC as mediators of cardiac iron overload and cellular damage by using specific Calcium channel blockers as potential suppressors of labile Fe(II) and Fe(III) ingress in cultured cardiomyocytes and ensuing apoptosis.

Methods: Fe(II) and Fe(III) uptake was assessed by exposing HL-1 cardiomyocytes to iron sources and quantitative real-time fluorescence imaging of cytosolic labile iron with the fluorescent iron sensor calcein while iron-induced apoptosis was quantitatively measured by flow cytometry analysis with Annexin V. The role of calcium channels as routes of iron uptake was assessed by cell pretreatment with specific blockers of LTCC and TTCC.

Results: Iron entered HL-1 cardiomyocytes in a time- and dose-dependent manner and induced cardiac apoptosis via mitochondria-mediated caspase-3 dependent pathways. Blockade of LTCC but not of TTCC demonstrably inhibited the uptake of ferric but not of ferrous iron. However, neither channel blocker conferred cardiomyocytes with protection from iron-induced apoptosis.

Conclusion: Our study implicates LTCC as major mediators of Fe(III) uptake into cardiomyocytes exposed to ferric salts but not necessarily as contributors to ensuing apoptosis. Thus, to the extent that apoptosis can be considered a biological indicator of damage, the etiopathology of cardiosiderotic damage that accompanies some forms of hemosiderosis would seem to be unrelated to LTCC or TTCC, but rather to other routes of iron ingress present in heart cells.

Editor: Alexander G. Obukhov, Indiana University School of Medicine, United States of America

Funding: This work was supported by Children's Thalassemia Foundation and Edward Sai Kim Hotung Pediatric Education and Research Fund (http://www.thalassaemia.org.hk). The funders had no role in study design, data collection and analysis, decision to publish, or preparation of the manuscript.

Competing Interests: The authors have declared that no competing interests exist.

* Email: gcfchan@hku.hk (GCFC); xfcheung@hku.hk (YFC)

Introduction

As an essential element for almost all living organisms, iron serves as a critical component in different metabolic processes including oxygen transport and storage, DNA, RNA and protein synthesis, and electron transport [1]. Tight regulation of iron concentrations is required for maintenance of cellular function, while excessive iron leads to generation of oxidative stress by increasing production of reactive oxygen species [2–4]. Of the different organs, the heart is particularly vulnerable to iron toxicity [5].

Iron overload cardiomyopathy (IOC) is well documented in patients with β-thalassemia major and is an important cause of morbidity and mortality [6–9]. Clinical manifestations include systolic and diastolic ventricular dysfunction, cardiac arrhythmias, and end-stage cardiomyopathy [5,8,10,11]. However, the mechanisms of iron-induced subclinical cardiac dysfunction and end-stage cardiomyopathy remain unclear. Progressive loss of cardiomyocytes, albeit at a low level, through apoptosis is believed to contribute to the remodeling process and ventricular dysfunction in heart failure [12–17]. There is, however, a paucity of data on the phenomenon of cardiomyocyte apoptosis and the pathway involved in the setting of iron overload.

Under physiologic condition, iron uptake into cardiomyocytes is mediated through transferrin-transferrin receptor-mediated endocytosis with negative feedback regulatory mechanisms [18]. However, under iron overloading conditions, transferrin becomes saturated and excess plasma iron will present as non-transferrin-bound iron (NTBI), which contributes to the intracellular labile iron pool and the generation of reactive oxygen species [9]. Reported mechanisms of NTBI entry into cardiomyocytes are nonetheless controversial [19]. While some studies have proposed L-type calcium channels (LTCC) to be a major pathway for NTBI entry [20–22], others suggest that T-type calcium channel (TTCC) may be the alternative portal of entry [23,24]. However, direct evidence for possible protective effects of calcium channel blockers against iron-induced cardiomyocyte apoptosis is lacking.

Using HL-1 cardiomyocytes, a spontaneously contracting cardiomyocyte cell line that expresses both LTCC and TTCC molecularly and functionally [25–27], together with the real-time technique tracing cellular iron uptake and flow cytometry, we explored (i) the phenomenon of and mechanisms involved in cardiomyocyte apoptosis induced by iron overload, (ii) the effects of LTCC and TTCC blockers on Fe(II) and Fe(III) entry into cardiomyocytes, and (iii) the potential protective effect on iron-induced cardiomyocyte apoptosis by calcium channel blockade.

Materials and Methods

Cell culture

HL-1 cardiomyocytes were kindly provided by Prof. W.C. Claycomb (Louisiana State University Health Science Center, New Orleans, LA, USA) who created the cell line [25]. HL-1 cells were established from the AT-1 mouse atrial cardiomyocyte tumor, and can be serially passaged while maintaining contractile phenotype. The cells were grown in culture vessels pre-coated with 0.02% gelatin (Difco, Fisher Scientific, Suwanee, GA, USA) - 5 μg/ml fibronectin (Sigma, St Louis, MO, USA) solution at 37°C in a humidified 5% CO_2 incubator, maintained in Claycomb Medium (SAFC Biosciences, Sigma) supplemented with 10% fetal bovine serum (Sigma), 0.1 mM norepinephrine (Sigma), 2 mM L-glutamine (Invitrogen, Life Technologies, Grand Island, NY, USA) and penicillin/streptomycin (100 U/ml:100 μg/ml) (Invitrogen). The medium was changed approximately 5 days per week.

Iron treatment and calcium channel blockade

For calcein green-acetomethoxy (CALG-AM) fluorescent assay, HL-1 cells were seeded at 6×10^4 cells/well in gelatin-fibronectin coated 96-well black CulturPlate (PerkinElmer, Waltham, Massachusetts, USA). Cells reached around 90% confluence after 24 hr culture. L-type calcium channel blockers including amlodipine (Cipla, India) and verapamil (Abbott, Ludwigshafen, Germany) and TTCC blocker, efonidipine (Sigma), were loaded at 0.1, 1, 10, 100 μM in assay buffer, which consisted of HEPES-buffered saline, pH 7.4 (HBS) supplemented with 0.5 mM probenecid (Sigma), 30 min before iron challenge, and the concentrations were maintained during the assay. $FeCl_3$ was loaded at 150, 300, 600 μM with and without 1 mM ascorbic acid in assay buffer, which has been indicated to represent Fe(II) and Fe(III) respectively [24,28,29]. Controls (with and without ascorbate) was defined as the conditions without calcium channel blockers and iron.

For flow cytometric assay, HL-1 cells were seeded at a density of 1.5×10^5 cells/ml in gelatin-fibronectin coated plates. After 24 hr incubation, culture medium was changed into norepinephrine-free medium containing 2% fetal bovine serum, 2 mM L-glutamine and penicillin/streptomycin (100 U/ml:100 μg/ml), and also 150, 300, 600 μM $FeCl_3$ with and without 1 mM ascorbic acid for test groups. Calcium channel blockers were pre-loaded at 1 μM 60 min before iron challenge without media change before treatment endpoint. For treatments with iron chelator deferiprone (Apotex, Toronto, Canada), 10 or 100 μM deferiprone was loaded 20 min after iron loading. Blank controls (with and without ascorbate) was defined as the conditions without calcium channel blockers, chelator and iron loading. After 72 hr of incubation, cells in the control group had confluency at around 90%, while cells in iron treatment groups had less. Cells were gently detached by 0.05% Trypsin-EDTA (Invitrogen) for flow cytometric assays.

CALG-AM fluorescent assay

To trace iron transport in live HL-1 cells, CALG-AM fluorescent assay was used [30]. Non-fluorescent CALG-AM is converted to green-fluorescent calcein once diffuses into live cells, going through acetoxymethyl ester hydrolysis by intracellular esterases. Cells were exposed to 0.25 μM CALG-AM (Molecular Probes, Life Technologies, Grand Island, NY, USA) at 37°C for 30 min in Claycomb Medium containing 10 mM Na-HEPES (Sigma). Cells were then rinsed with HBS, followed by the perfusion of assay buffer, HBS supplemented with 0.5 mM probenecid, which prevented leakage of anionic fluorescent probes from cells. Calcium channel blockers and ascorbic acid were added simultaneously under the conditions mentioned. Fluorescent intensity was measured using fluorescent plate reader Fusion (Packard, Perkin Elmer Life Sciences, Boston, MA, USA) at excitation/emission wavelength 485 nm/520 nm. Local average reading at 10 min after assay buffer loading was set as initial fluorescence level. $FeCl_3$ was loaded at 20 min after the first plate reading (Figure 1A). Calcein was quenched by intracellular labile iron, and hence, the fluorescence intensity was inversely proportional to the level of labile intracellular iron. Iron entry was terminated by adding 100 μM impermeant chelator diethylene-triamine-pentaacetic acid (DTPA) at 115 min after assay buffer loading. Identification of intracellular labile iron was verified by 100 μM permeant iron chelator deferasirox (Exjade, ICL670) at 136 min after assay buffer loading to reverse the calcein-Fe quenching. Control was defined as treatments without addition of calcium channel blockers, iron, DTPA and ICL670. Experiments were performed in triplicate. Each reading at any given time was normalized to the local initial fluorescence level.

Annexin V/PI assay

Fluorescein isothiocyanate (FITC) Annexin V Apoptosis Detection Kit (Becton Dickinson, Franklin Lakes, NJ, USA) was used according to manufacturer's instructions. Briefly, cells from cultures were collected and washed with cold PBS and then resuspended in annexin V binding buffer. After staining with annexin V-FITC and PI for 15 min at room temperature in the dark, cells suspended in annexin V binding buffer were tested by LSR II flow cytometer (Becton Dickinson). For each measurement, at least 10,000 cells were counted. Flow data were analyzed by FlowJo 8.8.4 (Tree Star). Only single cell events were gated out for analysis.

Activated caspase-3 assay

FITC Active Caspase-3 Apoptosis Kit (Becton Dickinson) was used according to manufacturer's instructions. Briefly, cells from culture were collected and washed with cold PBS, then fixed and permeabilized in BD Cytofix/Cytoperm solution for 20 min on ice. After washing with BD Perm/Wash buffer, cells were stained with FITC-conjugated anti- active caspase-3 antibody for 30 min at room temperature. With further wash with Perm/Wash buffer, cells suspended in Perm/Wash buffer were tested by LSR II flow cytometer. Flow cytometry was performed as aforementioned.

JC-1 assay

The mitochondrial membrane potential ($\Delta\psi$) of HL-1 cardiomyocytes was evaluated by Flow Cytometry Mitochondrial Membrane Potential Detection Kit (Becton Dickinson). JC-1 (5,5′,6,6′-tetrachloro-1,1′,3,3′-tetraethylbenzimidazolcarbocyanine iodide) is a fluorochrome widely used to evaluate the status of $\Delta\psi$. Mitochondria with normal $\Delta\psi$ increases JC-1 uptake, which leads to the formation of JC-1 aggregates that emit red

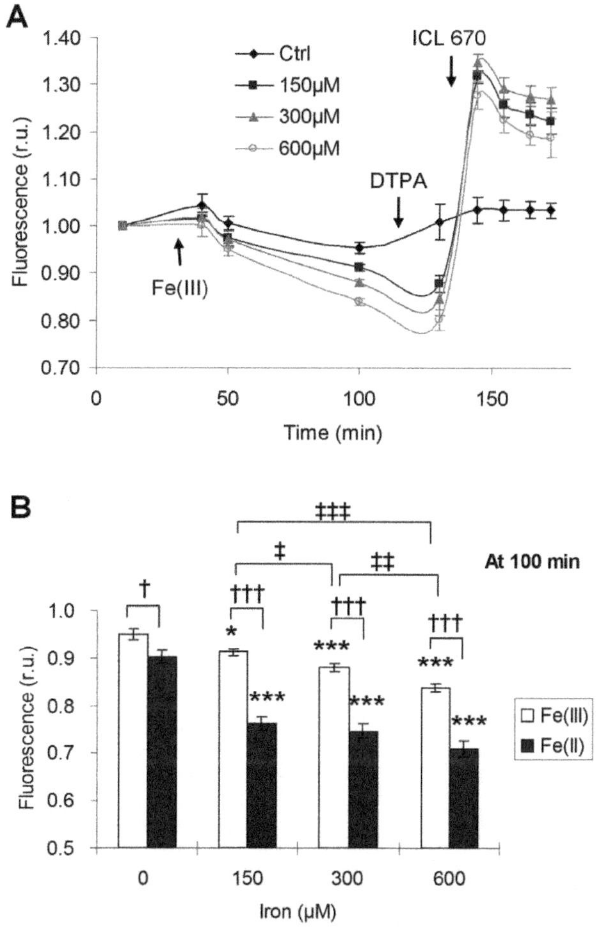

Figure 1. Exogenous iron entered cardiomyocytes in a time- and dose- dependent manner. (**A**) Fe(III) uptake by live HL-1 cells treated at 3 indicated doses, detected by CALG-AM fluorescent assay. Fluorescence intensity was carried out by fluorescent plate reader Fusion. Local average reading at 10 min was set as initial fluorescence level. Each reading at any given time was normalized to the local initial fluorescence level. $FeCl_3$ was load at 30 min. Impermeant chelator DTPA was loaded at 115 min; permeant iron chelator ICL 670 was loaded at 136 min. Control was defined as treatment without addition of iron, DTPA and ICL670. (**B**) Fe(III) and Fe(II) uptake at 100 min of the assessment time point indicated in (A), i.e. 70 min after iron loading. $FeCl_3$ loaded with ascorbate represented Fe(II) treatment. Both controls with and without ascorbate were shown. *, †, ‡, $p<0.05$; **, ††, ‡‡, $p<0.01$; ***, †††, ‡‡‡, $p<0.001$; * versus respective controls. The results represented as mean ± SEM of five independent triplicate experiments.

fluorescence at 590 nm. In depolarized mitochondria, low concentration of JC-1 inside would stay at monomer form, emitting green fluorescence maximally at 527 nm. The staining protocol followed manufacturer's instructions. Briefly, cells were collected and incubated in JC-1 solution for 15 min at 37°C in CO_2 incubator. After subsequent washes with Assay Buffer, cells were resuspended in Assay Buffer for flow cytometry by LSR II as aforementioned.

Statistical analysis

Data are presented as mean ± SEM. Statistical analysis was performed using one-way analysis of variance (ANOVA) with post test for multiple comparisons, and unpaired t test for comparisons of two groups by GraphPad Instat 3 (GraphPad Software, Inc., San Diego, CA, USA). A $p<0.05$ was regarded as statistically significant.

Results

Exogenous iron entered cardiomyocytes in a time- and dose- dependent manner

To detect intracellular labile iron, iron influx was visualized in real time by tracking the gradual decrease of fluorescence signals in the live HL-1 cardiomyocytes. Within the detection period from 10 to 70 min after iron loading (Figure 1A), we observed iron entering HL-1 cells in a time-dependent manner. With elimination of extracellular iron by addition of the impermeable chelator DTPA, the subsequent addition of permeable chelator ICL670 restored the calcein fluorescence quenched by labile iron significantly, confirming that CALG-AM assay could assess intracellular iron in HL-1 cardiomyocytes effectively.

Based on the difference of uptake rate at 70 min after iron challenge with or without ascorbate, Fe(II) was found to be significantly more permeable than Fe(III) ($p<0.001$) (Figure 1B). Fe(III) showed a dose-dependent acquisition at 150, 300, 600 µM loading. In contrast, Fe(II) achieved a near plateau loading at 150 µM (Figure 1B).

Iron loading induced cardiomyocyte apoptosis

Annexin V/Propidium Iodide (PI) flow cytometric assay was used to quantify the amount of apoptosis. Cells positive for annexin V but negative for PI represented those undergoing early apoptosis, while cells stained positive for both annexin V and PI represented the population undergoing late apoptosis or necrosis [31,32]. By quantifying the percentage of total annexin V positive cells (lower and upper right quadrant in the representative flow cytometry charts as shown in Figure 2A), we found a dose-dependent increase in apoptotic cell population when HL-1 cells were treated with $FeCl_3$ with or without ascorbic acid for 72 hr (Figure 2A) (pH of each condition changed within 7.4–7.8). Such increase in apoptosis was noted in cells treated with concentrations of $FeCl_3$ at ≥300 µM ($p<0.001$). At the concentration of 600 µM, Fe(II) induced significantly more apoptosis than Fe(III) ($p<0.01$).

To further define the underlying apoptotic mechanism of iron overload on HL-1 cardiomyocytes, caspase-3 activity and mitochondrial membrane potential change were also assessed. In line with the findings of annexin V/PI assay, iron overload induced a dose-dependent activation of caspase-3 (Figure 2B) and alteration of mitochondrial membrane potential (Figure 2C), which suggested an involvement of the intrinsic apoptotic pathway.

High-dose LTCC but not TTCC ameliorated Fe(III) entry under condition of iron load

The potential roles of LTCC and TTCC for iron entry into HL-1 cardiomyocytes were evaluated using CALG-AM fluorescent assay, with treatments with LTCC blockers, amlodipine and verapamil, and TTCC blocker, efonidipine, at 30 min prior to iron loading. The blockade effects for Fe(III) (Figure 3A) and Fe(II) (Figure 3B) treated at 150, 300, 600 µM were assessed at logarithmic increments of calcium channel blocker concentrations from 0.1 to 100 µM. The time point of assay was at 70 min after iron loading, which was approximately 100 min after administration of different calcium channel blockers. Fluorescent signal changes were normalized to respective negative controls of each treatment arm.

Compared with the increase in iron entry into cells under Fe(III) treatment alone with decreased fluorescent signals, pretreatment with 10 to 100 µM of amlodipine and verapamil significantly

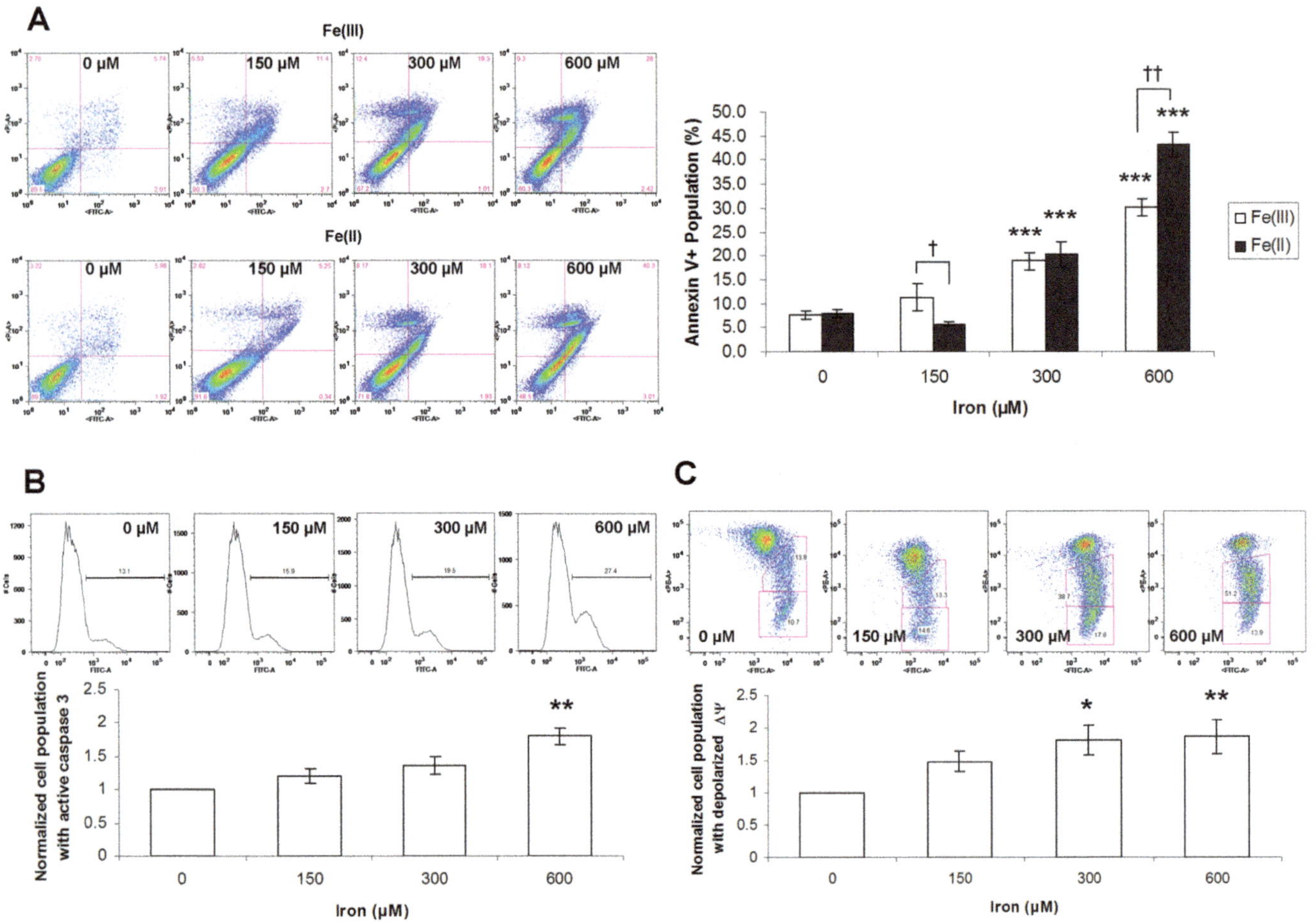

Figure 2. Iron overload induced cardiomyocyte apoptosis. HL-1 cells were treated with Fe(III) and Fe(II) for 72 hr, followed by (**A**) annexin V/PI flow cytometry assay, (**B**) active caspse-3 flow cytometry assay, and (**C**) JC-1 flow cytometry assay. *, †, $p<0.05$; **, ††, $p<0.01$; ***$p<0.001$; * versus respective controls. The results represented as mean ± SEM of five to six independent experiments.

increased normalized fluorescent signals (Figure 3A). The effect was more pronounced with 300 μM and 600 μM than 150 μM of Fe(III) load. These findings suggested blockade of Fe(III) entry by both LTCC blockers. However, efonidipine did not exert significant blocking effect on iron entry in Fe(III) overload.

Trend of LTCC and TTCC blockade of Fe(II) entry

With regard to Fe(II) loading condition, increased trends of fluorescent signals were observed with increased LTCC and TTCC blockade (Figure 3B). However, statistical significance was only found with pretreatment using 100 μM amlodipine.

Calcium channel blockers could not salvage cardiomyocytes from iron-induced apoptosis

To further explore whether calcium channel blockade could reduce HL-1 cardiomyocyte apoptosis induced by iron overload, annexin V/PI assay was performed on HL-1 cells loaded with Fe(III) and Fe(II) at different concentrations, with pretreatment of LTCC and TTCC blockers at a concentration of 1 μM. There was no significant decrease in apoptotic cell population, whether loaded with Fe(III) (Figure 4A) or Fe(II) (Figure 4B).

The findings suggested that calcium channel blockers at this concentration had no protective effects on HL-1 cells against iron-induced apoptosis. However, at the doses of 10 μM or 100 μM, amlodipine or verapamil, which showed significant iron blockade effect on HL-1 cells (Figure 3), appeared to have high cellular toxicity (Figure 5A). Pretreatment of TTCC blockers in iron treated HL-1 cells led to similar or even worse effects.

By contrast, the commonly-used iron chelator deferiprone induced less toxic effect under non-iron overloaded condition (Figure 5A) and further showed protective effect on iron-induced apoptosis of cardiomyocytes (Figure 5B).

Discussion

The present study shows that i) iron induces apoptosis of HL-1 cardiomyocytes via the mitochondria-mediated caspase-3 dependent pathway, ii) blockade of LTCC but not TTCC prevented Fe(III) but not Fe(II) entry under iron overload condition and (iii) blockade of neither LTCC nor TTCC could salvage the cultured cardiomyocytes from iron overload induced apoptosis.

Iron-induced cardiomyocyte apoptosis

The levels of plasma NTBI in thalassemia patients under iron overload are variable, with an estimation suggested to be 0-25 μM [33]. For the proof of principle, comparable iron concentrations as previously reported were used in the current *in vitro* study [24,34]. The apoptotic effect of iron overload on HL-1 cells and its involvement of mitochondria-dependent pathway were suggested

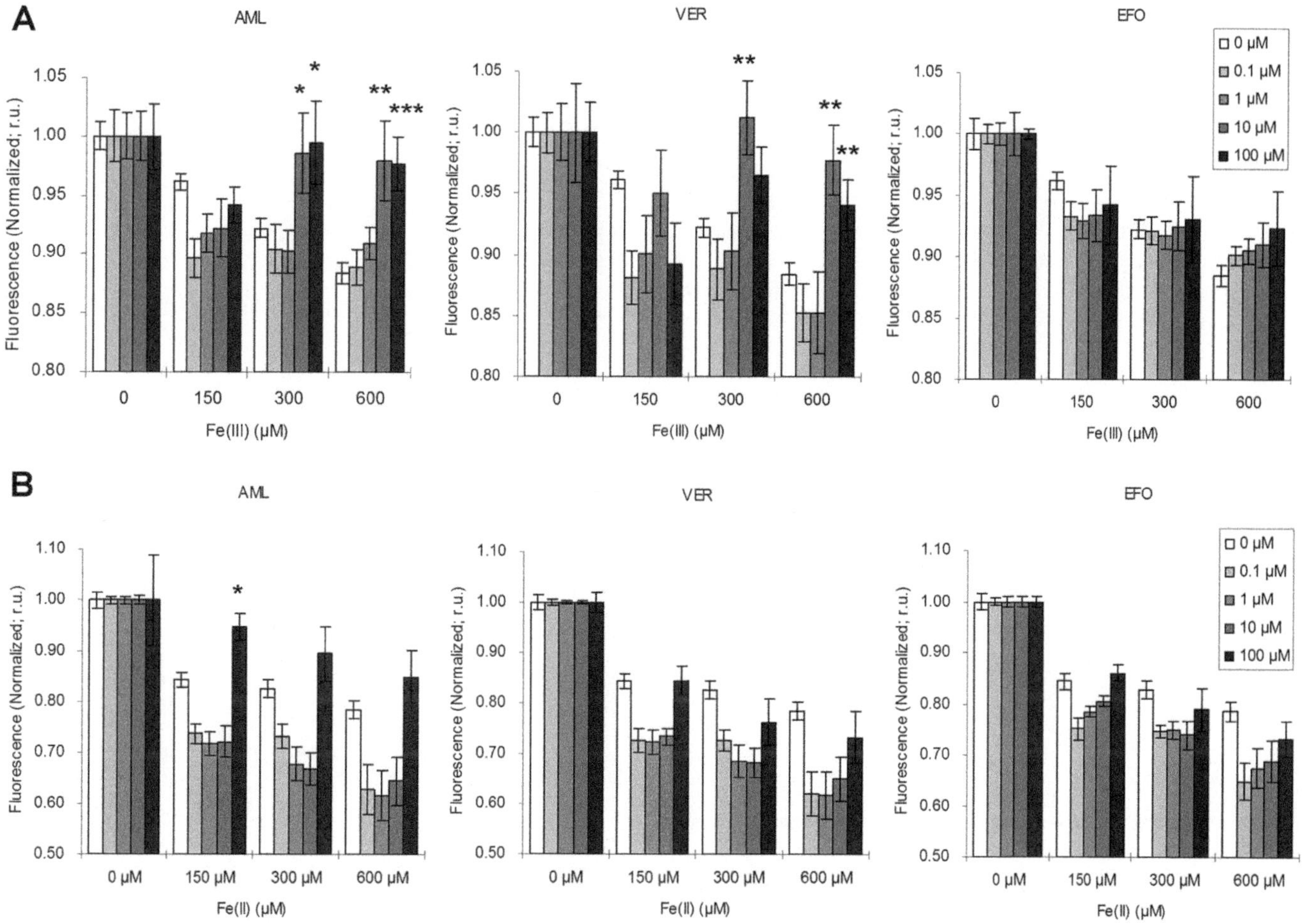

Figure 3. Iron blockade effects of LTCC and TTCC blockers on iron-overloaded cardiomyocytes. In this CALG-AM fluorescent assay, HL-1 cells were pretreated with LTCC blockers, amlodipine (AML) and verapamil (VER), and TTCC blocker, efonidipine (EFO), at logarithmic scale from 0.1 to 100 μM. 3 indicated doses of Fe(III) (**A**) and Fe(II) (**B**) were loaded 30 min after blocker treatment. Fluorescence readings were at 70 min after iron loading. Fluorescence signal changes were normalized to respective negative controls of each treatment arm. **p*<0.05; ***p*<0.01; ****p*<0.001. The results represented as mean ± SEM of four independent triplicate experiments.

by the findings of increase in phosphatidylserine exposure, increased caspase-3 activity, and a dose-dependent drop on mitochondrial membrane potential in iron-overloaded HL-1 cells. Our results are in agreement with the *in vivo* studies suggesting the cardiac apoptotic effect of iron overload on mice [20] and gerbils [35] as revealed by increased nucleic DNA fragmentation and caspase activity. Although other study suggesting the necrotic effect of iron overload on cardiomyocytes [36], more evidences will be of interest to the further mechanism behind, including the postulated cross link between apoptosis and necrosis in series or parallel [37], as well as the differences among experimental models.

Fe(II) and Fe(III) entry into cardiomyocytes

As both redox states of iron have been shown to form cardiac iron deposit [28], our study explored both ferric and ferrous irons. The results agree with those reported previously regarding the more permeative nature of ferrous iron, which is maintained with ascorbate as a reducing agent [24], as evaluated by kinetic parameters [28,38]. Previous studies have implicated either the LTCC or TTCC as the main candidate for NTBI entry into cardiomyocytes. The controversies have in part been related to different models and methods used.

The effect of LTCC blockade on iron entry

Calcium channels play an important role in myocardial contractility and remain open for long duration (>400 ms) in each contraction cycle [39]. Except for the primary transport of Ca^{2+}, LTCC also facilitate transport for many other divalent cations including Fe^{2+}, Co^{2+} and Zn^{2+} [22,40,41]. Previous studies suggest that LTCC is the major portal for iron uptake into cardiomyocytes in IOC [20–22]. For a further mechanism, we assessed the role of LTCC in iron-overloaded cardiomyocytes by the real-time approach.

Our results showed significant reduction of ferric iron ingress by both LTCC blockers at higher doses of iron treatment, 300 μM and 600 μM, but not at lower dose of iron at 150 μM. This phenomenon implicated the classic concept of iron delivery through transferrin at lower dose of iron treatment [9], while confirming the blockade effect from LTCC blockers toward excessive iron, Fe(III) from this result, uptake into cardiomyocytes [20–22]. It is worth noting, however, that LTCC blockers displayed their iron blockade effect only at concentrations of 10 and 100 μM, higher than the therapeutic serum levels of 0.1 to 1 μM [42,43]. Hence, the clinical translation of the use of LTCC blockers to prevent iron-induced cardiotoxicity remains uncertain.

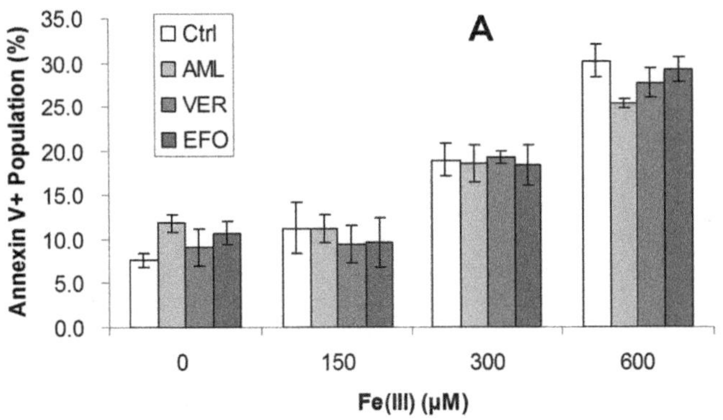

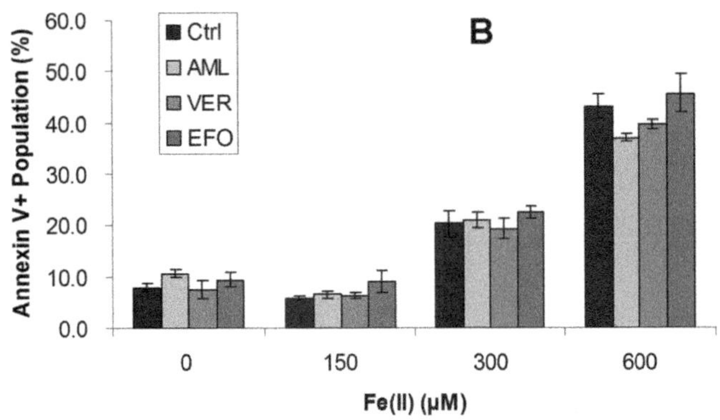

Figure 4. Calcium channel blockers could not salvage HL-1 cells from iron overload induced apoptosis. HL-1 cells were pretreated with LTCC blockers AML or VER, and TTCC blocker EFO for 1 hr, followed by Fe(III) (**A**) and Fe(II) (**B**) loading for 72 hr. Controls were defined as treatments without blockers. Apoptosis was determined by annexin V/PI flow cytometry assay. Total annexin V positive cell portion was counted. The results represented as mean ± SEM of three independent experiments.

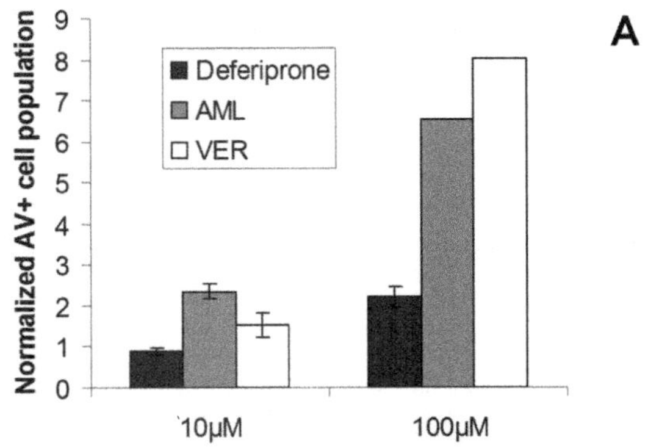

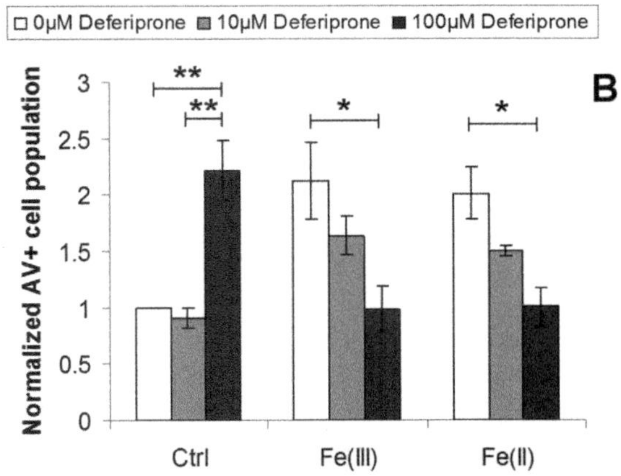

Figure 5. Cellular toxicity of LTCC blockers and the comparison with deferiprone. (A) Apoptotic effects of 10 or 100 μM AML, VER and deferiprone on HL-1 cells for 72 hr were assessed by annexin V/PI flow cytometry assay. (B) HL-1 cells were challenged with 300 μM Fe(III) or Fe(II), followed by treatments of 10 or 100 μM deferiprone 20 min after iron loading. Apoptosis was determined after 72 hr incubation by annexin V/PI assay. Data were shown as total annexin V positive cell portion with normalization to respective negative controls. * $p<0.05$; ** $p<0.01$. The results represented as mean ± SEM of three independent experiments, except 100 μM AML and VER (n = 1).

It is widely recognized that the promiscuous property of LTCC for the transport of other metals is limited to divalent, but not trivalent cations [22,40,41,44]. Interestingly, our data indicated a significant reduction of Fe(III) uptake, but only a trend to reduce Fe(II) uptake, at the presence of LTCC blockers. Together with evidence that a reduction of Fe(III) is required for NTBI uptake into cardiomyocytes [22,28], it raised the possibility that LTCC blockers achieve the effect on NTBI blockade not by stopping Fe(II) entry directly but through alternative mechanism. Recent studies provide an alternative explanation on the role of LTCC in NTBI entry. LTCC has been shown to contribute to the activation of endocytotic machinery in neuronal cells [45]. Interestingly, endocytosis has also been demonstrated to be a possible pathway for macromolecule-associated NTBI uptake into various cell types including cardiomyocytes [38,46]. As LTCC blockade interferes calcium-induced endocytosis, a subsequent interruption of Fe(III) uptake via such pathway can be a possible speculation.

The effect of TTCC blockade on iron entry

With abundant expression in embryonic cardiomyocytes, and subsequent suppression shortly after birth [47], TTCC has been shown to reappear in murine hearts with pathological abnormalities including hypertrophy [48], myocardial infarction [49] and also thalassemia [23,24]. Using efonidipine, the TTCC blocker, Kumfu et al. shows effective blockade of iron uptake both *in vitro* and *in vivo* using the thalassemic mice model, together with the protection effects as assessed *in vivo*, while LTCC blockers appeared inferior [23,24]. However, in our present experimental model, with pretreatment of efonidipine, uptake of neither Fe(II) nor Fe(III) was significantly decreased in iron-overloaded HL-1 cardiomyocytes, implicating an insignificant role of TTCC in HL-1 cells for excessive iron uptake.

Differences in study models

The mechanisms and portal of iron entry into cardiomyocytes under iron overload condition have been controversial, in part being related to differences in experiment approaches, types of iron load models, and the nature of cardiomyocytes explored. In the present study, immortalized HL-1 atrial myocytes were used, which have the advantages of being the only cardiomyocyte cell line currently available that continuously divides and spontaneously contracts while retaining a differentiated adult cardiac phenotype [25,26]. Apart from the superior cardiac properties and cell purity compared with isolated primary cardiomyocytes, HL-1 cells express, from molecular and functional regards, both LTCC and TTCC *in vitro* [27]. In addition, atrial myocytes may provide a model for the study of cardiac iron toxicity, given that atrial dilation and dysfunction have been reported to be earlier markers than depressed ventricular function of cardiac iron toxicity in patients with thalassemia major [50].

LTCC blockade and cardiomyocyte apoptosis

For the therapy of IOC, protection of iron overload induced cardiac apoptosis is apparently crucial beyond the maintenance of regular iron metabolism. Such protection effect was presented in our *in vitro* study by deferiprone, the effective iron chelator commonly used in current clinical practice [51]. However, in our assessment, none of the calcium channel blockers showed significant protection effect on iron overload induced apoptosis, though LTCC blockers, in particular amlodipine, presented slight protection at 600 μM of ferric or ferrous iron challenge. This result is to a certain extent contrary to the previous finding that amlodipine and verapamil attenuate cardiac apoptosis in iron-overloaded mice evaluated by TUNEL assay [20]. One possible explanation is that NTBI initiates apoptosis of cardiomyocyte prior to its entry through cell membrane; and for the *in vivo* model, apart from the effect on NTBI blockade, it cannot rule out the possible contribution from the impacts of LTCC blockers on other physiological conditions which subsequently reduce such iron induced apoptosis. Furthermore, the different susceptibility to iron overload between atrial and ventricular cardiomyocytes should also be taken into consideration [52]. Despite the demonstrable ability of LTCC blockers to inhibit iron ingress into the cytosol of cardiomycytes, their apparent failure to protect them from apoptosis might be due to various properties associated with iron traffic within cells, particularly between cytosol and into mitochondria. As shown earlier [38,53,54], a major fraction of exogenously added iron can access mitochondria, by mechanism that seemingly by-pass the labile iron pool, which is sensed by the calcein probe, and it can even be refractory to some intracellular chelators [38]. While those features would imply that LTCC might provide a path for NTBI entry into cardiomyocytes, they also indicate that such paths might not be relevant for trafficking iron across cytosol to mitochondria, particularly in the pathophysiological context. Consequently, although the prevention of iron ingress into cardiomyocytes was observed in treatment with LTCC blockers at higher doses, due to their toxicity, at least shown *in vitro*, further studies would be of importance for their protective roles in iron-overloaded cardiomyocytes, and also for a better understanding of the etiology of IOC.

Clinical implications

Apoptosis is rare in normal human heart. In all reported cases, including those in failing hearts, apoptosis levels are substantially lower than 1% as revealed by TUNEL assay [55]. Due to the poor regenerative capacity of cardiomyocytes, a constant, albeit low, level of apoptosis can have serious consequence. Apart from limited studies showing the potential anti-apoptotic effect of deferasirox [35] and taurine [4] in myocardium of iron-overloaded murine model, little is known about the anti-apoptotic approach for iron overload. Further studies on the mechanism of iron induced apoptosis would provide novel targets for advanced therapy against IOC.

Limitations

Several limitations to this study warrant discussion. Firstly, the findings of the present *in vitro* study may reflect perhaps a relatively acute effect of iron load on cardiomyocytes. Ideally, the experimental protocols should be extended to longer duration with lower iron levels. However, given the technical constraints including the confounding influence of cell proliferation with prolonged culture on fluorescent assay of iron entry and the need for medium change with alteration in iron concentrations, we have elected to adopt the current methodology. With regard to animal studies, previous works have been done on mouse [20,24] and gerbil [35], which mimic the effect of chronic iron overload better, although results remained controversial. Secondly, we have not assessed the effects of calcium channel blockade on cellular beating in the present study. Calcium channel blockade may reduce beating rate or cause cessation of cardiomyocyte contraction *in vitro* [56,57]. Nonetheless, LTCC and TTCC have been shown to remain functional in HL-1 cells without apparent contraction [27,58]. The effect of cardiomyocytes beating rate on iron uptake, however, requires further studies for its clarification. Thirdly, although HL-1 cells are the only cardiomyocyte cell line that retains contractile phenotype with differentiated cardiac characteristics [25,26], they are established from AT-1 mouse atrial cardiomyocyte tumor lineage. The different electrical properties, including calcium kinetics, between atrial and ventricular myocytes [52] may potentially lead to differences in response to iron overload between HL-1 cells and ventricular cardiomyocytes merit further studies. With advances in the induced pluripotent stem cell technology, the use of human ventricular cardiomyocytes may be a better model to study the effects of iron cardiotoxicity. Finally, we have not assessed the detailed pro-apoptotic signaling pathways in the present study. In mesenchymal stem cells [59,60], hepatocytes [61], neuroblastoma cells [62] and gerbil [63], p38 and JNK are activated under iron overload conditions. This would undoubtedly be important when designing future studies.

Conclusions

In summary, our current study illustrated the patterns of iron entry in HL-1 atrial myocytes under ferric or ferrous iron overload condition. The blockade of LTCC but not TTCC was identified to prevent labile ferric iron entry. The uptake of ferrous iron probably involves other mechanism. As expected, iron overload was shown to induce cardiac apoptosis via mitochondria-mediated caspase-3 dependent pathways. However, LTCC blockers have very limited protective effect toward iron induced apoptosis. Our study provided a better understanding to the role of LTCC and TTCC on NTBI uptake into cardiomyocytes, contributing to the conceptual framework in the development of advanced therapeutic strategy for IOC in combination with the current chelation therapy.

Acknowledgments

We thank Prof. William C. Claycomb (Louisiana State University Health Science Center, LA, USA) for HL-1 cardiomyocytes, and Dr. Wing Keung Chan (St. Jude Children's Research Hospital, TN, USA) for technical advice on experiments and data analysis. We also thank Prof. George J. Kontoghiorghes (Postgraduate Research Institute, Limassol,Cyprus) for comment on iron chelation experiment.

Author Contributions

Conceived and designed the experiments: MPC GCFC YFC. Performed the experiments: MPC. Analyzed the data: MPC GCFC YFC. Contributed reagents/materials/analysis tools: MPC ZIC SC. Wrote the paper: MPC ZIC GCFC YFC.

References

1. Lieu PT, Heiskala M, Peterson PA, Yang Y (2001) The roles of iron in health and disease. Mol Aspects Med 22: 1–87.
2. Esposito BP, Breuer W, Sirankapracha P, Pootrakul P, Hershko C, et al. (2003) Labile plasma iron in iron overload: redox activity and susceptibility to chelation. Blood 102: 2670–2677.
3. Hershko CM, Link GM, Konijn AM, Cabantchik ZI (2005) Iron chelation therapy. Curr Hematol Rep 4: 110–116.
4. Oudit GY, Trivieri MG, Khaper N, Husain T, Wilson GJ, et al. (2004) Taurine supplementation reduces oxidative stress and improves cardiovascular function in an iron-overload murine model. Circulation 109: 1877–1885.
5. Gujja P, Rosing DR, Tripodi DJ, Shizukuda Y (2010) Iron overload cardiomyopathy: better understanding of an increasing disorder. J Am Coll Cardiol 56: 1001–1012.
6. Kremastinos DT, Tiniakos G, Theodorakis GN, Katritsis DG, Toutouzas PK (1995) Myocarditis in beta-thalassemia major. A cause of heart failure. Circulation 91: 66–71.
7. Kremastinos DT, Flevari P, Spyropoulou M, Vrettou H, Tsiapras D, et al. (1999) Association of heart failure in homozygous beta-thalassemia with the major histocompatibility complex. Circulation 100: 2074–2078.
8. Muhlestein JB (2000) Cardiac abnormalities in hemochromatosis. In: Barton JC, Edwards CQ, editors. Hemochromatosis: genetics, pathophysiology, diagnosis, and treatment Cambridge University Press. pp. 297–310.
9. Murphy CJ, Oudit GY (2010) Iron-overload cardiomyopathy: pathophysiology, diagnosis, and treatment. J Card Fail 16: 888–900.
10. Olivieri NF, Nathan DG, MacMillan JH, Wayne AS, Liu PP, et al. (1994) Survival in medically treated patients with homozygous beta-thalassemia. N Engl J Med 331: 574–578.
11. Horwitz LD, Rosenthal EA (1999) Iron-mediated cardiovascular injury. Vasc Med 4: 93–99.
12. Narula J, Haider N, Virmani R, DiSalvo TG, Kolodgie FD, et al. (1996) Apoptosis in myocytes in end-stage heart failure. N Engl J Med 335: 1182–1189.
13. Olivetti G, Abbi R, Quaini F, Kajstura J, Cheng W, et al. (1997) Apoptosis in the failing human heart. N Engl J Med 336: 1131–1141.
14. Kang PM, Izumo S (2000) Apoptosis and heart failure: A critical review of the literature. Circ Res 86: 1107–1113.
15. Wencker D, Chandra M, Nguyen K, Miao W, Garantziotis S, et al. (2003) A mechanistic role for cardiac myocyte apoptosis in heart failure. J Clin Invest 111: 1497–1504.
16. Foo RS, Mani K, Kitsis RN (2005) Death begets failure in the heart. J Clin Invest 115: 565–571.
17. Lee Y, Gustafsson AB (2009) Role of apoptosis in cardiovascular disease. Apoptosis 14: 536–548.
18. Hentze MW, Muckenthaler MU, Andrews NC (2004) Balancing acts: molecular control of mammalian iron metabolism. Cell 117: 285–297.
19. Chattipakorn N, Kumfu S, Fucharoen S, Chattipakorn S (2011) Calcium channels and iron uptake into the heart. World J Cardiol 3: 215–218.
20. Oudit GY, Sun H, Trivieri MG, Koch SE, Dawood F, et al. (2003) L-type Ca2+ channels provide a major pathway for iron entry into cardiomyocytes in iron-overload cardiomyopathy. Nat Med 9: 1187–1194.
21. Oudit GY, Trivieri MG, Khaper N, Liu PP, Backx PH (2006) Role of L-type Ca2+ channels in iron transport and iron-overload cardiomyopathy. J Mol Med (Berl) 84: 349–364.
22. Tsushima RG, Wickenden AD, Bouchard RA, Oudit GY, Liu PP, et al. (1999) Modulation of iron uptake in heart by L-type Ca2+ channel modifiers: possible implications in iron overload. Circ Res 84: 1302–1309.
23. Kumfu S, Chattipakorn S, Chinda K, Fucharoen S, Chattipakorn N (2012) T-type calcium channel blockade improves survival and cardiovascular function in thalassemic mice. Eur J Haematol 88: 535–548.
24. Kumfu S, Chattipakorn S, Srichairatanakool S, Settakorn J, Fucharoen S, et al. (2011) T-type calcium channel as a portal of iron uptake into cardiomyocytes of beta-thalassemic mice. Eur J Haematol 86: 156–166.
25. Claycomb WC, Lanson NA Jr, Stallworth BS, Egeland DB, Delcarpio JB, et al. (1998) HL-1 cells: a cardiac muscle cell line that contracts and retains phenotypic characteristics of the adult cardiomyocyte. Proc Natl Acad Sci U S A 95: 2979–2984.
26. White SM, Constantin PE, Claycomb WC (2004) Cardiac physiology at the cellular level: use of cultured HL-1 cardiomyocytes for studies of cardiac muscle cell structure and function. Am J Physiol Heart Circ Physiol 286: H823–829.
27. Xia M, Salata JJ, Figueroa DJ, Lawlor AM, Liang HA, et al. (2004) Functional expression of L- and T-type Ca2+ channels in murine HL-1 cells. J Mol Cell Cardiol 36: 111–119.
28. Parkes JG, Olivieri NF, Templeton DM (1997) Characterization of Fe2+ and Fe3+ transport by iron-loaded cardiac myocytes. Toxicology 117: 141–151.
29. Randell EW, Parkes JG, Olivieri NF, Templeton DM (1994) Uptake of non-transferrin-bound iron by both reductive and nonreductive processes is modulated by intracellular iron. J Biol Chem 269: 16046–16053.
30. Glickstein H, El RB, Shvartsman M, Cabantchik ZI (2005) Intracellular labile iron pools as direct targets of iron chelators: a fluorescence study of chelator action in living cells. Blood 106: 3242–3250.
31. Lecoeur H, Melki MT, Saidi H, Gougeon ML (2008) Analysis of apoptotic pathways by multiparametric flow cytometry: application to HIV infection. Methods Enzymol 442: 51–82.
32. Oancea M, Mazumder S, Crosby ME, Almasan A (2006) Apoptosis assays. Methods Mol Med 129: 279–290.
33. Kontoghiorghes GJ (2006) Iron mobilization from transferrin and non-transferrin-bound-iron by deferiprone. Implications in the treatment of thalassemia, anemia of chronic disease, cancer and other conditions. Hemoglobin 30: 183–200.
34. Nday CM, Malollari G, Petanidis S, Salifoglou A (2012) In vitro neurotoxic Fe(III) and Fe(III)-chelator activities in rat hippocampal cultures. From neurotoxicity to neuroprotection prospects. J Inorg Biochem 117: 342–350.
35. Wang Y, Wu M, Al-Rousan R, Liu H, Fannin J, et al. (2011) Iron-induced cardiac damage: role of apoptosis and deferasirox intervention. J Pharmacol Exp Ther 336: 56–63.
36. Munoz JP, Chiong M, Garcia L, Troncoso R, Toro B, et al. (2010) Iron induces protection and necrosis in cultured cardiomyocytes: Role of reactive oxygen species and nitric oxide. Free Radic Biol Med 48: 526–534.
37. Whelan RS, Kaplinskiy V, Kitsis RN (2010) Cell death in the pathogenesis of heart disease: mechanisms and significance. Annu Rev Physiol 72: 19–44.
38. Shvartsman M, Kikkeri R, Shanzer A, Cabantchik ZI (2007) Non-transferrin-bound iron reaches mitochondria by a chelator-inaccessible mechanism: biological and clinical implications. Am J Physiol Cell Physiol 293: C1383–1394.
39. Catterall WA, Striessnig J (1992) Receptor sites for Ca2+ channel antagonists. Trends Pharmacol Sci 13: 256–262.
40. Winegar BD, Kelly R, Lansman JB (1991) Block of current through single calcium channels by Fe, Co, and Ni. Location of the transition metal binding site in the pore. J Gen Physiol 97: 351–367.
41. Atar D, Backx PH, Appel MM, Gao WD, Marban E (1995) Excitation-transcription coupling mediated by zinc influx through voltage-dependent calcium channels. J Biol Chem 270: 2473–2477.
42. Hamann SR, Blouin RA, McAllister RG Jr (1984) Clinical pharmacokinetics of verapamil. Clin Pharmacokinet 9: 26–41.
43. Mak IT, Weglicki WB (1990) Comparative antioxidant activities of propranolol, nifedipine, verapamil, and diltiazem against sarcolemmal membrane lipid peroxidation. Circ Res 66: 1449–1452.
44. Lansman JB, Hess P, Tsien RW (1986) Blockade of current through single calcium channels by Cd2+, Mg2+, and Ca2+. Voltage and concentration dependence of calcium entry into the pore. J Gen Physiol 88: 321–347.
45. Rosa JM, Nanclares C, Orozco A, Colmena I, de Pascual R, et al. (2012) Regulation by L-Type Calcium Channels of Endocytosis: An Overview. J Mol Neurosci.
46. Sohn YS, Ghoti H, Breuer W, Rachmilewitz E, Attar S, et al. (2012) The role of endocytic pathways in cellular uptake of plasma non-transferrin iron. Haematologica 97: 670–678.
47. Yasui K, Niwa N, Takemura H, Opthof T, Muto T, et al. (2005) Pathophysiological significance of T-type Ca2+ channels: expression of T-type Ca2+ channels in fetal and diseased heart. J Pharmacol Sci 99: 205–210.
48. Martinez ML, Heredia MP, Delgado C (1999) Expression of T-type Ca(2+) channels in ventricular cells from hypertrophied rat hearts. J Mol Cell Cardiol 31: 1617–1625.
49. Huang B, Qin D, Deng L, Boutjdir M, Nabil ES (2000) Reexpression of T-type Ca2+ channel gene and current in post-infarction remodeled rat left ventricle. Cardiovasc Res 46: 442–449.
50. Li W, Coates T, Wood JC (2008) Atrial dysfunction as a marker of iron cardiotoxicity in thalassemia major. Haematologica 93: 311–312.
51. Kolnagou A, Kleanthous M, Kontoghiorghes GJ (2011) Efficacy, compliance and toxicity factors are affecting the rate of normalization of body iron stores in thalassemia patients using the deferiprone and deferoxamine combination therapy. Hemoglobin 35: 186–198.
52. Grandi E, Pandit SV, Voigt N, Workman AJ, Dobrev D, et al. (2011) Human atrial action potential and Ca2+ model: sinus rhythm and chronic atrial fibrillation. Circ Res 109: 1055–1066.
53. Shvartsman M, Fibach E, Cabantchik ZI (2010) Transferrin-iron routing to the cytosol and mitochondria as studied by live and real-time fluorescence. Biochem J 429: 185–193.
54. Shvartsman M, Ioav Cabantchik Z (2012) Intracellular iron trafficking: role of cytosolic ligands. Biometals 25: 711–723.
55. Chiong M, Wang ZV, Pedrozo Z, Cao DJ, Troncoso R, et al. (2011) Cardiomyocyte death: mechanisms and translational implications. Cell Death Dis 2: e244.
56. Wang T, Hu N, Cao J, Wu J, Su K, et al. (2013) A cardiomyocyte-based biosensor for antiarrhythmic drug evaluation by simultaneously monitoring cell growth and beating. Biosens Bioelectron 49: 9–13.
57. Jonsson MK, Wang QD, Becker B (2011) Impedance-based detection of beating rhythm and proarrhythmic effects of compounds on stem cell-derived cardiomyocytes. Assay Drug Dev Technol 9: 589–599.
58. Rao F, Deng CY, Wu SL, Xiao DZ, Huang W, et al. (2013) Mechanism of macrophage migration inhibitory factor-induced decrease of T-type Ca(2+) channel current in atrium-derived cells. Exp Physiol 98: 172–182.

59. Lu WY, Zhao MF, Chai X, Meng JX, Zhao N, et al. (2013) [Reactive oxygen species mediate the injury and deficient hematopoietic supportive capacity of umbilical cord derived mesenchymal stem cells induced by iron overload]. Zhonghua Yi Xue Za Zhi 93: 930–934.
60. Lu WY, Zhao MF, Sajin R, Zhao N, Xie F, et al. (2013) [Effect and mechanism of iron-catalyzed oxidative stress on mesenchymal stem cells]. Zhongguo Yi Xue Ke Xue Yuan Xue Bao 35: 6–12.
61. Dai J, Huang C, Wu J, Yang C, Frenkel K, et al. (2004) Iron-induced interleukin-6 gene expression: possible mediation through the extracellular signal-regulated kinase and p38 mitogen-activated protein kinase pathways. Toxicology 203: 199–209.
62. Salvador GA, Oteiza PI (2011) Iron overload triggers redox-sensitive signals in human IMR-32 neuroblastoma cells. Neurotoxicology 32: 75–82.
63. Al-Rousan RM, Paturi S, Laurino JP, Kakarla SK, Gutta AK, et al. (2009) Deferasirox removes cardiac iron and attenuates oxidative stress in the iron-overloaded gerbil. Am J Hematol 84: 565–570.

24

Measuring the Chemical and Cytotoxic Variability of Commercially Available Kava (*Piper methysticum* G. Forster)

Amanda C. Martin[1], Ed Johnston[2], Chengguo Xing[3*¤a], Adrian D. Hegeman[1*¤b]

1 The Department Horticultural Science, Plant Biological Sciences Graduate Program, Department of Plant Biology, and Microbial and Plant Genomics Institute, University of Minnesota, St Paul, MN, United States of America, **2** Alia Point 'Awa Nursery, Pepe'ekeo, Hawai'i, United States of America, **3** Department of Medicinal Chemistry, University of Minnesota, Minneapolis, MN, United States of America

Abstract

Formerly used world-wide as a popular botanical medicine to reduce anxiety, reports of hepatotoxicity linked to consuming kava extracts in the late 1990s have resulted in global restrictions on kava use and have hindered kava-related research. Despite its presence on the United States Food and Drug Administration consumer advisory list for the past decade, export data from kava producing countries implies that US kava imports, which are not publicly reported, are both increasing and of a fairly high volume. We have measured the variability in extract chemical composition and cytotoxicity towards human lung adenocarcinoma A549 cancer cells of 25 commercially available kava products. Results reveal a high level of variation in chemical content and cytotoxicity of currently available kava products. As public interest and use of kava products continues to increase in the United States, efforts to characterize products and expedite research of this potentially useful botanical medicine are necessary.

Editor: Stephen J. Polyak, University of Washington, United States of America

Funding: This work was funded by the National Science Foundation Plant Genome Research Program grants IOS-0923960 and IOS-1238812 (www.nsf.gov)(ACM, ADH), the National Science Foundation Graduate Research Fellowship Program 00006595 (www.nsf.gov) (ACM), and the National Cancer Institute, National Institutes of Health grant R01 CA142649 (www.cancer.gov) (CX). The funders had no role in study design, data collection and analysis, decision to publish, or preparation of the manuscript.

Competing Interests: Ed Johnston is affiliated with Alia Point 'Awa Nursery. The authors would like to clarify that Alia Point 'Awa Nursery is not a commercial 'awa nursery. Alia Point 'Awa Nursery provides 'awa cuttings and plants to research and educational institutions including University of Hawaii, Mano'a, University of Saskatchewan, Saskatoon, Canada, and Como Park Zoo and Conservatory, Saint Paul, MN, free of charge or for the cost of shipping only. Recorded 'awa costs for 2012 and 2013 were $60 and $50, respectively. The role of Alia Point 'Awa Nursery is described by Vincent Lebot in Ethnobotany Research & Applications Vol 2/i as a site for preservation of Hawaiian cultivars of 'awa / kava. The first recorded DNA analysis of Hawaiian 'awa was performed at Alia Point 'Awa Nursery and reported in Lebot, et. al. (1999) Morphological Phytochemical and genetic variation in Hawaiian cultivars of 'awa. Economic Botany, 53 (4): 407–418. Please let the authors know if you would like any additional information about the status of Alia Point 'Awa Nursery.

* Email: xingx009@umn.edu (CX); hegem007@umn.edu (ADH)

¤a Current address: University of Minnesota, College of Pharmacy, Minneapolis, MN, United States of America
¤b Current address: University of Minnesota, Department of Horticultural Science, Saint Paul, MN, United States of America

Introduction

Kava (*Piper methysticum* G. Forster) is the name of a plant and drink that is prepared traditionally by macerating its roots in cool water or coconut water [1]. It has been used for many centuries in the South Pacific and Hawaii for social ceremonies, relaxation, medicine, and a multitude of other purposes [1]. More recently, standardized kava extracts, containing 30% active constituents, have been used globally as an anxiolytic [2,3]. Additionally, a tight inverse correlation between high rates of kava consumption and low incidences of cancer for populations in the South Pacific has been reported [4]. Subsequent studies have shown that kava displays cancer preventive properties [5–8].

There are about 200 different cultivated varieties of kava [9], each with a unique chemotype that produces specific physiological and psychoactive effects [10–13]. The active constituents are chemically classified as kavalactones and six (kawain, dihydrokawain, methysticin, dihydromethysticin, yangonin, and desmethoxyyangonin) constitute the primary chemicals that are responsible for individual cultivars' unique chemotypes [1,14–16].

A 2002 the Kava Act passed in Vanuatu established four classes of kava cultivars: noble, which have a long history of safe use as traditional drink; medicinal, which have long been used by traditional herbalists in the South Pacific and are banned as export commodities; 'Tu dei', which have a very strong effect that lasts two days; and 'Wichmanni' or wild varieties [9,17,18]. Cultivars from the noble class are typically used to prepare kava extract as they have the optimal therapeutic chemotype. Cultivars belonging to other classes have been reported to have overpowering and unpredictable effects causing symptoms such as nausea and headaches [19,20]. Kava's active constituents are primarily located in its roots; other plant parts such as stems and leaves should not be used in extract preparations [1]. Traditional kava is prepared using a 100% aqueous solvent, which results in a drink containing an average of 0.3–20% kavalactone content [1]. Commercial manufacturers use up to 100% ethanol or acetone in the extraction process resulting in up to 70% kavalactone content in the final product [1,20]. Studies have shown a difference between traditionally prepared extracts and those prepared with ethanol both in cytotoxicity and chemical composition [21–25]. In addition to those described above, other potential sources of variation in kava products include, contamination of raw kava materials, impurities, post-harvest handling and storage procedures (drying, whole vs. ground material, humidity, temperature), age of harvested kava plants, mixture and quality of cultivar(s) used [17,26–28] Kava is distributed in variable forms, including dry powder, freeze-dried, liquid tincture, and capsule, making it difficult to know exactly which cultivar(s), plant part(s), extraction solvent(s), and other factors were used in the preparation [26].

Due to reports linking modern kava consumption to individual cases of hepatotoxicity, kava was banned in the European Union and Canada in 2003, voluntarily recalled in Australia in 2003, and included on the United States Food and Drug Administration (US FDA) consumer advisory list in March of 2002 [14,29–32]. These bans and advisories have hindered research on kava as an alternative anti-anxiety and cancer preventive medicine [17]. Despite its presence on the US FDA consumer advisory list for the past decade, the extrapolation of export data from the kava producing nations Fiji, The Republic of Vanuatu, and Tonga to the US (Table 1) indicates that kava imports to the US, which are not publicly reported, are presumably both increasing and of a fairly high volume [33–36].

There are many hypothesized mechanisms potentially linking kava consumption to hepatotoxicity [3,14,25,26,32,37–43]. We intended to measure the overall variation in cellular toxicity and chemical composition among the large volume of diverse kava products currently available. Only six kavalactones have been intensively studied [2,16,18,44,45] making it necessary to assess the complete pool of extracted compounds. We performed metabolic fingerprinting; a metabolomics technique that facilitates comparisons based on global metabolite patterns of whole extracts [46]. We used ultra-performance liquid chromatography-electrospray ionization-time-of-flight-mass spectrometry (UPLC-ESI-TOF-MS) to fingerprint replicate aqueous and 95% ethanolic extracts of 25 commercial kava products (Table S1). We also quantified six compounds found in kava that may be associated with either the medicinal or negative cytotoxic effects of modern kava usage: kawain (K); dihydrokawain (DHK); methysticin (M); dihydromethysticin (DHM); flavokawain A (FLK A); and flavokawain B (FLK B) [24,37–39] (Figure 1). Absolute quantification was performed using pure standards and a UPLC-single quadrupole mass spectrometer (MS). Finally, we determined the cytotoxicity levels of each extract in cell viability assays towards human lung adenocarcinoma A549 cancer cell line.

Results and Discussion

Metabolic fingerprinting experiments measured three aspects of chemical variation: reproducibility of replicate extractions of individual products; differences between using 100% water or 95% ethanol as the extraction solvent; and the overall variation among the set of kava products tested. Similarly to previous quantitative studies of compounds from kava, we observed a high level of reproducibility of replicate extractions of material from individual kava sources [45]. Principal component analysis showed that replicate extractions from the same kava source are tightly clustered (Figure S1). Moreover, there were small standard errors (average standard error 12.6%) from the absolute quantification measurements of K, DHK, M, DHM, FLKA, and FLKB (Tables S2 and S3). These results provide evidence that there is consistency in the material contained within a single batch of kava from any given source.

Extract chemical composition was strongly influenced by extraction solvent. Metabolic fingerprints from aqueous and ethanolic extracts plotted in principal component space formed two distinct groups driven by extraction solvent where the use of either 100% water or 95% ethanol was responsible for 71.1% of the variation among all samples explained by PC1 (Figure S2). The detected ion m/z, retention time pairs that contribute the most to the loadings for PC1 were 315.1132 m/z, 9.1823 min and 285.1021 m/z, 9.4699 min, which correspond to the masses and retention times of FLKA and FLKB, respectively. Compound quantification showed that extracts prepared with 95% ethanol resulted in higher yields and greater consistency among replicates, compared with extracts prepared with 100% water. This result is similar to previous studies that found water produced kava extracts with decreased compound concentrations compared to extracts prepared with ethanol [22,23]. Specifically, K, DHK, M, and DHM concentrations were 1.5–5x higher in samples extracted with 95% ethanol than in those extracted with 100% water. The concentrations of FLKA & FLKB were up to fifty times higher in

Table 1. Kava exports from Fiji, Tonga and Vanuatu: 2008 through 2013[a].

	Total metric tons exported (subset exp. to US)		
Year	**from Fiji[33]**	**from Tonga[34]**	**from Vanuatu[35]**
2008	184 (93)	27[36]	356[36]
2009	212 (123)	38.9 (0.7)	485[36]
2010	244 (91)	61.6 (29.2)	498
2011	276 (95)[b]	68.6 (42)	734
2012	NA	117 (80)	643
2013	NA	NA	558[c]

[a]Kava exports are reported in metric tons where available from 2008 through 2013. The subset of exports to the United States is given parenthetically next to each total export figure where available. NA indicates that the data were not available for that year from the sources cited.
[b]Represents exports for January through November of 2011.
[c]Represents exports for January through August 2013.

samples extracted by 95% ethanol than in those extracted with 100% water although a significant number of the water extracts contained concentrations of FLKA or FLKB that were below detectable limits (Figure 1). Extracts prepared with 95% ethanol consistently contained greater quantities of FLKA and FLKB than corresponding water extracts, and were highly variable across kava products (ranging from undetectable concentrations up to 14.7 ppm; Figure 1).

While extraction solvent was the most influential variable affecting the observed chemical composition, significant variation in the concentrations of K, DHK, M, and DHM for identically prepared extracts was observed from different source materials. This variation was even more dramatic in regard to the concentration of FLKA and FLKB. The variation in chemical composition was further reflected by the differences in cytotoxicity observed for each commercial kava product extract.

Cytotoxicity assays against human lung adenocarcinoma A549 cancer cell line with aqueous extracts from all 25 commercial kavas showed no toxicity at any concentration measured up to 500 μg/mL. This result is similar to previous studies indicating that aqueous extracts have low to no cytotoxic effect [20]. In contrast, identically prepared ethanol extracts from different commercial sources varied greatly in their relative cytotoxicity at all concentrations measured 37.5, 75 (shown in Figure 2 top), and 150 μg/mL. Ethanol extracts prepared from commercial kava sources K, M, N, O, W, Y, DD, and EE exhibited very low cell toxicity at all concentrations, indicated by a relative cell viability level of greater than 90%. In contrast, ethanolic extracts from G, H, P, Q, R, S, V, Z, and BB, displayed the highest levels of toxicity, with a relative cell viability level of less than 30%. Cytotoxicity levels at these three discreet extract concentrations varied over a wide range similarly to the variation observed in extract chemical composition, especially in regard to FLKA and FLKB.

We observed a moderate correlation between the concentrations of FLKA & FLKB (log_2 normalized) and the relative cytotoxicity across the sampled kava products. High concentrations of the flavokawains generally mirrored lower relative cell viability (Figure 2). Some samples, however, deviated from this correlation, specifically N and BB, which have medium concentrations of FLKA and FLKB and display low and high cytotoxcity, respectively. Scatterplots of these data revealed that concentrations of FLKA and FLKB correlated similarly to cytotoxicity level with R^2 values equal to 0.68 and 0.69, respectively for extracts prepared at 75 μg/mL and 0.78 and 0.77, respectively for extracts prepared at 150 μg/mL (Figures S3 and S4). Based on these correlation values, clearly, FLKA and FLKB are likely the major but not the solely compounds responsible for the extract toxicity. However, additional reports of flavokawain toxicity, including measured IC_{50} levels for FLKA and FLKB of 13±1.1 and 6.6±0.1 ppm, respectively against Hepa 1c1c7 liver cells [24], and 57% growth inhibition of bladder T24 tumor cells by FLKA [8] warrant further exploration of the link between flavokawains and kava's cytotoxicity [21,22].

We also generated a non-polar flavokawain enriched kava fraction to determine the IC_{50} values of 48-hour toxicities against hepatocytes from mouse, rat, and monkey. We measured IC_{50} values for this FLK rich fraction of 57±9, 45±4 and 49±6 μg/mL, for mouse, rat, and monkey hepatocytes, respectively. In contrast polar and medium polarity fractions and whole traditionally prepared kava had non-detectable IC_{50} values greater than 400 μg/mL in all three cell types.

For each compound, K, DHK, M, and DHM there was no obvious association between concentration and relative cell viability, although extracts with higher overall concentrations of all six compounds resulted in lower relative cell viability. This trend suggests that K, DHK, M, and DHM are less likely to be involved in the specific mechanism(s) of cytotoxicity.

Additional reports of cytotoxic compounds found in kava describe one additional flavokawain, distinct from FLKA and FLKB with the following chemical formula and exact mass $C_{17}H_{16}O_5$, 300.0998 (FLK C) [35] and three alkaloids found in kava leaves with the following chemical formulae and exact masses: $C_{14}H_{17}NO_2$, 231.1259 (awaine); $C_{16}H_{17}NO4$, 287.1157 (pipermethystine); $C_{16}H_{17}NO_5$, 303.1106 (3α, 4α-epoxy-5β-pipermethystine) [47]. These alkaloids may have been present in commercial kava products produced by European companies leading up to the European ban of kava in 2003 [18] We looked for patterns between cytotoxicity level and the presence of these potentially toxic compounds using M^+H extracted ion chromatograms. Overall, only a peak corresponding to FLKC was detectable above the limit of detection ($s/n>3$), where the relative intensity of this peak was higher in extracts with higher toxicities. Additional experiments are necessary to understand how these compounds interact with FLKA and FLKB to produce extract cytotoxicity [17,18,26,28]. Regardless of the precise cause of cytotoxicity it is clear that tremendous variation exists in the chemical composition and resulting toxicity of commercially available kava products.

Conclusions

Kava export data show that in spite of bans and warnings, consumption of unregulated kava products appears to be increasing. Our analysis shows that the assortment of commercially available kava products vary widely in chemical composition and cytotoxicity level. Certain kava cultivars and preparation methods may produce products that vary broadly in both their toxicity and their efficacy and thus a rapid and easily applied method to characterize and classify kava products would be beneficial to the consumer. Disregarding kava and its potential use as an anxiolytic or for cancer preventive ignores the great potential societal benefits of the rational and informed medicinal use of this plant.

Materials and Methods

Solvents and reagents

HPLC grade solvents from Sigma Aldrich (St. Louis, MO, USA) were used including: acetonitrile, dimethyl sulfoxide (DMSO), formic acid, ethyl acetate, 95% ethanol, and hexanes. Reverse osmosis deionized glass distilled water was obtained in house using a Thermo Scientific Barnstead B-pure filter and Distinction water still model D4000 (Bibby Scientific Limited, Stone, Staffordshire ST15 0SA, UK). Standard kava compounds were purified from Gaia Herb (Brevard, NC, USA) commercial Kava extract. Commercial kava samples were obtained from a variety of sources (Table S1).

Kava extraction

Kava samples from 25 different sources were classified as either powder (P) or liquid (L) (Table S1). Four extraction methods were used; methods I and II for powder samples and methods III and IV for liquid samples. The method details are as follows: Method I: 10 mL of room temperature water was added to 5 grams of powdered kava, shaken for 2 hrs, centrifuged to remove insoluble material and the supernatant evaporated to dryness and re-dissolved in water at a concentration of 1.5 mg of residue per mL.

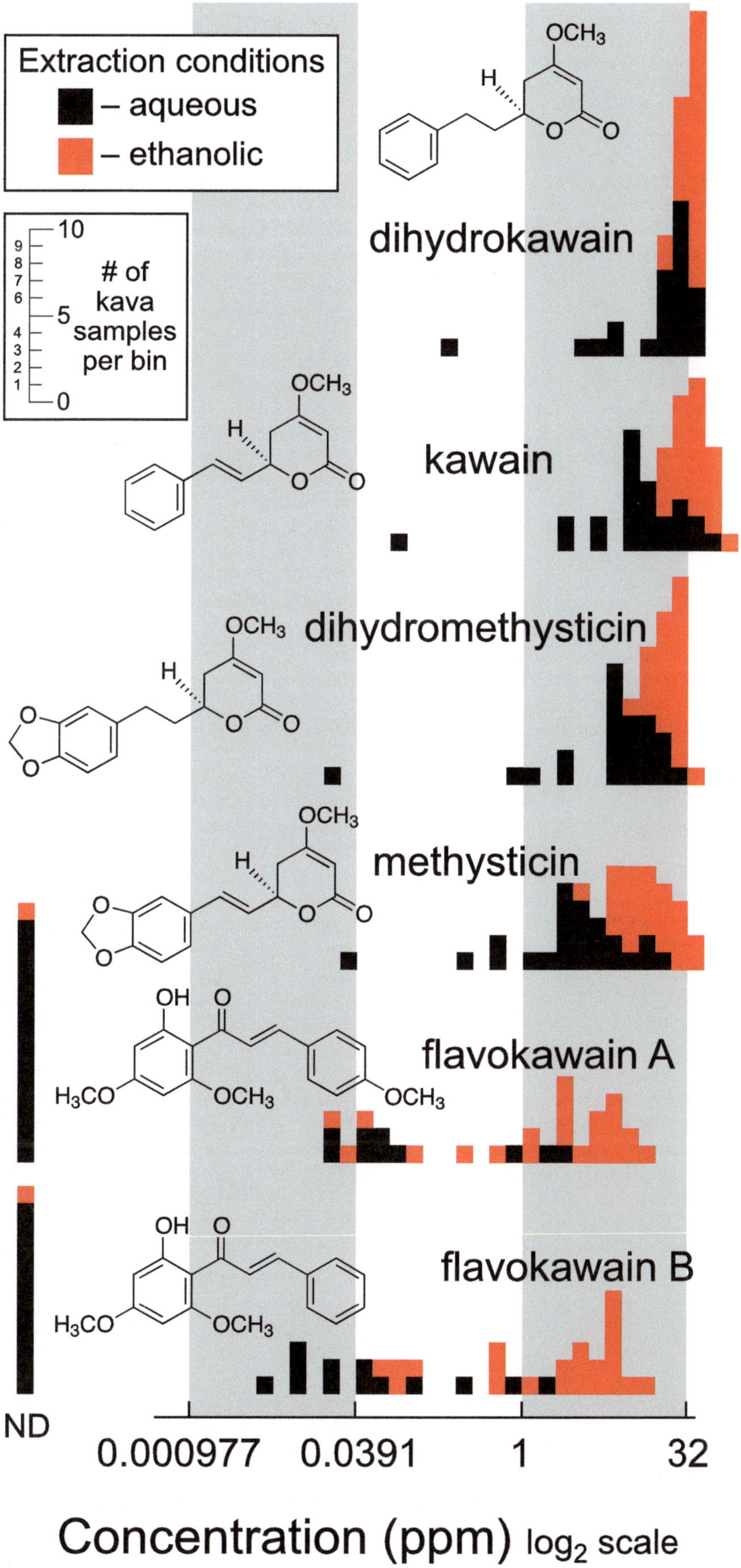
Extraction conditions
– aqueous
– ethanolic
10
9
8
7
6
5
4
3
2
1
0
# of kava samples per bin
OCH3
H
O
O
dihydrokawain
kawain
dihydromethysticin
methysticin
OH
H3CO
flavokawain A
flavokawain B
ND
0.000977
0.0391
1
32
Concentration (ppm) log2 scale

Figure 1. Histograms showing the distribution of concentrations of compounds found in commercial kava preparations. Kava samples were extracted with both 100% water (black) and 95% ethanol (red). The six compounds shown were quantified in each extract by LC-MS and the resulting concentrations in part per million are displayed histogramatically. Distributions were normalized by display on a $\log_2$ scale. The inset y-axis scale indicates the numbers of kava samples in each bin. Measurements designated as not detected (ND) were below the limits of detection (LOD ($s/n<3$) for each analysis, which were typically 0.0005 to 0.001 ppm depending on variation in signal to noise from sample to sample.

Method II: the same as I, except 95% ethanol was used in place of water. Method III: 200 µL of liquid kava sample was dried *in vacuo*, reconstituted in 500 µL of water and then adjusted to a concentration of 1.5 mg of residue per mL with additional water. Method IV: samples were directly diluted to 1.5 mg/mL with 95% ethanol. For each extraction method four replicates per sample were prepared for analysis. Extract yield was determined gravimetrically by evaporating 500 µL of extract to dryness using a Savant model SVC-200 H SpeedVac concentrator (Farmingdale, NY, USA). Extracts were normalized to 1.5 mg/mL for experiments and stored at 4°C in the dark for no more than a week prior to LC/MS analysis. All extractions were carried out at room temperature (approximately 25°C).

Metabolic fingerprinting

Metabolic fingerprints were generated using C_{18}-reversed-phase ultra-performance liquid chromatography-positive electrospray ionization-time-of-flight mass spectrometry (UPLC-ESI(+)-TOF-MS) carried out on a UPLC-TOF LCT Premier XE mass spectrometer fitted with an autosampler with a sample vial block maintained at 4°C (Acquity, Waters, Milford MA, USA). The following MS conditions were used: full scan mass scan range: 100–1000 *m/z*, W analyzer mode, extended dynamic range, 0.1 s scan time, desolvation temperature 350°C, desolvation nitrogen flow rate: 7.0 L/min, capillary voltage: 2900 V, sample cone voltage: 30 V, source temperature: 120°C. Separations were carried out on a reversed-phase C_{18} HSS T_3 1.8 µm particle size, 2.1×100 mm column (Waters). Column temperature was 50°C, solvent flow rate 0.3 mL/min, injection volume 5 µL. A 14-minute gradient using mobile phases A: 0.1% formic acid in water and B: 0.1% formic acid in acetonitrile was run according to the following gradient elution profile: initial, 10%; 3 minutes, 50% B; 8 minutes, 60% B; 13 minutes, 98% B; 14 minutes, 98% B. A 7-minute wash cycle was run between every sample and monitored for the absence of carryover. MassLynx version 4.1 (Waters) was used for data collection and visualization. Sample analysis order was randomized across the entire sample set.

Feature detection and multivariate statistical analysis

LC-MS files were processed using MarkerLynx version 4.1 software (Waters) for feature detection using the following parameters: mass tolerance: 0.01 Da; peak width at 5% height: 0.2 s; intensity threshold: 2000 counts; mass window: 0.05 Da; retention time window: 0.20 s. Following feature detection the feature lists were imported into *Analyst* version 7.5 software (Genedata, Lexington, MA, USA). Feature lists were inspected and a feature was considered to be real if it was present in greater than 75% of replicate samples with similar intensity in all replicates. Once highly confident feature lists were obtained, principle component analysis (PCA) was performed.

Absolute quantification

Absolute quantification was performed using a UPLC-single quadrupole mass spectrometer (Waters). Independent standard curves were generated for six compounds (K, DHK, M, DHM, FLKA, and FLKB) found in kava. Mixtures of pure standards were made in seven concentrations from 0.05 ppm to 100 ppm; four technical replicates were completed to account for chromatographic drift and ionization variability. After LC method optimization, standard curves were generated in selected ion recording (SIR) mode with the following retention time windows: 0–3.5 min: scan 220–700 *m/z*; 3.5–6.5 min: 275±2 *m/z*; 3.5–6.5 min: 277±2 *m/z*; 4.0–7.0 min: 231±2; 4.5–7.5 min: 233±2 *m/z*; 7.5–8.0 min scan 220–700 *m/z*; 8–11 min: 315±2 *m/z*; 9–12: 285±2 *m/z*; 12.0–14.0 min: scan 220–700 *m/z*. A cone voltage of 40 V was used to disfavor non-covalent compound dimerization in ESI^+ mode. The liquid chromatography and column parameters are identical to those used for the metabolic fingerprinting. Standard curves were linear up to 50 ppm. Samples were analyzed using the same LC-MS method with three or four replicates in most cases, although for six samples (N, IV; X, III; Y, IV; BB, III&IV; CC, II) and four samples (J, IV; M, IV; O, IV; X, IV) only two or one replicates were suitable for the final quantification, respectively. The limit of detection was set at concentrations corresponding to a signal-to-noise ratio of 3 to 1; peaks occurring below this threshold are not detected (ND). The limit of quantification was set at a signal-to-noise ratio of 10 to 1. The MassLynx™ application manager QuanLynx™ (Waters) was used to assist with automatic integration and of this large dataset. All integrations were visually inspected and manually adjusted to ensure consistent and accurate quantification.

Kava fractionation and characterization

Fractionation of commercial kava purchased from Gaia Herb (Brevard, NC, USA) was performed as described previously [7] with normal phase silica gel chromatography generating three modalities – fraction A (hydrophilic), B (medium polarity), and C (lipophilic). Briefly, 300 mL, net weight, of kava residue was mixed with silica gel (300 g). Ethanol and water were removed by vacuum. This silica gel with adsorbed kava residue was subjected to coarse chromatographic separation using a 750-gram pre-packed silica gel cartridge. The elution method was 28% ethyl acetate and 72% hexane 5 column volumes, followed by 90% ethyl acetate and 10% hexane, 4.1 column volumes, and then 35% methanol and 65% ethyl acetate, 5.5 column volumes. Different eluents were analyzed by TLC and the desired eluents were combined with solvent removed to generate fractions A, B, and C. The fractionation process was automated and monitored by the Biotag Separation System. Each individual fraction was analyzed by ^1H NMR and HPLC to confirm the success of fractionation.

Cytotoxicity test

Cytotoxicity tests were performed on fractions A, B, & C and different kava extracts. Extracts were dried and reconstituted in DMSO at a concentration of 10 mg/mL. From these stock solutions working solutions of 150, 75, and 37.5 µg/mL were prepared from each extract. Their cytotoxicity against human lung adenocarcinoma A549 cancer cells (American Type Culture Collection CCL-185) were evaluated by following our established procedures [48]. Briefly, A549 cells were plated in a 96-well plate (2.5×103 cells/well). The cells were treated with kava extracts with 0.5% DMSO in the final cell media (cells treated with media containing 0.5% DMSO served as a control). After 48 h of treatment, the relative cell viability in each well was determined by

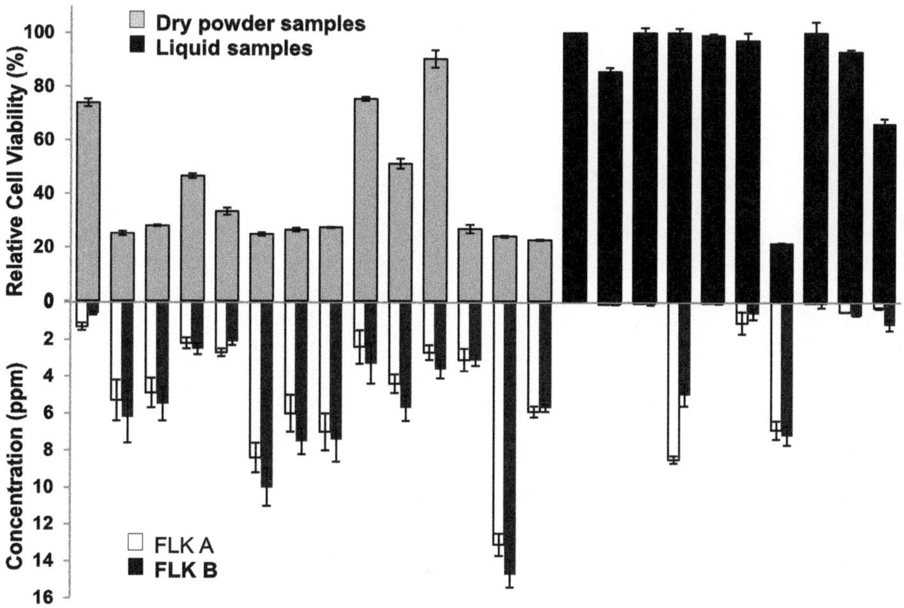

Figure 2. Comparison of relative cell viability to flavokawain (FLK) A and B concentrations. Top. Relative cell viability of human lung adenocarcinoma A549 cancer cell line after 48-hour incubation with ethanolic kava extracts at 75 μg/mL. Samples are organized according to kava preparation type with the gray bars representing the dry power samples organized from coarse grind on the left to very fine grind on the right with the last three dry powder samples (P, Z, and V) being instant freeze-dried kava. Black bars represent liquid samples. **Bottom.** Concentration of two potentially cytotoxic compounds found in kava (white bars: FLKA and black bars: FLKB respectively). Error bars represent standard error of 3–4 replicates.

using CellTiter-Blue cell viability assay kit (Promega, CA). Two biological repeats with three replicates per experiment were performed.

Supporting Information

Figure S1 Principal components analysis (PCA) of commercial kava preparations. Dry ground kava was extracted with water. Replicate extractions of the same material form clusters identified by drawn circles. The percent of variation explained by each principal component is shown along the appropriate axis.

Figure S2 Principal components analysis (PCA) of commercial kava preparations. Dry ground kava was extracted with water (blue) and ethanol (red). Replicate extractions of the same material form tight clusters. Secondary groups identified by drawn circles are formed based on the extraction solvent used; where the large amount of variation explained by Eigenrow 1 (PC1) is due to the use of either water or ethanol. The percent of variation explained by each principal component is shown along the appropriate axis.

Figure S3 Correlation between relative cell viability and Flavokawain A concentration. Relative cell viability of human cancer cells after 48-hour incubation with kava extracts at 75 μg/mL (red, circles) and 150 μg/mL (blue squares) is plotted to sample FLK A concentrations ($\log_2$ normalized) with R^2 values shown at the top for each extract concentration.

Figure S4 Correlation between relative cell viability and Flavokawain B concentration. Relative cell viability of human cancer cells after 48-hour incubation with kava extracts at 75 μg/mL (red, circles) and 150 μg/mL (blue squares) is plotted to sample FLK B concentrations ($\log_2$ normalized) with R^2 values shown at the top for each extract concentration.

Table S1 Commercial Kava Sources.

Table S2 Average concentration (ppm) of compounds from dry powder commercial kava sources.

Table S3 Average concentration (ppm) of compounds from liquid commercial kava sources.

Acknowledgments

The authors would like to thank the Center for Mass Spectrometry and Proteomics and the Minnesota Supercomputing Institute at the University of Minnesota for software and computational support.

Author Contributions

Conceived and designed the experiments: ACM EJ CX ADH. Performed the experiments: ACM CX. Analyzed the data: ACM EJ CX ADH. Contributed reagents/materials/analysis tools: ACM EJ CX ADH. Contributed to the writing of the manuscript: ACM. Edited manuscript and approved final version to be submitted: ACM EJ CX ADH. Generated figures and tables: ACM ADH.

References

1. E Johnston, H Rogers (2006) Introduction, Hawaiian Cultivars, and Non-Hawaiian Cultivars Grown in Hawai'i Today. In: E Johnston, H Rogers, editors. Hawaiian 'Awa: Views of an ethnobotanical treasure. Hilo, Hawaii: Association for Hawaiian 'Awa, pp. 1–7, 30–63.
2. He X, Lin L, Lian L (1997) Electrospray high performance liquid chromatography-mass-spectrometry in phytochemical analysis of kava (Piper methysticum) extract. Planta Med 63: 70–74.
3. Sarris J, Teschke R, Stough C, Scholey A, Schweltzer I (2011) Re-introduction of kava (Piper methysticum) to the EU: Is there a way forward? Planta Med 77: 107–110.
4. Steiner G (2000) The correlation between cancer incidence and kava consumption. Hawaii Med J 59: 420–422.
5. Johnson TE, Kassie F, O'Sullivan MG, Negia M, Hanson TE, et al. (2008) Johnson, Chemopreventive effect of kava on 4-(Methylnitrosamino)-1-(3-pryridyl)-1-butanone plus Benzo[a]pyrene-induced lung tumorigenesis in A/J mice. Can Prev Res 1: 430–438.
6. Johnson TE, Hermanson D, Wang L, Kassie F, Upadhyaya P, et al. (2011) Lung tumorigenesis suppressing effects of commercial kava extract and its selected compounds in A/J mice. Am J Chinese Med 39: 727–742.
7. Leitzman P, Narayanapillai SC, Balbo S, Zhou B, Upadhyaya P, et al. (2014) Kava blocks 4-(methylnitrosamino)-1-(3-pyridyl)-1-butanone–Induced lung tumorigenesis in association with reducing O^6-methylguanine DNA adduct in A/J Mice. Cancer Prev Res 7: 86–96.
8. Zi X, Simoneau AR (2005) Flavokawain A, a novel chalcone from kava extract, induces apoptosis in bladder cancer cells by involvement of bax protein-dependent and mitochondria-dependent apoptotic pathway and suppresses tumor growth in mice. Cancer Res, 65: 3479–3486.
9. Teschke R, Sarris J, Glass X, Schulze J (2011) Kava, the anxiolytic herb: back to basics to prevent liver injury? British Journal of Clinical Pharmacology 71: 445–448.
10. Dinh LD, Simmen U, Bueter KB, Bueter B, Lundstrom K, et al. (2001) Interaction of various Piper methysticum cultivars with CNS receptors in vitro. Planta Medica 67: 306–311.
11. Jokhan AD and McLenachan PA (2004) Genetic fingerprinting of Fijian kava. S Pac J Nat Sci 22: 43–45.
12. Lebot V, Siméoni P (2004) Is the quality of kava (Piper methysticum Forst f.) responsible for different geographical patterns. Ethnobotany Res App 2: 19–28.
13. Lebot V, Johnston E, Zheng QZ, McKern D, McKenna DJ (1999) Morphological, phytochemical, and genetic variation in Hawaiian cultivars of 'awa (Kava, Piper methysticum, Piperaceae). Econ Bot 53: 407–418.
14. Dasgupta A, Hammett-Stabler CA (2011) Abnormal Liver Function Tests Due to Hepatotoxic Herbs. In: Dasgupta A, Hammett-Stabler C A, editors. Herbal supplements: Efficacy, toxicity, interactions with western drugs, and effects on clinical tests. New York: Wiley, New York. pp. 155–167.
15. Shao Y, He K, Zheng B, Zheng Q (1998) Reversed-phase high-performance liquid chromatographic method for quantitative analysis of the six major kavalactones in Piper methysticum. J Chromatogr A 825: 1–8.
16. Siméoni P, Lebot V (2002) Identification of factors determining kavalactone content and chemotype in kava (Piper methysticum Forst. F.). Biochem Syst Ecol 30: 413–424.
17. Teschke R, Sarris J, Lebot V (2011) Kava hepatotoxicity solution: A six-point plan for new kava standardization. Phytomedicine; 18: 96–103.
18. Lebot V, Do TKT, Legendre L (2014) Detection of Flavokavins (A, B, C) in cultivars of kava (*Piper methysticum*) using high performance thin layer chromatography (HPTLC). Food Chem 151: 554–560.
19. Lebot V (2006) The quality of kava consumed in the South Pacific. HerbalGram. 71: 34–37.
20. Teschke R, Genthner A, Wolff A (2009) Kava hepatotoxicity: Comparison of aqueous, ethanolic, acetonic kava extracts and kava-herbs mixtures. J Ethnopharmacol 123: 378–384.
21. Jhoo JW, Freeman JP, Heinze TM, Moody JD, Schnackenberg LK, et al. (2006) In vitro cytotoxicity of nonpolar constituents from different parts of kavaplant (Piper methysticum). J Agr Food Chem 54: 3157–3162.
22. Johansen NH, Glitse V, Knudsen KEB (1996) Influence of extraction solvent and temperature on the quantitative determination of oligosaccharides from plant materials by high-performance liquid chromatography. J Agric Food Chem 44: 1470–1474.
23. Lapornik B, Prošek M, Wondra G A (2005) Comparison of extracts prepared from plant by-products using different solvents and extraction time, J Food Eng 71: 214–222.
24. Shaik AA, Hermanson DL, Xing C (2009) Identification of methysticin as a potent and non-toxic NF-κB inhibitor from kava, potentially responsible for kava's chemopreventive actvity. Bioorg. Med Chem Lett 19: 5732–5736.
25. Zhou P, Gross S, Liu J, Yu B, Feng L, et al, (2010) Flavokawain B, the hepatotoxic constituent from kava root, induces GSH-sensitive oxidative stress through modulation of IKK/NF-κB and MAPK signaling pathways. FASEB J 24: 4722–4732.
26. Teschke R, Qiu SX, Xuan TD, Lebot V (2011) Kava and kava hepatotoxicity: Requirements for novel experimental, ethnobotanical, and clinical studies based on a review of the evidence. Phytother Res, 25: 1263–1274.
27. Teschke R, Qiu SX, Lebot V (2011) Herbal hepatotoxicity by kava: Update on pipermethystine, flavokavain B, and mould hepatotoxins as primarily assumed culprits. Digestive and Liver Disease, 43: 676–681.
28. Teschke R, Lebot V (2011) Proposal for a kava quality standard code. Food and Chem Toxicol. 49: 2503–2516.
29. Teschke R (2011) Kava and the risk of liver toxicity: past, current, and future. Am Herb Prod Assoc Rep 26: 8–17.
30. Teschke R, Gaus W, Loew D (2003) Kava extracts: Safety and risks including rare hepatotoxicity. Phytomedicine 10: 440–446.
31. U.S. Food and Drug Administration (2002) Consumer advisory: kava-containing dietary supplements may be associated with severe liver injury (FDA Publication, www.fda.gov/Food/ResourcesForYou/Consumers/ucm085482.htm, date accessed: January 11, 2014).
32. Zhang LY, Rowe A, Ramzan I (2011) Does inflammation play a role in kava hepatotoxicity. Phytother Res 25: 629–630.
33. Fiji Bureau of Statistics (2013) 2011 Export Statistics. Fiji Bureau of Statistics Publicatio. Available: http://www.statsfiji.gov.fj/index.php/document-library/cat_view/12-economic-statistics/17-international-trade?start=25. Accessed: March 18, 2014.
34. Tonga Statistics Department (2008–2012) Kingdom of Tonga annual foreign trade report. Nuku'alofa Statistics Department Publication. Available: www.spc.int/prism/tonga/index.php/economic/merchandise-trade/export. Accessed: March 18, 2014.
35. Vanuatu National Statistics Office (2013) Statistics Update: Overseas Trade, August 2013 Highlights. Vanuatu National Statistics Office Publication. Available: http://www.vnso.gov.vu/index.php/economic-statistics/economic-summary. Accessed: March 18, 2014.
36. Discussion paper on the development of a standard for kava products (2012) Codex Alimentarius Commission (FAO/WHO), Food Standards Programme, 12th session, agenda item 6.
37. Johnson BM, Qiu S, Zhang S, Burdette JE, Yu L, et al. (2003) Identification of novel electrophilic metabolites of Piper methysticum Forst. (Kava). Chem Res Toxicol 16: 733–740.
38. Whitton PA, Lau A, Salisbury A, Whitehouse J, Evans CS (2003) Kava lactones and the kava-kava controversy. Phytochemistry 64: 673–679.
39. Olsen LR, Grillo MP, Skonberg C (2011) Constituents in kava extracts potentially involved in hepatotoxicity: A Review. Chem Res Toxicol 24: 992–2002.
40. Anke J, Ramzan I (2004) Pharmacokinetic and pharmacodynamic drug interactions with Kava (Piper methysticum Forst. f.). J Ethnopharmacol 93: 153–160.
41. Behl M, Nyska A, Chhabra RS, Travlos GS, Fomby LM, et al. (2011) Liver toxicity and carcinogenicity in F344/N rats and B6C3F1 mice exposed to Kava Kava. Food Chem Toxicol 49: 2820–2829.
42. Zou L, Henderson GL, Harkey MR, Sakai Y, Li A (2004) Effects of kava (Kava-kava, 'Awa, Yaqona, Piper methysticum) on c-DNA-expressed cytochrome P450 enzymes and human cryopreserved hepatocytes. Phytomedicine 11: 285–294.
43. Yang X, Salminen W F (2011) Kava extract, an herbal alternative for anxiety relief, potentiates acetaminophen-induced cytotoxicity in rat hepatic cells. Phytomedicine 18: 592–600.
44. Smith RM, Thakrar H, Arowolo TA, Shafi AA (1984) High-performance liquid chromatography of kava lactones from Piper methysticum. J Chromatogr 283: 303–308.
45. Meissmer O and Haberlein H (2005) HPLC analysis of flavokavins and kavapyrones from *Piper methysticum* Forst. Journal of Chromatography B, 826: 46–49.
46. Hegeman AD (2010) Plant metabolomics–meetings the analytical challenges of comprehensive metabolite analysis. Brief Funct Gen 9: 139–148.
47. Dragull K, Yoshida WY, Tang C (2003) Piperidine alkaloids from Piper methysticum. Phytochemistry 63: 193–198.
48. Warmka JK, Solberg EL, Zeliadt NA, Srinivasan B, Charlson AT, et al. (2012) Inhibition of mitogen activated protein kinases increases the sensitivity of A549 lung cancer cells to the cytotoxicity induced by a kava chalcone analog. Biochem Biophys Res Comm 424: 488–492.

HIV Cure Strategies: How Good Must They Be to Improve on Current Antiretroviral Therapy?

Paul E. Sax[1,3]*, Alexis Sypek[5,7], Bethany K. Berkowitz[5,7], Bethany L. Morris[5,7], Elena Losina[2,3,5,7,8], A. David Paltiel[10], Kathleen A. Kelly[5,7], George R. Seage III[4], Rochelle P. Walensky[1,3,6,7], Milton C. Weinstein[4], Joseph Eron[11], Kenneth A. Freedberg[3,4,5,6,7,9]

1 Division of Infectious Diseases, Brigham and Women's Hospital, Boston, Massachusetts, United States of America, **2** Department of Orthopedic Surgery, Brigham and Women's Hospital, Boston, Massachusetts, United States of America, **3** Harvard University Center for AIDS Research, Harvard University, Boston, Massachusetts, United States of America, **4** Harvard School of Public Health, Harvard University, Boston, Massachusetts, United States of America, **5** Division of General Medicine, Department of Medicine, Massachusetts General Hospital, Boston, Massachusetts, United States of America, **6** Division of Infectious Diseases, Department of Medicine, Massachusetts General Hospital, Boston, Massachusetts, United States of America, **7** Medical Practice Evaluation Center, Department of Medicine, Massachusetts General Hospital, Boston, Massachusetts, United States of America, **8** Department of Biostatistics, Boston University School of Public Health, Boston, Massachusetts, United States of America, **9** Department of Epidemiology, Boston University School of Public Health, Boston, Massachusetts, United States of America, **10** Yale School of Public Health, New Haven, Connecticut, United States of America, **11** Division of Infectious Disease, School of Medicine, University of North Carolina at Chapel Hill, Chapel Hill, North Carolina, United States of America

Abstract

Background: We examined efficacy, toxicity, relapse, cost, and quality-of-life thresholds of hypothetical HIV cure interventions that would make them cost-effective compared to life-long antiretroviral therapy (ART).

Methods: We used a computer simulation model to assess three HIV cure strategies: Gene Therapy, Chemotherapy, and Stem Cell Transplantation (SCT), each compared to ART. Efficacy and cost parameters were varied widely in sensitivity analysis. Outcomes included quality-adjusted life expectancy, lifetime cost, and cost-effectiveness in dollars/quality-adjusted life year ($/QALY) gained. Strategies were deemed cost-effective with incremental cost-effectiveness ratios <$100,000/QALY.

Results: For patients on ART, discounted quality-adjusted life expectancy was 16.4 years and lifetime costs were $591,400. Gene Therapy was cost-effective with efficacy of 10%, relapse rate 0.5%/month, and cost $54,000. Chemotherapy was cost-effective with efficacy of 88%, relapse rate 0.5%/month, and cost $12,400/month for 24 months. At $150,000/procedure, SCT was cost-effective with efficacy of 79% and relapse rate 0.5%/month. Moderate efficacy increases and cost reductions made Gene Therapy cost-saving, but substantial efficacy/cost changes were needed to make Chemotherapy or SCT cost-saving.

Conclusions: Depending on efficacy, relapse rate, and cost, cure strategies could be cost-effective compared to current ART and potentially cost-saving. These results may help provide performance targets for developing cure strategies for HIV.

Editor: Nicolas Sluis-Cremer, University of Pittsburgh, United States of America

Funding: This research was funded by the National Institute of Allergy and Infectious Diseases (http://www.niaid.nih.gov/Pages/default.aspx): R37 AI042006 (PS, AS, BB, BM, EL, ADP, KK, RW, MW, KF), AI051966 (GS), R01 AI093269 (RW, KF, ADP, MW), U01 AI 069472 (PS, EL, KF, RQ, ADP, AS, BB, BM, KK) (Partners/Harvard AIDS Clinical Trials Unit), U01 AI068636 (PS, EL, KF, RQ, ADP, AS, BB, BM, KK) (Central AIDS Clinical Trials Group), and UM1 AI 069423 (JE) (UNC Global Clinical Trials Unit). The funders had no role in study design, data collection and analysis, decision to publish, or preparation of the manuscript.

Competing Interests: Dr. Sax has served as a consultant to AbbVie, BMS, Gilead, GSK, Merck and Janssen, and has received grant support from BMS, Gilead, and GSK. Dr. Weinstein serves as a consultant to OptumInsight for work unrelated to the submitted research. Dr. Eron has served as a consultant to Abbvie, BMS, Gilead, GSK, Merck, Janssen and ViiV, and has received grant support from BMS, ViiV and Merck. All other authors report no conflicts of interest.

* Email: psax@partners.org

Introduction

Combination antiretroviral therapy (ART) durably controls HIV replication and halts progression of clinical HIV disease in the vast majority of patients who receive and continue treatment [1]. Projected survival for people with HIV is now estimated to be several decades. Some reports suggest that survival for people with HIV on successful therapy approaches that of those without infection if therapy is initiated early and HIV suppression is sustained [2].

Despite the remarkable success of treatment, ART nonetheless has many limitations. Although much less toxic than earlier regimens, current treatment still may be associated with cardiovascular, renal, bone, and other complications [3,4]. The inflammation and immune activation that persist in many patients on suppressive ART may have long-term negative consequences [5]. Therapy in the US and Europe remains costly, and, because not curative, it must be continued indefinitely [6,7]. Successful ART also does not eliminate the stigma associated with HIV infection [8].

The first report of successful HIV cure after allogeneic stem cell transplant for acute leukemia demonstrated that eradicating HIV from an individual is viable [9]. While allogeneic transplant in the absence of usual indications carries substantial risk, cost, and post-transplant consequences of chronic immunosuppression, other strategies are being studied that could potentially cure HIV and be practically deployed [10–12]. In this analysis we aim to establish thresholds of efficacy, toxicity, durability, cost, and quality of life necessary for a cure strategy to compare favorably with current antiretroviral therapy in the United States.

Methods

Analytic Overview

To analyze the potential life expectancy and cost-effectiveness of HIV cure strategies under study, we utilized the Cost-Effectiveness of Preventing AIDS Complications (CEPAC) model, a Monte-Carlo microsimulation of HIV disease and treatment [13]. We completed a 'what if' analysis, in order to understand the possible role of HIV cure strategies as they are developed. Model outputs included life expectancy, quality-adjusted life expectancy, and lifetime costs (2012 USD), all discounted to present value at 3% annually [14]. Incremental cost-effectiveness ratios (ICERs) were calculated by comparing each hypothetical cure strategy to the standard of care, lifelong ART. We determined parameter thresholds at which potential cure strategies were either cost-effective, defined as ICERs <$100,000/quality-adjusted life year (QALY), or cost-saving compared to current ART [15].

Strategies Evaluated

We evaluated three hypothetical HIV cure strategies: a "low efficacy," "low risk" gene therapy approach (Gene Therapy); a "moderate efficacy," "moderate risk" chemotherapy approach (Chemotherapy); and a "high efficacy," "high risk" allogeneic stem cell transplant (SCT). Costs of these strategies would likely vary widely and are currently uncertain.

The Gene Therapy strategy was modeled after the use of zinc finger nucleases to modify the CCR5 receptor on the surface of CD4 cells [12]. Patients undergo pheresis, their cells are modified using zinc finger nucleases, and re-infused with the goal of establishing a CCR5-negative cell population that is resistant to HIV infection. Based on preliminary reports, this type of procedure would have lower risk and toxicity than Chemotherapy and SCT and, we assumed, lower likelihood of achieving cure [16–19]. Simulated patients were modeled to receive the benefit of cure one month after Gene Therapy, if effective. Input parameters for all strategies were varied widely in sensitivity analysis, as described below.

The Chemotherapy intervention was derived from both in vitro and in vivo experiments using histone deacetylase inhibitors (such as vorinostat) to stimulate and eliminate the HIV viral reservoir [10]. Simulated patients received ART combined with Chemotherapy for 96 weeks, after which, if effective, they had the benefit of cure. There was increased cost and toxicity for the chemotherapy-based administration of vorinostat [17,20].

SCT had the highest assumed risk of mortality and toxicity, but was assumed the most effective. Simulated SCT patients received the benefit of cure in the first month after successful transplant.

The Cost-Effectiveness of Preventing AIDS Complications (CEPAC) Model

Simulations were performed using the CEPAC model, a widely-published, validated state-transition microsimulation of HIV disease [13]. HIV natural history is modeled as a series of monthly transitions between health states characterized by CD4 count and HIV RNA. Without treatment, patients' CD4 counts decline according to a viral load-dependent trajectory [21]. Patients are also subject to age- and sex-specific non-HIV-related mortality [22].

Once patients initiate ART, the probability of virologic suppression and subsequent CD4 count increases, with the greatest CD4 gain occurring in the first two months [23]. CD4 count gains are associated with reduced risk of developing opportunistic infections and HIV-related death. Patients' HIV RNA and CD4 counts are routinely monitored to detect treatment failure. Upon virologic rebound, patients switch to the next available ART regimen. Costs of HIV treatment and care are from the health system perspective and derived from HIV Research Network data and the Medicare fee schedule [24–27].

Cure Simulation

This analysis focused on patients who had received fully suppressive first-line ART for one year and were thereby eligible for a cure strategy, as is the case in planned or ongoing cure trials [28]. We maintained the CD4 benefit associated with virologic suppression for each cure strategy. With each cure regimen, patients faced strategy-specific probabilities of achieving cure as well as toxicity, quality of life (QOL) decrements and increases (associated with both toxicity and the regimen itself), and monthly probabilities of relapse. Additionally, patients accrued strategy-specific intervention costs. Cured patients were no longer subject to monthly probabilities of opportunistic infections and AIDS-related death, but were subject to monthly probabilities of relapse and subsequent return to ART. After cure, patients faced monthly probabilities of non-AIDS mortality and accrued monthly costs for routine care and continued HIV RNA monitoring for relapse. Patients who failed cure, or later relapsed after cure, resumed first-line ART, followed by additional ART regimens if virologic failure occurred later.

Model Inputs and Analysis

We used the CEPAC model itself to determine the distribution of CD4 counts in the eligible population by simulating a cohort of patients entering the model with the age, sex, and CD4 count distribution of HIV-infected patients in North America at care presentation. Patients were given a first-line ART regimen of efavirenz, tenofovir, and emtricitabine for one year [29]. Per current guidelines, all patients received ART, regardless of CD4 count [30]. Following one year on suppressive ART, patients became eligible for a cure intervention, beginning these cure strategies with mean CD4 count of 564/μl (SD 250/μl), based on this initialization.

Patients assigned to a cure intervention were subject to a strategy-specific probability of being cured (Table 1). All efficacies were hypothetical, since cure interventions do not currently exist. Cured patients had undetectable viremia for the duration of their lifetimes, unless they relapsed. We assumed relapse rates were highest during the first five years after a cure intervention (0.5%/

month); after five years the relapse rate was reduced by one half (0.25%/month). Relapse was detected through routine virologic monitoring. Both acute and chronic non-fatal toxicities resulted in a QOL decrement of 0.04, which lasted one month for acute non-fatal toxicities and until the patient failed the cure strategy for chronic toxicities [31]. Because the cohort was comprised only of patients virologically suppressed on first-line ART for one year, we assumed high rates of virologic re-suppression after a failed cure intervention. Those patients were also at risk for later virologic failure, at a rate of 0.13%/month [32]. Costs associated with each of the interventions and their associated toxicities were based on reported costs for similar procedures for other conditions (Table 1). In the base case, we assumed no additional QOL benefit related to achieving HIV cure compared to being on effective ART. In sensitivity analyses, we considered scenarios in which cured patients had an increase in their QOL from the base case. Any QOL benefit was suspended if the patient relapsed and re-initiated ART.

Gene Therapy was assumed to have an efficacy of 10.0% with no risk of fatal toxicity [16]. Patients incurred a 25.0% risk of acute, non-fatal toxicity (e.g., headache or oropharyngeal pain) lasting for one month [16]. While receiving Gene Therapy, patients incurred an immediate cost of $100,000, based on current estimates for gene therapies, plus $2,000 for continued ART (from weighted average of current drug prices) during the month they received Gene Therapy [20,33,34]. This intervention cost was based on ivacaftor, an oral cystic fibrosis medication that acts on the genetic mutation causing the disease [20].

Chemotherapy was assumed to have an efficacy of 20.0%, and 1.2% probability of fatal toxicity [17]. Patients incurred a 6.0% risk of acute non-fatal toxicity and 5.8% risk of chronic non-fatal toxicity [17,18]. Chemotherapy was modeled as a 96-week course (24 months) with monthly costs of $12,400; $2,000/month was included for maintenance ART [17]. At any point in the 96-weeks patients could fail ART and experience HIV virologic rebound. Patients who had not experienced ART failure during the 96 weeks could be cured at the end of that period (assumed efficacy 20.0%).

SCT was assumed to have an efficacy of 70.0%, with 5.0% mortality from the procedure [35]. Patients had a 47.3% probability of acute graft-versus-host disease and 37.2% probability of chronic graft-versus-host-disease [19]. The initial cost of the transplant was assumed to be $150,000 with monthly costs of $1,000 for six months for immunosuppressive medications [36,37].

Sensitivity Analysis

Because the focus of this analysis was on strategies under research and development, we conducted extensive sensitivity analysis on all cure parameters to identify those most important in changing the main conclusions. For each cure strategy and parameter, we determined thresholds at which the strategy would become cost-effective at a threshold of $100,000/QALY, as well as become cost-saving compared to ART. For sensitivity analyses involving relapse rates, early (≤5 years) and late (>5 years) relapse rates were varied together. Recognizing the impact a cure might have on patients' well-being (physical, emotional, and social), we

Table 1. Parameter inputs for a model-based analysis of potential HIV cure strategies.

Variable: Base Case (Range)	Gene Therapy	Chemotherapy	Stem Cell Transplant	References
Cohort Characteristics				
CD4 count, mean cells/μl (SD)	564 (250)	564 (250)	564 (250)	See Methods[a]
Age, mean years (SD)	44 (12)	44 (12)	44 (12)	[29]
Percent male	84	84	84	[29]
Cure Characteristics				
Efficacy (%)	10.0 (10.0–90.0)	20.0 (10.0–90.0)	70.0 (10.0–90.0)	Assumptions
Monthly relapse rate (%), early/late	0.50/0.25 (0.0–2.0)	0.50/0.25 (0.0–2.0)	0.50/0.25 (0.0–2.0)	Assumptions
Initial cost ($)	100,000 (50,000–200,000)	12,400/month[b] (6,200–24,800)	150,000 (75,000–300,000)	Assumptions based on [20,33,34,36]/[20]/[36]
Additional cost ($, while on cure regimen only)	2,000/month[c]	2,000/month[c]	1,000/month[d] (for 6 months)	[20,34]/[20,34]/[37]
Fatal Toxicity				
Probability (%)	0.0	1.2	5.0	Assumption based on [16]/[17]/[35]
Cost ($)	–	63,110	63,110	Derived from [24,25,27,46]
Acute Non-fatal Toxicity				
Probability (%)	25.0	6.0	47.3	Assumption based on Ivacaftor package insert [16]/[18]/[19]
Cost ($)	50	3,100	18,700	[25]/[47]/Derived from [48]
Chronic Non-fatal Toxicity				
Probability (%)	0.0	5.8	37.2	Assumption based on [16]/[18]}/[19]
Cost ($)	–	1,040	1,900	[49]/Derived from [50]

SD: standard deviation; **QOL:** quality-of-life.
[a]Determined through initialization run of simulated cohort; [b]For 24 months based on vorinostat; [c]For monthly antiretroviral therapy, derived from weighted averages of current therapies until gene- or chemo-therapy is complete; [d]For immunosuppressive agents, including methotrexate with tacrolimus.

also conducted sensitivity analysis on health-related QOL, both prior to and following HIV cure. Due to the major toxicity, including fatal toxicity, involved in SCT, we focused the QOL sensitivity analysis on the Gene Therapy and Chemotherapy strategies.

Ethics Statement

This study was reviewed and approved by the Partners Heath Care Human Research Committee (Protocol 2000P001927), Boston, Massachusetts, USA, as it was determined to meet the criteria for exemption from human studies. A waiver for written informed consent from participants was not necessary because only secondary data were used in this study and no human subjects were involved. Secondary patient data that serve as our model inputs were anonymized and de-identified prior to analysis.

Results

Base Case Scenarios

The standard of care (lifelong ART) had a discounted projected life expectancy of 19.0 years (16.4 QALYs) and discounted lifetime cost of $591,400. Undiscounted life expectancy with standard of care was 32.3 years, compared to 32.8, 32.3, and 32.6 years, for Gene Therapy, Chemotherapy, and SCT under the base case set of assumptions. Gene Therapy (10% efficacy) resulted in a discounted life expectancy of 19.3 years (16.6 QALYs) and increased discounted lifetime costs to $658,700, for an ICER of $330,600/QALY gained compared to continued ART. Chemotherapy (20% efficacy) led to a discounted life expectancy of 19.0 years (16.4 QALYs) and discounted lifetime cost of $807,300, and was more expensive and less effective than ART. SCT resulted in a discounted life expectancy of 19.0 years (16.3 QALYs) and increased costs to $607,400; it was also more expensive and less effective than ART (Table 2).

One-way Sensitivity Analyses

With efficacy increased to 22% and other inputs remaining the same, Gene Therapy had an ICER <$100,000/QALY, and at an efficacy of 34% became cost-saving, relative to ART (Table 3). With a reduced cost of $54,000, Gene Therapy achieved an ICER<$100,000/QALY gained even at 10% efficacy; it was cost-saving at $34,000. Chemotherapy was not cost-effective unless efficacy increased to 88% and was not cost-saving at any efficacy. Varying any other single parameter within reasonable limits did not result in Chemotherapy reaching thresholds for cost-effectiveness or cost savings (Table 3). The efficacy threshold for SCT was 79% to achieve cost-effectiveness and 80% to achieve cost savings. Reducing fatal toxicity to 3.0% from 5.0% also led to SCT becoming cost-effective (Table 3).

Multiway Sensitivity Analyses

With no relapse risk, Gene Therapy was cost-saving with efficacy of at least 30%. With increasing relapse rates, higher efficacy was required to achieve cost savings. At a decreased cost of $50,000, Gene Therapy became cost-effective at the base case values for relapse and efficacy and cost-saving with lower relapse rates or higher efficacies (Figure 1). At increased cost of $200,000, the intervention was not cost-effective compared to standard of care ART for almost all combinations of input parameters (Figure 1).

For Chemotherapy, at the base case cost and relapse rate of greater than 0.5%/month, the intervention was never cost-effective (Figure 2). With no relapse risk, the intervention was not cost-effective at efficacies of 20–50% but was cost-saving at efficacies above 60%. If the cost was halved ($6,200/month), Chemotherapy was cost-saving at substantially lower efficacies and higher relapse rates than in the base case. For example, at this decreased cost, Chemotherapy was cost-saving with relapse rate of 0.5%/month with efficacy 60%. If the cost of Chemotherapy was doubled to $24,800/month, it was not cost-effective with any combination of efficacy (20–90%) and relapse rate (0.0–2.0%). The window for cost-effectiveness was narrow; with most parameter combinations, Chemotherapy was either cost-saving or not cost-effective.

In most sensitivity analyses, SCT was not cost-effective. In selected cases where the cost was extremely low or efficacy very high, SCT became cost-saving (Figure 3). For one parameter combination, SCT was less effective and less expensive than ART, but it was not cost-effective because the ICER of ART was < $100,000/QALY compared to SCT. If the cost of SCT was halved ($75,000), the combinations where the intervention was cost-saving remained roughly the same, but several scenarios that were not cost-effective in the base case became less expensive and less effective than ART.

With an efficacy of 10% for Gene Therapy, improving QOL to a utility of 1.00 (i.e., the equivalent of perfect health) after successful cure would be insufficient to achieve an ICER < $100,000/QALY gained. With efficacy of 20%, however, an ICER <$100,000/QALY gained could be achieved if patient utility following cure increased from 0.85 to 0.88, or the equivalent of facing a 3% decreased risk of death every year. For efficacies of 30% or more, the Gene Therapy strategy would always be cost-effective, regardless of whether the cure had any impact on QOL. At the base-case QOL utility of 0.85, Chemotherapy was not cost-effective at any efficacy below 60%, even with the maximum QOL improvement. At an efficacy of 60% for Chemotherapy, cost-effectiveness could be achieved if patient utility following cure increased from 0.85 to 0.97. If the baseline QOL utility while living with HIV were 0.50, Chemotherapy would not reach the

Table 2. Base case results of an analysis of hypothetical HIV cure strategies*.

Strategy	Discounted Life Years (Undiscounted)	Discounted QALYs	Cost ($)	Incremental Cost-effectiveness compared to standard of care ($/QALY)
Standard of care ART	19.0 (32.3)	16.4	591,400	–
Gene Therapy	19.3 (32.8)	16.6	658,700	330,600
Chemotherapy	19.0 (32.3)	16.4	807,300	Dominated
Stem Cell Transplant	19.0 (32.6)	16.3	607,400	Dominated

*Based on assumptions for efficacy, durability, toxicity, and cost in Methods and Table 1. Life expectancy, QALYs, and costs all discounted at 3%/year. **ART:** antiretroviral therapy; **QALY:** Quality-adjusted life year; **Dominated:** Less effective and more costly than the standard of care ART strategy.

Table 3. Threshold which key parameters would need to reach for each type of HIV cure strategy to be cost-effective (ICER< $100,000/QALY gained) or cost-saving.

Parameter	Base case value	ICER<$100,000/QALY gained	Cost-saving
Gene Therapy (base case ICER: $330,600/QALY gained)			
Efficacy (%)	10	22	34
Fatal Toxicity (%)	0.0	None	None
Monthly relapse rate (%), early (late)	0.5/0.25	None	None
Intervention cost ($)	100,000, one-time	54,000, one-time	34,000, one-time
Chemotherapy (base case ICER: Dominated)			
Efficacy (%)	20	88	None
Fatal Toxicity (%)	1.2	None	None
Monthly relapse rate (%), early (late)	0.5/0.25	None	None
Intervention cost ($)	12,400/month, for 24 months	*	*
Stem Cell Transplant (base case ICER: Dominated)			
Efficacy (%)	70	79	80
Fatal Toxicity (%)	5.0	3.0	None
Monthly relapse rate (%), early (late)	0.5/0.25	None	0.25/0.125
Intervention cost ($)	150,000, one-time	*	*

ICER: incremental cost-effectiveness ratio; **QALY:** quality-adjusted life year; **QOL:** quality of life; **Dominated:** strategy was less effective and more expensive than current ART.
*Cost reductions led to the strategy being less effective and less expensive than current ART. One could calculate an ICER for ART compared to Chemotherapy or Stem Cell Transplant, but it is not clinically plausible that these strategies would be used if they resulted in worse outcomes than standard of care with ART, even if they saved money by avoiding the costs of lifelong ART.

cost-effectiveness threshold of <$100,000/QALY at cure efficacies below 40%. At cure efficacy of 40%, Chemotherapy would achieve an ICER below $100,000/QALY gained with improvement in QOL utility to 0.88. If we used ICER thresholds below $150,000 or $200,000 per QALY gained to define cost-effectiveness, there were no appreciable changes in results [15].

Discussion

With intense pre-clinical investigation underway towards finding a cure for HIV, we sought to evaluate the cost-effectiveness of three potential HIV cure approaches, each compared to standard of care ART. We used a variety of assumptions, anchored in published data on gene-targeted therapy, chemotherapy, and stem cell transplant for diseases other than HIV. By doing extensive sensitivity analyses on efficacy, toxicity, relapse rates, and cost, we defined a range of benchmarks that might justify the adoption of a cure strategy, and identified combinations of parameters under which these could potentially be cost-effective or cost-saving. For a Gene Therapy approach, modest increases in efficacy (above 10%) or moderate decreases in cost (below $100,000), led to this strategy being cost-saving compared to ART. For Chemotherapy and SCT, the inventions became cost-saving with very high efficacies and low relapse rates.

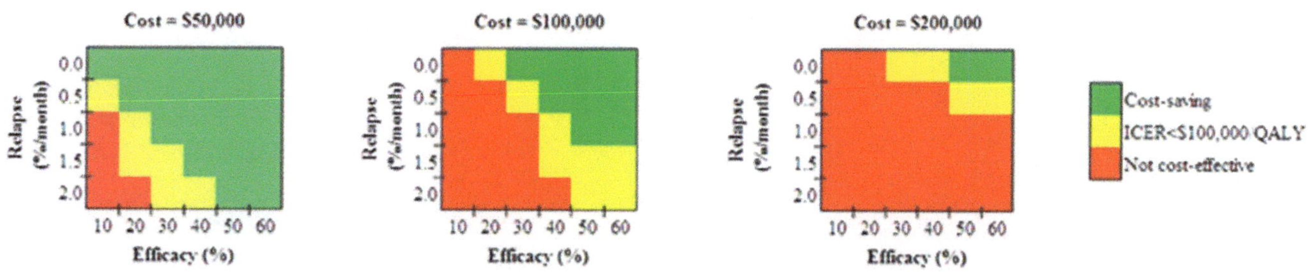

Figure 1. Gene Therapy compared to standard of care ART. The figure depicts the cost-effectiveness of Gene Therapy compared to standard of care ART as a function of the three influential parameters identified via the one-way sensitivity analysis in Table 3: cost, relapse rate, and efficacy. In each panel, the horizontal axis denotes efficacy while the vertical axis denotes the relapse rate. Inside each panel, the shading denotes the resultant cost-effectiveness finding, ranging from cost-saving (green), through cost-effective (with an ICER<$100,000/QALY, yellow), to not cost-effective (≥ $100,000/QALY or more expensive and less effective than ART, red). **ART:** antiretroviral therapy; **ICER:** incremental cost-effectiveness ration; **QALY:** quality-adjusted life year.

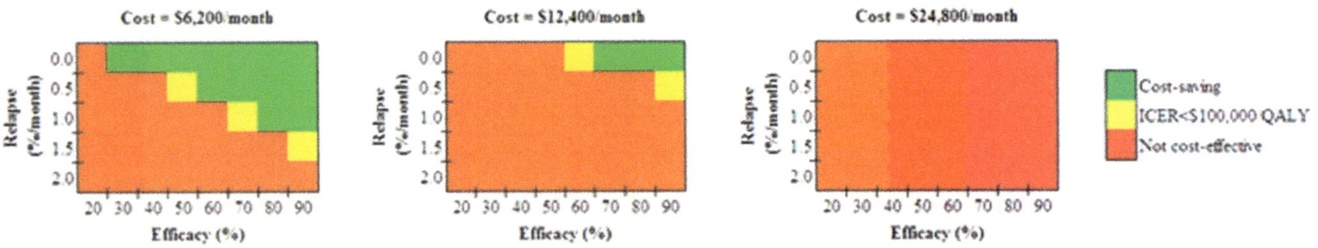

Figure 2. Chemotherapy compared to standard of care ART. The figure depicts the cost-effectiveness of Chemotherapy compared to standard of care ART as a function of the three influential parameters identified via the one-way sensitivity analysis in Table 3: cost, relapse rate, and efficacy. In each panel, the horizontal axis denotes efficacy while the vertical axis denotes the relapse rate. Inside each panel, the shading denotes the resultant cost-effectiveness finding, ranging from cost-saving (green), through cost-effective (with an ICER<$100,000/QALY, yellow), to not cost-effective (≥$100,000/QALY or more expensive and less effective than ART, red). **ART:** antiretroviral therapy; **ICER:** incremental cost-effectiveness ration; **QALY:** quality-adjusted life year.

We found that changes in efficacy, relapse rates, and/or cost rapidly moved the strategies from being worse than ART to being cost-saving – that is, to being both equally or more effective and less costly. The range in which any strategy would be cost-effective but not cost-saving is narrow (Figures 1–3, yellow area). High initial costs of cure strategies could be justified, and would save money, if (and essentially only if) the strategy eliminates the lifetime cost of ART. For example, with an initial cost of $100,000 and an efficacy of 34%, the Gene Therapy strategy is cost-saving compared to ART, even if all other assumptions remain the same. In such a scenario, identification of conditions that could theoretically increase the likelihood of cure – such as ART started during acute infection, or heterozygosity of the CCR5delta32 gene – would make a cure strategy even more attractive [38]. Alternatively a substantial decrease in the cost of lifelong ART would make these interventions less cost-effective.

It is possible that combination approaches to cure may be needed to improve efficacy [39]. These would, nonetheless, each have some combination of efficacy, toxicity, and cost. The value in terms of cost-effectiveness, compared to ART, can be inferred from those combinations as shown in Figures 1–3. Further, some lower-risk interventions, such as zinc finger nucleases, could also have higher efficacy than other interventions. If so, then they would both be more effective and less costly, and thus 'dominant' from a cost-effectiveness perspective, compared to those other interventions, such as HDAC inhibitors.

No published studies to date have examined the cost-effectiveness of hypothetical HIV cure strategies in comparison to ART. Similar model-based analyses have, however, been done for other previously unproven strategies in HIV, including therapeutic and preventive HIV vaccines and pre-exposure prophylaxis (PrEP) [40–42]. These analyses have been used to design subsequent vaccine and PrEP research. In the case of PrEP, modeled results before proven efficacy closely matched the outcome of some later trials [43].

At present, strategies to cure HIV have only progressed to the proof of concept stage. Given this early stage, current complexity, anticipated cost, and possible risks, a cure strategy will not be ready for implementation anytime soon. However, this analysis suggests that potential HIV cure strategies must be moderately effective and have low toxicity and low relapse rates to compare favorably to standard of care ART. The optimal cost threshold for such strategies will depend on both the likelihood of durable cure (initial efficacy and subsequent relapse rate) and the cost of ART. As initial efforts at cure are developed, this work can help investigators determine the efficacy and toxicity targets which would make the strategies attractive. Further, if any cure strategies are proven effective, the results of this analysis can help inform policymakers as to their appropriate role. This issue has recently been highlighted by the high efficacy and cost of new HCV cures [44].

From a societal and quality-of-life perspective, with a base case utility of 0.85 for patients doing well on ART, improvements in

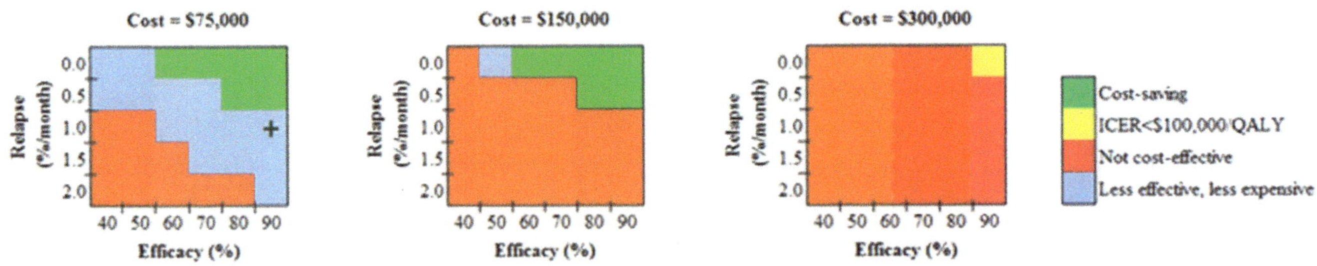

Figure 3. Stem Cell Transplantation compared to standard of care ART. The figure depicts the cost-effectiveness of Stem Cell Transplantation compared to standard of care ART as a function of the three influential parameters identified via the one-way sensitivity analysis in Table 3: cost, relapse rate, and efficacy. In each panel, the horizontal axis denotes efficacy while the vertical axis denotes the relapse rate. Inside each panel, the shading denotes the resultant cost-effectiveness finding, ranging from cost-saving (green), through cost-effective (with an ICER<$100,000/QALY, yellow), to not cost-effective (≥$100,000/QALY or more expensive and less effective than ART, red). Instances where the intervention is both less expensive and less effective than ART are denoted in blue, but most were not cost-effective because the ICER of ART was <$100,000/QALY compared to SCT. The plus sign indicates a strategy that had an ICER for ART compared to SCT >$100,000/QALY gained. **ART:** antiretroviral therapy; **ICER:** incremental cost-effectiveness ration; **QALY:** quality-adjusted life year.

quality of life after cure do not have a major impact on cost-effectiveness. However, many might argue that there is an important psychological, social, and emotional distinction to be drawn between curing HIV and controlling it via therapy.

Our study has several limitations. The most important is that HIV cure interventions do not yet exist, so model parameters such as efficacy, mortality, cost, and relapse rates were assumed using specific data wherever possible and then varied widely. The effect of cure strategies on the incidence and severity of "non-HIV" complications, such as malignancies, heart disease, and other chronic non-communicable diseases was not included; one might anticipate either an increase or decrease in these complications, based on the strategy employed. If non-AIDS events are driven primarily by HIV-mediated immune activation and inflammation, then curing HIV would presumably ameliorate these processes. In addition, adverse effects of antiretroviral drugs would also be eliminated. By contrast, some of the treatments proposed for HIV cure may themselves increase risks of non-AIDS events. For example, some are analogous to cancer chemotherapy, and such treatments may increase the risk of secondary malignancies; radiation used for stem cell transplant could also raise cardiovascular risk; and alteration in stem cells could also increase the long-term risk of cancers. The demographics of the suppressed patients eligible for cure interventions were based on the demographics of the population presenting to care in the United States and may not be completely representative of those who achieve suppression after one year. Since we modeled only patients virologically suppressed after a year, this represents the most adherent subset of patients. If cure strategies were utilized in a broader group of patients, such as those with early infection, the strategies might be more or less effective and cost-effective compared to ART, depending on the requirements of the particular cure strategy. Gene therapy may require stem cell modification to achieve cure, which could increase the risk of rare but substantial toxicity of cancer induction; this risk was not included. Although we did include relapse rates – indicating a later chance of HIV viral rebound after initial cure – we did not include the possibility of re-infection among cured patients, which has been documented after successful HCV cure [45]. Adding this possibility would make any cure strategy less attractive. Increased use of newer, more effective branded therapies, however, may keep the costs of ART in their current range [20].

In summary, the key determinants of the cost-effectiveness of HIV cure strategies, compared to current antiretroviral therapy, are initial efficacy, toxicity, relapse rate, and cost. Potential cure strategies must have moderate efficacy, low toxicity, and relatively low risk of relapse to be cost-effective and, in combination, would likely be cost-saving.

Author Contributions

Conceived and designed the experiments: PES AS BKB BLM EL ADP KAK GRS RPW MCW JE KAF. Analyzed the data: PES AS BKB BLM EL ADP KAK GRS RPW MCW JE KAF. Contributed to the writing of the manuscript: PES AS BKB BLM EL ADP KAK GRS RPW MCW JE KAF. Performed model analyses: AS BKB BLM KAK.

References

1. Moore RD, Bartlett JG (2011) Dramatic decline in the HIV-1 RNA level over calendar time in a large urban HIV practice. Clin Infect Dis 53: 600–604.
2. Rodger AJ, Lodwick R, Schechter M, Deeks S, Amin J, et al. (2013) Mortality in well controlled HIV in the continuous antiretroviral therapy arms of the SMART and ESPRIT trials compared with the general population. AIDS 27: 973–979.
3. Sabin CA, Worm SW, Weber R, Reiss P, El-Sadr W, et al. (2008) Use of nucleoside reverse transcriptase inhibitors and risk of myocardial infarction in HIV-infected patients enrolled in the D:A:D study: a multi-cohort collaboration. Lancet 371: 1417–1426.
4. Martin A, Bloch M, Amin J, Baker D, Cooper DA, et al. (2009) Simplification of antiretroviral therapy with tenofovir-emtricitabine or abacavir-Lamivudine: a randomized, 96-week trial. Clin Infect Dis 49: 1591–1601.
5. Lederman MM, Funderburg NT, Sekaly RP, Klatt NR, Hunt PW (2013) Residual immune dysregulation syndrome in treated HIV infection. Advances in immunology 119: 51–83.
6. Farnham PG, Gopalappa C, Sansom SL, Hutchinson AB, Brooks JT, et al. (2013) Updates of lifetime costs of care and quality-of-life estimates for HIV-infected persons in the United States: late versus early diagnosis and entry into care. J Acquir Immune Defic Syndr 64: 183–189.
7. Finzi D, Hermankova M, Pierson T, Carruth LM, Buck C, et al. (1997) Identification of a reservoir for HIV-1 in patients on highly active antiretroviral therapy. Science 278: 1295–1300.
8. Andrinopoulos K, Clum G, Murphy DA, Harper G, Perez L, et al. (2011) Health related quality of life and psychosocial correlates among HIV-infected adolescent and young adult women in the US. AIDS Educ Prev 23: 367–381.
9. Allers K, Hutter G, Hofmann J, Loddenkemper C, Rieger K, et al. (2011) Evidence for the cure of HIV infection by CCR5Delta32/Delta32 stem cell transplantation. Blood 117: 2791–2799.
10. Archin NM, Liberty AL, Kashuba AD, Choudhary SK, Kuruc JD, et al. (2012) Administration of vorinostat disrupts HIV-1 latency in patients on antiretroviral therapy. Nature 487: 482–485.
11. Margolis DM, Hazuda DJ (2013) Combined approaches for HIV cure. Curr Opin HIV AIDS 8: 230–235.
12. Tebas P, Stein D, Tang WW, Frank I, Wang SQ, et al. (2014) Gene editing of CCR5 in autologous CD4 T cells of persons infected with HIV. N Engl J Med 370: 901–910.
13. Walensky RP, Sax PE, Nakamura YM, Weinstein MC, Pei PP, et al. (2013) Economic savings versus health losses: the cost-effectiveness of generic antiretroviral therapy in the United States. Ann Intern Med 158: 84–92.
14. Siegel JE, Weinstein MC, Russell LB, Gold MR (1996) Recommendations for reporting cost-effectiveness analyses. Panel on Cost-Effectiveness in Health and Medicine. JAMA 276: 1339–1341.
15. Ubel PA, Hirth RA, Chernew ME, Fendrick AM (2003) What is the price of life and why doesn't it increase at the rate of inflation? Arch Intern Med 163: 1637–1641.
16. Vertex Pharmaceuticals Incorporated (2012) Kalydeco (ivacaftor) package insert. Available: http://pi.vrtx.com/files/uspi_ivacaftor.pdf. Accessed 18 August 2014.
17. Merck Sharp & Dohme Corp. (2006) Zolinza (vorinostat) [package insert]. Available: http://www.merck.com/product/usa/pi_circulars/z/zolinza/zolinza_pi.pdf. Accessed 18 August 2014.
18. Kavanaugh SM, White LA, Kolesar JM (2010) Vorinostat: A novel therapy for the treatment of cutaneous T-cell lymphoma. Am J Health Syst Pharm 67: 793–797.
19. Stamatovic D, Balint B, Tukic L, Elez M, Tarabar O, et al. (2011) Impact of stem cell source on allogeneic stem cell transplantation outcome in hematological malignancies. Vojnosanit Pregl 68: 1026–1032.
20. RED BOOK Online (2013) Product Information for antiretroviral drug prices, Kalydeco, Zolinza, and Tivicay. Micromedex 2.0, Truven Health Analytics.
21. Mellors JW, Muñoz A, Giorgi JV, Margolick JB, Tassoni CJ, et al. (1997) Plasma viral load and CD4+ lymphocytes as prognostic markers of HIV-1 infection. Ann Intern Med 126: 946–954.
22. United Nations, Department of Economic and Social Affairs and Population Division (2009) World Population Prospects: The 2008 Revision, Highlights, Working Paper No. ESA/P/WP.210 Available: http://www.un.org/esa/population/publications/wpp2008/wpp2008_highlights.pdf. Accessed 10 June 2013.
23. Pozniak AL, Gallant JE, DeJesus E, Arribas JR, Gazzard B, et al. (2006) Tenofovir disoproxil fumarate, emtricitabine, and efavirenz versus fixed-dose zidovudine/lamivudine and efavirenz in antiretroviral-naive patients: virologic, immunologic, and morphologic changes-a 96-week analysis. J Acquir Immune Defic Syndr 43: 535–540.
24. Centers for Medicare and Medicaid Services (2012) Clinical Diagnostic Laboratory Fee Schedule 2012. Available: https://www.cms.gov/Medicare/Medicare-Fee-for-Service-Payment/ClinicalLabFeeSched/clinlab.htm. Accessed 18 August 2014.
25. Centers for Medicare and Medicaid Services (2012) Medicare Physician Fee Schedule 2012. Available: http://www.cms.gov/apps/physician-fee-schedule/overview.aspx. Accessed 20 January 2014.
26. University HealthSystems Consortium (2008) CDP Online Report. Available: www.uhc.edu. Accessed 28 January 2014.
27. Bozzette SA, Berry SH, Duan N, Frankel MR, Leibowitz AA, et al. (1998) The care of HIV-infected adults in the United States. HIV Cost and Services Utilization Study Consortium. N Engl J Med 339: 1897–1904.
28. Mellors J, McMahon D (2014) Evaluating the safety and efficacy of single-dose romidepsin in combination with antiretroviral therapy in HIV-infected adults

with suppressed viral load. NCT01933594. Available: http://clinicaltrials.gov/ct2/show/NCT01933594?term=romidepsin+hiv&rank=1. Accessed 23 Janaury 2014.
29. Althoff KN, Gange SJ, Klein MB, Brooks JT, Hogg RS, et al. (2010) Late presentation for human immunodeficiency virus care in the United States and Canada. Clin Infect Dis 50: 1512–1520.
30. Panel on Antiretroviral Guidelines for Adults and Adolescents (2013) Guidelines for the prevention and treatment of opportunistic infections in HIV-infected adults and adolescents: recommendations from the Centers for Disease Control and Prevention, the National Institutes of Health, and the HIV Medicine Association of the Infectious Diseases Society of America. Department of Health and Human Services. Available: http://aidsinfo.nih.gov/contentfiles/lvguidelines/adult_oi.pdf. Accessed 18 August 2014.
31. Pepper PV, Owens DK (2002) Cost-effectiveness of the pneumococcal vaccine in healthy younger adults. Med Decis Making 22: S45–57.
32. Messou E, Chaix ML, Gabillard D, Minga A, Losina E, et al. (2011) Association between medication possession ratio, virologic failure and drug resistance in HIV-1-infected adults on antiretroviral therapy in Côte d'Ivoire. J Acquir Immune Defic Syndr 56: 356–364.
33. Dotinga R (2010) Gene therapy for HIV inches forward. Available: http://health.usnews.com/health-news/managing-your-healthcare/genetics/articles/2010/06/16/gene-therapy-for-hiv-inches-forward. Accessed 7 May 2014.
34. US Department of Health and Human Services (2005) Medicaid drug price comparisons: average manufacturer price to published prices. Available: http://oig.hhs.gov/oei/reports/oei-05-05-00240.pdf. Accessed 16 January 2014.
35. Shenoy S (2011) Hematopoietic stem cell transplantation for sickle cell disease: current practice and emerging trends. Hematology Am Soc Hematol Educ Program 2011: 273–279.
36. National Bone Marrow Transplant Link (2010) Bone marrow/stem cell transplant frequently asked questions. Available: http://nbmtlink.org/resources_support/faq/faq_question8.html. Accessed 18 August 2014.
37. Kasiske BL, Cohen D, Lucey MR, Neylan JF (2000) Payment for immunosuppression after organ transplantation. American Society of Transplantation. JAMA 283: 2445–2450.
38. Saez-Cirion A, Bacchus C, Hocqueloux L, Avettand-Fenoel V, Girault I, et al. (2013) Post-treatment HIV-1 controllers with a long-term virological remission after the interruption of early initiated antiretroviral therapy ANRS VISCONTI Study. PLoS Pathog 9: e1003211.
39. Lewin SR (2014) Finding a Cure for HIV: Much Work to Do. Ann Intern Med 161: 368–369.
40. Paltiel AD, Freedberg KA, Scott CA, Schackman BR, Losina E, et al. (2009) HIV preexposure prophylaxis in the United States: impact on lifetime infection risk, clinical outcomes, and cost-effectiveness. Clin Infect Dis 48: 806–815.
41. Leelahavarong P, Teerawattananon Y, Werayingyong P, Akaleephan C, Premsri N, et al. (2011) Is a HIV vaccine a viable option and at what price? An economic evaluation of adding HIV vaccination into existing prevention programs in Thailand. BMC Public Health 11: 534.
42. Walensky RP, Paltiel AD, Goldie SJ, Gandhi RT, Weinstein MC, et al. (2004) A therapeutic HIV vaccine: how good is good enough? Vaccine 22: 4044–4053.
43. Grant RM, Lama JR, Anderson PL, McMahan V, Liu AY, et al. (2010) Preexposure chemoprophylaxis for HIV prevention in men who have sex with men. N Engl J Med 363: 2587–2599.
44. Petta S, Cabibbo G, Enea M, Macaluso FS, Plaia A, et al. (2014) Cost-effectiveness of sofosbuvir-based triple therapy for untreated patients with genotype 1 chronic hepatitis C. Hepatology 59: 1692–1705.
45. Lambers FA, Prins M, Thomas X, Molenkamp R, Kwa D, et al. (2011) Alarming incidence of hepatitis C virus re-infection after treatment of sexually acquired acute hepatitis C virus infection in HIV-infected MSM. AIDS 25: F21–27.
46. Gebo KA, Moore RD, Fleishman JA (2003) The HIV Research Network: a unique opportunity for real time clinical utilization analysis in HIV. Hopkins HIV Rep 15: 5–6.
47. Havrilesky LJ, Pokrzywinski R, Revicki D, Higgins RV, Nycum LR, et al. (2012) Cost-effectiveness of combination versus sequential docetaxel and carboplatin for the treatment of platinum-sensitive, recurrent ovarian cancer. Cancer 118: 386–391.
48. Dignan FL, Potter MN, Ethell ME, Taylor M, Lewis L, et al. (2013) High readmission rates are associated with a significant economic burden and poor outcome in patients with grade III/IV acute GvHD. Clin Transplant 27: E56–63.
49. Elting LS, Cantor SB, Martin CG, Hamblin L, Kurtin D, et al. (2003) Cost of chemotherapy-induced thrombocytopenia among patients with lymphoma or solid tumors. Cancer 97: 1541–1550.
50. Crespo C, Perez-Simon JA, Rodriguez JM, Sierra J, Brosa M (2012) Development of a population-based cost-effectiveness model of chronic graft-versus-host disease in Spain. Clin Ther 34: 1774–1787.

26

60-Day Chronic Exposure to Low Concentrations of $HgCl_2$ Impairs Sperm Quality: Hormonal Imbalance and Oxidative Stress as Potential Routes for Reproductive Dysfunction in Rats

Caroline S. Martinez[1], João Guilherme D. Torres[1], Franck M. Peçanha[1], Janete A. Anselmo-Franci[2], Dalton V. Vassallo[3], Mercedes Salaices[4], María J. Alonso[5], Giulia A. Wiggers[1]*

1 Postgraduate Program in Biochemistry, Universidade Federal do Pampa, Uruguaiana, Rio Grande do Sul, Brazil, **2** Department of Morphology, Stomatology and Physiology, School of Dentistry, Universidade de São Paulo, Ribeirão Preto, São Paulo, Brazil, **3** Department of Physiological Sciences, Universidade Federal do Espírito Santo, Vitória, Espírito Santo, Brazil, **4** Department of Pharmacology, School of Medicine, Universidad Autónoma de Madrid, Madrid, Spain, **5** Department of Biochemistry, Physiology and Molecular Genetics, Universidad Rey Juan Carlos, Alcorcón, Spain

Abstract

Mercury is a toxic and bio-accumulative heavy metal of global concern. While good deals of research have been conducted on the toxic effects of mercury, little is known about the mechanisms involved in the pathogenesis of male reproductive dysfunction induced by mercury. Therefore, the purpose of this study was to assess the effects and underlying mechanisms of chronic mercury exposure at low levels on male reproductive system of rats. Three-month-old male Wistar rats were divided into two groups and treated for 60 days with saline (i.m., Control) and $HgCl_2$ (i.m. 1st dose: 4.6 μg/kg, subsequent doses 0.07 μg/kg/day). We analyzed sperm parameters, hormonal levels and biomarkers of oxidative stress in testis, epididymis, prostate and vas deferens. Mercury treatment decreased daily sperm production, count and motility and increased head and tail morphologic abnormalities. Moreover, mercury treatment decreased luteinizing hormone levels, increased lipid peroxidation on testis and decreased antioxidant enzymes activities (superoxide dismutase and catalase) on reproductive organs. Our data demonstrate that 60-day chronic exposure to low concentrations of $HgCl_2$ impairs sperm quality and promotes hormonal imbalance. The raised oxidative stress seems to be a potential mechanism involved on male reproductive toxicity by mercury.

Editor: Stefan Schlatt, University Hospital of Münster, Germany

Funding: This work was supported by Coordenação de Aperfeiçoamento de Pessoal de Nível Superior; Conselho Nacional de Desenvolvimento Científico e Tecnológico; Fundação de Amparo a Pesquisa do Espírito Santo; Fundo Estadual de Ciência e Tecnologia [39767531/07]; Ministerio de Ciencia e Innovación [SAF 2012-36400 and PHB2011- 0001-PC]; Instituto de Salud Carlos III - ISCIII [RIC RD12/0042/0024]; Pró-reitoria de Pesquisa - Universidade Federal do Pampa [N° 10.0010.13]. The funders had no role in study design, data collection and analysis, decision to publish, or preparation of the manuscript.

Competing Interests: The authors have declared that no competing interests exist.

* Email: giuliawp@gmail.com

Introduction

Mercury is a persistent toxic and bio-accumulative heavy metal of global concern since their anthropogenic and natural emissions still pose high risk to human and environmental health. The main contribution to human exposure is the ingestion of fish and seafood (methylmercury); however, it can also occur from dental amalgam fillings (metallic mercury), vaccines (ethylmercury) and contaminated water and air (mercury chloride) [1,2].

Human exposure to mercury is still high even with the recommendations from international regulatory agencies. Within the US population, the chronic mercury exposure and average population of mercury concentrations rose sharply from 1999 to 2006 [3]. At Tapajós River Basin, Amazon, Brazil, known for mercury occupational exposure from gold mining and, consequently, bioaccumulation, exposure to mercury have decreased between 1999–2012 among women of reproductive age; however the potential risk is still an issue because 22% of the women shows high mercury levels. In the aforementioned cohort studies, the mercury levels are close to or above the tolerance limit established by WHO [4–6].

Mercury in a cumulative process can be deposited into the body and their toxic effects observed in several target organs [7–9]. Studies conducted in vitro and using animal models suggested male reproductive toxicity by mercury [10–12]. Thus, mercury, at high levels of exposure, affects germ cell development and promotes germ cell injury [13], alters reproductive behavior [11] and decreases sperm count and motility [14,15]. However, in humans results are conflicting and contradictory [16]; thus, it has been reported reproductive toxic effects of mercury among occupationally exposed workers such as sperm damage and hormonal imbalance [12,17,18] as well as an association between semen mercury concentrations and sperm abnormalities in

Table 1. Effect of a chronic treatment for 60 days at low concentrations of $HgCl_2$ on body weight (g), absolute (g or mg) and relative (g/100 g or mg/100 g) weights of reproductive organs of male rats.

Parameters	Experimental groups	
	Control (n = 10)	$HgCl_2$ (n = 10)
Initial body weight (g)	294.9±2.2	301.7±2.9
Final body weight (g)	384.8±18.7	400.7±9.4
Testis (g)	1.8±0.03	1.7±0.04
Testis (g/100 g)	0.4±0.02	0.4±0.01
Epididymis (mg)	671.9±16.1	651.7±17.4
Epididymis (mg/100 g)	177.1±6.3	163.0±3.8
Ventral prostate (mg)	546.0±37.2	507.2±22.4
Ventral prostate (mg/100 g)	142.7±7.7	127.0±5.6
Full seminal vesicle (g)	1.4±0.05	1.5±0.05
Full seminal vesicle (g/100 g)	0.4±0.02	0.4±0.01
Empty seminal vesicle (g)	0.4±0.02	0.4±0.02
Empty seminal vesicle (g/100 g)	0.1±0.01	0.1±0.01
Vesicular secretion (g)	0.9±0.04	1.0±0.04
Vas deferens (mg)	108.4±3.6	107.5±5.7
Vas deferens (mg/100 g)	28.5±1.1	26.8±1.4

Data are expressed as means ± SEM. The relative organ weight was calculated by use of the formula: organ weight/body weight ×100. Units: g: gram, mg: milligram.

subfertile men [18]; however, other studies did not observe any relationship between blood mercury levels and biomarkers of male reproductive health [19,20].

It has been proposed that heavy metals can interrupt the normal function of male reproductive system: a) at any level of hypothalamic – pituitary - testicular axis; b) directly, at reproductive organs, or c) by altering post- testicular events such as sperm capacitation, motility and function [21]. Although heavy metals are known as endocrine disruptors, the endocrine implications in the mercury toxicity remained unexplored [21]. Moreover, changes in the hormonal levels caused by mercury treatment at high doses were reported, with decreased testosterone, luteinizing hormone (LH) and follicle stimulating hormone (FSH) levels, suggesting an interference with neuroendocrine system [13,22].

Involvement of oxidative mechanisms in the toxic manifestation induced by mercury in the reproductive system has been recently

Table 2. Effect of a chronic treatment for 60 days at low concentrations of $HgCl_2$ on sperm counts in testis and epididymis of rats.

Parameters	Experimental groups	
Sperm count	Control (n = 10)	$HgCl_2$ (n = 10)
Testis		
Sperm number ($x10^6$)	184.1±7.4	138.6±5.2**
Sperm number ($x10^6$/g)	101.6±5.1	81.5±3.1**
DSP ($x10^6$/testis/day)	29.4±1.3	22.7±0.8**
DSPr ($x10^6$/testis/day/g)	16.6±0.8	13.3±0.5**
Epididymis		
Caput/Corpus		
Sperm number ($x10^6$)	109.0±2.6	90.3±2.2**
Sperm number ($x10^6$/g)	295.4±7.8	249.9±7.2**
Sperm transit time (days)	3.6±0.1	4.1±0.2
Cauda		
Sperm number ($x10^6$)	190.0±13.4	148.8±6.7*
Sperm number ($x10^6$/g)	713.3±37.0	640.5±15.1
Sperm transit time (days)	6.5±0.4	6.5±0.2

DSP: daily sperm production; DSPr: daily sperm production relative to testis weight. Units: g: gram, mg: milligram. Data are expressed as mean ± SEM. * $p<0.05$, ** $p<0.01$ (Student's t-test).

Table 3. Effect of a chronic treatment for 60 days at low concentrations of $HgCl_2$ on sperm morphology of rats.

Parameters	Experimental groups	
	Control (n = 10)	**$HgCl_2$ (n = 10)**
Normal	95.5 (92.5–96.7)	86.7 (80.1–90.1)**
Head Abnormalities		
Amorphous	1.5 (0.6–2.8)	2.7 (1.5–7.5)
Banana Head	0.2 (0.0–0.5)	4.5 (1.7–7.1) **
Detached Head	2.0 (0.5–2.8)	1.7 (1.0–3.8)
Total of Head Abnormalities	3.7 (3.5–5.1)	11.0 (6.7–13.6)**
Tail Abnormalities		
Bent Tail	0.5 (0.0–1.0)	2.2 (1.3–4.6)**
Broken Tail	0.5 (0.0–1.2)	0.7 (0.0–1.5)
Total of Tail Abnormalities	1.2 (0.2–1.5)	3.2 (1.8–5.1)**

Data are in percentage and expressed as median (Q_1–Q_3). **$p<0.01$ (Mann - Whitney).

investigated, but it is still quite insufficient [14,23]. Mercury toxicity in several organs has been related with increased reactive oxygen species (ROS) and inactivated antioxidant enzymes necessary for cellular protection against ROS attack [8,24].

While a good deal of research has been conducted on the toxic effects of mercury, little is known about the underlying mechanisms pathogenesis of male reproductive dysfunction induced by mercury. Recently, we observed decreases on sperm quality after 30 days of mercury exposure and its association with oxidative stress [25]; however, the effects and mechanisms involved after a long-term exposure, which best mimetize human exposure to this metal, are not explored. Therefore, the purpose of this study was to assess the effects of prolonged mercury exposure at low levels on sperm parameters (count, morphology and motility), hormonal levels as well as on biomarkers of oxidative stress on male reproductive system of rats.

Materials and Methods

Animals

Three-month-old male Wistar rats (280–320 g) were obtained from the Central Animal Laboratory of the Federal University of Santa Maria, Rio Grande do Sul, Brazil. During treatment, rats were housed at a constant room temperature, humidity, and light cycle (12:12 h light-dark), free access to tap water and fed with standard chow ad libitum. All experiments were conducted in compliance with the guidelines for biomedical research stated by the Brazilian Societies of Experimental Biology and approved by the Ethics Committee on Animal Use Experimentation of the Federal University of Pampa, CEUA, Uruguaiana, Rio Grande do Sul, Brazil (Process Number: 036/2012).

Rats were divided into two groups (n = 10 each) and treated for 60 days as follows: a) Control (saline solution, i.m.); b) Mercury ($HgCl_2$ - mercury chloride (1st dose 4.6 µg/kg, subsequent doses 0.07 µg/kg/day, i.m., to cover daily loss, using the model described by Wiggers *et al.* [26], with an extended exposure for 60 days). At the end of treatment, animals were euthanized by decapitation and the weights of prostate, seminal vesicle (empty, without coagulation gland), testis, epididymis and vas deferens (absolute and relative to body weights) were determined. Prostate, left testis and epididymis as well as right vas deferens were used for further analysis of oxidative stress; the right testis, epididymis and left vas deferens were used for sperm parameter analysis.

$HgCl_2$ was purchased from Sigma-Aldrich (St Louis, MO, USA). Salts and reagents were of analytical grade obtained from Sigma and Merk (Darmstadt, Germany).

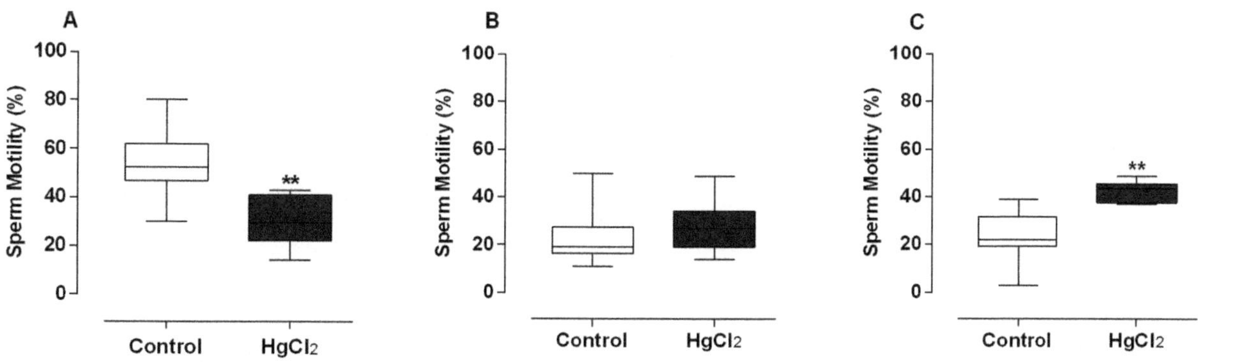

Figure 1. Effect of treatment for 60 days at low concentrations of $HgCl_2$ on sperm motility. Type A: motile with progressive movement (A), type B: motile without progressive movement (B) and type C: immotile (C). Data are expressed as median (Q1–Q3), (n = 10). **$p<0.01$ (Mann - Whitney).

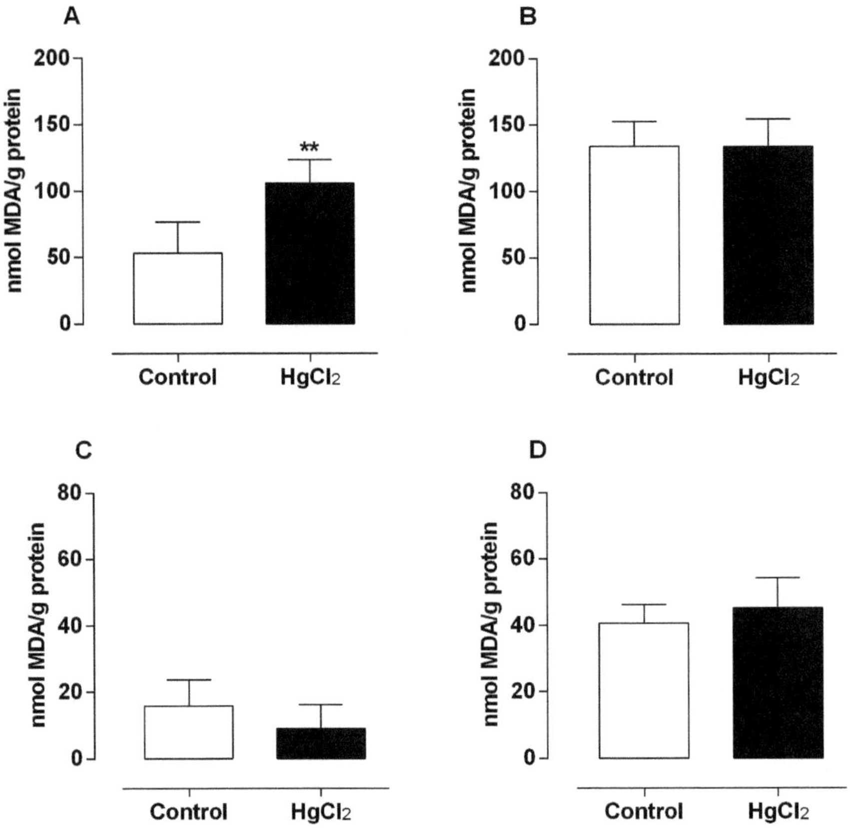

Figure 2. Effect of treatment for 60 days at low concentrations of $HgCl_2$ on lipid peroxidation. Values of TBARS on testis (A), epididymis (B), prostate (C) and vas deferens (D). Data are expressed as mean ± SEM (n = 6). ** $p<0.01$ (Student's t-test).

Sperm motility

Sperm were removed from the vas deferens by internal rising with 1 mL of Human Tubular Fluid (DMPBS-Nutricell-SP-Brazil) pre-warmed to 34°C. Then, a 10 µL aliquot was transferred to a histological slide. Under a light microscope (20X magnification, Binocular, Olympus CX31, Tokyo, Japan), 100 spermatozoa were analyzed and classified as type A: motile with progressive movement, type B: motile without progressive movement and type C: immotile. Sperm motility was expressed as % of total sperm [25].

Sperm morphology

Sperm were obtained from the vas deferens and stored with 1 mL of saline - formol until the analysis. For the analysis, smears were prepared on histological slides and 200 spermatozoa per animal were evaluated under 400X magnification (Binocular, Olympus CX31). Morphological abnormalities were classified into head (amorphous, banana and detached head) and tail morphology (bent and broken tail), according to Filler [27].

Daily sperm production per testis, sperm number and transit time in epididymis

Homogenization-resistant testicular spermatids (stage 19 of spermiogenesis) and sperm in the caput/corpus epididymis and cauda epididymis were counted as described by Robb et al. [28]. To calculate daily sperm production, the number of spermatids at stage 19 was divided by 6.1, which is the number of days these spermatids are present in the seminiferous epithelium. The sperm transit time through the epididymis was determined by dividing the number of sperm in each portion by the daily sperm production [28].

Biochemical assays

The reproductive organs (testis, epididymis, prostate and vas deferens) were immediately excised after euthanasia, washed on ice-cold saline and rapidly homogenized in 50 mM Tris–HCl, pH 7.5 (1/5, w/v), centrifuged at 2400×g for 15 min at 4°C and the resulting supernatant (S1) fraction was frozen at −80°C for further measurements of oxidative stress parameters.

Lipid peroxidation

The levels of lipid peroxidation were measured as malondialdehyde (MDA) using a colorimetric method, as previously described by Ohkawa *et al.* [29], with modifications. An aliquot of each tissue was incubated with thiobarbituric acid 0.8% (TBA), phosphoric acid buffer 1% (H_3PO_4), and sodium dodecil sulphate 0.8% (SDS) at 100°C for 60 min. The color reaction was measured at 532 nm against blanks (Spectrophotometer Femto 600 S, FEMTO, São Paulo, Brazil). The results were expressed as nanomoles of MDA per gram of protein.

Determination of non-protein thiols

The non-protein thiols (NPSH) levels were determined by the method of Ellman [30]; for that, homogenates were centrifuged at 4000×g at 4°C for 10 min and the supernatant was mixed (1:1) with 10% trichloroacetic acid. After centrifugation, the protein pellet was discarded and free -SH groups were determined in the clear supernatant. An aliquot of supernatant was added in

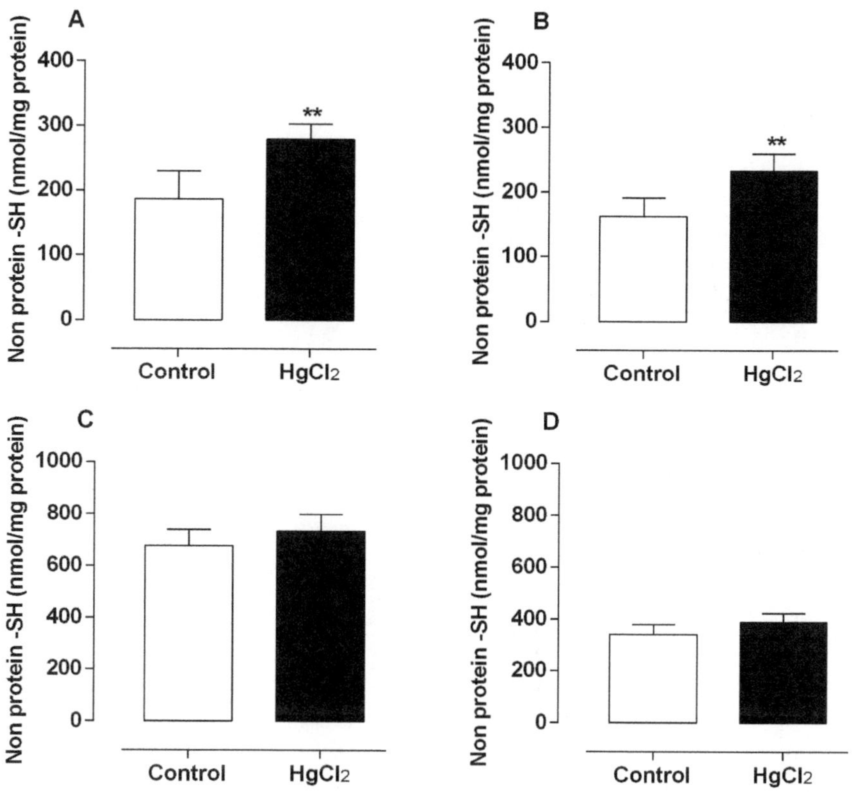

Figure 3. Effect of treatment for 60 days at low concentrations of $HgCl_2$ on non-protein thiol content. Values of non-protein thiol groups on testis (A), epididymis (B), prostate (C) and vas deferens (D). Data are expressed as mean ± SEM (n = 6). ** $p<0.01$ (Student's t-test).

potassium phosphate buffer 1 M, pH 7.4, and 5,5′- dithio-bis (2-nitrobenzoic acid) (DTNB) 10 mM. The color reaction was measured spectrophotometrically at 412 nm. The results were expressed as nmol of NPSH per milligram of protein.

Superoxide dismutase activity

Superoxide dismutase (SOD) activity was spectrophotometrically assayed as described by Misra and Fridovich [31]. This method is based on the capacity of SOD in inhibiting autoxidation of adrenaline to adrenochrome. The color reaction was measured at 480 nm. One unit of enzyme was defined as the amount of enzyme required to inhibit the rate of epinephrine autoxidation by 50% at 26°C. For analysis, tissues were diluted 1:10 (v/v) and aliquots were added in a glycine buffer 50 mM, pH 10.5. Enzymatic reaction was started by adding epinephrine, and results were expressed as Units (U) per milligram of protein.

Catalase activity

The catalase activity was spectrophotometrically assayed by the method of Aebi [32], which measures the disappearance of H_2O_2 in the presence of tissue homogenates at 240 nm. An aliquot of each tissue was added in potassium phosphate buffer 50 mM, pH 7.0, and the enzymatic reaction was initiated by adding H_2O_2. The enzymatic activity was expressed in Units (1 U decomposes 1 µmol H_2O_2/min at pH 7 at 25°C) per milligram of protein. Proteins for experiments were measured using bovine serum albumin as a standard.

Hormonal assay

Blood was collected (between 9:00 and 11:30 AM) and serum was obtained by centrifugation (1236xg for 20 min at 4°C). The concentration of LH and testosterone was determined by the technique of double antibody radioimmunoassay. The hormonal assay was performed using specific kits supplied by the National Institute of Arthritis, Diabetes and Kidney Diseases (NIADDK, USA). All samples were assayed in duplicate and in the same assay to avoid inter-assay errors. The intra-assay error was 3.4%.

Statistical analysis

Data are expressed as mean ± SEM of the number of animals used in each experiment. Statistical analysis was performed using Student's t test and Mann-Whitney test. Values of $p<0.05$ were considered statistically significant. Graphpad prism 5 software was used for statistical analysis and for plotting graphs.

Results

Long-term $HgCl_2$ exposure at low levels did not change the body weight as well as the absolute and relative organ weights (Table 1).

Sperm parameters

Mercuric chloride administration for 60 days resulted in a significant decrease in sperm count in testis and epididymis (caput/corpus and cauda). Furthermore, there was a reduction in daily sperm production per testis in $HgCl_2$ treated rats (Table 2).

Sperm morphology analysis revealed a significant increase in sperm abnormalities in rats treated for 60 days with mercury when compared with the control group. Within head phenotypes, more banana head as well as total head abnormalities were observed; concerning tail morphology, there were more bent tail and total tail abnormalities (Table 3).

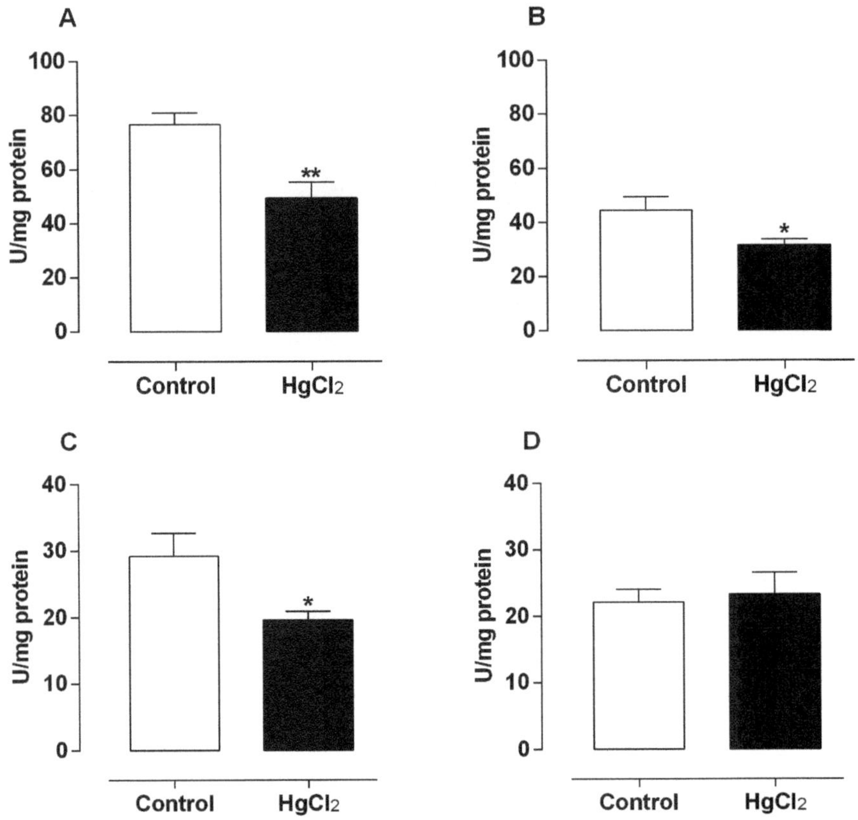

Figure 4. Effect of treatment for 60 days at low concentrations of $HgCl_2$ on superoxide dismutase activity. Values of superoxide dismutase activity on testis (A), epididymis (B), prostate (C) and vas deferens (D). Data are expressed as mean ± SEM (n = 6). ** $p<0.01$, *$p<0.05$ (Student's t-test).

Regarding the sperm motility, we observed decreased type A sperm (motile with progressive movement) (Figure 1A) accompanied by increases in type C sperm (immotile) in $HgCl_2$-treated rats (Figure 1C), without differences in type B sperm (motile without progressive movement) (Figure 1B).

Biochemical assays

In rats exposed to mercury for 60 days a significant increase on MDA levels was observed in testis (Figure 2A), whereas non-significant changes were observed in epididymis, prostate and vas deferens (Figure 2B, C, D).

With respect to NPSH levels, mercury treatment increased the testicular and epididymal SH levels (Figure 3A, B), while those in prostate and vas deferens were not statistically different between groups (Figure 3C, D).

The mercury exposure for 60 days decreased the enzymatic activity of SOD in testis, epididymis and prostate (Figure 4A, B, C), without change in rats vas deferens (Figure 4D). Moreover, the catalase activity after mercury treatment slightly decreased in epididymis and in a great extent in vas deferens of rats (Figure 5B, D); however, in testis and prostate there were not differences in the activity of this enzyme (Figure 5A, C).

Hormonal assay

There was a significant reduction in serum levels of LH in the mercury treated rats when compared with the control group (Figure 6A), with no differences in testosterone levels after mercury exposition (Figure 6B).

Discussion

In a previous report, we have shown that 30 days exposure to $HgCl_2$ induces male reproductive dysfunction associated to increased oxidative stress. The present study goes further in this field showing that when mercury exposure is prolonged for 60 days, this metal cumulatively affects the male reproductive system. The current study shows that marked changes after 30 days of mercury administration are maintained, although the defense mechanisms are differently affected after a prolonged mercury exposition. Long-term $HgCl_2$ administration impairs sperm function, as observed by suppression on sperm production and count reduction following by motility and morphological abnormalities mainly banana head and bent tail; and also caused hormonal disparity with decreased LH levels. These effects seem to be related with increased oxidative stress. Namely, this investigation demonstrates a fail from the adaptative response of SOD activity as well as an apparently improve from lipid damage when compared with 30 days of exposure.

Mercury is suspected to have negative impact on male fertility, although human studies are few and contradictory [12,16,18]. The United States Environmental Protection Agency (EPA) recommends a maximum blood concentration of mercury of 5.8 ng/ml [5,6]; however the human exposure to mercury is still increasing and sometimes are higher than allowed [4,33]. In the present study, we used a model of controlled low-dose administration of $HgCl_2$ for 60 days which was previously established in rats [26] and attains a blood mercury concentration of approximately 8 ng/

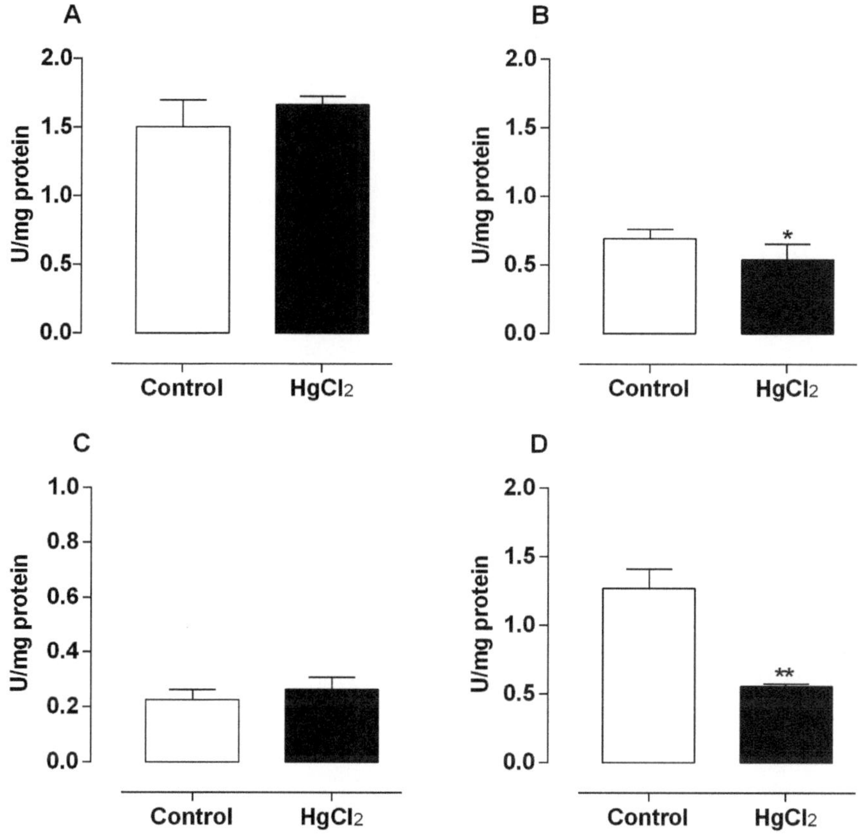

Figure 5. Effect of treatment for 60 days at low concentrations of $HgCl_2$ on catalase activity. Values of catalase activity on testis (A), epididymis (B), prostate (C) and vas deferens (D). Data are expressed as mean ± SEM (n = 6). ** *p*<0.01, **p*<0.05 (Student's t-test).

ml, similar to human exposure levels and near to those considered safe by the EPA [19,34].

Animal studies with 30, 60 and 90 days of mercury exposure at high levels had showed decreased testosterone, LH, FSH and prolactin levels [13,22,23]. Recently, within the US population, it was discovered a significant inverse relationship between chronic inorganic mercury and LH levels [3]. Our study raises a question about the association between chronic mercury exposure at a dose environmental relevant and LH as a causal mechanism for mercury associated diseases; the decreased serum LH levels highlights mercury as an endocrine disruptor. Endocrine disruptors are an exogenous agent that either mimic or block the effects of hormones at the target receptor or by directly stimulating or inhibiting production of hormones [21], being the complex hypothalamus–pituitary–gonad (HPG) axis within the domain of endocrine disruption [35].

The relationship between chronic mercury exposure and LH was first postulated considering that pituitary gland is one of the

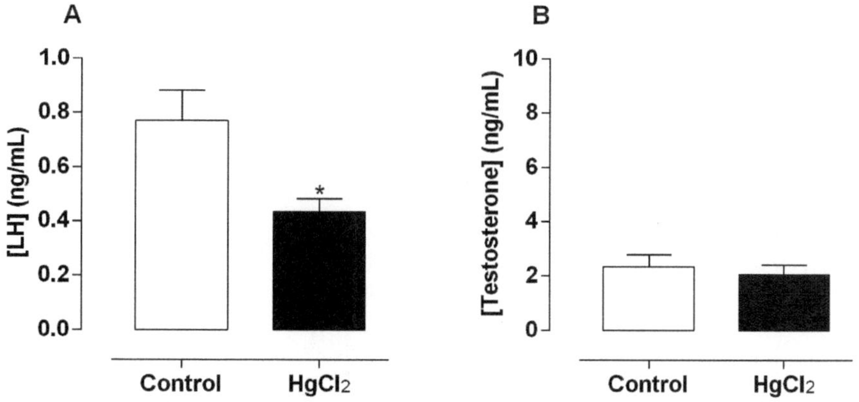

Figure 6. Effect of treatment for 60 days at low concentrations of $HgCl_2$ on serum hormonal levels. (A) Luteinizing hormone (LH) and testosterone levels (B). Data are expressed as mean ± SEM (n = 10). **p*<0.05 (Student's t-test).

target for mercury deposition within the brain [36]. Even that, relatively little mercury in inorganic form crosses the blood brain barrier; however, inorganic mercury can be formed as a metabolite of other forms of mercury in the brain where it can remain for years [36,37]. Moreover, the bran effect-response to mercury seems to be dependent on reaching a critical threshold concentration/deposition after prolonged exposure, since we did not find LH reduction after 30-days exposure. This hypothesis is supported by a recent publication of our group in which we observed that only exposure for 60 days to $HgCl_2$ in adulthood can cause deleterious effects on cognitive function and memory formation [38]. However, further experiments are still necessary to determine the time necessary for the effects we observed.

In the present study we did not find changes in either body weight or reproductive organs weight, similar to that previously found after 30 days exposure [25]. The fact that the testosterone levels were maintained supports these results because the reproductive system is directly under control of androgen stimulation [39,40].

The 60-day $HgCl_2$ exposure decreased sperm motility, daily sperm production and sperm quantity in testis and epididymis; in addition, we also observed a significantly increase in sperm abnormalities in exposed-animals. These impairments are similar to that found in our previous study in terms of percentage [25], except for sperm motility, in which a prolonged mercury exposure seems to be more deleterious to sperm movement. Our results are in agreement with other authors using $HgCl_2$ 3 times a week at high doses for 60 days [13]. Similarly, inorganic mercury at 0.25–1.00 mg/kg/day (21 days) produced adverse effects on mice reproductive performance with fertility and sperm survival reduction [11].

Other heavy metals such as lead, cadmium and arsenic are also known to induce reproductive toxicity, including abnormalities in sperm count, motility, morphology, testosterone production and spermatogenesis [21,41,42]. One of the target organs of heavy metals is testis and many authors have proposed that oxidative stress could be partially responsible to induce their toxicity [43,21].

In this sense, $HgCl_2$ has been reported to be one of the most pro–oxidant forms that induce oxidative stress in several organs [8,9]. Recently, this ability has been shown on male reproductive organs at high mercury levels [13,14] as well as at low levels in our previous study after 30 days of $HgCl_2$ exposition [25]. The current investigation suggests that oxidative stress might be one of the mechanisms explaining mercury reproductive toxicity such as hypospermatogenesis, as demonstrated by the lipid peroxidation enhancement in testis as well as the reduction in antioxidant enzymes activities in reproductive organs. The apparent discrepancies between increased lipid peroxidation in testis while not differences in other evaluated organs were found, even with reduction in antioxidant status, are expected in metal exposure due to differences in metal accumulation and tissue response to insults according to specific characteristic of this tissue [44].

ROS are important mediators of normal sperm function, but excessive production of ROS results in lipid peroxidation and membrane damage, inactivation of glycolytic enzymes, damage to the acrosomal membranes and DNA oxidation [45,46]. Moreover, mammalian spermatozoa are rich in polyunsaturated fatty acids (PUFAs) which are very susceptible to ROS attack. As a consequence of lipid peroxidation, PUFAs undergo degradation by a chain reaction leading to production of cytosolic aldehydes including MDA as well as loss of structural integrity required for sperm viability and motility [45,46]. Therefore, the suppression of spermatogenesis and sperm impairment found would be attributed to peroxidation of PUFAs in the sperm membrane.

Mercury compounds have high affinity to thiol (-SH) groups of biomolecules; depletion of intracellular thiols has been reported as a trigger for production of ROS by mercury [47]. The enhanced non protein -SH levels in testis and epididymis found after 60 days exposure could constitute an adaptive response to oxidative stress, since at 30-day of $HgCl_2$ exposure depletion of intracellular thiols in testis and epididymis of rats was observed [25].

Cells have several mechanisms to protect themselves from the toxic effect of ROS [48]. It is known the protective effect of the antioxidant enzymes SOD, CAT and GPx against toxicity induced by mercury [9,13,47]. Long-term exposure to mercury decreased SOD activity in testis, epididymis and prostate as well as catalase activity in epididymis and vas deferens; these reductions can raise the oxidative damage due to accumulation of superoxide radicals and hydrogen peroxide [14,23]. Catalase activity was also reduced after 30 days treatment, while SOD activity was increased in all reproductive organs, probably as a compensatory mechanism [25]; therefore, the lack of enzyme activity here found suggests a failure of antioxidant defense SOD after a prolonged mercury exposure. The reduction of SOD and CAT activities could be due either to a reduction on these enzymes expression, to a direct effect of ROS, or to a direct inhibition from mercury [48].

The present study raises concern about the toxicity of long-term $HgCl_2$ exposure, since another 30 days of mercury administration at environmental relevant levels was more deleterious and insufficient to a compensatory positive reaction from endogenous antioxidants at male reproductive system. Corroborating, Sharma et al. [49] after mercury administration of 1.25 mg/kg daily for 30 days observed only a partial recover after withdrawal of $HgCl_2$ for 45 days. Thus, further experiments are necessary to determine the time for recovery of the reproductive system effects.

Conclusions

Our results demonstrate that 60-day chronic exposure to low concentrations of $HgCl_2$ impairs sperm quality and promotes hormonal imbalance with decreased LH levels. The raised oxidative stress seems to be a potential mechanism involved in male reproductive toxicity by mercury.

Author Contributions

Conceived and designed the experiments: CSM FMP GAW. Performed the experiments: CSM JGDT JAA-F. Analyzed the data: CSM GAW JGDT. Contributed reagents/materials/analysis tools: GAW JAA-F DVV MJA MS. Wrote the paper: GAW CSM FMP DVV MJA MS.

References

1. Zhang Y, Gao D, Bolivar VJ, Lawrence DA (2011) Induction of autoimmunity to brain antigens by developmental mercury exposure. Toxicol Sci 119: 270–280.
2. Nyland JF, Fairweather D, Shirley DL, Davis SE, Rose NR, et al. (2012) Low dose inorganic mercury increases severity and frequency of chronic coxsackievirus- induced autoimmune myocarditis in mice. Toxicol Sci 125: 134–143.
3. Laks DR (2009) Assessment of chronic mercury exposure within the U.S. population, National Health and Nutrition Examination Survey, 1999–2006. Biometals 6: 1103–1114.
4. Corvelo TCO, Oliveira EA, de Parijós AM, de Oliveira CS, de Loiola RSP, et al. (2014) Monitoring mercury exposure in reproductive aged women inhabiting the tapajós river basin, Amazon. Bull Environ Contam Toxicol 1: 42–6.
5. Rice DC (2004) The US EPA reference dose for methylmercury: sources of uncertainty. Environ Res 95: 406–413.

6. National Academy of Sciences (2000) Toxicological effects of methylmercury. National Research Council, Washington, DC, USA.
7. Carmignani M, Boscolo P, Artese L, Del Rosso G, Porcinelli G, et al. (1992) Renal mechanisms in the cardiovascular effects of chronic exposure to inorganic mercury rats. Br J Indust Med 49: 226–232.
8. Perottoni J, Rodrigues OED, Paixão MW, Zeni G, Lobato LP, et al. (2004) Renal and hepatic ALA-D activity and selected oxidative stress parameters of rats exposed to inorganic mercury and organoselenium compounds. Food Chem Toxicol 42: 17–28.
9. Zemolin APP, Meinerz DF, de Paula MT, Mariano DOC, Rocha JBT, et al. (2012) Evidences for a role of glutathione peroxidase 4 (GPx4) in methylmercury induced neurotoxicity in vivo. Toxicology 302: 60–67.
10. Ghaffari MA, Motlagh B (2011) In vitro effect of lead, silver, tin, mercury, indium and bismuth on human sperm creatine kinase activity: a presumable mechanism for men infertility. Iran Biomed J 15, 1-2: 38–43.
11. Khan AT, Atkinson A, Graham TC, Thompson SJ, Ali S, et al. (2004) Effects of inorganic mercury on reproductive performance of mice. Food Chem Toxicol 42: 571–577.
12. Keck C, Bergman M, Ernst E, Muller C, Kliensch S, et al. (1993) Autometallographic detection of mercury in testicular tissue of an infertile man exposed to mercury vapour. Reprod Toxicol 7: 469–475.
13. El-Desoky GE, Bashandy SA, Alhazza IM, Al-Othman ZA, Aboul-Soud MAM, et al. (2013) Improvement of Mercuric Chloride-Induced Testis Injuries and Sperm Quality Deteriorations by Spirulina platensis in Rats. PLoS ONE 8(3): e59177. doi:10.1371/journal.pone.0059177.
14. Kalender S, Uzun FG, Demir F, Uzunhisarcıklı M, Aslanturk A (2013) Mercuric chloride-induced testicular toxicity in rats and the protective role of sodium selenite and vitamin E. Food Chem Toxicol 55: 456–462.
15. Mohamed MK, Barbacher TM, Mottet NK (1987) Effects of methylmercury on testicular functions in macaca fascicularis monkeys. Pharmacol Toxicol 60: 29–36.
16. Buck Louis GM (2014) Persistent environmental pollutants and couple fecundity: an overview. Reproduction 4: R97–R104.
17. Popescu HI (1978) Poisoning with alkylmercuric compounds. Br Med J 1: 23–47.
18. Choy CM, Lam CW, Cheung LT, Briton-Jones CM, Cheung LP, et al. (2002) Infertility, blood mercury concentrations and dietary seafood consumption: a case control study. BJOG 109: 1121–1125.
19. Mocevic E, Specht IO, Marott JL, Giwercman A, Jönsson BA, et al. (2013) Environmental mercury exposure, semen quality and reproductive hormones in Greenlandic Inuit and European men: a cross-sectional study. Asian J Androl 1: 97–104.
20. Erfurth EM, Schütz A, Nilsson A, Barregård L, Skerfving S (1990) Normal pituitary hormone response to thyrotrophin and gonadotrophin releasing hormones in subjects exposed to elemental mercury vapour. Br J Ind Med 9: 639–44.
21. Rana SV (2014) Perspectives in Endocrine Toxicity of Heavy Metals-A Review. Biol Trace Elem Res 1: 1–14.
22. Ramalingam V, Vimaladevi V, Rajeswary S, Suryavathi V (2003) Effect of mercuric chloride on circulating hormones in adult albino rats. J Environ Biol 24: 401–414.
23. Boujbiha MA, Hamdena K, Guermazib F, Bouslamac A, Omezzinec A, et al. (2009) Testicular toxicity in mercuric chloride treated rats: Association with oxidative stress. Reprod Toxicol 28: 81–89.
24. Carvalho CM, Chew EH, Hashemy SI, Lu J, Holmgren A (2008) Inhibition of the human thioredoxin system. A molecular mechanism of mercury toxicity. J Biol Chem 283: 11913–23.
25. Martinez CS, Escobar AG, Torres JGD, Brum DS, Santos FW, et al. (2014) Chronic exposure to low doses of mercury impairs sperm quality and induces oxidative stress in rats. J Toxicol Environ Health, Part A 77: 143–154.
26. Wiggers GA, Peçanha FM, Briones AM, Pérez-Girón JV, Miguel M, et al. (2008) Low mercury concentrations cause oxidative stress and endothelial dysfunction in conductance and resistance arteries. Am J Physiol Heart Circ Physiol 295: 1033–104.
27. Filler R (1993) Methods for evaluation of rats epididymal sperm morphology. In Male reproductive toxicology. Chapin RE, Heindel JH Eds. California: Academic Press. 334–343p.
28. Robb GW, Amman RP, Killian GJ (1978) Daily sperm production and epididymal sperm reserves of puberal and adult rats. J Reprod Fertil 54: 103–107.
29. Ohkawa H, Ohishi N, Yagi K (1979) Assay for lipid peroxides in animal tissues by thiobarbituric acid reaction. Anal Biochem 95: 351–358.
30. Ellman GL (1959) Tissue sulfhydryl groups. Arch Biochem 82: 70–77.
31. Misra HP, Fridovich I (1972) The role of superoxide anion in the autoxidation of epinephrine and simple assay for superoxide dismutase. J Biol Chem 247: 3170–3175.
32. Aebi H (1984) Catalase in vitro. Methods Enzymol 105: 121–126.
33. Birch RJ, Bigler J, Rogers JW, Zhuang Y, Clickner RP (2014) Trends in blood mercury concentrations and fish consumption among U.S. women of reproductive age, NHANES, 1999-2010. Environ Res 14: 00021–8.
34. McKelvey W, Gwynn RC, Jeffery N, Kass D, Thorpe LE, et al. (2007) A biomonitoring study of lead, cadmium, and mercury in the blood of New York city adults. Environ Health Perspect 115: 1435–1441.
35. Waye A, Trudeau VL (2011) Neuroendocrine disruption: more than hormones are upset. J Toxicol Environ Health, Part B 14: 270–291.
36. Vahter ME, Mottet NK, Friberg LT, Lind SB, Charleston JS, et al. (1995) Demethylation of methyl mercury in different brain sites of Macaca fascicularis monkeys during long-term subclinical methyl mercury exposure. Toxicol Appl Pharmacol 134: 273–84.
37. Cornett CR, Ehmann WD, Wekstein DR, Markesbery WR (1998) Trace elements in Alzheimer's disease pituitary glands. Biol Trace Elem Res 62: 107–14.
38. Mello-Carpes PB, Barros W, Borges S, Alves N, Rizzetti D, et al. (2013) Chronic exposure to low mercury chloride concentration induces object recognition and aversive memories deficits in rats. Int J Devl Neuroscience 31: 468–472.
39. Homma-Takeda S, Kugenuma Y, Iwamuro T, Kumagai Y, Shimojo N (2001) Impairment of spermatogenesis in rats by methylmercury: involvement of stage- and-cell specific germ cell apoptosis. Toxicology 169: 25–35.
40. Boujbiha MA, Hamden K, Guermazi F, Bouslama A, Omezzine A, et al. (2011) Impairment of spermatogenesis in rats by mercuric chloride: involvement of low 17β-estradiol level in induction of acute oxidative stress. Biol Trace Elem Res 3: 598–610.
41. Omura M, Tanaka A, Hirata M, Zhao M, Makita Y, et al. (1996) Testicular toxicity of galliumarsenide, indium arsenide and arsenic oxide in rats by repetitive intratracheal instillation. Fundam Appl Toxicol 32: 72–78.
42. Goyer RA, Liu J, Waalkes MP (2004) Cadmium and cancer of prostate and testis. Biometals 17: 555–558.
43. Kumar BA, Reddy AG, Kumar PR, Reddy YR, Rao TM, et al. (2013) Protective role of N-Acetyl L-Cysteine against reproductive toxicity due to interaction of lead and cadmium in male Wistar rats. J Nat Sci Biol Med 4: 414–9.
44. Santos FW, Zen G, Rocha JB, Weiss S, Fachineto JM, et al. (2005) Diphenyl diselenide reverses cadmium-induced oxidative damage on mice tissues. Chem Biol Interact 151: 159–165.
45. Aitken RJ (1995) Free radicals, lipid peroxidation and sperm function. Reprod Fertil Dev 7: 659–668.
46. Rao MV, Gangadharan B (2008) Antioxidative potential of melatonin against mercury induce intoxication in spermatozoa in vitro. Toxicol In Vitro 22: 935–42.
47. Farina M, Brandão R, de Lara FS, Pagliosa LB, Soares FA, et al. (2003) Profile of nonprotein thiols, lipid peroxidation and delta-aminolevulinate dehydratase activity in mouse kidney and liver in response to acute exposure to mercuric chloride and sodium selenite. Toxicol 184, 2-3: 179–87.
48. Bando I, Reus MIS, Andrés D, Cascales M (2005) Endogenous antioxidant defence system in rat liver following mercury chloride oral intoxication. J Biochem Mol Toxicol 19: 154–161.
49. Sharma AK, Kapadia AG, Fransis P, Rao MV (1996) Reversible effects of mercuric chloride on reproductive organs of the male mouse. Reprod Toxicol 10: 153–9.

Comparison of Long-Term Survival and Toxicity of Cisplatin Delivered Weekly versus Every Three Weeks Concurrently with Intensity-Modulated Radiotherapy in Nasopharyngeal Carcinoma

Chang-Juan Tao[1,2]ⓢ, Li Lin[1]ⓢ, Guan-Qun Zhou[1], Ling-Long Tang[1], Lei Chen[1], Yan-Ping Mao[1], Mu-Sheng Zeng[3], Tie-Bang Kang[3], Wei-Hua Jia[3], Jian-Yong Shao[4], Hai-Qiang Mai[5], Ai-Hua Lin[6], Jun Ma[1], Ying Sun[1]*

1 Department of Radiation Oncology, Sun Yat-sen University Cancer Center, State Key Laboratory of Oncology in South China, Collaborative Innovation Center for Cancer Medicine, Guangzhou, Guangdong Province, People's Republic of China, **2** Department of Radiation Oncology, Zhejiang Cancer Hospital, Hangzhou, Zhejiang Province, People's Republic of China, **3** Department of Experimental Research, Sun Yat-sen University Cancer Center, State Key Laboratory of Oncology in South China, Collaborative Innovation Center for Cancer Medicine, Guangzhou, Guangdong Province, People's Republic of China, **4** Department of Pathology, Sun Yat-sen University Cancer Center, State Key Laboratory of Oncology in South China, Collaborative Innovation Center for Cancer Medicine, Guangzhou, Guangdong Province, People's Republic of China, **5** Department of Nasopharyngeal Carcinoma, Sun Yat-sen University Cancer Center, State Key Laboratory of Oncology in South China, Collaborative Innovation Center for Cancer Medicine, Guangzhou, Guangdong Province, People's Republic of China, **6** Department of Medical Statistics and Epidemiology, School of Public Health, Sun Yat-sen University, Guangzhou, Guangdong Province, People's Republic of China

Abstract

Background: We aimed to compare the long-term survival outcomes and acute toxicity of cisplatin administered weekly versus every three weeks concurrently with intensity-modulated radiotherapy (IMRT) in patients with nasopharyngeal carcinoma (NPC).

Methods: This was a retrospective review of 154 patients with histologically proven, non-disseminated NPC who were treated using IMRT between January 2003 and December 2007. Seventy-three patients (47.4%) received 5–7 weeks of 30–40 mg/m^2 cisplatin weekly; 81 patients (52.6%) received two or three cycles of 80 mg/m^2 cisplatin every three weeks. IMRT was delivered at 68 Gy/30 fractions to the nasopharyngeal gross target volume and 60–66 Gy to the involved neck area.

Results: The clinical characteristics and treatment factors of the two groups were well-balanced. The median follow-up was 74 months (range, 6–123 months), and the 5-year overall survival, disease-free survival, locoregional relapse-free survival, and distant metastasis–free survival rates were 85.2% vs. 78.9% (P = 0.318), 71.6% vs. 71.0% (P = 0.847), 93.5% vs. 92.6% (P = 0.904), and 80.9% vs. 80.1% (P = 0.925) for the group treated every three weeks and weekly, respectively. Subgroup analyses indicated no significant differences in the survival rates of the two groups among patients with early- or advanced-stage disease. The incidence of acute toxicities was similar between groups.

Conclusion: IMRT with concurrent cisplatin administered weekly or every three weeks leads to similar long-term survival outcomes and acute toxicity in NPC regardless of whether patients have early- or advanced-stage disease.

Editor: Apar Kishor Ganti, University of Nebraska Medical Center, United States of America

Funding: This work was supported by grants from the National Natural Science Foundation of China (No. 8172409), the Sun Yat-Sen University Clinical Research 5010 Program (No. 2012011), the Natural Science Foundation of Guangdong Province (No. S2013010012220), the Science and Technology project of Guangzhou City, China (No. 132000507), the Health & Medical Collaborative Innovation Project of Guangzhou City, China (201400000001), and the National Science & Technology Pillar Program during the Twelfth Five-year Plan Period (2014BAI09B10). The funders had no role in study design, data collection and analysis, decision to publish, or preparation of the manuscript.

Competing Interests: The authors have declared that no competing interests exist.

* Email: sunying@sysucc.org.cn

ⓢ These authors contributed equally to this work.

Introduction

Nasopharyngeal carcinoma (NPC) is highly prevalent in southern China; incidence rates range between 15 and 50 per 100,000 individuals [1]. Due to the anatomical location and radiosensitivity of NPC, radiation therapy (RT) is the mainstay treatment. According to the 7th Edition of the American Joint Commission on Cancer (AJCC) staging system [2], 60–70% of patients present with stage III–IVB disease at diagnosis [3]. Several randomized clinical trials and meta-analyses have consistently demonstrated that concurrent chemoradiotherapy

(CCRT) confers survival benefit in locoregionally advanced NPC [4–11].

The National Comprehensive Cancer Network recommends that CCRT with cisplatin (CDDP) be delivered at an intermediate dose weekly or at a high dose at 3-week intervals for stage II–IVB NPC. The Intergroup 0099 trial (INT-0099) [3] and NPC-9901 trial [5] compared concurrent CDDP (100 mg/m^2) every three weeks with RT alone; compliance to the intensive regimen was poor, and CCRT was associated with a high incidence of acute toxicity. Two other phase III clinical trials [6–7] evaluated the efficiency of CCRT with weekly CDDP (40 mg/m^2) during RT; CCRT conferred survival benefit and weekly CDDP was well-tolerated. Therefore, it is of great importance to identify the optimal administration schedule for CDDP during CCRT in patients with NPC.

To the best of our knowledge, only one study [12] has compared the dose delivery and survival outcomes of weekly CDDP versus CDDP every three weeks delivered concurrently with RT in locally advanced NPC. However, in that study, the clinical characteristics and treatment factors were unevenly distributed, and the majority of patients were treated with 3-dimensional conformal RT. Radiation technology has evolved rapidly in recent years, especially with the introduction of intensity-modulated radiotherapy (IMRT). In NPC, this treatment modality has excellent local control and less late toxicities [13]. With the widespread application of IMRT, reevaluating the survival outcomes and toxicity of concurrent weekly CDDP versus CDDP every three weeks in patients with NPC undergoing CCRT would be worthwhile.

Therefore, we aimed to compare the long-term survival outcomes and acute toxicities of weekly CDDP versus CDDP every three weeks delivered concurrently with IMRT in patients with NPC. Our results will help guide clinical CCRT treatment strategies in NPC.

Materials and Methods

Patient characteristics

Between January 2003 and December 2007, we reviewed the records of 749 patients with newly diagnosed, untreated, non-disseminated NPC who were treated using IMRT at our center. We excluded 595 patients (79.4%): 214 (28.6%) who were treated with RT only, 274 (36.6%) who received neoadjuvant chemotherapy, 86 (11.5%) who received concurrent chemotherapy using a combination of CDDP and 5-fluorouracil, and 21 (2.8%) who received concurrent chemotherapy with taxol. We included the remaining 154 patients in our retrospective study; 115 were male and 39 were female (male:female ratio, 2.95:1) and the median age was 44.0 years (range, 13–73 years). One hundred and fifty patients (97.4%) had a pathological diagnosis of World Health Organization (WHO) type II or III NPC, only one patient (0.6%) had WHO type I NPC, and three patients (1.9%) had basaloid squamous cell carcinoma. The ethics committee of Sun Yat-sen University Cancer Center approved this study. Written consent was waived because this was a retrospective study; verbal consent was obtained from the patients via telephone and documented in the informed consent form if the patient agreed to participate in this study, as was consent on behalf of the children enrolled. The Institutional Review Board approved the use of verbal consent.

All patients underwent pre-treatment evaluation, which included complete patient history, physical and neurological examination, hematological and biochemical profiles, magnetic resonance imaging (MRI) scan of the neck and nasopharynx, chest radiography, abdominal sonography, and single-photon emission computed tomography whole-body bone scans. Positron emission tomography/computed tomography (CT) was performed on 41 patients (26.6%). All patients were restaged according to the 7th edition of the AJCC staging system. The stage distribution was as follows: stage II, 40/154 (26%); stage III, 66/154 (42.8%); stage IVA, 38/154 (24.7%); stage IVB, 10/154 (6.5%). Table 1 lists the characteristics of the patient cohort.

Radiotherapy

All patients were immobilized in the supine position with a head, neck, and shoulder thermoplastic mask. We obtained two sets of images, i.e., with and without contrast, from the CT simulator for treatment planning purposes. CT was performed after administering intravenous contrast medium, and we obtained 3-mm slices from the head to 2 cm below the sternoclavicular joint. The primary tumor and upper neck above the caudal edge of the cricoid cartilage were treated by IMRT. The target volumes were delineated using a previously described institutional treatment protocol [14] in accordance with the International Commission on Radiation Units and Measurements reports 50 and 62. The clinical target volumes (CTVs) were individually delineated based on the tumor invasion pattern [15]. The contoured images were transferred to a Corvus version 3.0 inverse IMRT planning system (Peacock; Nomos Corp., Deer Park, IL, USA). The prescribed radiation dose (as per the protocol) was total dose of 68 Gy in 30 fractions at 2.27 Gy/fraction to the planning target volume (PTV) of the primary gross tumor volume (GTV), 60–64 Gy to the nodal GTV PTV, 60 Gy to the CTV-1 PTV (i.e., high-risk regions), and 54 Gy to the CTV-2 PTV (i.e., low-risk regions) and CTV-N (i.e., neck nodal regions). Treatment was delivered by a dynamic, multileaf intensity-modulating collimator (MIMiC; Nomos Corp., Sewickley, PA, USA). A conventional anterior or anteroposterior opposing cervical technique was used for the lower neck. All patients were treated with one fraction daily over five days per week. All targets were treated simultaneously using the simultaneous integrated boost technique.

Concurrent chemotherapy

All patients were treated with CCRT; eight patients (5.2%) also received adjuvant chemotherapy using CDDP and 5-fluorouracil for two or three cycles. CCRT was initiated on the first day of RT and the choice of every three weeks or weekly CDDP was based on oncologists' opinions. Seventy-three patients (47.4%) received 30–40 mg/m^2 CDDP weekly for 5–7 planned weeks, and 81 patients (52.6%) received two or three cycles of 80 mg/m^2 CDDP every three weeks. In both groups, 5-hydroxytryptamine 3 antagonists and dexamethasone were administered as antiemetic prophylaxis.

Follow-up

The duration of follow-up was calculated from the first day of treatment to either the day of death or day of the last follow-up. Patients were examined at least every three months during the first two years, and every six months thereafter until death. At every follow-up, we assessed disease status with complete physical examination, nasopharyngoscopy, hematological and biochemical profiles, chest radiography, abdominal ultrasonography, and CT/MRI scans of the nasopharynx and cervical region. Acute toxicities were scored according to the Common Terminology Criteria for Adverse Events v3.0 (CTCAE v3.0).

Statistical analysis

We used Statistical Package for the Social Sciences version 16.0 (SPSS, Chicago, IL, USA) for statistical analysis. We compared

Table 1. Clinical characteristics of patients.

Characteristic	Weekly cisplatin (n = 73)	Cisplatin every three weeks (n = 81)	P*
Age (years)			.351
≤50	52	63	
>50	21	18	
≤18	0	3	.247
>18	73	78	
Sex			.581
Male	56	59	
Female	17	22	
Histology			.116
WHO I	1	0	
WHO II	7	3	
WHO III	65	75	
BSCC	0	3	
T classification†			.651
T1	7	13	
T2	17	17	
T3	31	30	
T4	18	21	
N classification†			.538
N0	14	13	
N1	42	55	
N2	12	8	
N3	5	5	
Staging†			.939
II	18	22	
III	33	33	
IVA	17	21	
IVB	5	5	
OTT (days)			.554
Median	43	44	
SD	4.3	4.0	
Response rate			.258
CR	60	61	
PR	11	19	

WHO: World Health Organization; BSCC: basaloid squamous cell carcinoma; CR: complete response; PR: partial response; SD: standard deviation; OTT: overall treatment time.
**P*-values were calculated using the chi-square test (or Fisher's exact test) and Mann–Whitney U test.
†According to the 7th AJCC/International Union against Cancer staging system.

and analyzed the clinical characteristics and toxicity rates of the two treatment groups using the chi-square test (or Fisher's exact test, if indicated). Overall treatment time and cumulative CDDP dose were compared using the Mann–Whitney U test. All events were measured from the start of treatment. The following end points (time to the first defined event) were assessed: overall survival (OS), disease-free survival (DFS), locoregional relapse–free survival (LRRFS), distant metastasis–free survival (DMFS). Actuarial rates were calculated using the Kaplan–Meier method [16]. Multivariate analyses using the Cox proportional hazards model

Table 2. Acute toxicity according to CTCAE v3.0.

Toxicity	Weekly cisplatin (n = 73)					Every three week cisplatin (n = 81)					P*
	Grade0	Grade1	Grade2	Grade3	Grade4	Grade0	Grade1	Grade2	Grade3	Grade4	
Hematological											
Leucopenia	30	17	20	5	1	23	28	25	5	0	0.312
Anemia	58	9	4	2	0	61	17	3	0	0	0.233
Thrombocytopenia	62	4	2	3	2	74	2	2	2	1	0.583
Non-hematological											
Dermatitis	0	59	13	1	0	0	54	24	3	0	0.121
Mucositis	0	29	21	23	0	0	23	34	23	1	0.139
Dysphagia	29	33	8	3	0	31	32	13	5	0	0.715
Nausea/Vomiting	32	24	17	0	0	36	27	15	3	0	0.295
Xerostomia	15	27	31	0	0	11	31	39	0	0	0.498
Ototoxicity	52	19	2	0	0	57	23	1	0	0	0.767
Neurotoxicity	70	2	1	0	0	77	3	1	0	0	0.632

**P*-values were calculated with the chi-square test (or Fisher's exact test, if indicated).

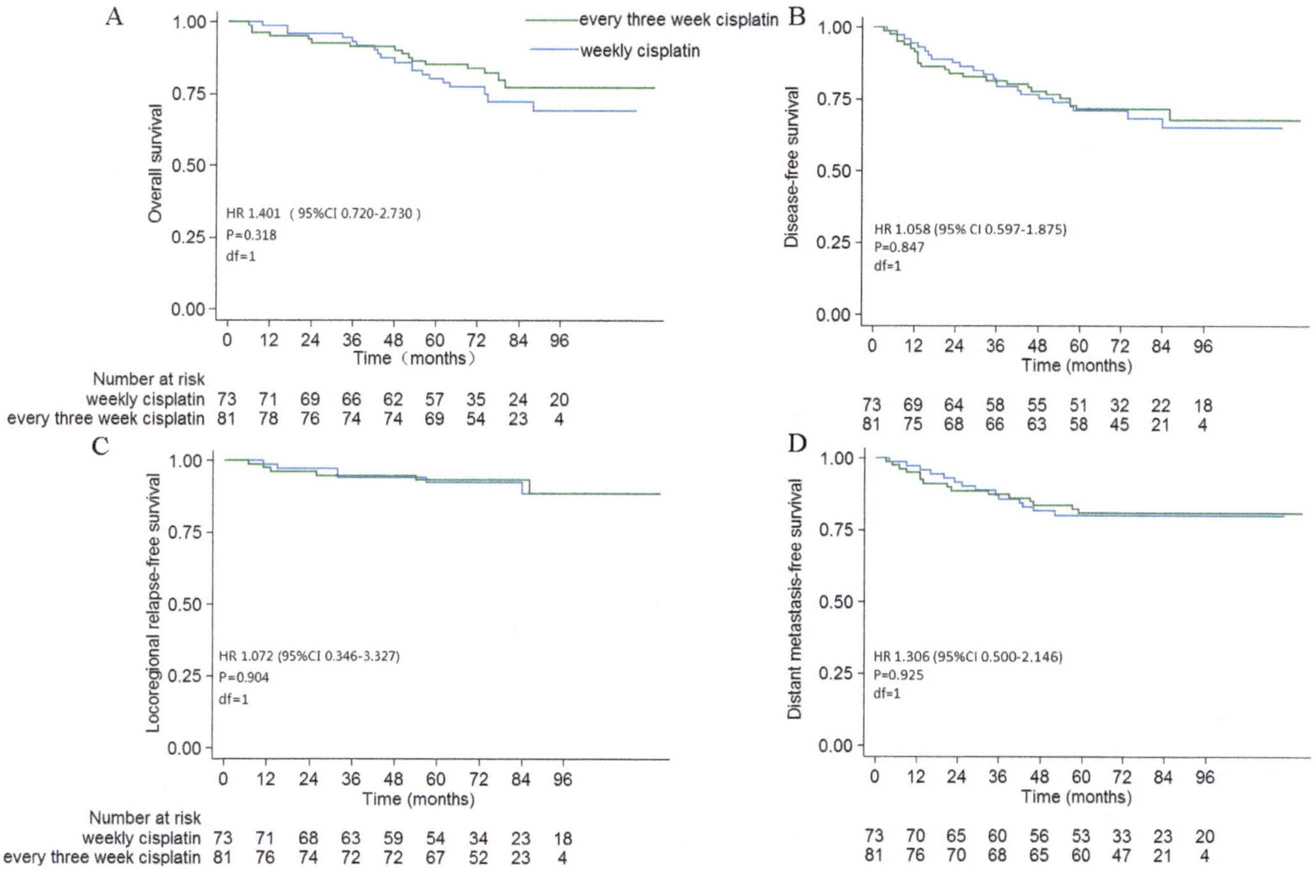

Figure 1. Survival curves for patients with NPC. (A) OS, (B) DFS, (C) LRRFS, and (D) DMFS. Hazard ratios (HRs) were calculated with an unadjusted Cox proportional hazards model; *P*-values were calculated with the unadjusted log-rank test.

were used to test for independent significance using backward elimination of insignificant explanatory variables of different parameters [17]. Host factors (age and sex) were included as covariates in all tests. The criterion for statistical significance was set at $a = 0.05$ and *P*-values were determined using 2-sided tests.

Results

Patient characteristics

Both groups were well-balanced in terms of baseline demographic and clinical characteristics and treatment factors (Table 1). Tumor response was evaluated via endoscopy and MRI scans of the neck and nasopharynx at three months after RT. The composite overall response rate for the primary site and neck region was 97.3% and 98.7% for the weekly and every three week CDDP group, respectively ($P = 0.258$).

Treatment compliance and acute toxicities

All 154 patients tolerated the treatment well and completed the prescribed dose of RT. The median overall treatment time was 43 days in the weekly group and 44 days in the every three weeks group ($P = 0.554$). In the weekly group, 66 patients (90.4%) received at least five weeks of CDDP and only four patients (5.5%) received seven weeks of CDDP. In the every three weeks group, 72 patients (88.9%) completed two cycles of CDDP and five patients (6.2%) received three cycles of CDDP. The median cumulative dose received during concurrent CCRT was 180 and 160 mg/m^2 CDDP for the weekly and every three weeks group, respectively ($P = 0.10$). Reductions in the number of chemotherapy cycles were mostly due to patient refusal, severe mucositis, or prolonged severe leucopenia.

Systemic toxicities were similar in both groups (Table 2). No treatment-related deaths were observed in either cohort. The most commonly recorded non-hematological adverse event was grade 3–4 mucositis in 23 patients (31.5%) in the weekly group and 24 patients (29.6%) in the every three weeks group ($P = 0.139$). The most common hematological adverse event was grade 3–4 leucopenia in six patients (8.2%) in the weekly group and five patients (6.2%) in the every three weeks group ($P = 0.312$).

Patterns of failure and survival rates

The overall median follow-up time was 74 months (range, 6–123 months). Twelve of the 154 patients (7.8%) developed locoregional relapse. There was distant metastasis in 29/154 patients (18.8%), and 35/154 patients (22.7%) died. The sites of relapse were local in 7/154 patients (4.5%), regional in 7/154 patients (4.5%), and both local and regional in 2/154 patients (1.3%). Three patients (1.9%) experienced both locoregional and distant failure.

The overall 5-year OS, DFS, LRRFS, and DMFS rates were 82.3%, 71.3%, 93.1%, and 80.5%, respectively. The 5-year OS, DFS, LRRFS, and DMFS rates were not significantly different between the two treatment groups (every three weeks vs. weekly: OS = 85.2% vs. 78.9%, $P = 0.318$; DFS = 71.6% vs. 71.0%, $P = 0.847$; LRRFS = 93.5% vs. 92.6%, $P = 0.904$; DMFS = 80.9%

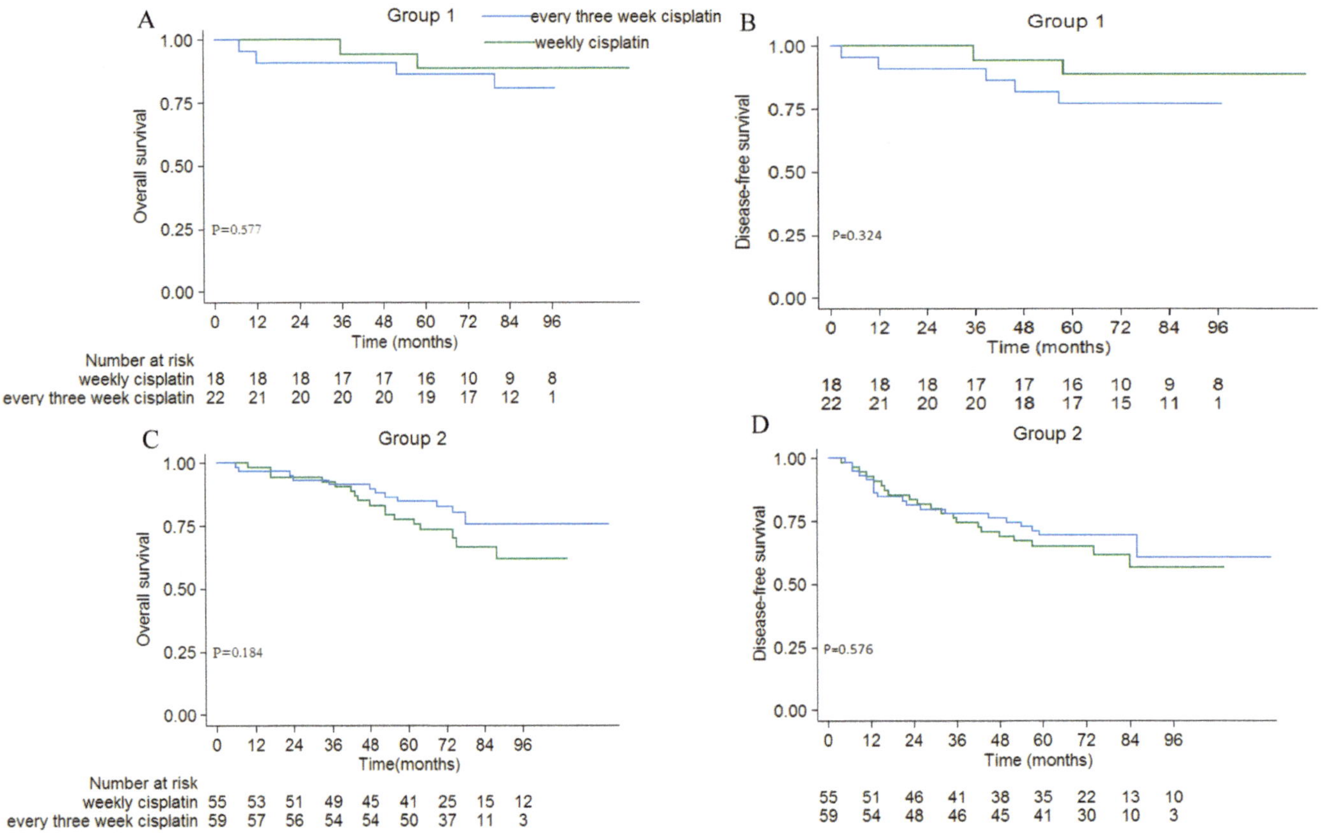

Figure 2. OS and DFS curves for NPC patients stratified by clinical stage. Group 1: patients with stage II disease; Group 2: patients with stage III–IVB disease. *P*-values were calculated with the unadjusted log-rank test.

vs. 80.1%, $P = 0.925$; Fig. 1). The 5-year OS and DFS rates for patients with early- or advanced-stage disease in the two treatment groups were not significantly different (Fig. 2).

Prognostic factors

Univariate analysis revealed that age and N classification were significant prognostic factors for OS ($P = 0.009$ and 0.011, respectively). The 5-year OS rate for patients who received

Table 3. Summary of multivariate analysis of prognostic factors in the 154 patients.

End point	HR (95% CI)	*P*-value[†]
OS		
Age,>50 years vs. ≤50 years	2.809 (1.404–5.620)	0.004
N classification (N0–1 vs. N2–3)	2.830 (1.388–5.767)	0.004
Chemotherapy regimen, weekly vs. every three weeks	0.862 (0.428–1.738)	0.679
DFS		
N classification (N0–1 vs. N2–3)	2.241 (1.213–4.141)	0.010
Chemotherapy regimen, weekly vs. every three weeks	1.099 (0.604–2.001)	0.757
DMFS		
N classification (N0–1 vs. N2–3)	2.752 (1.271–5.961)	0.010
Chemotherapy regimen, weekly vs. every three weeks	1.331 (0.607–2.917)	0.475
LRRFS		
Chemotherapy regimen, weekly vs. every three weeks	1.280 (0.382–4.288)	0.689

HR: hazard ratio; CI: confidence interval; OS: overall survival; DFS: disease-free survival; DMFS: distant metastasis-free survival; LRRFS: locoregional relapse-free survival.
[†]P-values were calculated using an adjusted Cox proportional hazards regression model adjusted for age (>50 years vs. ≤50 years), sex (male vs. female), T classification (T1–2 vs. T3–4), N classification (N0–1 vs. N2–3), chemotherapy regimen, (weekly vs. every three weeks) and cumulative cisplatin dose, (≥160 mg/m2 vs. <160 mg/m2). Only variables that were significantly associated with survival are presented except for the chemotherapy regimen.

cumulative CDDP≥160 mg/m^2 was 84.1% compared to 73.3% for patients who received cumulative CDDP<160 mg/m^2 ($P = 0.175$).

Multivariate analysis was performed to adjust for various prognostic factors. The following parameters were included in the Cox proportional hazards model by backward elimination of insignificant explanatory variables: age (>50 years vs. ≤50 years), sex, T classification (T1–2 vs. T3–4), N classification (N0–1 vs. N2–3), concurrent chemotherapy regimen (CDDP weekly vs. every three weeks), and cumulative CDDP dose (≥160 mg/m^2 vs. <160 mg/m^2). Multivariate analysis demonstrated that the CCRT regimen was not a significant prognostic factor for any end point. Age and N classification were significant predictive factors for OS; furthermore, N classification was the only significant prognostic factor for DFS and DMFS (Table 3).

Discussion

NPC is a highly chemosensitive tumor and CCRT has long been a standard treatment option for locoregionally advanced NPC. Treatment regimens delivering medium doses of CDDP weekly or high CDDP doses every three weeks have been evaluated. Weekly medium doses are efficient and simple to administer, especially for outpatients; high doses every three weeks are more effective for reducing distant metastasis, but may lead to more severe acute toxicities [4–7].

In this retrospective study, we observed that CDDP-based (both weekly and every three weeks) CCRT led to similar long-term survival outcomes and acute toxicities in patients with NPC receiving IMRT regardless of whether the patients had early- or advanced-stage disease. Our results are in agreement with that of Jagdis et al. [12], which showed that 40 mg/m^2 CDDP weekly and 100 mg/m^2 CDDP every three weeks had similar deliverability, toxicity profiles, and comparable survival outcomes in locally advanced NPC.

In our study, the overall 5-year OS, DFS, LRRFS, and DMFS rates were 82.3%, 71.3%, 93.1%, and 80.5%, respectively. CDDP delivered both weekly or every three weeks concurrently with IMRT achieved excellent OS. Distant metastasis was the predominant mode of failure. One approach to improving the strategy tested in this study would be changing the sequence to induction–concurrent, as substantially better tolerance and compliance rates have been reported for induction–concurrent regimens [18]. The meta-analysis by OuYang et al. in 2013 [19] also showed that neoadjuvant chemotherapy effectively enhances OS and reduces the rate of distant metastasis in NPC.

With respect to dose delivery, both treatment groups achieved similar cumulative dose intensities; however, a low proportion of patients completed seven weeks of weekly CDDP or three cycles of every three week CDDP. The compliance rates in the present study are lower than that of previously published trials [4–7]. This may be because the weekly CDDP regimen was mostly administered to outpatients and the regimen given every three weeks was mainly administered to inpatients at our center. Poor outpatient compliance and inadequate numbers of inpatient beds led to delays in administering chemotherapy, which often resulted in patients missing the final cycle of chemotherapy.

We adopted an every three week CDDP dose of 80 mg/m^2 in our study, which was effective with low toxicity. The INT-0099 regimen is widely used, where 100 mg/m^2 CDDP is administered on days 1, 22, and 43 with radical RT, given its impressive results compared to RT alone when treating NPC. Regrettably, only 63% of patients in the chemoradiation arm of INT-0099 completed all three cycles of concurrent chemotherapy as planned. Ho et al. [20] compared the dose intensity and toxicity of CDDP administered weekly and every three weeks concurrently to 52 patients with locally advanced squamous head and neck cancer. They found that 100 mg/m^2 CDDP every three weeks with RT was less well tolerated than 40 mg/m^2 CDDP weekly or 80 mg/m^2 CDDP every three weeks and resulted in fewer patients achieving a cumulative dose of >200 mg/m^2. Therefore, the optimal CDDP dose in CCRT regimes for NPC warrants further exploration.

The dose intensity of CDDP is an important prognostic factor in NPC patients receiving CCRT [21–22]. Using pooled data from three prospective trials (in which most patients were treated with conventional RT), Loong et al. [21] explored the prognostic significance of the total CDDP dose delivered during CCRT and reported that patients with stage II or III NPC who received more than five weeks of 40 mg/m^2 CDDP weekly had significantly better OS than patients who received less than five weeks. In our study, a cumulative dose of ≥160 mg/m^2 led to obviously better OS, although the difference was not significant (84.1% vs. 73.3%, $P = 0.175$); this may be related to the small number of patients who received a cumulative dose of <160 mg/m^2. Additionally, the use of IMRT has improved the long-term outcome for patients with NPC, which may render detection of the survival benefit of a higher total CDDP dose difficult.

The toxicity profiles of both groups were similar. However, there was a slightly higher incidence of grade 3–4 hematological toxicity and oropharyngeal mucositis in the weekly CDDP group, although the difference was not significant. The main reasons for this observation may be the inadequate monitoring of blood counts and poor oral care in outpatients in the weekly CDDP group. It is worth noting that there was a lower incidence of grade 3–4 dermatitis, dysphagia, and gastrointestinal reactions in patients who received CDDP weekly even though they received a relatively high median cumulative dose.

To our knowledge, this is the first single-institution study to review the difference between CDDP administered weekly and every three weeks concurrently with IMRT in NPC. The principal limitation of this study is its retrospective nature. However, the clinical characteristics of the patients in the two groups were well-balanced, which may reduce potential selection bias. To derive a definitive conclusion, a randomized phase II study of CCRT with CDDP administered every three weeks versus weekly in patients with locally advanced NPC is ongoing in Korea. The findings of this study will aid in the standardization of CCRT regimens for NPC.

Conclusions

IMRT in conjunction with concurrent CDDP chemotherapy administered weekly or every three weeks leads to similar long-term survival outcomes and acute toxicities in NPC regardless of whether the patient has early- or advanced-stage disease. The results of this study require further validation in prospective, multicenter controlled trials.

Author Contributions

Conceived and designed the experiments: C-JT LL YS. Performed the experiments: C-JT LL YS G-QZ L-LT LC Y-PM M-SZ T-BK W-HJ J-YS H-QM JM. Analyzed the data: C-JT LL YS A-HL. Contributed reagents/materials/analysis tools: H-QM A-HL JM. Wrote the paper: C-JT LL YS G-QZ.

References

1. Jemal A, Bray F, Center MM, Ferlay J, Ward E, et al. (2011) Global cancer statistics. CA Cancer J Clin 61: 69–90.
2. Edge SB, Byrd DR, Compton CC, Fritz AG, Greene FL, et al. (2009) American Joint Committee on Cancer Manual for Staging of Cancer [M]. 7th ed. Philadelphia: JB Lippincott. 649 p.
3. Chen L, Mao YP, Xie FY, Liu LZ, Sun Y, et al. (2012) The seventh edition of the UICC/AJCC staging system for nasopharyngeal carcinoma is prognostically useful for patients treated with intensity-modulated radiotherapy from an endemic area in China. Radiat Oncol Biol Phys 104: 331–337.
4. Al-Sarraf M, LeBlanc M, Giri PG, Fu KK, Cooper J, et al. (1998) Chemoradiotherapy versus radiotherapy in patients with advanced nasopharyngeal cancer: phase III randomized Intergroup Study 0099. J Clin Oncol 16: 1310–1317.
5. Lee AW, Lau WH, Tung SY, Chua DT, Chappell R, et al. (2005) Preliminary results of a randomized study on therapeutic gain by concurrent chemotherapy for regionally-advanced nasopharyngeal carcinoma: NPC-9901 trial by the Hong Kong Nasopharyngeal Cancer Study Group. J Clin Oncol 23: 6966–6975.
6. Lee AW, Tung SY, Chua DT, Ngan RK, Chappell R, et al. (2010) Randomized trial of radiotherapy plus concurrent-adjuvant chemotherapy vs. radiotherapy alone for regionally advanced nasopharyngeal carcinoma. J Natl Cancer Inst 102: 1188–1198.
7. Chen Y, Sun Y, Liang SB, Zong JF, Li WF, et al. (2013) Progress report of a randomized trial comparing long-term survival and late toxicity of concurrent chemoradiotherapy with adjuvant chemotherapy versus radiotherapy alone in patients with stage III to IVB nasopharyngeal carcinoma from endemic regions of China. Cancer 119: 2230–2238.
8. Chan AT, Leung SF, Ngan RK, Teo PM, Lau WH, et al. (2005) Overall survival after concurrent cisplatin-radiotherapy compared with radiotherapy alone in locoregionally advanced nasopharyngeal carcinoma. J Natl Cancer Inst 97: 536–539.
9. Lin JC, Jan JS, Hsu CY, Liang WM, Jiang RS, et al. (2003) Phase III study of concurrent chemoradiotherapy versus radiotherapy alone for advanced nasopharyngeal carcinoma: positive effect on overall and progression-free survival. J Clin Oncol 21: 631–637.
10. Chen L, Hu CS, Chen XZ, Hu GQ, Cheng ZB, et al. (2012) Concurrent chemoradiotherapy plus adjuvant chemotherapy versus concurrent chemoradiotherapy alone in patients with locoregionally advanced nasopharyngeal carcinoma: a phase 3 multicentre randomised controlled trial. Lancet Oncol 13: 163–171.
11. Baujat B, Audry H, Bourhis J, Chan AT, Onat H, et al. (2006) Chemotherapy in locally advanced nasopharyngeal carcinoma: an individual data meta-analysis of eight randomized trials and 1753 patients. Int J Radiat Oncol Biol Phys 64: 47–56.
12. Jagdis A, Laskin J, Hao D, Hay J, Wu J, et al. (2012) Dose delivery analysis of weekly versus 3-weekly cisplatin concurrent with radiation therapy for locally advanced nasopharyngeal carcinoma (NPC). Am J Clin Oncol Oct 29.
13. Peng G, Wang T, Yang KY, Zhang S, Zhang T, et al. (2012) A prospective, randomized study comparing outcomes and toxicities of intensity-modulated radiotherapy vs. conventional two-dimensional radiotherapy for the treatment of nasopharyngeal carcinoma. Radiother Oncol 104: 286–293.
14. Zhao C, Han F, Lu LX, Huang SM, Lin CG, et al. (2004) Intensity modulated radiotherapy for local-regional advanced nasopharyngeal carcinoma. Ai Zheng 23: 1532–1537.
15. Liang S, Sun Y, Liu L, Chen Y, Chen L, et al. (2009) Extension of local disease in nasopharyngeal carcinoma detected by magnetic resonance imaging: improvement of clinical target volume delineation. Int J Radiat Oncol Biol Phys 75: 742–750.
16. Kaplan EL, Meier P (1958) Nonparametric estimation from incomplete observation. J Am Stat Assoc 53: 457–481.
17. Cox DR (1972) Regression models and life tables. J R Stat Soc B 34: 187–220.
18. Lee AWM, Ng WT, Chan OSH, Sze HCK (2012) If concurrent-adjuvant chemoradiotherapy is beneficial for locoregionally advanced nasopharyngeal carcinoma, would changing the sequence to induction-concurrent achieve better outcome? J Radiat Oncol 1: 107–115.
19. OuYang PY, Xie C, Mao YP, Zhang Y, Liang XX, et al. (2013) Significant efficacies of neoadjuvant and adjuvant chemotherapy for nasopharyngeal carcinoma by meta-analysis of published literature-based randomized, controlled trials. Ann Oncol 24: 2136–2146.
20. Ho KF, Swindell R, Brammer CV (2008) Dose intensity comparison between weekly and 3-weekly Cisplatin delivered concurrently with radical radiotherapy for head and neck cancer: A retrospective comparison from New Cross Hospital, Wolverhampton, UK. Acta Oncol 47: 1513–1518.
21. Loong HH, Ma BB, Leung SF, Mo F, Hui EP, et al. (2012) Prognostic significance of the total dose of cisplatin administered during concurrent chemoradiotherapy in patients with locoregionally advanced nasopharyngeal carcinoma. Radiother Oncol 104: 300–304.
22. Chen QY, Wen YF, Guo L, Liu H, Huang PY, et al. (2011) Concurrent chemoradiotherapy vs radiotherapy alone in stage II nasopharyngeal carcinoma: phase III randomized trial. J Natl Cancer Inst 103: 1761–1770.

Activation of the NLRP3 Inflammasome Complex is Not Required for Stress-Induced Death of Pancreatic Islets

Jibran A. Wali[1,2], Esteban N. Gurzov[1,2], Stacey Fynch[1], Lorraine Elkerbout[1], Thomas W. Kay[1,2], Seth L. Masters[3], Helen E. Thomas[1,2*]

1 St Vincent's Institute, Fitzroy, Victoria, Australia, **2** The University of Melbourne Department of Medicine, St. Vincent's Hospital, Fitzroy, Victoria, Australia, **3** The Walter and Eliza Hall Institute of Medical Research, Parkville, Victoria, Australia

Abstract

Loss of pancreatic beta cells is a feature of type-2 diabetes. High glucose concentrations induce endoplasmic reticulum (ER) and oxidative stress-mediated apoptosis of islet cells *in vitro*. ER stress, oxidative stress and high glucose concentrations may also activate the NLRP3 inflammasome leading to interleukin (IL)-1β production and caspase-1 dependent pyroptosis. However, whether IL-1β or intrinsic NLRP3 inflammasome activation contributes to beta cell death is controversial. This possibility was examined in mouse islets. Exposure of islets lacking functional NLRP3 or caspase-1 to H_2O_2, rotenone or thapsigargin induced similar cell death as in wild-type islets. This suggests that oxidative or ER stress do not cause inflammasome-mediated cell death. Similarly, deficiency of NLRP3 inflammasome components did not provide any protection from glucose, ribose or gluco-lipotoxicity. Finally, genetic activation of NLRP3 specifically in beta cells did not increase IL-1β production or cell death, even in response to glucolipotoxicity. Overall, our results show that glucose-, ER stress- or oxidative stress-induced cell death in islet cells is not dependent on intrinsic activation of the NLRP3 inflammasome.

Editor: David M. Ojcius, University of California Merced, United States of America

Funding: This study was supported by a National Health and Medical Research Council of Australia (NHMRC; http://www.nhmrc.gov.au/) and Juvenile Diabetes Research Foundation (JDRF; http://www.jdrf.org/) joint special program grant in type 1 diabetes (APP466658), an NHMRC project grant (APP1032610), fellowships from the NHMRC (HET and SLM) and a University of Melbourne Viola Edith Reid Bequest Scholarship (JAW). This study was supported in part by the Victorian Government's Operational Infrastructure Support Program (http://www.business.vic.gov.au/grants-and-assistance/programs/medical-research-operational-infrastructure-program). The funders had no role in study design, data collection and analysis, decision to publish, or preparation of the manuscript.

* Email: hthomas@svi.edu.au

These authors contributed equally to this work.

Introduction

There is evidence from human studies and animal models that loss of pancreatic beta cell mass occurs in type 2 diabetes [1–5]. Histological examination of pancreatic specimens from type 2 diabetic individuals showed a reduction in beta cell mass and an increase in the number of terminal deoxynucleotidyl transferase dUTP nick end labelling (TUNEL) positive beta cells compared to non-diabetic individuals [1–3]. Elevated plasma glucose is a hallmark of diabetes, and chronic exposure to high concentrations of glucose *in vitro* causes apoptosis (glucose toxicity) of islet cells [6–9]. We have demonstrated that apoptosis induced by glucose is due to activation of the intrinsic apoptosis pathway [6]. The pro-apoptotic BH3-only proteins BIM and PUMA, and downstream effector molecule BAX are important mediators of glucose toxicity [6]. Expression of pro-apoptotic molecules including BIM, PUMA and BAX was observed in islets isolated from subjects with type 2 diabetes [10,11].

It has been reported that exposure of mouse or human islets to high glucose concentrations induces production of IL-1β that could be toxic for islet cells [9,12–14]. IL-1β is produced as a result of activation of the NLRP3 inflammasome. This protein complex comprises of NLRP3, the adaptor protein ASC and caspase-1 [15]. Activation of the NLRP3-inflammasome requires two signals. Signal-1 increases intracellular pro-IL-1β concentration in response to binding of ligands to toll-like receptors (TLR). For most *in vitro* studies, lipopolysaccharide (LPS) is used to provide signal-1 whereas substances such as minimally modified LDL and free fatty acids could possibly serve this role *in vivo*. Signal-2 causes activation of the NLRP3-inflammasome resulting in generation of active caspase-1. Caspase-1 cleaves pro-IL-1β to IL-1β that can then be secreted by the cell. Substances such as cholesterol crystals, nigericin, alum and uric acid crystals have been shown to activate the inflammasome [14,16–18].

Some substances can provide both signal-1 and 2. These include glucose, ER stress-inducing drugs such as thapsigargin and tunicamycin, and the mitochondrial oxidative stress-inducing drug rotenone [13,19,20]. It has been suggested that increased reactive oxygen species (ROS) accumulation inside the cell causes dissociation of the thioredoxin-interacting protein (TXNIP) from the antioxidant protein thioredoxin resulting in its activation. TXNIP then binds to NLRP3 to stimulate its activation [13,20]. In

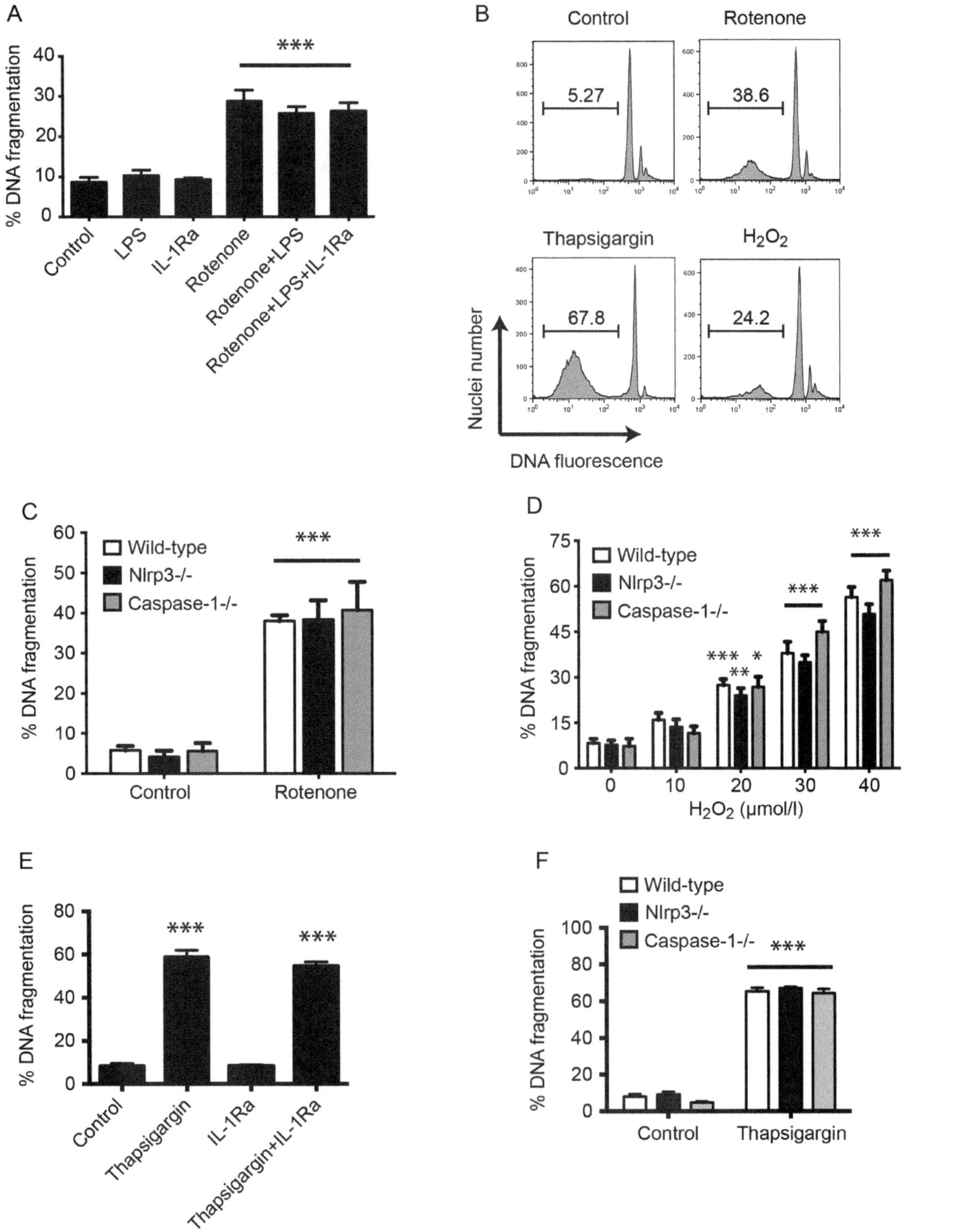

Figure 1. Oxidative and ER stress induced islet cell death is not mediated by inflammasome activation. (**A**) NIT cells were cultured in control medium, or medium containing 100 ng/mL LPS, 5 μg/mL IL-1Ra or 100 nM rotenone for 2 days. The frequency of cells undergoing DNA fragmentation was measured by flow cytometry. Results are mean+SEM of n≥3 independent experiments. ***p<0.001 Rotenone treated NIT cell groups vs controls without rotenone. (**B**) Representative FACS profiles of DNA fragmentation in islets after treatment with 100 nM rotenone for 2 days, 20 μM H_2O_2 for 2 days or 5 μM thapsigargin for 5 days. The percentage of islet cells with fragmented nuclei is indicated. (C) C57BL/6, Nlrp3$^{-/-}$ or caspase-1$^{-/-}$ islets were cultured in control medium or medium containing 100 nM rotenone for 2 days. Control islets were incubated in medium containing an equal volume of solvents. Results are mean+SEM of n≥3 independent experiments. ***p<0.001 rotenone treated islets vs controls of

same genotype. (**D**) DNA fragmentation was measured by flow cytometry after incubation of wild-type C57BL/6, Nlrp3$^{-/-}$ or caspase-1$^{-/-}$ islets with 0–40 µM H_2O_2 for 2 days. Results are mean+SEM of n≥3 independent experiments. *p<0.05, **p<0.01, ***p<0.001 H_2O_2 treated islets vs untreated islets of same genotype. (**E, F**) Wild-type islets (E), or wild-type, Nlrp3$^{-/-}$ or caspase-1$^{-/-}$ islets (F) were cultured in control medium, or medium containing 5 µg/mL IL-1Ra or 5 µM thapsigargin for 5 days. Control islets were incubated in medium containing an equal volume of solvents. The frequency of cells undergoing DNA fragmentation was measured by flow cytometry. Results are mean+SEM of n≥3 independent experiments. ***p< 0.001 Thapsigargin treated islets vs controls without thapsigargin of same genotype. No significant differences were observed between the different genotypes.

addition to IL-1β mediated apoptosis, inflammasome activation can lead to caspase-1-dependent death of the IL-1β-producing cell called pyroptosis. Pyroptosis is characterised by DNA fragmentation, cellular swelling and formation of pores in the plasma membrane [15].

Activation of ER stress molecules IRE1α and PERK, as well as high concentrations of glucose can also activate the inflammasome via TXNIP upregulation [9,12,13,19,21]. However, the involvement of TXNIP in inflammasome activation is controversial. While suppression of TXNIP in the INS-1 beta cell line reduced thapsigargin toxicity [21], mouse islets lacking TXNIP were not protected across a range of concentrations tested [22].

Macrophages express inflammasome components including NLRP3, ASC and caspase-1, and show inflammasome activation, in response to pathogens [18,23,24]. However, while their expression in pancreatic islets has been demonstrated, it is at much lower levels when compared with bone marrow derived macrophages [13,25]. In addition, although the expression of TLR4 and IL-1β has been reported previously in human and mouse islets and MIN6 beta cells [25–29], inflammasome activation has not convincingly been demonstrated in beta cells. In addition, a recent study reported absence of TLR4 in rat beta cells, making the understanding of the role for these factors in beta cells more complicated [29]. Therefore, we decided to directly examine whether deletion of inflammasome components in islets results in protection from cell death. Our results show that deletion of inflammasome components does not affect the apoptosis of islet cells in response to glucose or chemicals that induce oxidative or ER stress, and that activation of NLRP3 in beta cells does not contribute to IL-1β production and islet cell death.

Methods

Mice

Mice deficient in caspase-1 were obtained from Jackson Laboratories on a NOD/Lt background (NOD.129S2(B6)-*Casp1^{tm1Sesh}*/LtJ) and backcrossed onto the C57BL/6 genetic background for 10 generations by Dr Ben Croker (WEHI). NLRP3 hypomorphic mutant mice (Nlrp3^{A350VneoR}) were generated by Dr Susannah Brydges and Dr Hal Hoffman (UCSD) [30]. These mice are effectively a loss of function mutation when bred as homozygous mutants without Cre recombinase. For simplicity of nomenclature, these mice have been called NLRP3$^{-/-}$ in this article. NLRP3$^{A350V/A350V}$ mutant mice were bred with RIP-Cre from Jackson Laboratories (B6.Cg-Tg(Ins2-cre)^{25Mgn}/J) to generate mice with gain of function NLRP3 activity in the pancreatic beta cells. All animal experiments were carried out in compliance with the 'Australian code for the care and use of animals for scientific purposes' and were approved by the St Vincent's Hospital Animal Ethics Committee (AEC protocol number 016/11).

Reagents

D-glucose (used at 33.3 mM) and D-ribose (50 mM) were purchased from Invitrogen (Invitrogen Corporation, Grand Island, New York) and Sigma-Aldrich (St Louis, MO) respectively. Similar concentrations of L-glucose or L-ribose were used as hyperosmolarity controls (data not shown) [6]. Thapsigargin (used at 5 µM) was purchased from Calbiochem (Darmstadt, Germany) and IL-1Ra (used at 5 µg/mL) was obtained from Amgen (Thousand Oaks, CA) [31]. Hydrogen peroxide (H_2O_2, Sigma-Aldrich) was used at 0–40 µM, Rotenone (Sigma-Aldrich) at 100 nM and lipopolysaccharide (LPS, Sigma-Aldrich L2630, from *Escherichia coli* 0111:B4) at 100 nM. These reagents were tested across a range of concentrations in wild-type islets to determine the optimal concentration to use for further experiments. Control conditions contained the same concentration of diluent (DMSO for rotenone and ethanol for thapsigargin) in complete medium. Palmitate (Sigma-Aldrich) was used at 1 mM coupled to fatty acid free 1% BSA (Roche, Mannheim, Germany) and was added to 10% serum containing medium for islet treatment.

Islet isolation, culture and DNA fragmentation assay

Islets of Langerhans were isolated as described previously [32]. Islets were washed, hand-picked, and cultured overnight at 37°C in 5% CO_2 in CMRL medium-1066 (Invitrogen) supplemented with 100 units/mL penicillin, 100 µg/mL streptomycin, 2 mmol/L glutamine, and 10% FCS (JRH Biosciences, Lenexa, KS) (referred to below as complete CMRL). CMRL medium contains 5.5 mM glucose that is similar to the glucose concentrations islets are exposed to *in vivo*. Following incubation with different reagents, 100 islets per sample were trypsinized and DNA fragmentation was analyzed by staining with propidium iodide and flow cytometric analysis as previously described [6]. NIT-1 cells are derived from transgenic NOD mice expressing the SV40 large T antigen in pancreatic beta cells under control of the rat insulin promoter [33]. Cells were maintained in 10% CO_2 at 37°C in DMEM (Invitrogen) supplemented with non-essential amino acids, antibiotics and 10% FCS.

PCR genotyping

DNA was prepared from purified islets, and subjected to PCR with the following primers: Nlrp3^{A350V}F: CCCTGCATTTTGTTGTTGTTG, Nlrp3^{A350V}R: CCTGCTTCTCACATGTCGTC, NeoF: GAAGCGGGAAGG-GACTGGCTGCTA and NeoR: CGGGAGCGGCGATACCG-TAAAGC.

ELISA

After isolation, islets were cultured overnight at 37°C in complete CMRL medium. Next day, 400 islets/sample were hand-picked and cultured in 1 mL of complete CMRL medium for 2.5 days. Alternatively, macrophages were prepared by culturing mouse bone marrow in M-CSF for 7 days, re-plated at 10^5 per well in 96 well plates, followed by stimulation with palmitate conjugated to BSA overnight. At the conclusion of culture, the supernatant was removed and stored at −80°C for later use. ELISA for IL-1β was performed according to manufacturers instructions (R & D Systems, Minneapolis, MN).

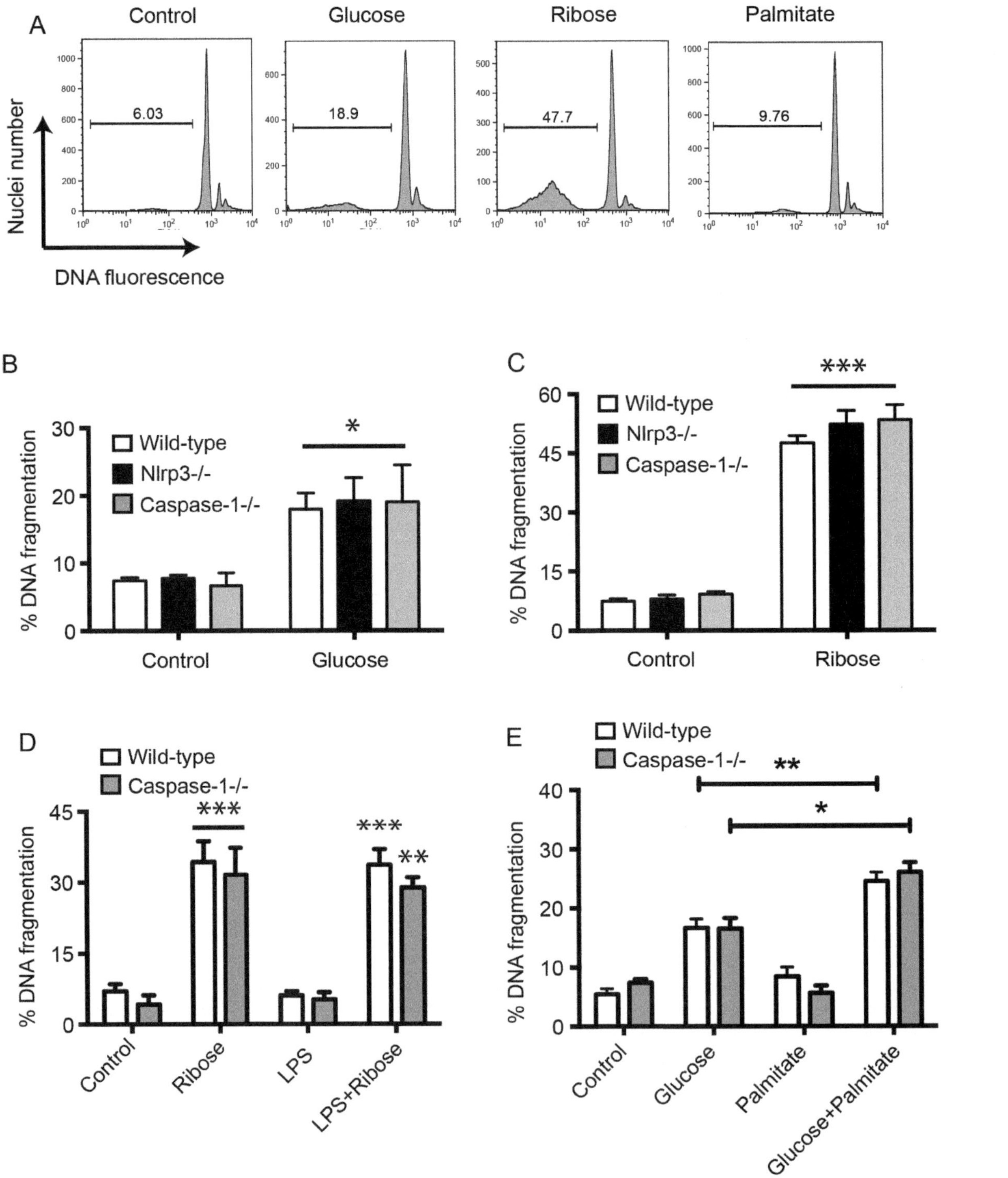

Figure 2. Glucose toxicity of islet cells is not mediated by inflammasome activation. (A) Representative FACS profiles of DNA fragmentation in islets after treatment with 33.3 mM glucose for 6 days, 50 mM ribose for 4 days or 1 mM palmitate conjugated to 1% BSA for 6 days. The percentage of islet cells with fragmented nuclei is indicated. (B, C) DNA fragmentation was measured by flow cytometry after incubation of wild-type C57BL/6, Nlrp3$^{-/-}$ or caspase-1$^{-/-}$ islets with 33.3 mM glucose for 6 days (B) or 50 mM ribose for 4 days (C). Control islets were incubated in medium containing 5.5 mM glucose. Results are mean+SEM of n≥4 independent experiments. *$p<0.05$ glucose treated islets vs control islets of same genotype. ***$p<0.001$ ribose treated islets vs control islets of same genotype. (D) DNA fragmentation was measured by flow cytometry after incubation of wild-type or caspase-1$^{-/-}$ islets with 50 mM ribose or 100 ng/mL LPS for 3 days. Results are mean+SEM of n = 3 independent experiments. **$p<0.01$, ***$p<0.001$ ribose treated islets vs controls of the same genotype without ribose. (E) DNA fragmentation was measured by flow cytometry after incubation of wild-type or caspase-1$^{-/-}$ islets with 33.3 mM glucose or 1 mM palmitate conjugated to 1% BSA for 6 days. Control islets were incubated in a medium containing 5.5 mM glucose and 1% BSA. Results are mean+SEM of n = 3 independent experiments. *$p<0.05$ glucose vs glucose+palmitate-treated caspase-1$^{-/-}$ islets, **$p<0.01$ glucose vs glucose+palmitate-treated wild-type islets.

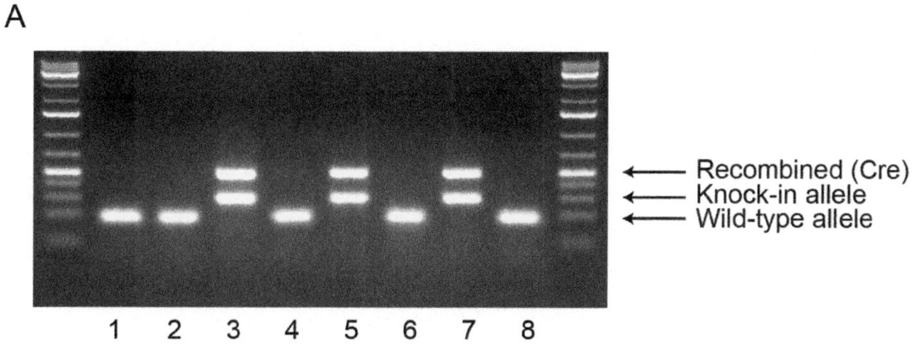

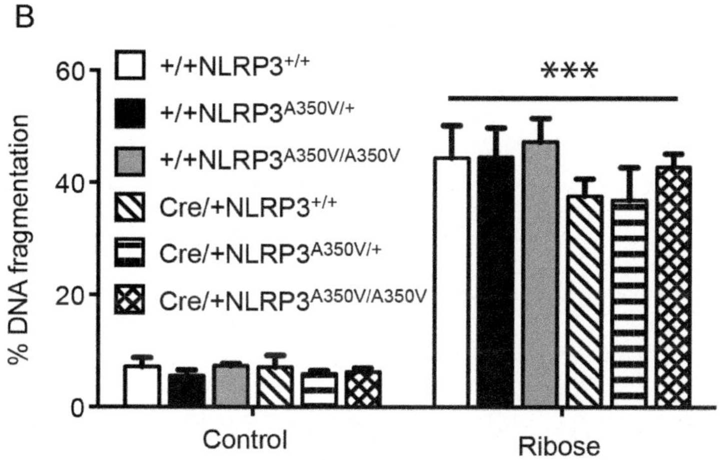

Figure 3. NLRP3 activating mutation does not increase islet glucose toxicity. (**A**) DNA was isolated from 300–500 islets of the indicated genotypes and PCR was performed using appropriate primers. Islets isolated from wild-type (+/+ $NLRP3^{+/+}$) mice were used for lanes 1, 2, 4, 6 and 8, while lanes 3, 5 and 7 show genotyping results for islets expressing mutant NLRP3 (Cre/+ $NLRP3^{A350V/A350V}$). (**B**) Islets from wild-type mice, floxed mice with activating NLRP3 mutation knocked into NLRP3 locus ($NLRP3^{A350V/+}$ or $NLRP3^{A350V/A350V}$), and mice expressing Cre recombinase in beta cells (Cre/+) were treated with 50 mM ribose for 4 days. DNA fragmentation was measured by flow cytometry. Results are mean+SEM of n = 3 independent experiments. ***$p<0.001$ ribose-treated islets vs control islets of the same genotype.

Western blotting

Four hundred islets/sample were cultured in 1 mL of 1% BSA containing complete CMRL medium and treated with a combination of LPS (100 nM), glucose (33.3 mM) and palmitate (1 mM) for 2.5 days. Control islets were cultured in the same medium containing 100 nM LPS. Islet lysates were then prepared in RIPA buffer as described previously [34]. Macrophages were also treated with the same concentration of LPS, glucose and palmitate for 24 h then lysed directly in reducing sample buffer. Western blotting was performed using anti-caspase-1 (p20) from Adipogen (Casper-1, Adipogen International, San Diego, CA).

Lactate dehydrogenase (LDH) assay

Islets were cultured and treated as described above for the western blotting experiments. Supernatants were obtained at the end of culture and LDH concentrations were determined. A commercially available kit from Roche was used for LDH assay.

Statistical analysis

Statistical analysis was performed using GraphPad Prism Software (San Diego, CA). All data shown as bar graphs are mean + SEM. Data were analysed by one-way or two-way ANOVA with Bonferroni's or Dunnett's post test for comparison of multiple columns (as appropriate).

Results

Oxidative and ER stress do not cause NLRP3 inflammasome mediated islet cell death

Oxidative stress induced cell death is not inflammasome mediated. Oxidative stress has been shown to cause activation of the NLRP3 inflammasome [20]. In particular, there is evidence that mitochondrial ROS is a potent activator of the NLRP3 complex in non-islet cells [20]. We studied the role of inflammasome activation in mitochondrial ROS mediated cell death in both beta cell lines and primary islets. Oxidative stress-sensitive NIT-1 beta cells were treated with 100 nM rotenone to induce mitochondrial ROS production. Although rotenone is able to provide both signal 1 and 2, we nevertheless pre-treated cells with 100 ng/mL LPS to maximally increase cellular stores of pro-IL-1β. Cell death was quantified by DNA fragmentation, which is common to all known forms of programmed cell death [15]. Inhibition of IL-1β activity with the receptor antagonist for IL-1β (IL-1Ra, 5 μg/mL) did not provide any protection from rotenone induced DNA fragmentation. This shows that mitochondrial ROS does not cause IL-1R-dependent death of beta cells (Figure 1A).

It remains possible that inflammasome activation could occur in islet resident macrophages or other non-beta cells, resulting in IL-1β secretion and toxicity to beta cells. Therefore, islets isolated

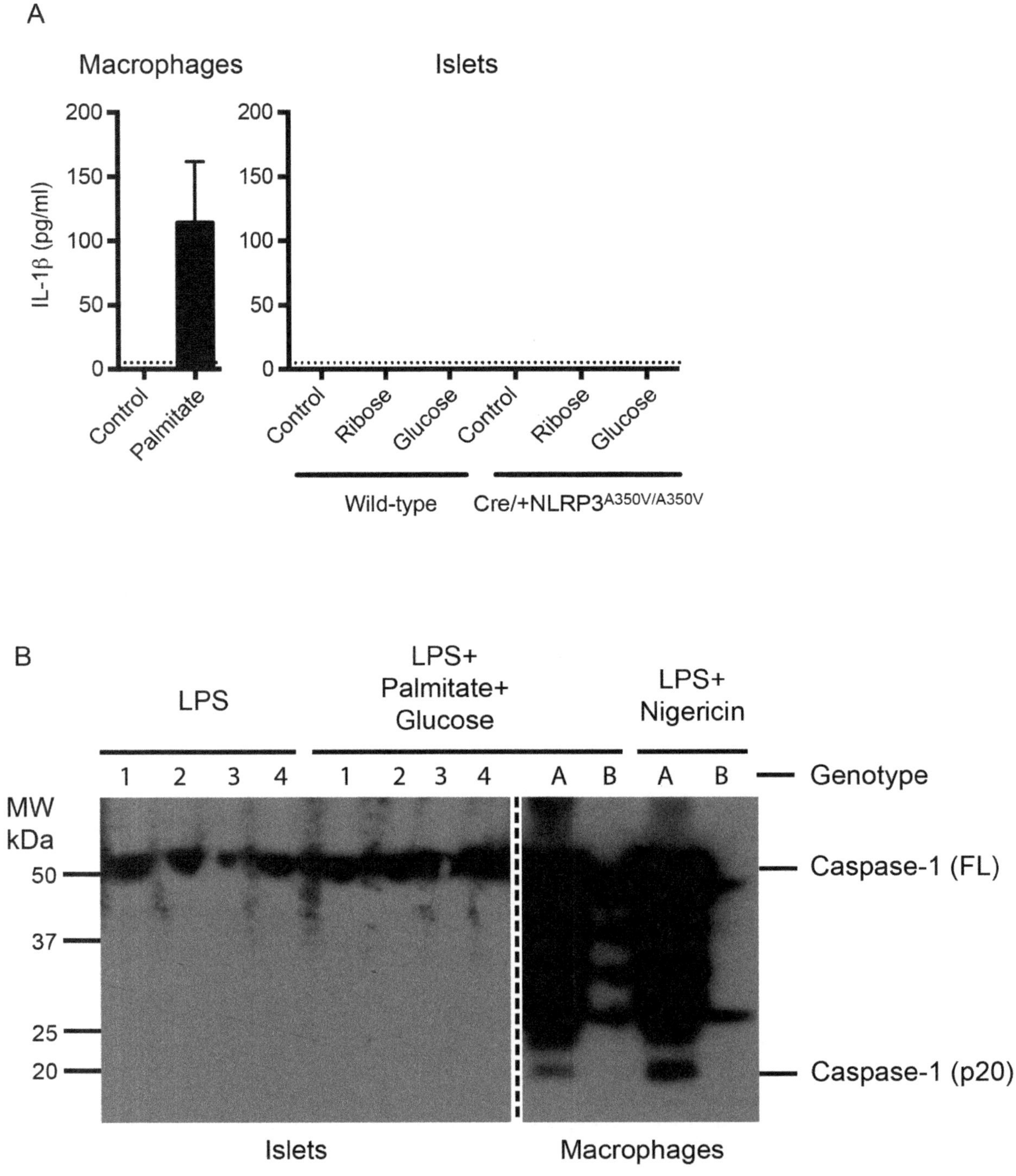

Figure 4. NLRP3 activating mutation does not induce caspase-1 cleavage in islets. (**A**) Four hundred islets isolated from wild-type or Cre/+ $NLRP3^{A350V/A350V}$ mice were cultured in 1 mL of medium containing 33.3 mM glucose or 50 mM ribose for 2.5 days and IL-1β concentration in the supernatant was quantified by ELISA. Control islets were incubated in a medium containing 5.5 mM glucose. Macrophages treated overnight with 1 mM palmitate were used as positive control. Results are mean+SEM of n = 3–4 independent experiments. (**B**) Four hundred islets/sample were cultured in control medium containing 1% BSA and 100 nM LPS or a medium containing 1% BSA, 100 nM LPS, 33.3 mM glucose and 1 mM palmitate for 2.5 days. Lysates were prepared in RIPA buffer and western blotting for caspase-1 (full length, FL or cleaved, p20) was performed. Genotypes 1: +/+ $NLRP3^{+/+}$, 2: Cre/+ $NLRP3^{+/+}$, 3: +/+ $NLRP3^{A350V/A350V}$ and 4: Cre/+ $NLRP3^{A350V/A350V}$. As a positive control, macrophages from wild-type (A) and caspase-$1^{-/-}$ (B) mice were treated with LPS+BSA+Glucose+Palmitate for 24 h or LPS+Nigericin for 1 h.

from wild-type, $Nlrp3^{-/-}$ and caspase $1^{-/-}$ mice were exposed to rotenone for 2 days. DNA fragmentation was similar in wild-type and NLRP3 or caspase-1 deficient islets suggesting that mitochondrial oxidative stress causes islet cell death through inflammmasome independent pathways (Figure 1B & C).

To further investigate the possible contribution of oxidative stress, islets from both $Nlrp3^{-/-}$ and Caspase-$1^{-/-}$ mice were exposed to cytosolic ROS by culturing them with hydrogen peroxide. Hydrogen peroxide-induced DNA fragmentation was similar in wild-type and knock-out islets, indicating that the NLRP3 inflammasome is not a mediator of islet cell death induced by cytosolic ROS (Figure 1D).

ER stress-induced death is not inflammasome mediated. To examine if ER stress could cause inflammasome mediated islet cell death, wild-type islets were treated with thapsigargin and IL-1Ra. The addition of IL-1Ra did not protect islets from thapsigargin toxicity (Figure 1B & E) indicating that ER stress induced islet cell death is not mediated by IL-1β secretion. For further confirmation, Nlrp3$^{-/-}$ and caspase-1$^{-/-}$ islets were incubated with 5 μm thapsigargin for 5 days. No difference in DNA fragmentation was observed between wild-type and Nlrp3$^{-/-}$ or caspase-1$^{-/-}$ islets, demonstrating that ER stress-induced islet toxicity does not involve activation of the NLRP3 inflammasome complex (Figure 1F).

A high concentration of glucose does not cause NLRP3 inflammasome mediated islet cell death

The possibility that glucose-induced death could be due to inflammasome activation in islets was studied. In addition to glucose, we tested toxicity of islets to the reducing sugar ribose, which is metabolized through the pentose phosphate pathway and glycolysis, and similar to glucose, induces islet cell death through glycation and formation of ROS [35]. Islets lacking functional NLRP3 or caspase-1 were not protected from glucose or ribose toxicity (Figure 2A, B & C). To rule out any possibility of a weak signal-1, islets were also cultured in the presence of LPS to increase the concentration of pro-IL-1β. However, addition of LPS did not increase ribose-induced islet cell death and compared with wild-type islets, we did not see any protection in LPS+ribose treated caspase1$^{-/-}$ islets (Figure 2D).

Whole islets were also treated with 1 mM palmitate (conjugated to 1% BSA in medium containing 10% serum) in the presence or absence of 33.3 mM glucose to study if the combined presence of glucose and palmitate (glucolipotoxicity) is required to cause significant IL-1β secretion. We did not see substantial cell death in whole islets after exposure to palmitate alone, however, exposure of islets to palmitate and glucose together caused greater cell death than each reagent alone. Deletion of caspase-1 from islets did not provide any protection from glucolipotoxicity (Figure 2E). These results, together with our previous findings that deficiency of IL-1 receptors did not protect islets from glucose toxicity [6], show that neither activation of the NLRP3-inflammasome complex, nor IL-1β production mediate islet cell death in response to high concentrations of glucose.

Beta-cell specific NLRP3 activating mutation does not result in IL-1β production or amplify glucose toxicity

To determine whether constitutive activation of NLRP3 in beta cells would exacerbate glucose toxicity, we isolated islets from mice with a human NLRP3 activating mutation knocked-in to the mouse NLRP3 locus (NLRP3$^{A350V/A350V}$) [30]. This mutation is controlled by the Cre enzyme, so by crossing the NLRP3$^{A350V/A350V}$ mutant mice to mice expressing Cre under control of the rat insulin promoter (RIP-Cre), we generated mice with mutant NLRP3 expressed at endogenous levels, but only in pancreatic beta cells, denoted NLRP3$^{A350V/A350V}$/RIP-cre. Successful expression of mutant NLRP3 in mouse beta cells was confirmed by PCR using DNA from purified islets (Figure 3A). This mutation is clearly active, as mice crossed to LysM-Cre do not survive beyond 14 days, however NLRP3$^{A350V/A350V}$/RIP-cre mice do not have any overt phenotype (data not shown).

If beta cells are able to produce significant IL-1β in response to glucose or ribose exposure, then islets with constitutive NLRP3 should produce even greater amounts of IL-1β resulting in amplification of glucose or ribose toxicity. Contrary to this expectation, we observed that islets with mutant NLRP3 did not show increased ribose toxicity when compared with controls (Figure 3B).

We measured the IL-1β produced by culturing 400 islets in 1 ml culture medium with 50 mM ribose or 33.3 mM glucose for 2.5 days. It has previously been shown that high concentrations of glucose neither induce IL-1β seretion nor affect its expression in cultured dendritic cells [18]. Therefore, wild-type macrophages cultured with palmitate were included as a positive control. Supernatant was collected and IL-1β concentration was determined by ELISA. Similar to our previous finding [6], we were unable to detect any IL-1β in supernatants from wild-type islets treated with ribose or glucose. In addition, the IL-1β concentration remained below the threshold for detection even in the supernatants from ribose or glucose treated NLRP3 mutant islets (Figure 4A). In contrast, 113.8±47.9 pg/mL of IL-1β (n = 4) was detected in the supernatants of palmitate treated macrophages (Figure 4A). To test glucolipotoxicity, islets were treated with 33.3 mM glucose and 1 mM palmitate conjugated to 1% BSA in the presence of 100 nM LPS for 2.5 days. Control islets were treated with LPS and 1% BSA. IL-1β was detectable at extrememly low levels in supernatants from wild-type islets. In NLRP3 mutant islets, this IL-1β concentration was unchanged (Figure S1A). These results indicate that beta-cells lack the ability to produce sufficient IL-1β that could make a noticeable contribution to glucose plus palmitate-mediated cell death.

Beta-cell specific NLRP3 activating mutation does not cause caspase-1 cleavage in islets

Finally, we examined if intrinsic NLRP3 activation could cause caspase-1 cleavage and, therefore, result in pyroptosis of beta cells. Pro-caspase-1 was detected in LPS treated islets, but addition of glucose and palmitate to the culture medium did not induce its cleavage in either wild-type islets or those expressing mutant NLRP3 (Figure 4B). In contrast, caspase-1 cleavage was observed in macrophages primed with LPS and then treated with glucose and palmitate, or nigericin as positive control (Figure 4B). DNA fragmentation is a feature of pyroptosis and we did not previously see a reduction in DNA fragmentation in caspase-1-deficient islets (Figure 2B–E). To confirm this we used an alternative cell death assay to measure glucolipotoxicity in islets by measuring lactate dehydrogenase (LDH) concentration. Treatment with glucose and palmitate for 2.5 days resulted in increased LDH concentration in the supernatant, but cell death was similar in wild-type as well as mutant NLRP3 islets (Figure S1B). These findings indicate that glucolipotoxicity does not induce pyroptosis of beta cells.

Discussion

In this study, we have investigated whether exposure of islets to stress could cause inflammasome mediated islet cell death. We did not find a significant role for the NLRP3 inflammasome or IL-1β in islet cell killing in response to high glucose concentrations or chemicals that induce ER or oxidative stress. Previous studies showed that exposure of human islets to 33.3 mM glucose for 5 days resulted in a very low level of IL-1β secretion [9,12]. Similarly in mouse islets, high glucose concentrations or the ER stress-inducing drug thapsigargin caused a modest amount of IL-1β production (20–40 pg/mL) that was abrogated in islets lacking NLRP3 or TXNIP [13,19]. These studies all suggested that the IL-1β produced as a result of incubation with glucose or thapsigargin was responsible for islet cell killing. However, the crucial experiment of incubating islets lacking inflammasome components or IL-1 receptors with the toxic stimulus was not done. When we performed this experiment, we found that loss of functional inflammasome factors, NLRP3 or caspase-1, did not inhibit glucose toxicity, ER stress or oxidative stress-mediated death of islet cells. These findings agree with previous data demonstrating that islets lacking IL-1 receptors are not protected from glucose toxicity [6]. Consistent with our results, another study showed that treatment of isolated human islets with high glucose concentration for up to 7 days neither induced IL-1β secretion nor affected IL-1β gene expression [25]. Our data are also in agreement with the concept that mouse islets are not susceptible to killing by IL-1β on its own, but require the addition of IL-1β+IFNγ to induce significant cell death [36–38].

Nevertheless, differences in the source (human vs. mouse) of islets, the number of islets, culture conditions, drug concentrations, differences between various beta-cell lines and assays used to measure cell death and inflammasome activity can not be ruled out as possible explanations for the differences between our data and results from some of the previous studies. For example, Maedler and colleagues cultured human islets for 2 days on extracellular matrix coated plates allowing the cells to attach and spread before treating islets with glucose that led to IL-1β secretion [12]. On the other hand we, and others, cultured islets in non-coated petri dishes and treated them with various reagents after 24 hours of isolation [25]. The DNA fragmentation assay we used in our experiments has the advantage that we analyze 10,000 cells per sample, which makes this a highly sensitive method for picking up subtle differences in treatments or genotypes. Although trypsinizing cells does induce some cell death (background DNA fragmentation always below 10%), the proportion of dead cells as a result of trypsinizing was consistent across all treatment groups. Nevertheless, we cannot rule out the possibility that small increases in islet cell death induced by IL-1β could have been masked by cell death induced by trypsinization. Another caveat is the use of LPS in our study to induce signal 1 in islets, where it is possible that other agents, such as Pam2CSK4 may provide a better signal [29]. The story may be further complicated during diabetes *in vivo* during which macrophages recuited to the islet could increase IL-1β production [24].

Compared to macrophages, the expression of NLRP3 inflammasome components is very low in islet beta cells [13], and beta cells produced only a modest amount of IL-1β after glucose treatment [13]. Even when we used islets with beta-cell specific expression of a NLRP3 activating mutation, IL-1β production was almost undetectable and glucose toxicity remained similar to wild-type islets. While islet resident macrophages may be a potential source of islet IL-1β, deficiency of either NLRP3 or caspase-1, both components of the NLRP3 inflammasome required for IL-1β processing in macrophages, did not affect *in vitro* glucose toxicity mediated killing of whole islets, suggesting the contribution of resident macrophage IL-1β is not significant in this context.

Animal and human studies have shown that high glucose concentration induces oxidative and ER stress in islets [10,11,35,39–41]. Increased mitochondrial ROS production in non-islet cells such as THP1 cells or bone marrow derived macrophages (BMDMs) caused inflammasome activation and IL-1β release [20]. Similarly, IL-1β release from glucose treated islets was inhibited by treatment with ROS inhibitors [13]. ER stress induced by reagents such as thapsigarin and tunicamycin also caused IL-1β release from non-islet cells, beta cell lines and primary islets [19,21]. In addition to IL-1β release, TXNIP was upregulated by these chemicals. These studies concluded that ER and oxidative stress induced inflamasome activation in islets is mediated by TXNIP [13,19,21], however, a direct link between TXNIP and inflammasome activation was not formally demonstrated in islets. Treatment of wild-type BMDMs with islet amyloid poplypeptide (IAPP) caused IL-1β production and this was inhibited by NLRP3 deletion but loss of TXNIP had no effect [18]. Our findings that islets lacking functional NLRP3 or caspase-1 are not protected from thapsigargin or rotenone, and previous findings that TXNIP-deficient islets were not protected from thapsigargin or palmitate [22] indicate that TXNIP dependent inflammasome activation may not be important for islet toxicity. TXNIP-deficiency did protect islets from glucose toxicity [22], but this may be due to increased antioxidant activity of thioredoxin in the absence of TXNIP, independent of any effects on the inflammasome.

In conclusion, our results clearly show that activation of the inflammasome does not mediate islet cell death in response to high glucose concentrations, ER or oxidative stress. It could be possible that islets do produce small amounts of IL-1β in response to these stimuli, but by itself this is not enough to cause substantial death *in vitro*. Small amounts of IL-1β may increase the expression of chemokines such as CXCL1 in beta cells, thereby facilitating the recruitment of immune cells in the islets [42]. More work will be required to determine if inflammasome activation downstream of glucose toxicity or ER stress has any adverse effects on beta cell function including insulin biosynthesis and glucose stimulated insulin release. In type 2 diabetes, there is an increased concentration of substances such as glucose, saturated and unsaturated fatty acids, IAPP, and circulating cytokines. It is likely that these are also able to induce IL-1β production by macrophages in adipose tissue and other tissues such as islets. However, our data show that while these circulating factors affect islet viability directly, it is less likely that they cause islet cell death through production of IL-1β in a beta cell autonomous manner.

Supporting Information

Figure S1 Activation of NLRP3 in beta cells does not alter IL-1β production and cell death. (A) Four hundred islets/sample were isolated from the mice of indiacted genotypes and cultured in 1 mL of medium containing 100 nM LPS, 33.3 mM glucose and 1 mM palmitate conjugated to 1% BSA for 2.5 days. IL-1β secretion into supernatant was quantified by ELISA. Control islets were incubated in a medium containing 5.5 mM glucose, 1% BSA and 100 nM LPS. n = 1 experiment. **(B)** Cells were treated as in A, then LDH concentration in the supernatant was quantified by LDH assay. Control islets were incubated in a medium containing 5.5 mM glucose, 1% BSA and 100 nM LPS. n = 1 experiment.

Acknowledgments

We thank Dr Ben Croker (WEHI) for caspase1$^{-/-}$ mice, Drs Susannah Brydges and Hal Hoffman (UCSD) for Nlrp3$^{-/-}$ mice and Mr Sam Thorburn, Ms Daniella Novembre-Cycon, Ms Rebecca Branch and Mr Anthony Gomes (St Vincent's Institute) for genotyping and animal husbandry.

Author Contributions

Conceived and designed the experiments: JAW TWK SLM HET. Performed the experiments: JAW ENG SF LE SLM. Analyzed the data:

JAW ENG SLM HET. Contributed to the writing of the manuscript: JAW ENG TWK SLM HET.

References

1. Sakuraba H, Mizukami H, Yagihashi N, Wada R, Hanyu C, et al. (2002) Reduced beta-cell mass and expression of oxidative stress-related DNA damage in the islet of Japanese Type II diabetic patients. Diabetologia 45: 85–96.
2. Clark A, Wells CA, Buley ID, Cruickshank JK, Vanhegan RI, et al. (1988) Islet amyloid, increased A-cells, reduced B-cells and exocrine fibrosis: quantitative changes in the pancreas in type 2 diabetes. Diabetes Res 9: 151–159.
3. Butler AE, Janson J, Bonner-Weir S, Ritzel R, Rizza RA, et al. (2003) Beta-cell deficit and increased beta-cell apoptosis in humans with type 2 diabetes. Diabetes 52: 102–110.
4. Oyadomari S, Koizumi A, Takeda K, Gotoh T, Akira S, et al. (2002) Targeted disruption of the Chop gene delays endoplasmic reticulum stress-mediated diabetes. J Clin Invest 109: 525–532.
5. Song B, Scheuner D, Ron D, Pennathur S, Kaufman RJ (2008) Chop deletion reduces oxidative stress, improves beta cell function, and promotes cell survival in multiple mouse models of diabetes. J Clin Invest 118: 3378–3389.
6. McKenzie MD, Jamieson E, Jansen ES, Scott CL, Huang DC, et al. (2010) Glucose induces pancreatic islet cell apoptosis that requires the BH3-only proteins Bim and Puma and multi-BH domain protein Bax. Diabetes 59: 644–652.
7. Mellado-Gil JM, Aguilar-Diosdado M (2004) High glucose potentiates cytokine- and streptozotocin-induced apoptosis of rat islet cells: effect on apoptosis-related genes. J Endocrinol 183: 155–162.
8. Federici M, Hribal M, Perego L, Ranalli M, Caradonna Z, et al. (2001) High glucose causes apoptosis in cultured human pancreatic islets of Langerhans: a potential role for regulation of specific Bcl family genes toward an apoptotic cell death program. Diabetes 50: 1290–1301.
9. Maedler K, Spinas GA, Lehmann R, Sergeev P, Weber M, et al. (2001) Glucose induces beta-cell apoptosis via upregulation of the Fas receptor in human islets. Diabetes 50: 1683–1690.
10. Wali JA, Rondas D, McKenzie MD, Zhao Y, Elkerbout L, et al. (2014) The proapoptotic BH3-only proteins Bim and Puma are downstream of endoplasmic reticulum and mitochondrial oxidative stress in pancreatic islets in response to glucotoxicity. Cell Death Dis 5: e1124.
11. Laybutt DR, Preston AM, Akerfeldt MC, Kench JG, Busch AK, et al. (2007) Endoplasmic reticulum stress contributes to beta cell apoptosis in type 2 diabetes. Diabetologia 50: 752–763.
12. Maedler K, Sergeev P, Ris F, Oberholzer J, Joller-Jemelka HI, et al. (2002) Glucose-induced beta cell production of IL-1beta contributes to glucotoxicity in human pancreatic islets. J Clin Invest 110: 851–860.
13. Zhou R, Tardivel A, Thorens B, Choi I, Tschopp J (2010) Thioredoxin-interacting protein links oxidative stress to inflammasome activation. Nat Immunol 11: 136–140.
14. Bergmann L, Kroncke KD, Suschek C, Kolb H, Kolb-Bachofern V (1992) Cytotoxic action of IL-1 beta against pancreatic islets is mediated via nitric oxide formation and is inhibited by NG-monomethyl-L-arginine. FEBS Lett 299: 103–106.
15. Kepp O, Galluzzi L, Zitvogel L, Kroemer G (2010) Pyroptosis - a cell death modality of its kind? Eur J Immunol 40: 627–630.
16. Ichinohe T, Pang IK, Kumamoto Y, Peaper DR, Ho JH, et al. (2011) Microbiota regulates immune defense against respiratory tract influenza A virus infection. Proc Natl Acad Sci U S A 108: 5354–5359.
17. Joosten LA, Netea MG, Mylona E, Koenders MI, Malireddi RK, et al. (2010) Engagement of fatty acids with Toll-like receptor 2 drives interleukin-1beta production via the ASC/caspase 1 pathway in monosodium urate monohydrate crystal-induced gouty arthritis. Arthritis Rheum 62: 3237–3248.
18. Masters SL, Dunne A, Subramanian SL, Hull RL, Tannahill GM, et al. (2010) Activation of the NLRP3 inflammasome by islet amyloid polypeptide provides a mechanism for enhanced IL-1beta in type 2 diabetes. Nat Immunol 11: 897–904.
19. Lerner AG, Upton JP, Praveen PV, Ghosh R, Nakagawa Y, et al. (2012) IRE1alpha induces thioredoxin-interacting protein to activate the NLRP3 inflammasome and promote programmed cell death under irremediable ER stress. Cell Metab 16: 250–264.
20. Zhou R, Yazdi AS, Menu P, Tschopp J (2011) A role for mitochondria in NLRP3 inflammasome activation. Nature 469: 221–225.
21. Oslowski CM, Hara T, O'Sullivan-Murphy B, Kanekura K, Lu S, et al. (2012) Thioredoxin-interacting protein mediates ER stress-induced beta cell death through initiation of the inflammasome. Cell Metab 16: 265–273.
22. Chen J, Fontes G, Saxena G, Poitout V, Shalev A (2010) Lack of TXNIP protects against mitochondria-mediated apoptosis but not against fatty acid-induced ER stress-mediated beta-cell death. Diabetes 59: 440–447.
23. Schroder K, Zhou R, Tschopp J (2010) The NLRP3 inflammasome: a sensor for metabolic danger? Science 327: 296–300.
24. Jourdan T, Godlewski G, Cinar R, Bertola A, Szanda G, et al. (2013) Activation of the Nlrp3 inflammasome in infiltrating macrophages by endocannabinoids mediates beta cell loss in type 2 diabetes. Nat Med 19: 1132–1140.
25. Welsh N, Cnop M, Kharroubi I, Bugliani M, Lupi R, et al. (2005) Is there a role for locally produced interleukin-1 in the deleterious effects of high glucose or the type 2 diabetes milieu to human pancreatic islets? Diabetes 54: 3238–3244.
26. Eguchi K, Manabe I, Oishi-Tanaka Y, Ohsugi M, Kono N, et al. (2012) Saturated Fatty Acid and TLR Signaling Link beta Cell Dysfunction and Islet Inflammation. Cell Metab 15: 518–533.
27. Goldberg A, Parolini M, Chin BY, Czismadia E, Otterbein LE, et al. (2007) Toll-like receptor 4 suppression leads to islet allograft survival. The FASEB Journal 21: 2840–2848.
28. Chan JY, Luzuriaga J, Bensellam M, Biden TJ, Laybutt DR (2013) Failure of the adaptive unfolded protein response in islets of obese mice is linked with abnormalities in beta-cell gene expression and progression to diabetes. Diabetes 62: 1557–1568.
29. Nackiewicz D, Dan M, He W, Kim R, Salmi A, et al. (2014) TLR2/6 and TLR4-activated macrophages contribute to islet inflammation and impair beta cell insulin gene expression via IL-1 and IL-6. Diabetologia 57: 1645–1654.
30. Brydges SD, Mueller JL, McGeough MD, Pena CA, Misaghi A, et al. (2009) Inflammasome-mediated disease animal models reveal roles for innate but not adaptive immunity. Immunity 30: 875–887.
31. Thomas HE, Angstetra E, Fernandes RV, Mariana L, Irawaty W, et al. (2006) Perturbations in nuclear factor-kappaB or c-Jun N-terminal kinase pathways in pancreatic beta cells confer susceptibility to cytokine-induced cell death. Immunol Cell Biol 84: 20–27.
32. McKenzie MD, Dudek NL, Mariana L, Chong MM, Trapani JA, et al. (2006) Perforin and Fas induced by IFNgamma and TNFalpha mediate beta cell death by OT-I CTL. Int Immunol 18: 837–846.
33. Hamaguchi K, Gaskins HR, Leiter EH (1991) NIT-1, a pancreatic beta-cell line established from a transgenic NOD/Lt mouse. Diabetes 40: 842–849.
34. Merry TL, Tran M, Stathopoulos M, Wiede F, Fam BC, et al. (2014) High-fat-fed obese glutathione peroxidase 1-deficient mice exhibit defective insulin secretion but protection from hepatic steatosis and liver damage. Antioxid Redox Signal 20: 2114–2129.
35. Tanaka Y, Tran PO, Harmon J, Robertson RP (2002) A role for glutathione peroxidase in protecting pancreatic beta cells against oxidative stress in a model of glucose toxicity. Proc Natl Acad Sci U S A 99: 12363–12368.
36. Gurzov EN, Germano CM, Cunha DA, Ortis F, Vanderwinden JM, et al. (2010) p53 up-regulated modulator of apoptosis (PUMA) activation contributes to pancreatic beta-cell apoptosis induced by proinflammatory cytokines and endoplasmic reticulum stress. J Biol Chem 285: 19910–19920.
37. Eizirik DL, Mandrup-Poulsen T (2001) A choice of death–the signal-transduction of immune-mediated beta-cell apoptosis. Diabetologia 44: 2115–2133.
38. Eizirik DL, Colli ML, Ortis F (2009) The role of inflammation in insulitis and beta-cell loss in type 1 diabetes. Nat Rev Endocrinol 5: 219–226.
39. Lipson KL, Ghosh R, Urano F (2008) The role of IRE1alpha in the degradation of insulin mRNA in pancreatic beta-cells. PLoS One 3: e1648.
40. Hou ZQ, Li HL, Gao L, Pan L, Zhao JJ, et al. (2008) Involvement of chronic stresses in rat islet and INS-1 cell glucotoxicity induced by intermittent high glucose. Mol Cell Endocrinol 291: 71–78.
41. Moore PC, Ugas MA, Hagman DK, Parazzoli SD, Poitout V (2004) Evidence against the involvement of oxidative stress in fatty acid inhibition of insulin secretion. Diabetes 53: 2610–2616.
42. Ribaux P, Ehses JA, Lin-Marq N, Carrozzino F, Boni-Schnetzler M, et al. (2007) Induction of CXCL1 by extracellular matrix and autocrine enhancement by interleukin-1 in rat pancreatic β-cells. Endocrinology 148: 5582–5590.

Permissions

All chapters in this book were first published in PLOS ONE, by The Public Library of Science; hereby published with permission under the Creative Commons Attribution License or equivalent. Every chapter published in this book has been scrutinized by our experts. Their significance has been extensively debated. The topics covered herein carry significant findings which will fuel the growth of the discipline. They may even be implemented as practical applications or may be referred to as a beginning point for another development.

The contributors of this book come from diverse backgrounds, making this book a truly international effort. This book will bring forth new frontiers with its revolutionizing research information and detailed analysis of the nascent developments around the world.

We would like to thank all the contributing authors for lending their expertise to make the book truly unique. They have played a crucial role in the development of this book. Without their invaluable contributions this book wouldn't have been possible. They have made vital efforts to compile up to date information on the varied aspects of this subject to make this book a valuable addition to the collection of many professionals and students.

This book was conceptualized with the vision of imparting up-to-date information and advanced data in this field. To ensure the same, a matchless editorial board was set up. Every individual on the board went through rigorous rounds of assessment to prove their worth. After which they invested a large part of their time researching and compiling the most relevant data for our readers.

The editorial board has been involved in producing this book since its inception. They have spent rigorous hours researching and exploring the diverse topics which have resulted in the successful publishing of this book. They have passed on their knowledge of decades through this book. To expedite this challenging task, the publisher supported the team at every step. A small team of assistant editors was also appointed to further simplify the editing procedure and attain best results for the readers.

Apart from the editorial board, the designing team has also invested a significant amount of their time in understanding the subject and creating the most relevant covers. They scrutinized every image to scout for the most suitable representation of the subject and create an appropriate cover for the book.

The publishing team has been an ardent support to the editorial, designing and production team. Their endless efforts to recruit the best for this project, has resulted in the accomplishment of this book. They are a veteran in the field of academics and their pool of knowledge is as vast as their experience in printing. Their expertise and guidance has proved useful at every step. Their uncompromising quality standards have made this book an exceptional effort. Their encouragement from time to time has been an inspiration for everyone.

The publisher and the editorial board hope that this book will prove to be a valuable piece of knowledge for researchers, students, practitioners and scholars across the globe.

List of Contributors

Suzanne C. Simmons
Department of Chemistry and Analytical Sciences, The Open University, Milton Keynes, United Kingdom

Dilson Silva and Celia M. Cortez
Institute of Mathematics and Statistics, Rio de Janeiro State University, Rio de Janeiro, Brazil

Edward A. McKenzie
Manchester Institute of Biotechnology, University of Manchester, Manchester, United Kingdom

Sirpa Salo and Sini Nurmenniemi
Department of Diagnostics and Oral Medicine, Institute of Dentistry, University of Oulu, Oulu, Finland

Juha Risteli
Institute of Diagnostics, Department of Clinical Chemistry, University of Oulu, Oulu, Finland

Paul E. C. Brenchley
Renal Research Group, University of Manchester, Manchester, United Kingdom

Sotiris Missailidis
Institute of Biophysics Carlos Chagas Filho, Federal University of Rio de Janeiro, Rio de Janeiro, Brazil

Hannaleena Jämsa, Carolina C. Bitu and Pia Nyberg
Department of Diagnostics and Oral Medicine, Institute of Dentistry, University of Oulu, Oulu, Finland
Medical Research Center and Oulu University Hospital, Oulu, Finland

Carlos E. B. deAlmeida
Department of Chemistry and Analytical Sciences, The Open University, Milton Keynes, United Kingdom
Laboratório de Radiobiologia, Instituto de Radioproteção e Dosimetria, Rio de Janeiro, Brazil

Tuula Salo
Department of Diagnostics and Oral Medicine, Institute of Dentistry, University of Oulu, Oulu, Finland
Medical Research Center and Oulu University Hospital, Oulu, Finland
Graduate Program in Estomatopatologia, Piracicaba Dental School, University of Campinas, Piracicaba, São Paulo, Brazil
Institute of Dentistry, University of Helsinki, Helsinki, Finland

Minjie Gao, Guang Yang, Lu Gao, Xiaosong Wu, Yi Tao, Xiaojing Hu, Jun Hou, Ying Han and Jumei Shi
Department of Hematology, Shanghai Tenth People's Hospital, Tongji University School of Medicine, Shanghai, China

Van S. Tompkins
Department of Pathology, University of Iowa Carver College of Medicine, Iowa City, Iowa, United States of America

Hongwei Xu and Fenghuang Zhan
Department of Internal Medicine, University of Iowa Carver College of Medicine, Iowa City, Iowa, United States of America

Laura B. Motta-Mena and Amanda Cordova
Dept. of Biological Sciences, University of Texas at El Paso, El Paso, Texas, United States of America

Oscar Ramirez and Kristine M. Garza
Dept. of Biological Sciences, University of Texas at El Paso, El Paso, Texas, United States of America
Border Biomedical Research Center, University of Texas at El Paso, El Paso, Texas, United States of America

Christelle Blanc-Bisson and Jessica Durrieu
CHU Bordeaux, Clinical Gerontology, Bordeaux, France

Christine Germain
CHU Bordeaux, Unité d'épidémiologie clinique, Bordeaux, France

Jérôme Dauba and Cyril Lahmar
Centre Hospitalier Layné-Mont de Marsan, Service Oncologie, Mont de Marsan, France

Eric Terrebonne
CHU Bordeaux, Service Hépato-gastro et oncologie digestive, Bordeaux, France

Cédric Lecaille
Polyclinique Bordeaux Nord, Service hépato gastroentérologie, Bordeaux, France

Joël Ceccaldi
Centre Hospitalier de Libourne, Libourne, France

Laurent Cany
Clinique Francheville, Périgueux, France

Sandrine Lavau-Denes
CHU Limoges, Unité de recherche clinique en oncologie médicale, Limoges, France

Nadine Houede, François Chomy and Mariane Fonck
Institut Bergonié, Departement d'oncologie médicale, Bordeaux, France

Adélaïde Doussau and Geneviéve Chene
CHU Bordeaux, Unité d'épidémiologie clinique, Bordeaux, France
INSERM, CIC-EC7, Bordeaux, France

Jean-Frédéric Blanc
CHU Bordeaux, Hépatologie, gastroentérologie and oncologie digestive, Bordeaux, France
Université Bordeaux, Bordeaux, France

Pierre Soubeyran
Université Bordeaux, Bordeaux, France
Institut Bergonié, Departement d'oncologie médicale, Bordeaux, France

Pierre Senesse
Universités Montpellier 1 et 3, Laboratoire Epsylon EA 4556, Montpellier, France
CLCC du Val d'Aurelle, Unité de Nutrition clinique et de gastroentérologie, Montpellier, France

Isabelle Bourdel-Marchasson
CHU Bordeaux, Clinical Gerontology, Bordeaux, France
CNRS, RMSB, UMR 5536, Bordeaux, France
Université Bordeaux, RMSB, UMR 5536, Bordeaux, France

Filippo Pietrantonio, Claudia Maggi, Francesca Ricchini, Filippo de Braud and Maria Di Bartolomeo
Department of Medical Oncology, Fondazione IRCCS Istituto Nazionale dei Tumori, Milan, Italy

Alessio Farcomeni
Department of Public Health and Infectious Diseases, Statistics Section, Sapienza University of Rome, Rome, Italy

Roberto Iacovelli and Antonella Palazzo
Department of Medical Oncology, Fondazione IRCCS Istituto Nazionale dei Tumori, Milan, Italy
PhD Program, Department of Radiology, Oncology and Human Pathology, Sapienza University of Rome, Rome, Italy

Zhen Li, Lianxian Cui, Yu Hu and Wei He
Department of Immunology, School of Basic Medicine, Peking Union Medical College; National Key Laboratory of Medical Molecular Biology, Institute of Basic Medical Sciences, Chinese Academy of Medical Sciences, Beijing, China

Wei Li, Yanmei Jiao, Hao Wu and Dexi Chen
Center for Infectious Diseases, Beijing You'an Hospital, Capital Medical University, Beijing, China

Ning Li
Beijing You'an Hospital, Capital Medical University, Beijing, China

M. Mar Díaz-Herrero, Pilar Carbonero-Aguilar and Juan Bautista
Departamento de Bioquímica y Biología Molecular, Facultad de Farmacia, Universidad de Sevilla, Sevilla, Spain

José A. del Campo and Manuel Romero-Gómez
Unidad de Gestión Clínica de Enfermedades Digestivas & Centro de Investigación Biomédica en Red de Enfermedades Hepáticas y Digestivas (CIBERehd), Hospital Universitario de Valme, Sevilla, Spain

José M. Vega-Pérez, Fernando Iglesias-Guerra and Ignacio Periñán
Departamento de Química Orgánica y Química Farmacéutica, Facultad de Farmacia, Universidad de Sevilla, Sevilla, Spain

Francisco J. Miñano
Unidad de Farmacología Experimental y Clínica (UFEC), Hospital Universitario de Valme, Universidad de Sevilla, Sevilla, Spain

Weiji Shi
Department of Epidemiology and Biostatistics, Memorial Sloan-Kettering Cancer Center, New York, NY, United States of America

Kim Salvaggio and Y. Pierre Gobin
Department of Neurosurgery, New York Presbyterian Hospital Weill Cornell Medical College, New York, NY, United States of America

Scott E. Brodie
Department of Ophthalmology, Mount Sinai Medical Center, New York, NY, United States of America

Ira J. Dunkel
Department of Pediatrics, Memorial Sloan-Kettering Cancer Center, New York, NY, United States of America
Department of Pediatrics, New York Presbyterian Hospital Weill Cornell Medical College, New York, NY, United States of America

Brian P. Marr and David H. Abramson
Department of Surgery, Memorial Sloan-Kettering Cancer Center, New York, NY, United States of America
Department of Ophthalmology, New York Presbyterian Hospital Weill Cornell Medical College, New York, NY, United States of America

Xuan Qiu, Yuanxin Liang, Roman Perez-Soler and Yiyu Zou
Departments of Medicine, Division of Medical Oncology, Albert Einstein College of Medicine, Bronx, NY, United States of America

Rani S. Sellers
Department of Pathology, Albert Einstein College of Medicine, Bronx, NY, United States of America

Alethéia L. Silveir and Demetrius A. M. Araújo
Departamento de Biotecnologia, Universidade Federal da Paraíba, João Pessoa, Paraíba, Brasil

Raquel C. Maia
Laboratório de Hemato-Oncologia Celular e Molecular, Instituto Nacional de Câncer (INCA), Rio de Janeiro, Rio de Janeiro, Brasil

Glaúcia V. Faheina-Martins
Departamento de Biotecnologia, Universidade Federal da Paraíba, João Pessoa, Paraíba, Brasil
Departamento de Biologia Molecular, Universidade Federal da Paraíba, João Pessoa, Paraíba, Brasil

Chih-Hsiang Chang and Che-Yu Hsu
Orthopaedic Research Center, Kaohsiung Medical University, Kaohsiung, Taiwan

Chau-Zen Wang and Mei-Ling Ho
Orthopaedic Research Center, Kaohsiung Medical University, Kaohsiung, Taiwan
Department of Physiology, College of Medicine, Kaohsiung Medical University, Kaohsiung, Taiwan

Je-Ken Chang
Orthopaedic Research Center, Kaohsiung Medical University, Kaohsiung, Taiwan
Department of Orthopaedics, College of Medicine, Kaohsiung Medical University Hospital, Kaohsiung Medical University, Kaohsiung, Taiwan
Department of Orthopaedics, Kaohsiung Municipal Ta-Tung Hospital, Kaohsiung, Taiwan

Justin L. McCarville, Heather J. Galipeau, Jennifer Jury and Elena F. Verdu
Farncombe Family Digestive Health Research Institute, McMaster University, Hamilton, Ontario, Canada

Yotam Nisemblat, Rinat Tabakman, Ad Cohen, Esmira Naftali, Bela Neiman and Efrat Halbfinger
BioLineRx, Ltd., Jerusalem, Israel

Joseph A. Murray
Division of Gastroenterology and Hepatology, Mayo Clinic, Rochester, Minnesota, United States of America

Arivarasu N. Anbazhagan and Pradeep K. Dudeja
Division of Gastroenterology and Hepatology, Department of Medicine, University of Illinois and Chicago and Jesse Brown VA Medical Center, Chicago, Illinois, United States of America

Alexander Varvak
Chromatography Unit, Scientific Equipment Center, The Mina and Everard Goodman Faculty of Life Sciences, Bar-Ilan University, Ramat-Gan, Israel

Jean-Christophe Leroux
Institute of Pharmaceutical Sciences, Department of Chemistry and Applied Biosciences, ETH Zürich, Zürich, Switzerland

Gábor Szalóki, Ágnes Tóth, Zolta´n Krasznai, Gábor Szabó and Katalin Goda
Department of Biophysics and Cell Biology, Faculty of Medicine, University of Debrecen, Debrecen, Hungary

Zoárd T. Krasznai
Department of Obstetrics and Gynecology, Faculty of Medicine, University of Debrecen, Debrecen, Hungary

Laura Vízkeleti and Margit Balázs
Department of Preventive Medicine, Faculty of Medicine, University of Debrecen, Debrecen, Hungary

Attila G. Szöllősi
Department of Physiology, Faculty of Medicine, University of Debrecen, Debrecen, Hungary

György Trencsényi, Imre Lajtos and Teréz Márián
Department of Nuclear Medicine, Faculty of Medicine, University of Debrecen, Debrecen, Hungary

István Juhász
Department of Dermatology, Faculty of Medicine, University of Debrecen, Debrecen, Hungary,
Department of Surgery and Operative Techniques, Faculty of Medicine, University of Debrecen, Debrecen, Hungary

Srikanth Mairpady Shambat, Axana Haggar, Mattias Svensson and Anna Norrby-Teglund
Karolinska Institutet, Centre for Infectious Medicine, Stockholm, Sweden

Francois Vandenesch and Gerard Lina
CIRI, International Center for Infectiology Research, LabEx Ecofect, Université Lyon 1, Inserm U1111, Ecole Normale Supérieure de Lyon, CNRS UMR5308, Centre National de Référence des Staphylocoques, Hospices civils de Lyon, Lyon, France

Willem J. B. van Wamel
Department of Medical Microbiology and Infectious Diseases, Erasmus Medical Centre, Rotterdam, Netherlands

Gayathri Arakere
Society for Innovation and Development, Indian Institute of Science Campus, Bangalore, India

Fu-Nan Cho
Department of Obstetrics and Gynecology, Kaohsiung Veterans General Hospital, Kaohsiung, Taiwan

Tsung-Hsien Chang and Chih-Wen Shu
Department of Medical Education and Research, Kaohsiung Veterans General Hospital, Kaohsiung, Taiwan

Ming-Chin Ko and Chien-Hui Hung
Department of Pediatrics, Kaohsiung Veterans General Hospital, Kaohsiung, Taiwan

Shuen-Kuei Liao
Graduate Institute of Cancer Biology and Drug Discovery and Center of Excellence for Cancer Research, Taipei Medical University, Taipei, Taiwan

Kang-Hsi Wu
Department of Pediatrics, Children's Hospital and School of Chinese Medicine, China Medical University Hospitals, Taichung, Taiwan

Ming-Sun Yu, Shyh-Jer Lin and Ying-Chung Hong
Haematology-Oncology Section, Department of Medicine, Kaohsiung Veterans General Hospital, Kaohsiung, Taiwan

Chien-Hsun Chen
Department of Radiation Oncology, Kaohsiung Veterans General Hospital, Kaohsiung, Taiwan

Yu-Hsiang Chang
Department of Pediatrics, Kaohsiung Veterans General Hospital, Kaohsiung, Taiwan
Faculty of Medicine, National Yang-Ming University, Taipei, Taiwan, 9 Department of Nursing, Tajen University, Yanpu Township, Pingtung County, Taiwan

Lukmanee Tradtrantip, Nithi Asavapanumas, Puay-Wah Phuan and A. S. Verkman
Departments of Medicine and Physiology, University of California San Francisco, San Francisco, California, United States of America

Cheng Gong, Deng Pan1, Fengwu Qiu1, Pei Sun and Yu-Hui Zhang
Britton Chance Center for Biomedical Photonics, Wuhan National Laboratory for Optoelectronics-Huazhong University of Science and Technology (HUST), Wuhan, Hubei, China

Key Laboratory of Biomedical Photonics of Ministry of Education, Department of Biomedical Engineering, Huazhong University of Science and Technology, Wuhan, Hubei, China

Marta Dossena, Gloria Bedini, Alessandra Canazza, Valentina Tosetti, Stefania Elena Navone, Giovanni Marfia, Giulio Alessandri and Eugenio Agostino Parati
Cellular Neurobiology Laboratory, Unit of Cerebrovascular Disease, Fondazione IRCCS Istituto Neurologico Carlo Besta, Milan, Italy

Paola Rusmini and Angelo Poletti
Dipartimento di Scienze Farmacologiche e Biomolecolari, Centro Interdipartimentale sulle Malattie Neurodegenerative, Università degli Studi di Milano, Milan, Italy

Ettore Salsano, Anna Sagnelli and Davide Pareyson
Clinic of Central and Peripheral Degenerative Neuropathies Unit, Department of Clinical Neurosciences, Fondazione IRCCS Istituto Neurologico Carlo Besta, Milan, Italy

Caterina Mariotti and Cinzia Gellera
Unit of Genetics of Neurodegenerative and Metabolic Diseases, Department of Diagnostic and Applied Technology, Fondazione IRCCS Istituto Neurologico Carlo Besta, Milan, Italy

Fabio Corsi
Surgery Division, Department of Clinical Sciences, University of Milan, "Luigi Sacco" Hospital, Milan, Italy

Elisa Giorgetti
Dipartimento di Scienze Farmacologiche e Biomolecolari, Centro Interdipartimentale sulle Malattie Neurodegenerative, Università degli Studi di Milano, Milan, Italy
Department of Pathology, University of Michigan, Ann Arbor, Michigan, 48109, United States of America

Yao Tan, Xiyan Wang, Lingling Ma, Duoming Wang, Yunhui Hu, Yonghui Qin, Kai Liu and Cheng Chang
Department of Radiation Oncology, The Affiliated Tumor Hospital of Xinjiang Medical University, Urumqi, Xinjiang Uygur Autonomous Region, People's Republic of China

Jinming Yu
Department of Radiation Oncology, Tumor Hospital of Shandong, Jinnan, Shandong province, People's Republic of China

Ruozheng Wang
Tianjin Medical University, Tianjing, People's Republic of China
Department of Radiation Oncology, The Affiliated Tumor Hospital of Xinjiang Medical University, Urumqi, Xinjiang Uygur Autonomous Region, People's Republic of China

Hong Qin, Feng Pan, Jianjun Li, Xiaoli Zhang, Houjie Liang and Zhihua Ruan
Department of Oncology, Southwest Hospital, the Third Military Medical University, Chongqing, PR China

Jicheng Zhong and Helong Jiang
State Key Laboratory of Lake Science and Environment, Nanjing Institute of Geography and Limnology, Chinese Academy of Sciences, Nanjing, China

Jun Wu
Key Laboratory of Soil Environment and Pollution Remediation, Institute of Soil Science, Chinese Academy of Sciences, Nanjing, China

Liuyan Yang
State Key Laboratory of Pollution Control and Resources Reuse, School of the Environment, Xianlin Campus, Nanjing University, Nanjing, China

Huacheng Xu
State Key Laboratory of Pollution Control and Resources Reuse, School of the Environment, Xianlin Campus, Nanjing University, Nanjing, China
State Key Laboratory of Lake Science and Environment, Nanjing Institute of Geography and Limnology, Chinese Academy of Sciences, Nanjing, China

Guanghui Yu
State Key Laboratory of Lake Science and Environment, Nanjing Institute of Geography and Limnology, Chinese Academy of Sciences, Nanjing, China
College of Resources and Environmental Sciences, Nanjing Agricultural University, Nanjing, China

Mônica C. B. Martins, Monique C. Silva, Vera L. M. Lima and Nicácio Henrique da Silva
Departamento de Bioquímica, Universidade Federal de Pernambuco, Pernambuco, Brazil

Luanna R. S. Silva and Ana M. M. A. Melo
Departamento de Radiobiologia, Universidade Federal de Pernambuco, Pernambuco, Brazil

Eugênia C. Pereira
Departamento de Ciências Geográficas, Universidade de Pernambuco, Pernambuco, Brazil

Emerson P. S. Falcão
Centro Acadêmico de Vitória de Santo Antão, Universidade Federal de Pernambuco, Pernambuco, Brazil

Mei-pian Chen, Shing Chan, Godfrey Chi-fung Chan and Yiu-fai Cheung
Department of Pediatrics and Adolescent Medicine, The University of Hong Kong, Hong Kong, China

Z. Ioav Cabantchik
Department of Biological Chemistry, Alexander Silberman Institute of Life Sciences, Hebrew University of Jerusalem, Safra Campus at Givat Ram, Jerusalem, Israel

Amanda C. Martin and Adrian D. Hegeman
The Department Horticultural Science, Plant Biological Sciences Graduate Program, Department of Plant Biology, and Microbial and Plant Genomics Institute, University of Minnesota, St Paul, MN, United States of America

Ed Johnston
Alia Point 'Awa Nursery, Pepe'ekeo, Hawai'i, United States of America

Chengguo Xing
Department of Medicinal Chemistry, University of Minnesota, Minneapolis, MN, United States of America

A. David Paltiel
Yale School of Public Health, New Haven, Connecticut, United States of America

George R. Seage III and Milton C. Weinstein
Harvard School of Public Health, Harvard University, Boston, Massachusetts, United States of America

Joseph Eron
Division of Infectious Disease, School of Medicine, University of North Carolina at Chapel Hill, Chapel Hill, North Carolina, United States of America

Paul E. Sax
Division of Infectious Diseases, Brigham and Women's Hospital, Boston, Massachusetts, United States of America
Harvard University Center for AIDS Research, Harvard University, Boston, Massachusetts, United States of America

Alexis Sypek, Bethany K. Berkowitz and Bethany L. Morris
Division of General Medicine, Department of Medicine, Massachusetts General Hospital, Boston, Massachusetts, United States of America
Medical Practice Evaluation Center, Department of Medicine, Massachusetts General Hospital, Boston, Massachusetts, United States of America

Rochelle P. Walensky
Division of Infectious Diseases, Brigham and Women's Hospital, Boston, Massachusetts, United States of America
Harvard University Center for AIDS Research, Harvard University, Boston, Massachusetts, United States of America
Division of Infectious Diseases, Department of Medicine, Massachusetts General Hospital, Boston, Massachusetts, United States of America
Medical Practice Evaluation Center, Department of Medicine, Massachusetts General Hospital, Boston, Massachusetts, United States of America

Elena Losina
Department of Orthopedic Surgery, Brigham and Women's Hospital, Boston, Massachusetts, United States of America
Harvard University Center for AIDS Research, Harvard University, Boston, Massachusetts, United States of America
Division of General Medicine, Department of Medicine, Massachusetts General Hospital, Boston, Massachusetts, United States of America
Medical Practice Evaluation Center, Department of Medicine, Massachusetts General Hospital, Boston, Massachusetts, United States of America
Department of Biostatistics, Boston University School of Public Health, Boston, Massachusetts, United States of America

Kenneth A. Freedberg
Harvard University Center for AIDS Research, Harvard University, Boston, Massachusetts, United States of America
Harvard School of Public Health, Harvard University, Boston, Massachusetts, United States of America
Division of General Medicine, Department of Medicine, Massachusetts General Hospital, Boston, Massachusetts, United States of America
Division of Infectious Diseases, Department of Medicine, Massachusetts General Hospital, Boston, Massachusetts, United States of America
Medical Practice Evaluation Center, Department of Medicine, Massachusetts General Hospital, Boston, Massachusetts, United States of America
Department of Epidemiology, Boston University School of Public Health, Boston, Massachusetts, United States of America

Caroline S. Martinez, João Guilherme D. Torres, Franck M. Peçanha and Giulia A. Wiggers
Postgraduate Program in Biochemistry, Universidade Federal do Pampa, Uruguaiana, Rio Grande do Sul, Brazil

Janete A. Anselmo-Franci
Department of Morphology, Stomatology and Physiology, School of Dentistry, Universidade de São Paulo, Ribeirão Preto, São Paulo, Brazil

Dalton V. Vassallo
Department of Physiological Sciences, Universidade Federal do Espírito Santo, Vitória, Espírito Santo, Brazil

Mercedes Salaices
Department of Pharmacology, School of Medicine, Universidad Autónoma de Madrid, Madrid, Spain

María J. Alonso
Department of Biochemistry, Physiology and Molecular Genetics, Universidad Rey Juan Carlos, Alcorcón, Spain

Li Lin, Guan-Qun Zhou, Ling-Long Tang, Lei Chen, Yan-Ping Mao, Jun Ma, Jun Ma and Ying Sun
Department of Radiation Oncology, Sun Yat-sen University Cancer Center, State Key Laboratory of Oncology in South China, Collaborative Innovation Center for Cancer Medicine, Guangzhou, Guangdong Province, People's Republic of China

Mu-Sheng Zeng, Tie-Bang Kang and Wei-Hua Jia
Department of Experimental Research, Sun Yat-sen University Cancer Center, State Key Laboratory of Oncology in South China, Collaborative Innovation Center for Cancer Medicine, Guangzhou, Guangdong Province, People's Republic of China

Jian-Yong Shao
Department of Pathology, Sun Yat-sen University Cancer Center, State Key Laboratory of Oncology in South China, Collaborative Innovation Center for Cancer Medicine, Guangzhou, Guangdong Province, People's Republic of China

Hai-Qiang Mai
Department of Nasopharyngeal Carcinoma, Sun Yat-sen University Cancer Center, State Key Laboratory of Oncology in South China, Collaborative Innovation Center for Cancer Medicine, Guangzhou, Guangdong Province, People's Republic of China

Ai-Hua Lin
Department of Medical Statistics and Epidemiology, School of Public Health, Sun Yatsen University, Guangzhou, Guangdong Province, People's Republic of China

Chang-Juan Tao
Department of Radiation Oncology, Sun Yat-sen University Cancer Center, State Key Laboratory of Oncology in South China, Collaborative Innovation Center for Cancer Medicine, Guangzhou, Guangdong Province, People's Republic of China
Department of Radiation Oncology, Zhejiang Cancer Hospital, Hangzhou, Zhejiang Province, People's Republic of China

Stacey Fynch and Lorraine Elkerbout
St Vincent's Institute, Fitzroy, Victoria, Australia

Seth L. Masters
The Walter and Eliza Hall Institute of Medical Research, Parkville, Victoria, Australia

Jibran A. Wali, Esteban N. Gurzov, Thomas W. Kay and Helen E. Thomas
St Vincent's Institute, Fitzroy, Victoria, Australia
The University of Melbourne Department of Medicine, St. Vincent's Hospital, Fitzroy, Victoria, Australia

Index